FINITE FIELDS
AND
GALOIS RINGS

FINITE FIELDS
AND
GALOIS RINGS

Zhe-Xian Wan

Chinese Academy of Sciences, China

NEW JERSEY • LONDON • SINGAPORE • BEIJING • SHANGHAI • HONG KONG • TAIPEI • CHENNAI

Published by

World Scientific Publishing Co. Pte. Ltd.

5 Toh Tuck Link, Singapore 596224

USA office: 27 Warren Street, Suite 401-402, Hackensack, NJ 07601

UK office: 57 Shelton Street, Covent Garden, London WC2H 9HE

British Library Cataloguing-in-Publication Data
A catalogue record for this book is available from the British Library.

FINITE FIELDS AND GALOIS RINGS

For photocopying of material in this volume, please pay a copying fee through the Copyright Clearance Center, Inc., 222 Rosewood Drive, Danvers, MA 01923, USA. In this case permission to photocopy is not required from the publisher.

ISBN-13 978-981-4366-34-2
ISBN-10 981-4366-34-X

Printed in Singapore.

Preface

The present book is a revision and addendum of a former one entitled Lectures on Finite Fields and Galois Rings, which will be referred as "Lectures" in the following. The "Lectures" is based on a course on finite fields I gave at Nankai University, Tianjin and a seminar on Galois rings I conducted at Suzhou University, Suzhou, both in 2002. It has been used as a textbook for students in mathematics as well as those majoring in computer science or communication engineering. Of course, it has also been used for self-study.

The first five chapters and Chapter 12 of the "Lectures" are prerequisites for studying finite fields and Galois rings, respectively. They are prepared for students who had no background on abstract algebra, for instance, students in computer science or communication engineering. For those who already took a course on abstract algebra these chapters may be skipped. Chapter 6 to Chapter 11 of the "Lectures" are the main contents of finite fields; they are: structure theorems, automorphisms, norms and traces, various bases, factoring polynomials, constructing irreducible polynomials, and quadratic forms over (or of) finite fields. Galois rings are treated in Chapter 14, and Hensel's Lemma and Hensel lift, which are needed in studying Galois rings, are contained in Chapter 13.

In the present book many typos of the "Lectures" are corrected, some changes are made for clarity, and many exercises are added. The most significant additions to the "Lectures" are a section on optimal normal basis (i.e., section 8.5) and an expansion of the original section 12.4 so that a simple proof of Theorem 8.34 is included.

People are interested in finite fields and Galois rings, mainly because they have important applications in science and technology; for instances: shift register sequences, algebraic coding theory, cryptography, design theory, algebraic system theory, etc. However, in order to make this book not too thick, these applications are not included.

In preparing the book I benefitted a lot from the existing literature on finite fields and Galois rings; in particular, the encyclopedia of Lidl and Niederreiter, the books of Jungnickel, McDonald, McEliece, and Menezes et als.

The author is most grateful to Professor Qinghu Hou of Nankai University and Drs. Yuping Deng, Ruoxia Du, Mei Fu, Jun-Wei Guo, Qiu-Min Guo, Xia Jiang, De-Gang Liang, Chun-Lin Liu, Yan-Ping Mu, Yun Qin, Chao Wang, Limin Yang, Ling-Ling Yang, Yiting Yang, Bao-Yin Zhang, Jing-Yu Zhao who went through the tedious work of making a fair copy of the book during their attending my lectures at Nankai University. The author thanks Ms Jinling Chang who helped him with Latex problems and the compiling of the index. Finally, the author is also indebted to Professor Zhongming Tang of Suzhou University who read the whole text and gave valuable comments.

Thanks are due to those readers who took the trouble to point out typos of the "Lectures".

Beijing, 2010   Zhe-Xian Wan

Contents

Chapter 1

Sets and Integers

In this chapter some preparatory knowledge for studying this book is given. At first we give the fundamental concepts of set and map, then the factorization of integers, and finally the equivalence relation.

1.1   Sets and Maps

Set and map are two fundamental concepts in modern mathematics. We give a brief introduction to them in what follows.

A *set* is a collection of objects which is considered in its totality. For example, the totality of rational numbers forms a set which is denoted by $\mathbb{Q}$, the totality of real numbers also forms a set which is denoted by $\mathbb{R}$, and $\mathbb{Z}_p$ denotes a set consisting of $\bar{0}, \bar{1}, \bar{2}, \ldots, \overline{p-1}$. The members that form a set are called the *elements* of the set. We denote by

$$a \in S$$

that a is an element of the set S, read as *a belongs to S*. Denote by

$$a \notin S$$

that a is not an element of the set S, read as *a does not belong to S*.

Usually there are two ways to define a set, one is to list all its elements and the other is to specify the property which is possessed by all the elements and only the elements in the set. For example,

$$\mathbb{Z}_p = \{\bar{0}, \bar{1}, \bar{2}, \ldots, \overline{p-1}\},$$

1

which means listing all the elements in $\mathbb{Z}_p$. Also, $\mathbb{Q}[\sqrt{2}]$ is the set of all the real numbers of the form $a + b\sqrt{2}$, where a and b are rational numbers and this set is denoted by

$$\mathbb{Q}[\sqrt{2}] = \{a + b\sqrt{2} : a, b \in \mathbb{Q}\},$$

which means specifying the property of the elements in $\mathbb{Q}[\sqrt{2}]$. In the parentheses $\{\cdots\}$, $a + b\sqrt{2}$ which appears before the symbol : indicates the form of the elements in $\mathbb{Q}[\sqrt{2}]$, but what a, b really are is indicated by the property a, $b \in \mathbb{Q}$ after the symbol :.

For convenience, we introduce the concept of the *empty set*, which is denoted by $\emptyset$, and regard it as a set consisting of no elements. The other sets are then called *nonempty sets*, hence a nonempty set is one which indeed consists of elements.

Denote by $|S|$ the number of elements in S, which is called the *cardinal* of S. If $|S|$ is a finite number, we say that S is a *finite set*. We regard the empty set as a finite set consisting of 0 elements, i.e., $|\emptyset| = 0$. If S is not a finite set, then it is called an *infinite set*.

If two sets S and T consist of exactly the same elements, that is, $a \in S$ if and only if $a \in T$, then we say that they are *equal*, denoted by

$$S = T.$$

If all the elements of a set S are elements of a set T, that is, from $a \in S$ we deduce $a \in T$, then we say that S is a *subset* of T, denoted by

$$S \subset T \quad \text{or} \quad T \supset S.$$

For example, let $\mathbb{Q}$, $\mathbb{R}$, and $\mathbb{C}$ be the set of rational, real, and complex numbers, respectively, then $\mathbb{Q} \subset \mathbb{R}$, $\mathbb{R} \subset \mathbb{C}$. By definition, every set is a subset of itself. We also say that the empty set is a subset of any set.

Let S be a set and T be a subset of S. We denote by $S \setminus T$ the set which consists of those elements in S that do not belong to T, that is,

$$S \setminus T = \{a : a \in S \text{ and } a \notin T\},$$

and call $S \setminus T$ the *complementary set of T in S*. For example, if $\mathbb{Z}$ is the set consisting of all the integers and $2\mathbb{Z}$ is the set consisting of all the even integers (including 0), then $\mathbb{Z} \supset 2\mathbb{Z}$ and $\mathbb{Z} \setminus 2\mathbb{Z}$ is the set consisting of all the odd integers.

Let S and T be two sets. The set consisting of all the elements which belong to both S and T is called the *intersection* of S and T, denoted by

$$S \cap T.$$

That is,
$$S \cap T = \{a : a \in S \text{ and } a \in T\}.$$

For example, when S is the set of all the even integers and T is the set of all the positive integers which are less than seven, then

$$S \cap T = \{2, 4, 6\}.$$

It is clear that
$$S \cap T \subset S, \quad S \cap T \subset T.$$

When $S \cap T = \emptyset$, they are said to be *disjoint*.

Again let S and T be two sets. The set consisting of all the elements which belong to S or T is called the *union* of S and T, denoted by

$$S \cup T.$$

That is,
$$S \cup T = \{a : a \in S \text{ or } a \in T\}.$$

For example,
$$\{1, 2, 3, 4\} \cup \{2, 4, 5\} = \{1, 2, 3, 4, 5\}.$$

Clearly,
$$S \cup T \supset S, \quad S \cup T \supset T.$$

When $S \cap T = \emptyset$, $S \cup T$ is called a *disjoint union*.

The two concepts, the intersection and union of sets, can be generalized to the case of any number (finite or infinite) of sets. Let I be an index set which may be finite or infinite. Assume that for any $\alpha \in I$, there exists a set S_α, then the set of all the elements which belong to each S_α ($\alpha \in I$) is called the *intersection* of these S_α's, denoted by

$$\cap_{\alpha \in I} S_\alpha.$$

Similarly, the set of all the elements which belong to any of S_α ($\alpha \in I$) is called the *union* of these S_α's, denoted by

$$\cup_{\alpha \in I} S_\alpha.$$

Clearly,
$$\cap_{\alpha \in I} S_\alpha \subset S_\alpha,$$
$$\cup_{\alpha \in I} S_\alpha \supset S_\alpha.$$

When S_α's ($\alpha \in I$) are pairwise disjoint, $\cup_{\alpha \in I} S_\alpha$ is called a *disjoint union*.

Let S and T be two sets which need not be distinct. We define the *Cartesian product set $S \times T$* as follows:

$$S \times T = \{(x, y) : x \in S, \ y \in T\}.$$

The elements (x, y) and (x', y') of $S \times T$ are regarded as equal if and only if $x = x'$ and $y = y'$. Thus if S consists of m elements $x_1, x_2, \ldots, x_m$ and T consists of n elements $y_1, y_2, \ldots, y_n$, then $S \times T$ consists of mn elements (x_i, y_j), $i = 1, 2, \ldots, m$; $j = 1, 2, \ldots, n$.

More generally, let $S_1, S_2, \ldots, S_r$ be sets. Define the *Cartesian product set $S_1 \times S_2 \times \cdots \times S_r$* to be

$$S_1 \times S_2 \times \cdots \times S_r = \{(x_1, x_2, \ldots, x_r) : x_i \in S_i \ (i = 1, 2, \ldots, r)\}.$$

Next we introduce the concept of a map.

Let S and S' be two sets. A *map* from the set S to the set S' is a rule by which every element a in S has a definite element in S' corresponding to it. Sometimes we also call a map from S to S' a *function* which is defined on S and takes values in S'. Let σ be a map from S to S', which we denote by

$$\sigma : S \to S'.$$

If the element $a' \in S'$ corresponds to the element $a \in S$ according to the rule, then we denote by

$$\sigma : a \mapsto a',$$

or

$$\sigma(a) = a'.$$

We call a' the *image* of a and a an *original* of a' under the map σ. The set of all originals of a' under σ is denoted by $\sigma^{-1}(a')$, i.e.,

$$\sigma^{-1}(a') = \{a \in S : \sigma(a) = a'\},$$

and is called the *complete inverse image* of a' under the map σ. More generally, let T' be any subset of S', the set of elements of S which are mapped to elements of T' under σ is denoted by $\sigma^{-1}(T')$, i.e.,

$$\sigma^{-1}(T') = \{a \in S : \sigma(a) \in T'\},$$

and is called the *complete inverse image* of T' under the map σ.

Let σ be a map from a set S to a set S'. We denote by

$$\sigma(S)$$

the set of all the images of the elements of S under the map σ, that is,

$$\sigma(S) = \{\sigma(a) : a \in S\}.$$

$\sigma(S)$ is called the *image* of σ and is usually denoted by $\operatorname{Im}\sigma$. It is clear that

$$\sigma(S) \subset S'.$$

If $\sigma(S) = S'$, then the map σ is called a *surjection* or is said to be surjective. If, under the map σ, the images of different elements in S are also different, that is, from $a_1 \neq a_2$, we must have $\sigma(a_1) \neq \sigma(a_2)$, then we say that σ is an *injection* or is said to be injective. A map which is both injective and surjective is called a *bijection* or is said to be bijective.

Let S be a set, we call the map which maps every element of S to itself

$$a \mapsto a, \ a \in S,$$

the *identity map* of S, denoted by 1_S, and when there is no confusion, it can be also denoted simply by 1. Clearly, the identity map is a bijection.

We give some examples to elucidate the above concepts.

Example 1.1 $S = \{1, 2, 3, 4\}$, $S' = \{1', 2', 3'\}$. Define $\sigma(1) = \sigma(2) = 1'$, $\sigma(3) = \sigma(4) = 3'$. The map σ is not surjective since $2'$ does not have an original. σ is not injective either, since the images of 1 and 2 are the same and the images of 3 and 4 are the same. Moreover, $\sigma^{-1}(1') = \{1, 2\}$, $\sigma^{-1}(\{1', 2'\}) = \{1, 2\}$, and $\sigma^{-1}(\{1', 3'\}) = \sigma^{-1}(S') = S$.

Example 1.2 $S = \mathbb{Z}$ and S' is the set of all the nonnegative integers. Define a map σ from S to S' by

$$\sigma(a) = |a|, \ a \in S.$$

Then σ is surjective but not injective.

Example 1.3 Let S be the set of all the positive integers and

$$S' = \{x, x^2, x^3, \ldots\},$$

that is, S' is the set of all positive powers of an indeterminate x. Define a map from S to S' by

$$\sigma(n) = x^n, \ n \in S.$$

Then σ is both surjective and injective, hence it is a bijection.

Now let σ be a map from a set S to a set S' and σ' be a map from S' to a set S''. We define the *composite* of σ and σ', denoted by $\sigma' \circ \sigma$, to be the map from S to S'' by

$$\sigma' \circ \sigma : \ S \to S''$$

$$a \mapsto \sigma'(\sigma(a)).$$

Example 1.4 Let $S = S' = S'' = \mathbb{Z}$, σ be the map

$$\sigma : \ \mathbb{Z} \to \mathbb{Z}$$

$$a \mapsto a^3$$

and σ' be the map

$$\sigma' : \ \mathbb{Z} \to \mathbb{Z}$$

$$a \mapsto a^2.$$

Then

$$\sigma' \circ \sigma : \ \mathbb{Z} \to \mathbb{Z}$$

$$a \mapsto a^6.$$

Moreover, if σ'' is a map from S'' to S''', then

$$\sigma'' \circ (\sigma' \circ \sigma) = (\sigma'' \circ \sigma') \circ \sigma.$$

In fact, for any $a \in S$, we have

$$(\sigma'' \circ (\sigma' \circ \sigma))(a) = \sigma''((\sigma' \circ \sigma)(a)) = \sigma''(\sigma'(\sigma(a))),$$

$$((\sigma'' \circ \sigma') \circ \sigma)(a) = (\sigma'' \circ \sigma')(\sigma(a)) = \sigma''(\sigma'(\sigma(a))).$$

Let σ be a bijective map from a set S to itself. Define a map τ from S to itself by $\tau(a) = b$ if $\sigma(b) = a$. Then τ is also bijective and $\sigma \circ \tau = \tau \circ \sigma = 1_S$ (Exercise 1.5). τ is called the *inverse map* of σ and denoted usually by σ^{-1}. (It is important to distinguish the complete inverse image $\sigma^{-1}(a')$ and $\sigma^{-1}(T')$ defined previously with the inverse map σ^{-1} just defined.)

Let σ be a map from a set S to a set S' and T be a subset of S. Define a map from T to S', which will be denoted by $\sigma|_T$, by

$$\sigma|_T : \ T \to S'$$

$$a \mapsto \sigma(a), \ a \in T.$$

$\sigma|_T$ is called the *restriction* of σ to T.

Let S be a set. A *binary operation* or, simply, an *operation* in S is defined to be a map ϕ from the Cartesian product set $S \times S$ to the set S

$$\phi : \ S \times S \to S$$
$$(a, b) \mapsto \phi(a, b).$$

We also say that $\phi(a, b)$ is the result of performing the operation ϕ on a and b. We do not assume $\phi(a, b) = \phi(b, a)$ for all $a, b \in S$. If we denote

$$\phi(a, b) = a + b,$$

then ϕ (or $+$) is called an *addition* in S and $a + b$ the *sum* of a and b; if we denote

$$\phi(a, b) = a \cdot b,$$

then ϕ (or $\cdot$) is called a *multiplication* in S and $a \cdot b$ or, simply, ab the *product* of a and b.

For example, let $S = \mathbb{Q}$, then the usual addition in $\mathbb{Q}$

$$+ : \ \mathbb{Q} \times \mathbb{Q} \to \mathbb{Q}$$
$$(a, b) \mapsto a + b$$

is a binary operation in $\mathbb{Q}$ and the usual multiplication in $\mathbb{Q}$

$$\cdot : \ \mathbb{Q} \times \mathbb{Q} \to \mathbb{Q}$$
$$(a, b) \mapsto a \cdot b$$

is another binary operation in $\mathbb{Q}$.

1.2 The Factorization of Integers

Denote by $\mathbb{Z}$ the set of all integers (positive integers, negative integers, and 0). We know that addition, substraction, and multiplication can be performed in $\mathbb{Z}$. For division, we have

Theorem 1.1 *For any two integers a and b with $b \neq 0$, there exist a unique pair of integers q and r such that*

$$a = qb + r, \quad 0 \leq r < |b|, \tag{1.1}$$

where $|b|$ denotes the absolute value of b.

Proof. First we prove the existence part. If $a = 0$ then we set $q = r = 0$. For $a > 0$, we use induction on a. If $a < |b|$ then we take $q = 0$ and $r = a$. If $a \geq |b|$ let $a_1 = a - |b|$. Then $0 \leq a_1 < a$. By induction hypothesis, there is q_1 and r_1, such that

$$a_1 = q_1 b + r_1, \quad 0 \leq r_1 < |b|.$$

Substituting $a_1 = a - |b|$ into the above equation and transforming, we obtain

$$a = q_1 b + |b| + r_1, \quad 0 \leq r_1 < |b|.$$

If $b > 0$, we take $q = q_1 + 1$ and if $b < 0$, we take $q = q_1 - 1$. In both cases, we take $r = r_1$. Then we obtain

$$a = qb + r, \quad 0 \leq r < |b|.$$

There remains to consider the case $a < 0$. Then $-a > 0$. By the above case, there are q_1 and r_1 such that

$$-a = q_1 b + r_1, \quad 0 \leq r_1 < |b|.$$

Then $a = -q_1 b - r_1$. If $b > 0$, we take $q = -(q_1 + 1)$ and $r = b - r_1$. If $b < 0$, we take $q = -(q_1 - 1)$ and $r = -b - r_1$. In both cases, we have (1.1).

To prove the uniqueness we assume that there exist q, r such that (1.1) holds and that there exist q', r' such that

$$a = q'b + r', \quad 0 \leq r' < |b|. \tag{1.2}$$

Without loss of generality, we can assume that $r' \geq r$. Substracting (1.2) from (1.1), we obtain

$$(q - q')b + r - r' = 0,$$

or, equivalently,

$$(q - q')b = r' - r \geq 0.$$

But $r' - r \leq r' < |b|$. Thus $|(q - q')| \cdot |b| = r' - r < |b|$, which implies $q = q'$ and, hence, $r = r'$. □

Formula (1.1) is called the *division algorithm* in $\mathbb{Z}$, q in (1.1) the *quotient* of dividing a by b, and r in (1.1) the *remainder*, and we write

$$r = (a)_b.$$

When $r = 0$, we say that b is a *divisor* (or a *factor*) of a, a is a *multiple* of b, a is *divisible* by b, or b divides a, denoted by

$$b \mid a.$$

We also denote by $b \nmid a$ that b does not divide a. Clearly, if b is a divisor of a and $a \neq 0$, then

$$|b| \leq |a|.$$

Let a, b, c be integers and $c \neq 0$. If c is a divisor of both a and b, then we say that c is a *common divisor* of a and b. When a and b are not both equal to 0, then there is a maximum among the common divisors of a and b; we call it the *greatest common divisor* of a and b and denote it by

$$\gcd(a, b).$$

Clearly, $\gcd(a, b) > 0$. When both a and b are equal to 0, any integer is their common divisor, hence in this case a and b have no greatest common divisor and the symbol $\gcd(0,0)$ is meaningless.

Let a and b both be integers which are not equal to 0. Let us review the *Euclidean algorithm* to calculate $\gcd(a, b)$. Assume that $|a| > |b|$. By the division algorithm, we have the sequence of formulas

$$a = q_1 b + r_1, \quad 0 < r_1 < |b|, \tag{1.3}$$

$$b = q_2 r_1 + r_2, \quad 0 < r_2 < r_1, \tag{1.4}$$

$$r_1 = q_3 r_2 + r_3, \quad 0 < r_3 < r_2, \tag{1.5}$$

$$\dots\dots\dots\dots$$

$$r_{n-3} = q_{n-1} r_{n-2} + r_{n-1}, \quad 0 < r_{n-1} < r_{n-2}, \tag{1.6}$$

$$r_{n-2} = q_n r_{n-1} + r_n, \quad 0 < r_n < r_{n-1}, \tag{1.7}$$

$$r_{n-1} = q_{n+1} r_n. \tag{1.8}$$

Then

$$\gcd(a, b) = r_n.$$

In fact, from equation (1.8) we know that $r_n \mid r_{n-1}$, from (1.7) $r_n \mid r_{n-2}$, from (1.6) $r_n \mid r_{n-3}$, ... Continuing in this way, we finally deduce that $r_n \mid a$ from $r_n \mid r_1$, $r_n \mid b$, and (1.3). Hence r_n is a common divisor of a and b. Conversely, assume that d is a common divisor of a and b, that is, $d \mid a$, $d \mid b$. Then by (1.3) $d \mid r_1$, by (1.4) $d \mid r_2$, by (1.5) $d \mid r_3, \dots$. Continuing in this way, finally we deduce that $d \mid r_n$ from $d \mid r_{n-2}$, $d \mid r_{n-1}$, and (1.7). Hence $r_n = \gcd(a, b)$.

Further, we can rewrite equations (1.3) through (1.7) as

$$r_1 = a + (-q_1)b, \tag{1.9}$$

$$r_2 = b + (-q_2)r_1, \tag{1.10}$$

$$r_3 = r_1 + (-q_3)r_2, \tag{1.11}$$

$$\cdots\cdots\cdots\cdots$$

$$r_{n-1} = r_{n-3} + (-q_{n-1})r_{n-2}, \tag{1.12}$$

$$r_n = r_{n-2} + (-q_n)r_{n-1}. \tag{1.13}$$

(1.13) implies that r_n is a linear combination of r_{n-2} and r_{n-1} with integer coefficients. Substituting (1.12) into (1.13) we have

$$r_n = (-q_n)r_{n-3} + (1 + q_n q_{n-1})r_{n-2}.$$

That is, r_n is a linear combination of r_{n-3} and r_{n-2} with integer coefficients. By successive substitutions, we can express the greatest common divisor r_n of a and b as a linear combination of a and b with integer coefficients, that is,

$$r_n = ca + db,$$

where both c and d are integers. Then from the above equation, we deduce again that any common divisor of a and b is a divisor of r_n.

If $a \neq 0$ and $b = 0$, then $\gcd(a, b) = |a| = ca + db$, where $c = 1$ when $a > 0$, $c = -1$ when $a < 0$, and $d = 0$. Similarly, if $a = 0$ and $b \neq 0$, then $\gcd(a, b) = |b| = ca + db$, where $c = 0$, and $d = 1$ when $b > 0$, $d = -1$ when $b < 0$.

The above argument proves the following

Theorem 1.2 *Let a, $b \in \mathbb{Z}$ and are not both 0. Then there exist integers c and d such that*

$$\gcd(a, b) = ca + db.$$

Example 1.5 Let $a = 49$ and $b = 36$. By the division algorithm, we obtain the following sequence of formulas

$$49 = 1 \cdot 36 + 13, \quad 13 < 36$$

$$36 = 2 \cdot 13 + 10, \quad 10 < 13$$

$$13 = 1 \cdot 10 + 3, \quad 3 < 10$$

$$10 = 3 \cdot 3 + 1, \quad 1 < 3$$

$$3 = 3 \cdot 1.$$

This indicates that
$$\gcd(49, 36) = 1.$$
Rewrite the first four equations as
$$13 = 49 - 1 \cdot 36$$
$$10 = 36 - 2 \cdot 13$$
$$3 = 13 - 1 \cdot 10$$
$$1 = 10 - 3 \cdot 3.$$

Substitute the right-hand side of the third equation for 3 in the fourth equation, then substitute the right-hand side of the second equation for 10 in the equation just obtained and finally substitute the right-hand side of the first equation for 13 in the equation just obtained, we have

$$1 = 10 - 3 \cdot (13 - 1 \cdot 10)$$
$$= (-3) \cdot 13 + 4 \cdot 10$$
$$= (-3) \cdot 13 + 4 \cdot (36 - 2 \cdot 13)$$
$$= 4 \cdot 36 + (-11) \cdot 13$$
$$= 4 \cdot 36 + (-11) \cdot (49 - 1 \cdot 36)$$
$$= (-11) \cdot 49 + 15 \cdot 36.$$

Let $a, b \in \mathbb{Z}$. If 1 can be expressed as a linear combination of a and b with integer coefficients, that is,

$$1 = ca + db,$$

where both c and d are integers, then a and b are said to be *coprime*. For example, by the previous calculation, we know that 49 and 36 are coprime.

Theorem 1.3 *Let $a, b \in \mathbb{Z}$. Then a and b are coprime if and only if* $\gcd(a, b) = 1$.

Proof. If $\gcd(a, b) = 1$, then a and b are not both 0. By Theorem 1.2, $1 = ca + db$, where $c, d \in \mathbb{Z}$, i.e., a and b are coprime. The converse is trivial. □

Let a and b be nonzero integers. If c is a multiple of both a and b, then we say that c is a *common multiple* of a and b. It is clear that $|ab|$ is a positive common multiple of a and b. Hence among the positive common multiples of a and b, there must be a minimum; we call it the *least common multiple* of a and b and denote it by

$$\mathrm{lcm}[a, b].$$

Clearly, for any finite number of integers $a_1, a_2, \ldots, a_n$ which are not all equal to 0, we can also define their common divisor and their greatest common divisor. We say that c is a *common divisor* of them if c is a divisor of every $a_i (i = 1, 2, \ldots, n)$; the largest one among their common divisors is called their *greatest common divisor* and is denoted by $\gcd(a_1, a_2, \ldots, a_n)$. For any finite number of integers $a_1, a_2, \ldots, a_n$, no one of which is equal to 0, we can also define their common multiple and their least common multiple. We say that c is a *common multiple* of them if c is a multiple of every a_i $(i = 1, 2, \ldots, n)$; the smallest one among their positive common multiples is called their *least common multiple* and is denoted by $\operatorname{lcm}[a_1, a_2, \ldots, a_n]$. By definition, it is clear that

$$\gcd(a_1, a_2, \ldots, a_n) = \gcd(a_1, \gcd(a_2, \ldots, a_n))$$

$$= \gcd(a_1, \gcd(a_2, \ldots, \gcd(a_{n-1}, a_n) \ldots)), \quad (1.14)$$

$$\operatorname{lcm}[a_1, a_2, \ldots, a_n] = \operatorname{lcm}[a_1, \operatorname{lcm}[a_2, \ldots, a_n]]$$

$$= \operatorname{lcm}[a_1, \operatorname{lcm}[a_2, \ldots, \operatorname{lcm}[a_{n-1}, a_n] \ldots]]. \quad (1.15)$$

Moreover, Theorem 1.2 can be generalized as follows.

Theorem 1.4 *Let $a_1, a_2, \ldots, a_n$ be integers which are not all 0. Then there exist integers $c_1, c_2, \ldots, c_n$ such that*

$$\gcd(a_1, a_2, \ldots, a_n) = c_1 a_1 + c_2 a_2 + \cdots + c_n a_n.$$

The proof is left as an exercise.

Let p be an integer greater than 1. We say that p is a *prime* if the positive divisors of p are 1 and p only. Integers greater than 1 which are not primes are called *composite numbers*. We know that the *number of primes is infinite*. To obtain all the primes $\leq$ a given positive integer, we may use the *sieve method*. For example, to obtain all primes ≤ 100, we can first list all the integers from 2 through 100 in ascending order:

2	3	4	5	6	7	8	9	10	11	12	13	14	15	16
17	18	19	20	21	22	23	24	25	26	27	28	29	30	31
32	33	34	35	36	37	38	39	40	41	42	43	44	45	46
47	48	49	50	51	52	53	54	55	56	57	58	59	60	61
62	63	64	65	66	67	68	69	70	71	72	73	74	75	76
77	78	79	80	81	82	83	84	85	86	87	88	89	90	91
92	93	94	95	96	97	98	99	100						

2 is the smallest number among them and it is itself a prime. Cross all the multiples of 2 (not including 2 itself) off the list. Among the uncrossed numbers which are greater than 2, 3 is the smallest number and hence a prime. Cross all the multiples of 3 among the remaining numbers (3 not included) off the list. Repeatedly doing this until the smallest number amongst the remaining uncrossed numbers is 11. Since every composite number ≤ 100 must have a prime divisor $\leq \sqrt{100} = 10$, the remaining numbers are all primes. In such a way, we obtain the table for primes ≤ 100:

$$2 \quad 3 \quad 5 \quad 7 \quad 11 \quad 13 \quad 17 \quad 19 \quad 23 \quad 29 \quad 31 \quad 37 \quad 41$$

$$43 \quad 47 \quad 53 \quad 59 \quad 61 \quad 67 \quad 71 \quad 73 \quad 79 \quad 83 \quad 89 \quad 97$$

The so-called *fundamental theorem of arithmetic* is the following

Theorem 1.5 *Any integer greater than 1 can be expressed as a product of primes. If we disregard the order of these primes in the product, then this expression is unique.*

Proof. (First part). Let n be an integer greater than 1. If n is a prime, then it is naturally a product of primes, that is, $n = n$. If n is not a prime, then among the divisors of n which are greater than 1, there is a smallest one. Let this smallest one be p_1. Since the divisors of p_1 are also divisors of n, p_1 must be a prime. Assume that

$$n = p_1 n_1, \ 1 < n_1 < n.$$

If n_1 is a prime, n is already expressed as a product of primes. If n_1 is not a prime, let p_2 be the smallest divisor of n_1 which is greater than 1, then p_2 must be a prime. Assume that

$$n = p_1 p_2 n_2, \ 1 < n_2 < n_1 < n.$$

Continuing in this way, we obtain a sequence of decreasing positive integers

$$n > n_1 > n_2 > \cdots > 1.$$

This process can not be repeated more than n times. Hence finally we must have

$$n = p_1 p_2 p_3 \cdots p_r,$$

where $p_1, p_2, \ldots, p_r$ are all primes. This proves the first part of the fundamental theorem of arithmetic.

In order to prove the second part of this theorem, we first prove the following lemma.

Lemma 1.6 *Let p be a prime and $p \mid ab$, where a and b are both integers, then $p \mid a$ or $p \mid b$.*

Proof. If $p \nmid a$, then $\gcd(p, a) = 1$. By Theorem 1.2 there exist integers c and d such that

$$1 = cp + da.$$

Multiplying both sides of the above equation by b, we have

$$b = bcp + bda.$$

Since $p \mid ab$, $p \mid b$. $\qquad\square$

From Lemma 1.6, we deduce

Lemma 1.7 *Let p be a prime and $p \mid a_1 a_2 \cdots a_m$, then $p \mid a_1$, or $p \mid a_2, \ldots$, or $p \mid a_m$.*

Proof of Theorem 1.5 (Last part). Let n be an integer greater than 1 and assume that there are two ways to express n as a product of primes:

$$n = p_1 p_2 \cdots p_r = q_1 q_2 \cdots q_s, \tag{1.16}$$

where $p_1, p_2, \ldots, p_r$ and $q_1, q_2, \ldots, q_s$ are all primes. We use induction on r to prove the uniqueness of this expression.

When $r = 1$, $n = p_1$ is a prime. Hence $s = 1$, $n = p_1 = q_1$, so the expression is unique.

Now let us assume that for any integer greater than 1, if it can be expressed as a product of $r - 1$ primes, then the expression is unique. We are going to prove that the expression of n into a product of primes is also unique. We have $p_1 \mid n$, that is, $p_1 \mid q_1 q_2 \cdots q_s$. Hence by Lemma 1.7, $p_1 \mid q_1$, or $p_1 \mid q_2, \ldots$, or $p_1 \mid q_s$. Then after rearrangement of $q_1, q_2, \ldots, q_s$, we can assume that $p_1 \mid q_1$. Since p_1 and q_1 are both primes, $p_1 = q_1$. Then from (1.16) we deduce

$$p_2 p_3 \cdots p_r = q_2 q_3 \cdots q_s.$$

By the induction hypothesis, we must have $r - 1 = s - 1$ and after rearranging $q_2, q_3, \ldots, q_s$ we must have $p_2 = q_2$, $p_3 = q_3, \ldots, p_r = q_r$. This proves that the expression of n into a product of primes, disregarding the order of these primes, is unique. $\qquad\square$

By the fundamental theorem of arithmetic, any nonzero integer a can be expressed in the following form

$$a = \pm p_1^{e_1} p_2^{e_2} \cdots p_r^{e_r}, \tag{1.17}$$

where $p_1, p_2, \ldots, p_r$ are distinct primes, $e_1, e_2, \ldots, e_r$ are some positive integers and the sign "$+$" or "$-$" is determined by whether $a > 0$ or $a < 0$, respectively. (1.17) is called the *prime factorization* of a and also called the *complete factorization* of a.

Let $a, b, \ldots, c$ be a finite number of nonzero integers. Express them as

$$a = \pm p_1^{e_1} p_2^{e_2} \cdots p_m^{e_m},$$

$$b = \pm p_1^{f_1} p_2^{f_2} \cdots p_m^{f_m},$$

$$\cdots\cdots\cdots$$

$$c = \pm p_1^{g_1} p_2^{g_2} \cdots p_m^{g_m},$$

where $p_1, p_2, \ldots, p_r$ are distinct primes, e_i, f_i, g_i $(i = 1, 2, \ldots, m)$ are all integers ≥ 0, then the greatest common divisor $\gcd(a, b, \ldots, c)$ and the least common multiple $\mathrm{lcm}[a, b, \ldots, c]$ of $a, b, \ldots, c$ are, respectively,

$$\gcd(a, b, \ldots, c) = p_1^{\min\{e_1, f_1, \ldots, g_1\}} p_2^{\min\{e_2, f_2, \ldots, g_2\}} \cdots p_m^{\min\{e_m, f_m, \ldots, g_m\}},$$
(1.18)

and

$$\mathrm{lcm}[a, b, \ldots, c] = p_1^{\max\{e_1, f_1, \ldots, g_1\}} p_2^{\max\{e_2, f_2, \ldots, g_2\}} \cdots p_m^{\max\{e_m, f_m, \ldots, g_m\}},$$
(1.19)

where $\min\{e_i, f_i, \ldots, g_i\}$ and $\max\{e_i, f_i, \ldots, g_i\}$ denote the minimum and maximum of $e_i, f_i, \ldots, g_i$, respectively. From these two formulas, it can be deduced immediately that for any two positive integers a and b,

$$\gcd(a, b)\mathrm{lcm}[a, b] = ab$$
(1.20)

always holds.

1.3   Equivalence Relation and Partition

Let S be a nonempty set and $R = \{0, 1\}$. A map $\sim$ from the Cartesian product set $S \times S$ to the set R

$$\sim: \ S \times S \to R$$

$$(a, b) \mapsto \sim (a, b)$$

is sometimes called a *relation* on S. When $\sim (a, b) = 1$ we say that a is in the relation $\sim$ to b and write $a \sim b$; when $\sim (a, b) = 0$ we say that a is not in the relation $\sim$ to b and write $a \not\sim b$. For example, let $S = \mathbb{R}$ and we can define a relation $<$ on $\mathbb{R}$ by

$$< (a, b) = 1 \quad \text{if and only if} \quad a < b,$$

$$< (a, b) = 0 \quad \text{if and only if} \quad a \geq b.$$

Let $\sim$ be a relation on S. The relation $\sim$ is called an *equivalence relation* on S if it satisfies the following three conditions:

1. $a \sim a$ for all $a \in S$ (reflexive property),

2. $a \sim b$ implies $b \sim a$ (symmetric property),

3. $a \sim b$ and $b \sim c$ imply $a \sim c$ (transitive property).

Example 1.6 Let $S = \mathbb{Z}$ and m be a positive integer. For any two integers a and b, define

$$a \sim b \quad \text{if and only if } a - b = cm \text{ for some } c \in \mathbb{Z}. \tag{1.21}$$

Then we have

1. $a \sim a$ for all $a \in \mathbb{Z}$, since $a - a = 0 = 0 \cdot m$.

2. $a \sim b \Rightarrow a - b = cm$ for some $c \in \mathbb{Z} \Rightarrow b - a = (-c)m$ and $-c \in \mathbb{Z} \Rightarrow b \sim a$.

3. $a \sim b$ and $b \sim c \Rightarrow a - b = cm$ and $b - c = dm$, where $c, d \in \mathbb{Z} \Rightarrow a - c = (a - b) + (b - c) = (c + d)m$ and $c + d \in \mathbb{Z} \Rightarrow a \sim c$.

Therefore $a \sim b$ defined by (1.21) is an equivalence relation on $\mathbb{Z}$. We usually write $a \equiv b \pmod{m}$ and read "a is *congruent* to b modulo m" for $a \sim b$.

Example 1.7 Let $S = \mathbb{R} \times \mathbb{R}$. Define

$$(a, b) \sim (a', b') \text{ if and only if } a = a'.$$

It is easy to verify that this is an equivalence relation on $\mathbb{R} \times \mathbb{R}$. However, if we define

$$(a, b) \sim (a', b') \text{ if and only if } a = a' \text{ and } b < b',$$

this is not an equivalence relation; in fact, the symmetric property is obviously not satisfied.

Let $\sim$ be an equivalence relation on a nonempty set S. A subset T of S is called an *equivalence class* if $a \sim b$ for any two elements a and b of T and for any element $c \in S$ $c \sim a$ for some $a \in T$ implies $c \in T$.

Let S be a nonempty set and I be an index set. A set of nonempty subsets S_α ($\alpha \in I$) of S is called a *partition* of S if $\cup_{\alpha \in I} S_\alpha = S$ and $S_\alpha \cap S_\beta = \emptyset$ whenever $\alpha \neq \beta$.

Now we shall prove that any equivalence relation on a nonempty set gives rise to a partition of the set into equivalence classes and that any partition of a nonempty set defines an equivalence relation on the set by defining $a \sim b$ when a and b belong to the same subset of the partition.

Theorem 1.8 *Let $\sim$ be an equivalence relation on a nonempty set S. For any $a \in S$ define $S_a = \{x \in S : a \sim x\}$. Then S_a is an equivalence class and for any two elements a, $b \in S$ either $S_a \cap S_b = \emptyset$ or $S_a = S_b$. Moreover, there is a subset I of S such that $S_a \cap S_b = \emptyset$ for any two distinct elements a, b of I and $\cup_{a \in I} S_a = S$, that is, $\{S_a : a \in I\}$ is a partition of S into equivalence classes; moreover for any two elements a, $b \in S$, $a \sim b$ if and only if a, $b \in S_c$ for some $c \in I$.*

Proof. First we prove that $b \sim c$ for any two elements b, $c \in S_a$. By definition of S_a, we have $a \sim b$ and $a \sim c$. By the symmetric property of $\sim$, $a \sim b$ implies $b \sim a$, and by the transitive property, $b \sim a$ and $a \sim c$ imply $b \sim c$. Then we prove that for any element $d \in S$, $d \sim b$ for some $b \in S_a$ implies $d \in S_a$. In fact, $d \sim b$ implies $b \sim d$ and $b \in S_a$ means $a \sim b$. By the transitive property, $a \sim b$ and $b \sim d$ imply $a \sim d$. Therefore $d \in S_a$. This proves that S_a is an equivalence class.

Assume that $S_a \cap S_b \neq \emptyset$ and let $c \in S_a \cap S_b$. Then $a \sim c$ and $b \sim c$. But $b \sim c$ implies $c \sim b$, and $a \sim c$ and $c \sim b$ imply $a \sim b$. For any $x \in S_b$ we have $b \sim x$, then $a \sim b$ and $b \sim x$ imply $a \sim x$, thus $x \in S_a$. Similarly, $x \in S_a$ implies $x \in S_b$. Therefore, $S_a = S_b$.

Choose any $a \in S$, then we have an equivalence class S_a. If $S_a = S$, let $I = \{a\}$. If $S_a \neq S$, there is a $b \in S \setminus S_a$ and $S_a \cap S_b = \emptyset$. Let I be a maximal set of S such that $S_a \cap S_b = \emptyset$ for any two distinct elements of I. Clearly, $\cup_{a \in I} S_a = S$. Thus $\{S_a : a \in I\}$ is a partition of S into equivalence classes. Moreover, it is clear that for any two elements a, $b \in S$, $a \sim b$ if and only if a, $b \in S_c$ for some $c \in I$. $\qquad\square$

Conversely, we have

Theorem 1.9 *Let S be a nonempty set and $\{S_\alpha : \alpha \in I\}$ be a partition of S, where I is an index set. For any a, $b \in S$ define $a \sim b$ if and only if both a and b belong to an S_γ for some $\gamma \in I$. Then $\sim$ is an equivalence relation on S and for any $\alpha \in I$ $a \in S_\alpha$ implies $S_\alpha = S_a = \{x \in S : a \sim x\}$.*

The proof is left as an exercise.

Example 1.8 Let m be a fixed positive integer. For any two integers a and b we defined in Example 1.6 that

$$a \equiv b \,(\text{mod}\, m) \quad \text{if and only if} \quad m \,|\, (a - b)$$

and proved that $a \equiv b \,(\text{mod}\, m)$ is an equivalence relation in $\mathbb{Z}$. By Theorem 1.8, $\mathbb{Z}$ is partitioned into equivalence classes. The equivalence classes are called the *residue classes* mod m or, simply, *residue classes*. The residue class containing the integer a will be denoted by $\bar{a}$. Clearly, $\bar{a} = \bar{b}$ if and only if $m \,|\, (a - b)$ or $b = a + km$ for some integer k.

For example, for $m = 4$, we have residue classes:

$$\bar{0} = \{\ldots, -8, -4, 0, 4, 8, \ldots\},$$

$$\bar{1} = \{\ldots, -7, -3, 1, 5, 9, \ldots\},$$

$$\bar{2} = \{\ldots, -6, -2, 2, 6, 10, \ldots\},$$

$$\bar{3} = \{\ldots, -5, -1, 3, 7, 11, \ldots\}.$$

Notice that $\bar{0} = \bar{4} = \bar{8} = \overline{-4}$, etc. Clearly, $\mathbb{Z} = \bar{0} \cup \bar{1} \cup \bar{2} \cup \bar{3}$ and $\bar{0}, \bar{1}, \bar{2}, \bar{3}$ are pairwise disjoint.

We return to the general case. For any $a \in \mathbb{Z}$, dividing a by m, we have

$$a = qm + r, \quad \text{where } q, r \in \mathbb{Z} \text{ and } 0 \leq r < m.$$

Then $\bar{a} = \bar{r}$. Clearly, $0, 1, 2, \ldots, m - 1$ are pairwise not congruent modulo m, and $\bar{0}, \bar{1}, \bar{2}, \ldots, \overline{m - 1}$ are all the residue classes modulo m. For any a, $0 \leq a \leq m - 1$, we have

$$\bar{a} = \{a + km : k \in \mathbb{Z}\}$$

$$= \{\ldots, a - 2m, a - m, a, a + m, a + 2m, \ldots\}.$$

Clearly, $\mathbb{Z} = \bar{0} \cup \bar{1} \cup \bar{2} \cup \cdots \cup \overline{m - 1}$ and $\bar{0}, \bar{1}, \bar{2}, \ldots, \overline{m - 1}$ are pairwise disjoint. We introduce the notation

$$\mathbb{Z}_m = \{\bar{0}, \bar{1}, \bar{2}, \ldots, \overline{m - 1}\}.$$

1.4 Exercises

1.1 Let M, N, and P be subsets of a set S. Prove the distributive laws

$$M \cap (N \cup P) = (M \cap N) \cup (M \cap P),$$

$$M \cup (N \cap P) = (M \cup N) \cap (M \cup P).$$

1.2 Let M and N be subsets of a set S. Prove that

$$S \setminus (M \cup N) = (S \setminus M) \cap (S \setminus N),$$
$$S \setminus (M \cap N) = (S \setminus M) \cup (S \setminus N).$$

1.3 Let S be a finite set and ϕ be a map from S to itself. Prove that ϕ is injective if and only if it is surjective. Give examples showing that this statement is not necessarily true if S is infinite.

1.4 Let S and T be sets and ϕ be a map from S to T. Prove that

(i) ϕ is injective if and only if there is a map ψ from T to S such that $\psi \circ \phi = 1_S$. Moreover, if ψ is unique, then ϕ is bijective.

(ii) ϕ is surjective if and only if there is a map ψ from T to S such that $\phi \circ \psi = 1_T$. Moreover, if ψ is unique, then ϕ is bijective.

1.5 Let σ be a bijective map from a set S to itself. Define a map τ by $\tau(a) = b$ if $\sigma(b) = a$. Prove that τ is also bijective and $\sigma \circ \tau = \tau \circ \sigma = 1_S$.

1.6 Compute $\gcd(187, 221)$ and express it as a linear combination of 187 and 221 with integer coefficients.

1.7 Do the same for $\gcd(6188, 4709)$.

1.8 Prove (1.14) and (1.15).

1.9 Prove Theorem 1.4.

1.10 Prove (1.18) and (1.19).

1.11 Prove (1.20).

1.12 Let a and b be integers which are not both zero, prove that $\gcd(a, b)$ is the least positive integer of the form $ra + sb$, where $r, s \in \mathbb{Z}$. More generally, let $a_1, a_2, \ldots, a_n$ be integers which are not all zero, prove that $\gcd(a_1, a_2, \ldots, a_n)$ is the least positive integer of the form $c_1 a_1 + c_2 a_2 + \cdots + c_n a_n$, where $c_1, c_2, \ldots, c_n \in \mathbb{Z}$.

1.13 Let $a, b \in \mathbb{Z}$, which are not both zero and m be a positive integer. Prove that

$$\gcd(ma, mb) = m \gcd(a, b).$$

More generally, let $a_1, a_2, \ldots, a_n \in \mathbb{Z}$, which are not all zero, and m be a positive integer. Prove that

$$\gcd(ma_1, ma_2, \ldots, ma_n) = m \gcd(a_1, a_2, \ldots, a_n).$$

Let a, b, m be nonzero integers and assume that $\gcd(a, m) = 1$ and $\gcd(b, m) = 1$. Prove that $\gcd(ab, m) = 1$.

1.14 Let a and b be coprime integers and (x_0, y_0) be a set of integer solution of the Diophantine equation $ax + by = 1$. Prove that any set of integer solution is of the form $x = x_0 + bt$, $y = y_0 - at$, where $t \in \mathbb{Z}$.

1.15 Prove Theorem 1.9.

Chapter 2

Groups

Algebra aims at the study of algebraic structures. By an algebraic structure we mean a set with one or several operations. Group is an algebraic structure with only one operation and is of fundamental importance and indispensable in the study of finite fields (Galois fields) and Galois rings, which are the main concern of this book. So we begin with the study of groups in this chapter.

2.1 The Concept of a Group and Examples

The concept of a group together with the concept of a field was introduced by E. Galois (1811–1832) in 1832. The following is the definition of a group.

Definition 2.1 *Let G be a set with a binary operation, denoted by $\cdot$ and called the multiplication in G. G is called a group with respect to the multiplication, if the following manipulation rules hold:*

G1 *The associative law. That is, for all $a, b, c \in G$, $(a \cdot b) \cdot c = a \cdot (b \cdot c)$.*

G2 *There is an element in G, denoted by e, such that $a \cdot e = a$ for all $a \in G$.*

G3 *For any element $a \in G$, there is an element in G, denoted by a^{-1}, such that $a \cdot a^{-1} = e$.*

When G is a group with respect to the binary operation $\cdot$, we also say that $(G, \cdot)$ is a group in which the binary operation $\cdot$ is specified. We call G1, G2, and G3 in Definition 2.1 the *axioms of groups*. $a \cdot b$ is called the product of a and b. Usually we omit the multiplication sign $\cdot$ in $a \cdot b$ and write ab instead. In some groups the binary operation is denoted by $+$ and is called the *addition*, and $a + b$ is called the *sum* of a and b.

Example 2.1 Denoted by $\mathbb{Q}$ the set of rational numbers and $\mathbb{Q}^*$ the set of nonzero rational numbers. Now we verify that $(\mathbb{Q}^*, \cdot)$ is a group. First, we know that the associative law of multiplication holds in $\mathbb{Q}$, and, in particular, in $\mathbb{Q}^*$. Thus $G1$ holds in $\mathbb{Q}^*$. Next, the number 1 has the property $a \cdot 1 = a$ for all $a \in \mathbb{Q}$ and, in particular, for all $a \in \mathbb{Q}^*$. Thus $G2$ holds in $\mathbb{Q}^*$. Finally, for all $a \in \mathbb{Q}^*$, the number $1/a \in \mathbb{Q}^*$ and $a \cdot (1/a) = 1$. Therefore $G3$ is also verified. Hence $(\mathbb{Q}^*, \cdot)$ is a group. We remark that the commutative law of multiplication also holds in $\mathbb{Q}$, and, in particular, in $\mathbb{Q}^*$, i.e., $ab = ba$ for any $a, b \in \mathbb{Q}^*$. Denoted by $\mathbb{R}$ and $\mathbb{C}$ the set of real numbers and the set of complex numbers, respectively, and let $\mathbb{R}^*$ and $\mathbb{C}^*$ be the corresponding sets of nonzero numbers in $\mathbb{R}$ and $\mathbb{C}$. Similarly, both $(\mathbb{R}^*, \cdot)$ and $(\mathbb{C}^*, \cdot)$ are groups and $ab = ba$ for all $a, b \in \mathbb{R}^*$ or $\mathbb{C}^*$, respectively.

In the above examples of groups the binary operation is commutative. More generally, we have the following definition.

Definition 2.2 *Let G be a group and the binary operation in G be denoted by $\cdot$. If the commutative law holds in G, i.e., for all $a, b \in G, ab = ba$, G is called an abelian (or a commutative) group.*

Example 2.2 All $(\mathbb{Q}^*, \cdot)$, $(\mathbb{R}^*, \cdot)$, and $(\mathbb{C}^*, \cdot)$ are abelian groups. Moreover, $\mathbb{Q}$ itself is an abelian group with respect to the addition of rational numbers. Similarly, both $(\mathbb{R}, +)$ and $(\mathbb{C}, +)$ are abelian groups.

Example 2.3 Denote by $\mathbb{Z}$ the set of integers, then $(\mathbb{Z}, +)$ is an abelian group. But $(\mathbb{Z}^*, \cdot)$ is not a group, where $\mathbb{Z}^*$ is the set of nonzero integers. In fact, $G2$ is satisfied only with $1 \in \mathbb{Z}$, i.e., $a \cdot 1 = a$ for all $a \in \mathbb{Z}^*$; but for any integer $a \neq \pm 1$, there does not exist an integer a^{-1} such that $aa^{-1} = 1$, i.e., $G3$ does not hold in $\mathbb{Z}^*$. However, $(\{\pm 1\}, \cdot)$ is a group. It is a group with only two elements.

Example 2.4 Let m be a fixed positive integer. For any two integers a and b we defined in Example 1.6 that

$$a \equiv b \ (\mathrm{mod}\, m) \quad \text{if and only if} \quad m \,|\, (a - b)$$

and proved that $a \equiv b \ (\mathrm{mod}\, m)$ is an equivalence relation on $\mathbb{Z}$. In Example 1.8 the equivalence classes are called residue classes $\mathrm{mod}\, m$, the residue class $\mathrm{mod}\, m$ containing the integer a is denoted by $\bar{a}$, and $\bar{a} = \{a + km : k \in \mathbb{Z}\}$. We saw that $\mathbb{Z}$ is partitioned into a disjoint union of residue classes $\mathrm{mod}\, m$: $\bar{0}, \bar{1}, \bar{2}, \ldots, \overline{(m - 1)}$. The totality of the residue classes $\mathrm{mod}\, m$ is denoted by $\mathbb{Z}_m$, i.e.,

$$\mathbb{Z}_m = \{\bar{0}, \bar{1}, \bar{2}, \ldots, \overline{(m - 1)}\}$$

and $|\mathbb{Z}_m| = m$.

Let $\bar{a}, \bar{b} \in \mathbb{Z}_m$. We define their *sum* by

$$\bar{a} + \bar{b} = \overline{a + b}. \tag{2.1}$$

If $a_1 \in \bar{a}$ and $b_1 \in \bar{b}$, then $a_1 = a + k_1 m$ and $b_1 = b + k_2 m$, where $k_1, k_2 \in \mathbb{Z}$. Thus $a_1 + b_1 = a + b + (k_1 + k_2)m$, hence $a_1 + b_1 \in \overline{(a + b)}$. Therefore Definition (2.1) is independent of the particular choices of the elements a and b in the residue classes $\bar{a}$ and $\bar{b}$, respectively. It follows that the map

$$\mathbb{Z}_m \times \mathbb{Z}_m \to \mathbb{Z}_m$$
$$(\bar{a}, \bar{b}) \mapsto \overline{a + b}$$

is a well-defined binary operation in $\mathbb{Z}_m$, which is called the modulo m *addition* or, simply, the *addition* in $\mathbb{Z}_m$. In performing the addition $\bar{a} + \bar{b}$ in $\mathbb{Z}_m$ we usually take the smallest nonnegative integer r in $\overline{a + b}$ and write $\overline{a + b} = \bar{r}$. For example, in $\mathbb{Z}_4$, $\bar{2} + \bar{2} = \bar{0}$, $\bar{2} + \bar{3} = \bar{1}$, etc.

Now we verify that $\mathbb{Z}_m$ is a group with respect to the modulo m addition. First, for all $\bar{a}, \bar{b}, \bar{c} \in \mathbb{Z}_m$

$$(\bar{a} + \bar{b}) + \bar{c} = \overline{a + b} + \bar{c} = \overline{(a + b) + c} = \overline{a + (b + c)} = \bar{a} + \overline{b + c} = \bar{a} + (\bar{b} + \bar{c}),$$

i.e., $G1$ holds in $\mathbb{Z}_m$. Second, for the residue class $\bar{0}$,

$$\bar{a} + \bar{0} = \overline{a + 0} = \bar{a} \quad \text{for any residue class } \bar{a},$$

i.e., $G2$ holds in $\mathbb{Z}_m$. Finally, for any residue class $\bar{a}$ we have the residue class $\overline{(-a)}$ such that

$$\bar{a} + \overline{(-a)} = \overline{a + (-a)} = \bar{0},$$

i.e., $G3$ also holds in $\mathbb{Z}_m$. Moreover, we also have

$$\bar{a} + \bar{b} = \overline{a + b} = \overline{b + a} = \bar{b} + \bar{a} \quad \text{for } \bar{a}, \bar{b} \in \mathbb{Z}_m.$$

Therefore $(\mathbb{Z}_m, +)$ is an abelian group.

For any two residue classes mod m, $\bar{a}$ and $\bar{b}$, define their *product* by

$$\bar{a} \cdot \bar{b} = \overline{ab}. \tag{2.2}$$

If $a_1 \in \bar{a}$ and $b_1 \in \bar{b}$, then $a_1 = a + k_1 m$ and $b_1 = b + k_2 m$, where $k_1, k_2 \in \mathbb{Z}$. Thus $a_1 b_1 = ab + (k_1 k_2 m + k_1 b + k_2 a)m$ and, hence, $a_1 b_1 \in \overline{ab}$. Therefore (2.2) is well-defined and the map $(\bar{a}, \bar{b}) \mapsto \bar{a} \cdot \bar{b}$ is called the modulo m *multiplication* or simply the *multiplication* in $\mathbb{Z}_m$. As in the case of addition, in performing the multiplication $\bar{a} \cdot \bar{b}$ we usually take the smallest nonnegative integer r in $\overline{ab}$ and write $\overline{ab} = \bar{r}$. For example, in $\mathbb{Z}_4$ $\bar{2} \cdot \bar{2} = \bar{0}$, $\bar{2} \cdot \bar{3} = \bar{2}$, etc. But $\mathbb{Z}_m$ is not a group with respect to the modulo m multiplication unless $m = 1$. When $m = 1$, $\mathbb{Z}_1 = \{\bar{0}\}$ is certainly a group with only one element. When $m > 1$, $\bar{1}$ is the only residue class such that $\bar{a} \cdot \bar{1} = \bar{a}$ for all $\bar{a} \in \mathbb{Z}_m$; and for the residue class $\bar{0}$ we have $\bar{0} \cdot \bar{a} = \bar{0}$ for all $\bar{a} \in \mathbb{Z}_m$, thus there exists no residue class $\bar{b}$ in $\mathbb{Z}_m$ such that $\bar{0} \cdot \bar{b} = \bar{1}$. Moreover, if we denote the set of residue classes $\neq \bar{0}$ by $\mathbb{Z}'_m$, $\mathbb{Z}'_m$ is, in general, not a group with respect to the modulo m multiplication. For example, let $m = 6$, then $\bar{2} \cdot \bar{3} = \bar{0}$. If there exists a residue class $\bar{b}$ in $\mathbb{Z}_m$ such that $\bar{3} \cdot \bar{b} = \bar{1}$, then $\bar{2} = \bar{2} \cdot \bar{1} = \bar{2} \cdot (\bar{3} \cdot \bar{b}) = (\bar{2} \cdot \bar{3}) \cdot \bar{b} = \bar{0} \cdot \bar{b} = \bar{0}$, a contradiction.

Define a residue class $\bar{a}$ to be coprime with m if there are integers c and d such that $ca + dm = 1$, which is equivalent to $\gcd(a, m) = 1$ by Theorem 1.3. Let $a_1 \in \bar{a}$, then $a_1 = a + km$, where $k \in \mathbb{Z}$ and $\gcd(a_1, m) = \gcd(a + km, m) = \gcd(a, m) = 1$. Therefore the definition of a residue class to be coprime with m is well-defined. Denote the set of residue classes coprime with m by $\mathbb{Z}_m^*$. We are going to verify that $\mathbb{Z}_m^*$ is an abelian group with respect to the modulo m multiplication. For any $\bar{a}, \bar{b} \in \mathbb{Z}_m^*$ we have $\gcd(a, m) = 1$ and $\gcd(b, m) = 1$, which imply $\gcd(ab, m) = 1$, hence, $\overline{ab} = \overline{ab} \in \mathbb{Z}_m^*$. That is, the modulo m multiplication is a binary operation in $\mathbb{Z}_m^*$. Moreover, $G1$ trivially holds. For $G2$, we have $\bar{1} \in \mathbb{Z}_m^*$ and $\bar{a} \cdot \bar{1} = \bar{a}$ for all $\bar{a} \in \mathbb{Z}_m^*$. Let $\bar{a} \in \mathbb{Z}_m^*$, i.e., $\gcd(a, m) = 1$, then there are integers c and d such that $ca + dm = 1$, which implies $\gcd(c, m) = 1$, $\bar{c} \in \mathbb{Z}_m^*$, $ac \equiv 1 \,(\mathrm{mod}\, m)$, and $\bar{a} \cdot \bar{c} = \bar{1}$. Therefore $G3$ is verified. Finally, for all $\bar{a}, \bar{b} \in \mathbb{Z}_m^*$, $\bar{a} \cdot \bar{b} = \overline{ab} = \overline{ba} = \bar{b} \cdot \bar{a}$. Therefore $(\mathbb{Z}_m^*, \cdot)$ is an abelian group. In particular, if p is a prime, $\mathbb{Z}_p^* = \{\bar{1}, \bar{2}, \ldots, \overline{p-1}\}$ is an abelian group.

The following is an example of a group which is not abelian.

Example 2.5 Let $\triangle A_1 A_2 A_3$ be a regular triangle with center O in the Euclidian plane P. Denote by $D_{2.3}$ the set of distance preserving maps of P which carry the triangle $\triangle A_1 A_2 A_3$ into itself. Note that each element of $D_{2.3}$ is a bijective map of P to itself. For $T_1, T_2 \in D_{2.3}$, define the product of T_1 and T_2 to be the composite $T_1 \circ T_2$ of T_1 and T_2, i.e., at first we perform T_2 and then perform T_1. It is clear that $T_1 \circ T_2$ is also an element of $D_{2.3}$. Thus a binary operation is introduced in $D_{2.3}$. Moreover, we can verify that the axioms $G1, G2$, and $G3$ hold in $D_{2.3}$. In fact, $G1$ is a special case of the associativity of the composition of maps. Denote by E the identity map of P, i.e., the map which leaves every point in P fixed. Clearly, E is a distance preserving map leaving $\triangle A_1 A_2 A_3$ invariant. E is called the *identity transformation*, and $T \circ E = T$ for all $T \in D_{2.3}$. Thus $G2$ holds. Finally, let $T \in D_{2.3}$ and denote the *inverse transformation* of T by T^{-1}, that is, if T carries a point x into y, then T^{-1} carries y into x, then T^{-1} is also a distance preserving map carrying $\triangle A_1 A_2 A_3$ into itself and $T \circ T^{-1} = E$. Therefore, $G3$ also holds. Hence $D_{2.3}$ is a group.

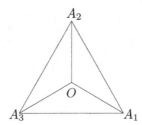

Figure 2.1: A regular triangle

$D_{2.3}$ is called the *symmetry group of the regular triangle*.

Let us give a concrete realization of $D_{2.3}$. Let $T \in D_{2.3}$, then T carries $\triangle A_1 A_2 A_3$ into itself, and, hence, leaves the center O invariant. Thus T carries OA_1 into OA_1, OA_2 or OA_3. If T carries OA_1 into OA_1, i.e., T leaves OA_1 invariant, then either T leaves also OA_2 and OA_3 invariant or T interchanges OA_2 and OA_3. In the former case, T is the identity transformation E and in the latter case T is the reflection with respect to OA_1. Denote these two elements by

$$
\begin{pmatrix} A_1 & A_2 & A_3 \\ A_1 & A_2 & A_3 \end{pmatrix} \text{ and } \begin{pmatrix} A_1 & A_2 & A_3 \\ A_1 & A_3 & A_2 \end{pmatrix},
$$

respectively. If T carries OA_1 into OA_2, then either T interchanges OA_1 and OA_2 and leaves OA_3 invariant or T carries OA_2 into OA_3 and OA_3 into OA_1. We have two elements of $D_{2.3}$ which can be represented by

$$
\begin{pmatrix} A_1 & A_2 & A_3 \\ A_2 & A_1 & A_3 \end{pmatrix} \text{ and } \begin{pmatrix} A_1 & A_2 & A_3 \\ A_2 & A_3 & A_1 \end{pmatrix},
$$

respectively. The former is the reflection with respect to OA_3, and the latter is the counterclockwise rotation around O through an angle $2\pi/3$. Finally, if T carries OA_1 into OA_3, then we have the following two possibilities

$$
\begin{pmatrix} A_1 & A_2 & A_3 \\ A_3 & A_2 & A_1 \end{pmatrix} \text{ and } \begin{pmatrix} A_1 & A_2 & A_3 \\ A_3 & A_1 & A_2 \end{pmatrix}.
$$

The former is the reflection with respect to OA_2, while the latter is the counterclockwise rotation around O through an angle $4\pi/3$.

Clearly,

$$
\begin{pmatrix} A_1 & A_2 & A_3 \\ A_2 & A_1 & A_3 \end{pmatrix} \circ \begin{pmatrix} A_1 & A_2 & A_3 \\ A_1 & A_3 & A_2 \end{pmatrix} = \begin{pmatrix} A_1 & A_2 & A_3 \\ A_2 & A_3 & A_1 \end{pmatrix},
$$

$$
\begin{pmatrix} A_1 & A_2 & A_3 \\ A_1 & A_3 & A_2 \end{pmatrix} \circ \begin{pmatrix} A_1 & A_2 & A_3 \\ A_2 & A_1 & A_3 \end{pmatrix} = \begin{pmatrix} A_1 & A_2 & A_3 \\ A_3 & A_1 & A_2 \end{pmatrix},
$$

and

$$\left(\begin{array}{ccc} A_1 & A_2 & A_3 \\ A_2 & A_3 & A_1 \end{array}\right) \neq \left(\begin{array}{ccc} A_1 & A_2 & A_3 \\ A_3 & A_1 & A_2 \end{array}\right).$$

Therefore, the commutative law does not hold in $D_{2 \cdot 3}$ and $D_{2 \cdot 3}$ is not an abelian group.

Example 2.6 Let $i_1, i_2, \ldots, i_n$ be an arrangement of $1, 2, \ldots, n$. We use the symbol

$$\left(\begin{array}{cccc} 1 & 2 & \cdots & n \\ i_1 & i_2 & \cdots & i_n \end{array}\right)$$

to denote a *permutation* on n elements $1, 2, \ldots, n$, which carries $1, 2, \ldots, n$ into $i_1, i_2, \ldots, i_n$, respectively. For instance,

$$\left(\begin{array}{cccc} 1 & 2 & 3 & 4 \\ 2 & 1 & 4 & 3 \end{array}\right)$$

is a permutation on four elements $1, 2, 3, 4$, which carries $1, 2, 3, 4$, respectively, into $2, 1, 4, 3$. Define the product of two permutations

$$\left(\begin{array}{cccc} 1 & 2 & \cdots & n \\ i_1 & i_2 & \cdots & i_n \end{array}\right) \text{ and } \left(\begin{array}{cccc} 1 & 2 & \cdots & n \\ j_1 & j_2 & \cdots & j_n \end{array}\right)$$

to be their composite

$$\left(\begin{array}{cccc} 1 & 2 & \cdots & n \\ j_1 & j_2 & \cdots & j_n \end{array}\right) \circ \left(\begin{array}{cccc} 1 & 2 & \cdots & n \\ i_1 & i_2 & \cdots & i_n \end{array}\right) = \left(\begin{array}{cccc} 1 & 2 & \cdots & n \\ j_{i_1} & j_{i_2} & \cdots & j_{i_n} \end{array}\right).$$

For instance,

$$\left(\begin{array}{cccc} 1 & 2 & 3 & 4 \\ 3 & 2 & 1 & 4 \end{array}\right) \circ \left(\begin{array}{cccc} 1 & 2 & 3 & 4 \\ 2 & 1 & 4 & 3 \end{array}\right) = \left(\begin{array}{cccc} 1 & 2 & 3 & 4 \\ 2 & 3 & 4 & 1 \end{array}\right).$$

There are altogether $n!$ arrangements of $1, 2, \cdots, n$ and, hence, altogether $n!$ permutations on n elements. It is easy to verify that they form a group with respect to the multiplication defined above. This group is called the *symmetric group* of degree n (or on n elements) and is denoted by S_n. It can be readily verified that S_n is not an abelian group, when $n \geq 3$ (Exercise 2.4).

In contrast to the groups $(\mathbb{Q}, +)$ and $(\mathbb{Q}^*, \cdot)$, which contain an infinite number of elements, each of the groups $\mathbb{Z}_m^*$, $D_{2 \cdot 3}$, and S_n contains only a finite number of elements. We have the following definition.

Definition 2.3 *Let G be any group. If G consists of an infinite number of elements, it is called an infinite group. If G consists of a finite number of elements, it is called a finite group. The cardinal $|G|$ of G is called the order of G.*

For example, the group $(\mathbb{Q}^*, \cdot)$ is an infinite abelian group; the symmetry group $D_{2\cdot3}$ of the regular triangle is a finite group of order 6; the symmetric group S_n is a finite group of order $n!$; and the group $(\{\pm1\}, \cdot)$ is a finite abelian group of order 2.

Theorem 2.1 *Let G be any group, then the following holds.*

(i) *The element e in G such that $ae = a$ for all $a \in G$ is unique and it satisfies also $ea = a$ for all $a \in G$; moreover, e is also the unique element in G which satisfies $ea = a$ for all $a \in G$.*

(ii) *For any $a \in G$, the element a^{-1} in G such that $aa^{-1} = e$ is unique and it satisfies also $a^{-1}a = e$; moreover, a^{-1} is also the unique element in G which satisfies $a^{-1}a = e$.*

(iii) $e^{-1} = e$.

Proof. (i) By $G2$ let $e \in G$ be such that $ae = a$ for all $a \in G$. Assume e_1 is another element in G such that $ae_1 = a$ for all $a \in G$. In particular, we have $e_1 e = e_1$ and $e_1 e_1 = e_1$. By group axiom $G3$, we have an $e_1^{-1} \in G$ such that $e_1 e_1^{-1} = e$. By the associative law we have

$$e_1 = e_1 e = e_1(e_1 e_1^{-1}) = (e_1 e_1)e_1^{-1} = e_1 e_1^{-1} = e.$$

This proves the uniqueness of e.

For any $a \in G$, by $G3$ we have $a^{-1} \in G$ such that $aa^{-1} = e$. By $G2$ $e = ee$. Thus

$$aa^{-1} = e = ee = e(aa^{-1}) = (ea)a^{-1}.$$

For $a^{-1} \in G$, by $G3$ there is an $(a^{-1})^{-1} \in G$ such that $a^{-1}(a^{-1})^{-1} = e$. Multiplying both sides of $aa^{-1} = (ea)a^{-1}$ by $(a^{-1})^{-1}$ from the right, we obtain $ea = a$.

Let e_1 be another element of G satisfying $e_1 a = a$ for all $a \in G$. From $ae = a$ for all $a \in G$, we obtain, in particular, $e_1 e = e_1$; and from $e_1 a = a$ for all $a \in G$, we obtain $e_1 e = e$. Therefore $e_1 = e$. Thus (i) is completely proved.

(ii) For any $a \in G$, let $a^{-1} \in G$ be such that $aa^{-1} = e$. Then

$$(a^{-1}a)a^{-1} = a^{-1}(aa^{-1}) = a^{-1}e = a^{-1}.$$

For $a^{-1} \in G$, there is an $(a^{-1})^{-1} \in G$ such that $a^{-1}(a^{-1})^{-1} = e$. Multiplying both sides of $(a^{-1}a)a^{-1} = a^{-1}$ by $(a^{-1})^{-1}$ from the right, we obtain $a^{-1}a = e$.

Now let a_1^{-1} be another element of G such that $aa_1^{-1} = e$. Then

$$a^{-1} = a^{-1}e = a^{-1}(aa_1^{-1}) = (a^{-1}a)a_1^{-1} = ea_1^{-1} = a_1^{-1}.$$

This proves that a^{-1} is the unique elements satisfying $aa^{-1} = e$. Similarly a^{-1} is also the unique element satisfying $a^{-1}a = e$.

(iii) By G2 $ee = e$. Then by (ii) $e^{-1} = e$.   □

Based on Theorem 2.1, we give the following definition.

Definition 2.4 *Let G be any group, the unique element e in G such that $ae = a$ for all $a \in G$ is called the identity of G. For any $a \in G$, the unique element a^{-1} in G such that $aa^{-1} = e$ is called the inverse of a.*

For instance, in Example 2.5 the identity of the symmetry group of the regular triangle is the identity transformation and

$$\left(\begin{array}{ccc} A_1 & A_2 & A_3 \\ A_2 & A_3 & A_1 \end{array} \right)^{-1} = \left(\begin{array}{ccc} A_1 & A_2 & A_3 \\ A_3 & A_1 & A_2 \end{array} \right),$$

$$\left(\begin{array}{ccc} A_1 & A_2 & A_3 \\ A_1 & A_3 & A_2 \end{array} \right)^{-1} = \left(\begin{array}{ccc} A_1 & A_2 & A_3 \\ A_1 & A_3 & A_2 \end{array} \right),$$

etc.; in Example 2.6 the identity permutation

$$\left(\begin{array}{cccc} 1 & 2 & \cdots & n \\ 1 & 2 & \cdots & n \end{array} \right)$$

is the identity of S_n, and in S_4

$$\left(\begin{array}{cccc} 1 & 2 & 3 & 4 \\ 2 & 3 & 1 & 4 \end{array} \right)^{-1} = \left(\begin{array}{cccc} 1 & 2 & 3 & 4 \\ 3 & 1 & 2 & 4 \end{array} \right),$$

$$\left(\begin{array}{cccc} 1 & 2 & 3 & 4 \\ 2 & 3 & 4 & 1 \end{array} \right)^{-1} = \left(\begin{array}{cccc} 1 & 2 & 3 & 4 \\ 4 & 1 & 2 & 3 \end{array} \right),$$

etc.

If the binary operation of a group is written as $+$, the identity is usually denoted by 0 and called the *zero* of the group and the inverse of an element a of the group is usually denoted by $-a$ and called the *negative* of a.

From the axioms of groups we can derive other manipulation rules which hold in any group. They are summarized in the following two theorems.

Theorem 2.2 *Let G be any group, then the following rules hold in G.*

(i) *The cancellation law. That is, for $a, b, c \in G$, if $ab = ac$, then $b = c$; and if $ba = ca$, then $b = c$.*

(ii) *Let a, b be any two elements in G, then $(ab)^{-1} = b^{-1}a^{-1}$.*

(iii) *Let a be any element in G, then $(a^{-1})^{-1} = a$.*

Proof. Let $ab = ac$. Multiplying both sides of the equation by a^{-1} from the left, we obtain $b = c$. Similarly, from $ba = ca$ we deduce $b = c$. Hence (i) holds.

Next we have
$$(ab)(ab)^{-1} = e.$$
On the other hand,
$$(ab)(b^{-1}a^{-1}) = a(b(b^{-1}a^{-1})) = a((bb^{-1})a^{-1})$$
$$= a(ea^{-1}) = aa^{-1} = e.$$

From the uniqueness of the inverse (i.e., Theorem 2.1(ii)), it follows that $(ab)^{-1} = b^{-1}a^{-1}$. So (ii) holds.

Finally we have $a^{-1}a = e$ and $a^{-1}(a^{-1})^{-1} = e$. Again from the uniqueness of the inverse it follows that $(a^{-1})^{-1} = a$. Hence, (iii) holds. $\square$

Let G be any group, a be any element in G, and n be a positive integer. Define
$$a^n = \underbrace{a\, a\, \cdots\, a}_{n\ a's},$$
$$a^0 = e,$$
$$a^{-n} = (a^{-1})^n.$$

Then we have

Theorem 2.3 *Let G be any group. Then the following rules hold in G.*

(i) *For any $a, b \in G$ such that $ab = ba$ and any integer n,*
$$(ab)^n = a^n b^n.$$

(ii) *For any $a \in G$ and integers m, n,*
$$a^{m+n} = a^m a^n,$$
$$a^{mn} = (a^m)^n.$$

As the proof is very simple, it will not be given here.

Finally, let us introduce the isomorphism of groups.

Definition 2.5 *Let G and G' be groups. If there is a bijection from G to G'*

$$\sigma : \quad G \to G'$$
$$a \mapsto \sigma(a),$$

which preserves the operations of the groups, i.e.,

$$\sigma(ab) = \sigma(a)\sigma(b) \quad \text{for all } a, b \in G,$$

then we say that G and G' are isomorphic, which is denoted by $G \simeq G'$, and that σ is an isomorphic map or, simply, an isomorphism from G to G'. An isomorphism from G to itself is called an automorphism of G.

Example 2.7 The map

$$\begin{pmatrix} 1 & 2 & 3 \\ i_1 & i_2 & i_3 \end{pmatrix} \longmapsto \begin{pmatrix} A_1 & A_2 & A_3 \\ A_{i_1} & A_{i_2} & A_{i_3} \end{pmatrix}$$

is an isomorphism from S_3 to $D_{2\cdot3}$.

Theorem 2.4 *Let σ be an isomorphism from a group G to a group G', then σ maps the identity e of G to that of G' and it also maps the inverse a^{-1} of any element a in G to the inverse $\sigma(a)^{-1}$ of the image $\sigma(a)$ of a.*

Proof. We have $e^2 = e$. Applying σ to it, we obtain $\sigma(e)^2 = \sigma(e)$. Let e' be the identity of G', then $\sigma(e)e' = \sigma(e)$. Thus $\sigma(e)^2 = \sigma(e)e'$. By the cancellation law, $\sigma(e) = e'$. For any $a \in G$, $aa^{-1} = e$. Applying σ to it, we obtain $\sigma(a)\sigma(a^{-1}) = \sigma(e) = e'$. Therefore $\sigma(a)^{-1} = \sigma(a^{-1})$.   □

We have also the following theorem.

Theorem 2.5 *Let G be a group and G' be a set with a binary operation, denoted by $\cdot$ and called the multiplication in G'. If there is a bijection $\sigma : G \to G'$ from G to G' which preserves the operations, i.e., $\sigma(ab) = \sigma(a) \cdot \sigma(b)$ for all $a, b \in G$, then G' is also a group and σ is an isomorphism.*

Proof. We verify that the group axioms are satisfied in G'.

For $a', b', c' \in G'$, since σ is a bijection, there are uniquely determined elements $a, b, c \in G$ such that $\sigma(a) = a', \sigma(b) = b'$, and $\sigma(c) = c'$. Since G is a group, we have $(ab)c = a(bc)$. Applying σ to this equation, we obtain $(\sigma(a) \cdot \sigma(b)) \cdot \sigma(c) = \sigma(a) \cdot (\sigma(b) \cdot \sigma(c))$, i.e., $(a' \cdot b') \cdot c' = a' \cdot (b' \cdot c')$. Therefore G1 holds in G'.

Let $e' = \sigma(e) \in G'$. Then for any $a' \in G'$, $a' \cdot e' = \sigma(a) \cdot \sigma(e) = \sigma(ae) = \sigma(a) = a'$. Therefore $G2$ also holds in G'.

Finally, for any $a' \in G'$ we have $a \in G$ such that $\sigma(a) = a'$. Applying σ to $aa^{-1} = e$, we obtain $\sigma(a) \cdot \sigma(a^{-1}) = \sigma(e)$, i.e., $a' \cdot \sigma(a^{-1}) = e'$. Therefore $G3$ also holds in G'. $\qquad\square$

2.2 Subgroups and Cosets

Definition 2.6 *Let G be a group and H a subset of G. If H is a group with respect to the operation of G, H is called a subgroup of G.*

Example 2.8 $(\mathbb{Q}^*, \cdot)$ is a subgroup of $(\mathbb{R}^*, \cdot)$, and $(\mathbb{R}^*, \cdot)$ is a subgroup of $(\mathbb{C}^*, \cdot)$. $(\{\pm 1\}, \cdot)$ is a subgroup of $(\mathbb{Q}^*, \cdot)$, also of $(\mathbb{R}^*, \cdot)$ and of $(\mathbb{C}^*, \cdot)$.

Example 2.9 Let m be a fixed positive integer and let

$$m\mathbb{Z} = \{mn : n \in \mathbb{Z}\}$$

be the set of integral multiples of m. Then $m\mathbb{Z}$ is a subgroup of the group $(\mathbb{Z}, +)$.

Example 2.10 The following two elements of S_3

$$\begin{pmatrix} 1 & 2 & 3 \\ 1 & 2 & 3 \end{pmatrix}, \begin{pmatrix} 1 & 2 & 3 \\ 2 & 1 & 3 \end{pmatrix}$$

form a subgroup of S_3, which is isomorphic to S_2 under the map

$$\begin{pmatrix} 1 & 2 & 3 \\ i_1 & i_2 & 3 \end{pmatrix} \longmapsto \begin{pmatrix} 1 & 2 \\ i_1 & i_2 \end{pmatrix}.$$

Example 2.11 The following four elements of S_4

$$\begin{pmatrix} 1 & 2 & 3 & 4 \\ 1 & 2 & 3 & 4 \end{pmatrix}, \begin{pmatrix} 1 & 2 & 3 & 4 \\ 2 & 1 & 4 & 3 \end{pmatrix}, \begin{pmatrix} 1 & 2 & 3 & 4 \\ 3 & 4 & 1 & 2 \end{pmatrix}, \begin{pmatrix} 1 & 2 & 3 & 4 \\ 4 & 3 & 2 & 1 \end{pmatrix}$$

form a subgroup of S_4, which is called the *Four group* and denoted by V_4. The following twelve elements

$$\begin{pmatrix} 1 & 2 & 3 & 4 \\ 1 & 2 & 3 & 4 \end{pmatrix}, \begin{pmatrix} 1 & 2 & 3 & 4 \\ 2 & 1 & 4 & 3 \end{pmatrix}, \begin{pmatrix} 1 & 2 & 3 & 4 \\ 3 & 4 & 1 & 2 \end{pmatrix}, \begin{pmatrix} 1 & 2 & 3 & 4 \\ 4 & 3 & 2 & 1 \end{pmatrix},$$

$$\begin{pmatrix} 1 & 2 & 3 & 4 \\ 2 & 3 & 1 & 4 \end{pmatrix}, \begin{pmatrix} 1 & 2 & 3 & 4 \\ 3 & 1 & 2 & 4 \end{pmatrix}, \begin{pmatrix} 1 & 2 & 3 & 4 \\ 2 & 4 & 3 & 1 \end{pmatrix}, \begin{pmatrix} 1 & 2 & 3 & 4 \\ 4 & 1 & 3 & 2 \end{pmatrix},$$

$$\begin{pmatrix} 1 & 2 & 3 & 4 \\ 3 & 2 & 4 & 1 \end{pmatrix}, \begin{pmatrix} 1 & 2 & 3 & 4 \\ 4 & 2 & 1 & 3 \end{pmatrix}, \begin{pmatrix} 1 & 2 & 3 & 4 \\ 1 & 3 & 4 & 2 \end{pmatrix}, \begin{pmatrix} 1 & 2 & 3 & 4 \\ 1 & 4 & 2 & 3 \end{pmatrix}$$

also form a subgroup of S_4, called the *alternating group* of degree 4 and denoted by A_4. Clearly, V_4 is also a subgroup of A_4.

Theorem 2.6 *Let G be a group and H be a subgroup of G. Then the identity of G belongs to H and is also the identity of H. The inverse of an element of H in G belongs to H and is also its inverse in H.*

Proof. Let e be the identity of G and e_H be that of H. Then $e_H e = e_H$ and $e_H e_H = e_H$. Thus $e_H e = e_H e_H$. By the cancellation law in G, $e = e_H$.

For any element $a \in H$, let a^{-1} be the inverse of a in G and a_H^{-1} be its inverse in H. Then $aa^{-1} = e$ and $aa_H^{-1} = e_H$. But $e = e_H$, therefore $aa^{-1} = aa_H^{-1}$. By the cancellation law in G, $a^{-1} = a_H^{-1}$.   □

The following theorem facilitates the verification of a subset of a group to be a subgroup.

Theorem 2.7 *Let G be a group and H be a nonempty subset of G. Then the following statements are equivalent:*

  (i) *H is a subgroup of G;*

 (ii) *For any $a, b \in H$ we have $ab \in H$ and for any $a \in H$ we have also $a^{-1} \in H$;*

(iii) *For any $a, b \in H$ we have $ab^{-1} \in H$.*

Proof. It is clear that (i) implies (ii) and (ii) implies (iii). Now let us prove that (iii) implies (i). Suppose that for any $a, b \in H$ we have $ab^{-1} \in H$. In particular, for any $a \in H$ we have $e = aa^{-1} \in H$. Thus for any $a \in H$, $a^{-1} = ea^{-1} \in H$. Consequently, for any $a, b \in H$ we have $b^{-1} \in H$ and $ab = a(b^{-1})^{-1} \in H$, that is, H is closed with respect to the operation of G. The group axiom $G1$ is automatically satisfied for H. The group axioms $G2$ and $G3$ were proved in the foregoing arguments. Therefore H is a subgroup of G.   □

Now let us come to the study of cosets.

Definition 2.7 *Let G be a group and H be a subgroup of G. Let a be an element of G. The set of elements*

$$\{ah : h \in H\}$$

is called a left coset of G relative to H, or more precisely, the left coset of the element a relative to H and is denoted by aH, i.e.,

$$aH = \{ah : h \in H\}.$$

Since $H = eH$, H is itself a left coset relative to H.

Example 2.12 Let $G = \mathbb{Q}^*$ and $H = \{\pm 1\}$. We know that $\mathbb{Q}^*$ is an abelian group with respect to the multiplication and H is a subgroup of $\mathbb{Q}^*$. For any $a \in \mathbb{Q}^*$, the set

$$aH = \{\pm a\}$$

is the left coset of a relative to H and $(-a)H = aH$. Moreover, $\mathbb{Q}^*$ is the disjoint union of left cosets aH, where a runs through the set of positive rational numbers.

Example 2.13 Let m be a fixed positive integer. By Example 2.9 $m\mathbb{Z}$ is a subgroup of the group $(\mathbb{Z}, +)$. For any $a \in \mathbb{Z}$, the set

$$a + m\mathbb{Z} = \{a + mk : k \in \mathbb{Z}\}$$

is the left coset of a relative to $m\mathbb{Z}$. As in Example 1.8, for any $a \in \mathbb{Z}$, denote by $\bar{a}$, the residue class mod m containing a, then $\bar{a} = a + m\mathbb{Z}$. Thus for any $a \in \mathbb{Z}$ the residue class mod m containing a coincides with the left coset of a relative to $m\mathbb{Z}$. Clearly, we have

$$(a + m\mathbb{Z}) \cap (b + m\mathbb{Z}) = \begin{cases} (a + m\mathbb{Z}) & \text{if } a \equiv b \pmod{m} \\ \emptyset & \text{if } a \not\equiv b \pmod{m}, \end{cases}$$

and $\mathbb{Z}$ is the disjoint union of the following m left cosets:

$$m\mathbb{Z},\, 1 + m\mathbb{Z},\, 2 + m\mathbb{Z},\, \ldots,\, (m-1) + m\mathbb{Z}.$$

By Example 2.4,

$$\mathbb{Z}_m = \{m\mathbb{Z},\, 1 + m\mathbb{Z},\, 2 + m\mathbb{Z},\, \ldots,\, (m-1) + m\mathbb{Z}\}.$$

Obviously, the map

$$\{0, 1, 2, \ldots, m-1\} \rightarrow \mathbb{Z}_m$$
$$a \mapsto a + m\mathbb{Z}$$

is a bijection. Sometimes, for simplicity, we also identity the element $\bar{a} = a + m\mathbb{Z} \in \mathbb{Z}_m$ ($a = 0, 1, 2, \ldots, m-1$) with the corresponding element $a \in \{0, 1, 2, \ldots, m-1\}$. Then

$$\mathbb{Z}_m = \{0, 1, 2, \ldots, m-1\}.$$

Example 2.14 Let

$$H = \left\{ \begin{pmatrix} 1 & 2 & 3 \\ 1 & 2 & 3 \end{pmatrix}, \begin{pmatrix} 1 & 2 & 3 \\ 2 & 1 & 3 \end{pmatrix} \right\}.$$

H is a subgroup of S_3. Clearly,

$$H, \begin{pmatrix} 1 & 2 & 3 \\ 1 & 3 & 2 \end{pmatrix} H = \left\{ \begin{pmatrix} 1 & 2 & 3 \\ 1 & 3 & 2 \end{pmatrix}, \begin{pmatrix} 1 & 2 & 3 \\ 3 & 1 & 2 \end{pmatrix} \right\},$$

$$\begin{pmatrix} 1 & 2 & 3 \\ 3 & 2 & 1 \end{pmatrix} H = \left\{ \begin{pmatrix} 1 & 2 & 3 \\ 3 & 2 & 1 \end{pmatrix}, \begin{pmatrix} 1 & 2 & 3 \\ 2 & 3 & 1 \end{pmatrix} \right\}$$

are three left cosets of S_3 relative to H. We can show that any left coset of S_3 relative to H coincides with one of the above three.

Example 2.15 We know that V_4 is a subgroup of A_4, thus V_4 is a left coset relative to V_4. Moreover,

$$\begin{pmatrix} 1 & 2 & 3 & 4 \\ 2 & 3 & 1 & 4 \end{pmatrix} V_4 =$$

$$\left\{ \begin{pmatrix} 1 & 2 & 3 & 4 \\ 2 & 3 & 1 & 4 \end{pmatrix}, \begin{pmatrix} 1 & 2 & 3 & 4 \\ 3 & 2 & 4 & 1 \end{pmatrix}, \begin{pmatrix} 1 & 2 & 3 & 4 \\ 1 & 4 & 2 & 3 \end{pmatrix}, \begin{pmatrix} 1 & 2 & 3 & 4 \\ 4 & 1 & 3 & 2 \end{pmatrix} \right\}$$

and

$$\begin{pmatrix} 1 & 2 & 3 & 4 \\ 3 & 1 & 2 & 4 \end{pmatrix} V_4 =$$

$$\left\{ \begin{pmatrix} 1 & 2 & 3 & 4 \\ 3 & 1 & 2 & 4 \end{pmatrix}, \begin{pmatrix} 1 & 2 & 3 & 4 \\ 1 & 3 & 4 & 2 \end{pmatrix}, \begin{pmatrix} 1 & 2 & 3 & 4 \\ 2 & 4 & 3 & 1 \end{pmatrix}, \begin{pmatrix} 1 & 2 & 3 & 4 \\ 4 & 2 & 1 & 3 \end{pmatrix} \right\}$$

are also left cosets relative to V_4. We can show that in A_4 any left coset relative to V_4 coincides with one of the above three left cosets and A_4 is the disjoint union of them.

In general we have

Lemma 2.8 *Let G be a group and H be a subgroup of G. Then any two left cosets of G relative to H either have no elements in common or coincide.*

Proof. Let aH and bH be any two left cosets relative to H, where $a, b \in G$. It suffices to prove that if aH and bH have an element in common then they must coincide. Let c be a common element of aH and bH, then there are elements $h_1, h_2 \in H$ such that

$$c = ah_1 = bh_2.$$

Thus $a = bh_2h_1^{-1}$ and $b = ah_1h_2^{-1}$. For any element $ah \in aH$, where $h \in H$, we have

$$ah = (bh_2h_1^{-1})h = b(h_2h_1^{-1}h) \in bH,$$

therefore $aH \subset bH$; similarly, we have $bH \subset aH$. Hence $aH = bH$.   □

From the above lemma we deduce

Theorem 2.9 *Let G be a group and H be a subgroup of G. Then G is the disjoint union of left cosets relative to H.*

Proof. Clearly, we have

$$G = \bigcup_{g \in G} gH.$$

Lemma 2.8 tells us that any two left cosets relative to H either have no elements in common or coincide. Therefore, by deleting the repeated ones from the set of cosets $\{gH : g \in G\}$ we obtain a set of mutually disjoint left cosets relative to H whose union is G. □

Moreover, we have

Lemma 2.10 *Let G be a finite group and H be a subgroup of G. Then any two left cosets of G relative to H contain the same number of elements.*

Proof. Let gH and $g'H$ be any two left cosets of G relative to H. It is clear that

$$gH \to g'H$$
$$gh \mapsto g'h \ (h \in H)$$

is a bijection. Hence the lemma follows. □

Example 2.16 S_3 is a disjoint union of the three left cosets relative to

$$H = \left\{ \left(\begin{array}{ccc} 1 & 2 & 3 \\ 1 & 2 & 3 \end{array} \right), \left(\begin{array}{ccc} 1 & 2 & 3 \\ 2 & 1 & 3 \end{array} \right) \right\},$$

i.e.,

$$S_3 = H \cup \left(\begin{array}{ccc} 1 & 2 & 3 \\ 1 & 3 & 2 \end{array} \right) H \cup \left(\begin{array}{ccc} 1 & 2 & 3 \\ 3 & 2 & 1 \end{array} \right) H,$$

and we have $|S_3| = 3|H|$.

Example 2.17 A_4 is a disjoint union of three left cosets relative to V_4

$$A_4 = V_4 \cup \left(\begin{array}{cccc} 1 & 2 & 3 & 4 \\ 2 & 3 & 1 & 4 \end{array} \right) V_4 \cup \left(\begin{array}{cccc} 1 & 2 & 3 & 4 \\ 3 & 1 & 2 & 4 \end{array} \right) V_4$$

and we have $|A_4| = 3|V_4|$.

Example 2.18 S_4 is a disjoint union of two left cosets relative to A_4

$$S_4 = A_4 \cup \left(\begin{array}{cccc} 1 & 2 & 3 & 4 \\ 2 & 1 & 3 & 4 \end{array} \right) A_4$$

and $|S_4| = 2|A_4|$.

Definition 2.8 *The number of distinct left cosets of a group G relative to its subgroup H is called the index of H in G and denoted by $G : H$.*

For example, $A_4 : V_4 = 3$, $S_4 : A_4 = 2$, $\mathbb{Z} : m\mathbb{Z} = m$, etc.

Theorem 2.11 (Lagrange) *The order of a subgroup of a finite group divides the order of the group. More precisely, let G be a finite group and H be a subgroup of G. Then $G : H$ is finite and*

$$|G| = (G : H)|H|.$$

Proof. It is an immediate consequence of Theorem 2.9 and Lemma 2.10. □

Definition 2.9 *Let G be a group and H be a subgroup of G. Any element in a left coset aH of G relative to H is called a left coset representative of the left coset aH. From each left coset of G relative to H we choose a left coset representative, then the set so obtained is called a complete system of left coset representatives of G relative to H.*

Clearly, if $\{a_\alpha | \alpha \in J\}$ is a complete system of left coset representatives of G relative to H, where J is an index set, then G is the disjoint union

$$G = \bigcup_{\alpha \in J} a_\alpha H.$$

For example, in A_4, the set

$$\left\{ \begin{pmatrix} 1 & 2 & 3 & 4 \\ 1 & 2 & 3 & 4 \end{pmatrix}, \begin{pmatrix} 1 & 2 & 3 & 4 \\ 2 & 3 & 1 & 4 \end{pmatrix}, \begin{pmatrix} 1 & 2 & 3 & 4 \\ 3 & 1 & 2 & 4 \end{pmatrix} \right\}$$

is a complete system of left coset representatives of A_4 relative to V_4; in S_4, the set

$$\left\{ \begin{pmatrix} 1 & 2 & 3 & 4 \\ 1 & 2 & 3 & 4 \end{pmatrix}, \begin{pmatrix} 1 & 2 & 3 & 4 \\ 2 & 1 & 3 & 4 \end{pmatrix} \right\}$$

is a complete system of left coset representatives of S_4 relative to A_4; in $\mathbb{Z}$, the set

$$\{0, 1, 2, \ldots, m-1\}$$

is a complete system of left coset representatives of $\mathbb{Z}$ relative to $m\mathbb{Z}$; etc.

Parallel to left cosets we can also introduce right cosets.

Definition 2.10 *Let G be a group and H be a subgroup of G. Let a be an element of G. The set of elements*

$$\{ha : h \in H\}$$

is called a right coset of G relative to H, or more precisely, the right coset of a relative to H and is denoted by Ha, i.e.,

$$Ha = \{ha : h \in H\}.$$

If G is abelian, for any element $a \in G$ the right coset Ha of a relative to a subgroup H coincides with the left coset aH of a relative to H. However, if G is not abelian, it may occur that $aH \neq Ha$.

Example 2.19 S_3 has the following three right cosets relative to

$$H = \left\{ \begin{pmatrix} 1 & 2 & 3 \\ 1 & 2 & 3 \end{pmatrix}, \begin{pmatrix} 1 & 2 & 3 \\ 2 & 1 & 3 \end{pmatrix} \right\},$$

namely,

$$H, H \begin{pmatrix} 1 & 2 & 3 \\ 1 & 3 & 2 \end{pmatrix}, H \begin{pmatrix} 1 & 2 & 3 \\ 3 & 2 & 1 \end{pmatrix},$$

and any right coset of S_3 relative to H coincides with one of the above three. But we have

$$\begin{pmatrix} 1 & 2 & 3 \\ 1 & 3 & 2 \end{pmatrix} H \neq H \begin{pmatrix} 1 & 2 & 3 \\ 1 & 3 & 2 \end{pmatrix}.$$

Example 2.20 It can be readily verified that A_4 has the following three right cosets relative to V_4:

$$V_4,$$

$$V_4 \begin{pmatrix} 1 & 2 & 3 & 4 \\ 2 & 3 & 1 & 4 \end{pmatrix} =$$

$$\left\{ \begin{pmatrix} 1 & 2 & 3 & 4 \\ 2 & 3 & 1 & 4 \end{pmatrix}, \begin{pmatrix} 1 & 2 & 3 & 4 \\ 1 & 4 & 2 & 3 \end{pmatrix}, \begin{pmatrix} 1 & 2 & 3 & 4 \\ 4 & 1 & 3 & 2 \end{pmatrix}, \begin{pmatrix} 1 & 2 & 3 & 4 \\ 3 & 2 & 4 & 1 \end{pmatrix} \right\},$$

and

$$V_4 \begin{pmatrix} 1 & 2 & 3 & 4 \\ 3 & 1 & 2 & 4 \end{pmatrix} =$$

$$\left\{ \begin{pmatrix} 1 & 2 & 3 & 4 \\ 3 & 1 & 2 & 4 \end{pmatrix}, \begin{pmatrix} 1 & 2 & 3 & 4 \\ 4 & 2 & 1 & 3 \end{pmatrix}, \begin{pmatrix} 1 & 2 & 3 & 4 \\ 1 & 3 & 4 & 2 \end{pmatrix}, \begin{pmatrix} 1 & 2 & 3 & 4 \\ 2 & 4 & 3 & 1 \end{pmatrix} \right\},$$

and any right coset of A_4 relative to V_4 coincides with one of the above three right cosets. Moreover, we have

$$\begin{pmatrix} 1 & 2 & 3 & 4 \\ 2 & 3 & 1 & 4 \end{pmatrix} V_4 = V_4 \begin{pmatrix} 1 & 2 & 3 & 4 \\ 2 & 3 & 1 & 4 \end{pmatrix}$$

and

$$\begin{pmatrix} 1 & 2 & 3 & 4 \\ 3 & 1 & 2 & 4 \end{pmatrix} V_4 = V_4 \begin{pmatrix} 1 & 2 & 3 & 4 \\ 3 & 1 & 2 & 4 \end{pmatrix}.$$

Example 2.21 S_4 has two right cosets relative to A_4, which are

$$A_4 \quad \text{and} \quad A_4 \begin{pmatrix} 1 & 2 & 3 & 4 \\ 2 & 1 & 3 & 4 \end{pmatrix}$$

and any right coset coincides with one of them. We have

$$\begin{pmatrix} 1 & 2 & 3 & 4 \\ 2 & 1 & 3 & 4 \end{pmatrix} A_4 = A_4 \begin{pmatrix} 1 & 2 & 3 & 4 \\ 2 & 1 & 3 & 4 \end{pmatrix}.$$

Parallel to Lemma 2.8, Theorem 2.9, Lemma 2.10 and Theorem 2.11 we have

Lemma 2.12 *Let G be a group and H be a subgroup of G. Then any two right cosets of G relative to H either have no elements in common or coincide.*

Theorem 2.13 *Let G be a group and H be a subgroup of G. Then G is the disjoint union of right cosets relative to H.*

Lemma 2.14 *Let G be a group and H be a subgroup of G. Then any two right cosets of G relative to H contain the same number of elements.*

Theorem 2.15 *Let G be a finite group and H be a subgroup of G. Denote the number of distinct right cosets of G relative to H by $(G : H)_r$. Then*

$$|G| = (G : H)_r |H|.$$

The proofs of them are parallel to the corresponding ones and will not be given here.

From Theorems 2.11 and 2.15 we deduce

Corollary 2.16 *Let G be a finite group and H be a subgroup of G. Then the number of left cosets of G relative to H is equal to the number of right cosets of G relative to H.*

Parallel to Definition 2.9 we can also define a *right coset representative* of a right coset of a group G relative to a subgroup H of G and a *complete system of right coset representatives*. The detailed definition will, however, be omitted.

2.3   Cyclic Groups

First we give the definition of the order of a group element.

Definition 2.11 *Let G be any group and a be an element in G. If for any positive integer n, we have $a^n \neq e$, then a is called an element of infinite order. If there exists a positive integer n such that $a^n = e$, then a is called an element of finite order and the smallest positive integer n such that $a^n = e$ is called the order of a. The order of a is denoted by $\mathrm{ord}\,(a)$.*

For example, in $D_{2\cdot3}$, 3 is the smallest positive integer such that

$$\begin{pmatrix} A_1 & A_2 & A_3 \\ A_2 & A_3 & A_1 \end{pmatrix}^3 = \begin{pmatrix} A_1 & A_2 & A_3 \\ A_1 & A_2 & A_3 \end{pmatrix}.$$

Thus

$$\begin{pmatrix} A_1 & A_2 & A_3 \\ A_2 & A_3 & A_1 \end{pmatrix}$$

is an element of order 3 in $D_{2.3}$. In $\mathbb{Q}^*$, $2^n \neq 1$ for any positive integer n. So ord $(2) = \infty$ in $\mathbb{Q}^*$.

Let G be any group and a be any element of G. Denote by $\langle a \rangle$ the set of all positive, negative, and 0-th powers of a. Clearly, the associative law holds in $\langle a \rangle$; $e = a^0 \in \langle a \rangle$; for any $a^n \in \langle a \rangle$, we have also $a^{-n} \in \langle a \rangle$ and $a^n a^{-n} = e$; and $a^n a^m = a^m a^n$ for all $m, n \in \mathbb{Z}$. Therefore $\langle a \rangle$ is an abelian subgroup of G. $\langle a \rangle$ is called the subgroup *generated* by a, and a is called a *generator* of the subgroup $\langle a \rangle$.

Theorem 2.17 *Let G be any group and $a \in G$. If* ord $(a) = n < \infty$, *then the following n elements*

$$a^0 = e, \ a^1 = a, \ a^2, \ a^3, \ldots, a^{n-1} \tag{2.3}$$

are distinct, any power of a is equal to an element in (2.3), and

$$\langle a \rangle = \{e, \ a, \ a^2, \ldots, a^{n-1}\} \tag{2.4}$$

is an abelian subgroup of order n of G. If ord $(a) = \infty$, *then the elements in the bi-infinite sequence*

$$\ldots, a^{-2}, \ a^{-1}, \ e, \ a, \ a^2, \ldots \tag{2.5}$$

are distinct and

$$\langle a \rangle = \{\ldots, a^{-2}, \ a^{-1}, \ e, \ a, \ a^2, \ldots\} \tag{2.6}$$

is an abelian subgroup of infinite order of G. Moreover, in both cases ord $(a) = |\langle a \rangle|$.

Proof. Let a be an element of finite order n in G. Suppose that $a^{m_1} = a^{m_2}, 0 \leq m_1 \leq m_2 < n$, then $a^{m_2 - m_1} = e$. Since ord $(a) = n$ and $0 \leq m_2 - m_1 < n$, we must have $m_2 - m_1 = 0$, i.e., $m_1 = m_2$. This proves that the n elements in (2.3) are distinct. Let m be any integer. Dividing m by n, we obtain

$$m = qn + r$$

where $q, r \in \mathbb{Z}$ and $0 \leq r < n$. Then

$$a^m = a^{qn+r} = (a^n)^q \cdot a^r = a^r.$$

Since a^r is in (2.3), a^m is also in (2.3). Therefore (2.4) holds and ord $(a) = |\langle a \rangle|$. That $\langle a \rangle$ is an abelian subgroup of G has already been proved in the paragraph above Theorem 2.17.

The proof of the case ord $(a) = \infty$ is left to the reader. $\square$

Corollary 2.18 *Let G be any finite group. Then any element of G is of finite order which is a divisor of the order of G.*

Proof. Let $a \in G$. If $\mathrm{ord}\,(a) = \infty$, by Theorem 2.17 all the elements in (2.5) are distinct and they are infinite in number, which contradicts $|G| < \infty$. Therefore $\mathrm{ord}\,(a) < \infty$. Again by Theorem 2.17 $\mathrm{ord}\,(a) = |\langle a \rangle|$. By Lagrange's Theorem, $\mathrm{ord}\,(a)\,|\,|G|$. $\square$

Now we give the definition of a cyclic group.

Definition 2.12 *Let G be any group. If there is an element $a \in G$ such that $G = \langle a \rangle$, G is called the cyclic group generated by a and a is called a generator of G. Moreover, if $|G| < \infty$, G is called a finite cyclic group of order $|G|$; if $|G| = \infty$, G is called an infinite cyclic group.*

Clearly, cyclic groups are abelian. We have the following consequence of Theorem 2.17.

Theorem 2.19 *Let G be the cyclic group generated by an element $a \in G$. If G is of finite order n, then $\mathrm{ord}\,(a) = n$ and*

$$G = \{e,\ a,\ a^2,\ \ldots, a^{n-1}\}.$$

If G is of infinite order, then $\mathrm{ord}\,(a) = \infty$ and

$$G = \{\ldots, a^{-2},\ a^{-1},\ e,\ a,\ a^2,\ \ldots\}.$$

As for the order of a group element we have the following theorems.

Theorem 2.20 *Let G be any group, $a \in G$, and $\mathrm{ord}\,(a) = n < \infty$. Then for any integer m, $a^m = e$ if and only if $n\,|\,m$.*

Proof. Let $n\,|\,m$. We can assume that $m = qn$, where $q \in \mathbb{Z}$. Then

$$a^m = a^{qn} = (a^n)^q = e^q = e.$$

Conversely, assume that $a^m = e$. Dividing m by n, let the quotient be q and the remainder be r, i.e.,

$$m = qn + r, \quad \text{where } 0 \le r < n.$$

Then

$$e = a^m = a^{qn+r} = a^{qn}a^r = ea^r = a^r.$$

Since n is the smallest positive integer such that $a^n = e$, $r = 0$. Therefore $n\,|\,m$. $\square$

Theorem 2.21 *Let G be any group, $a \in G$, $\operatorname{ord}(a) = n < \infty$, and $t \in \mathbb{Z}$. Then $\operatorname{ord}(a^t) = n/\gcd(t, n)$.*

Proof. Let $\gcd(t, n) = d$, then $(a^t)^{n/d} = (a^n)^{t/d} = e^{t/d} = e$. Thus $\operatorname{ord}(a^t) \mid (n/d)$. Assume that $\operatorname{ord}(a^t) = m$, then $a^{tm} = (a^t)^m = e$. By Theorem 2.20 $n \mid tm$. Consequently, $(n/d) \mid (t/d)m$. From $\gcd(t, n) = d$ we deduce $\gcd(t/d, n/d) = 1$. Therefore $(n/d) \mid m$. Hence $\operatorname{ord}(a^t) = n/d = n/\gcd(t, n)$. $\square$

Corollary 2.22 *Let $G = \langle a \rangle$ be a cyclic group of order n. For any integer k, a^k is a generator of G if and only if $\gcd(k, n) = 1$.*

Proof. a^k is a generator of G if and only if $\operatorname{ord}(a^k) = n$. $\square$

Let n be a positive integer. The number of integers k in $\{0, 1, 2, \ldots, n-1\}$ which are coprime with n is denoted by $\phi(n)$. We call the function $n \mapsto \phi(n)$ the Euler ϕ-*function*. Clearly, $|\mathbb{Z}_n^*| = \phi(n)$. From Corollary 2.22 we deduce

Theorem 2.23 *The number of generators of a finite cyclic group of order n is $\phi(n)$.*

The first few values of $\phi(n)$ are shown below.

n	1	2	3	4	5	6	7
$\phi(n)$	1	1	2	2	4	2	6

Now we proceed to give an explicit formula of the Euler ϕ-function. First, we give a group theoretical proof of the following number theoretical identity involving $\phi(n)$.

Theorem 2.24 *Let n be any positive integer. Then*

$$\sum_{d \mid n, d > 0} \phi(d) = n, \tag{2.7}$$

where the d under summation sign runs through all positive divisors of n.

Proof. Let G be a cyclic group of order n generated by $a \in G$. Then

$$G = \langle a \rangle = \{e, a, a^2, \ldots, a^{n-1}\}.$$

By Lagrange's Theorem, the order of each element $a^t (0 \le t \le n - 1)$ is a divisor of n. Thus to prove (2.7), it is enough to prove that for any $d > 0$ and $d \mid n$, the number of elements of order d in G is equal to $\phi(d)$.

Let $d \mid n$ and $d > 0$. By Theorem 2.21, $a^{n/d}$ is of order d and so are $a^{(n/d)t}$ for all t coprime to d. Let $a^k \in G$ ($0 \leq k \leq n - 1$) and assume that ord $(a^k) = d$. Then $a^{kd} = e$. By Theorem 2.20, $n \mid kd$ and then $(n/d) \mid k$. Let $k = (n/d)t$, where $t \in \mathbb{Z}$ and $0 \leq t < d$, then $a^k = a^{(n/d)t}$. By Theorem 2.21 $d = $ ord $(a^k) = $ ord $(a^{(n/d)t}) = d/\gcd(d, t)$, which implies $\gcd(d, t) = 1$. Therefore all the elements of order d in G are of the form $a^{(n/d)t}$, where $0 \leq t < d$ and t is coprime with d. Hence $\phi(d)$ is the number of elements of order d in G. $\qquad\square$

Corollary 2.25 *Let p be a prime and e a positive integer. Then*

$$\phi(p^e) = p^{e-1}(p - 1). \qquad (2.8)$$

Proof. Clearly, all 1, 2, 3, $\dots, p - 1$ are coprime with p. Therefore $\phi(p) = p - 1$. Thus, (2.8) is true for $e = 1$. Now assume that $e > 1$. By Theorem 2.24, we have

$$\phi(1) + \phi(p) + \phi(p^2) + \cdots + \phi(p^e) = p^e$$

and

$$\phi(1) + \phi(p) + \phi(p^2) + \cdots + \phi(p^{e-1}) = p^{e-1}.$$

Substracting, we obtain

$$\phi(p^e) = p^e - p^{e-1} = p^{e-1}(p - 1).$$

$\qquad\square$

We also need the *Chinese Remainder Theorem*.

Theorem 2.26 *Let m_1, m_2, $\dots, m_r$ be r integers greater than 1, which are pairwise coprime, and let $m = m_1 m_2 \cdots m_r$. Then for any given $n_i \in \mathbb{Z}$, $i = 1, 2, \dots, r$, there exists an $n \in \mathbb{Z}$ such that the following r congruences*

$$n \equiv n_i \pmod{m_i}, \ i = 1, 2, \dots, r \qquad (2.9)$$

hold simultaneously. Moreover, any $n' \in \mathbb{Z}$ is a solution of (2.9) if and only if $n \equiv n' \pmod{m}$. In particular, there is a unique $n \in \mathbb{Z}$ with $0 \leq n < m$ satisfying (2.9). Finally, if $\gcd(n_i, m_i) = 1$ for all $i = 1, 2, \dots, r$, then $\gcd(n, m) = 1$, and conversely.

Proof. Let $\hat{m}_i = m/m_i$, $i = 1, 2, \dots, r$. Then $\gcd(\hat{m}_1, \hat{m}_2, \dots, \hat{m}_r) = 1$. By Theorem 1.4 there exist r integers c_1, c_2, $\dots, c_r$ such that

$$1 = c_1 \hat{m}_1 + c_2 \hat{m}_2 + \cdots + c_r \hat{m}_r.$$

It is clear that

$$c_i \hat{m}_i \equiv 1 \pmod{m_i}, \ i = 1, 2, \dots, r,$$

$$c_i \hat{m}_i \equiv 0 \pmod{m_j}, \text{ for } j \neq i.$$

Let
$$n = c_1 \hat{m}_1 n_1 + c_2 \hat{m}_2 n_2 + \cdots + c_r \hat{m}_r n_r,$$

then
$$n \equiv n_i \pmod{m_i}, \ i = 1, 2, \ldots, r.$$

Moreover, assume that $n' \in \mathbb{Z}$ satisfies also

$$n' \equiv n_i \pmod{m_i}, \ i = 1, 2, \ldots, r.$$

Then
$$n - n' \equiv 0 \pmod{m_i}, \ i = 1, 2, \ldots, r.$$

Thus
$$m_i \mid (n - n'), \ i = 1, 2, \ldots, r.$$

Since $m_1, m_2, \ldots, m_r$ are pairwise coprime,

$$m \mid (n - n').$$

Therefore
$$n \equiv n' \pmod{m}.$$

Conversely, suppose that $n' \in \mathbb{Z}$ and $n \equiv n' \pmod{m}$. Clearly,

$$n' \equiv n_i \pmod{m_i}, \ i = 1, 2, \ldots, r.$$

Now assume $\gcd(n, m) = 1$. Then $\gcd(n, m_i) = 1$. But $n \equiv n_i \pmod{m_i}$, so $\gcd(n_i, m_i) = 1$. Conversely, assume $\gcd(n_i, m_i) = 1$. Since $n \equiv n_i \pmod{m_i}$, we have $\gcd(n, m_i) = 1$. Since $m_1, m_2, \ldots, m_r$ are pairwise coprime, we also have $\gcd(n, m) = 1$. $\qquad\square$

Corollary 2.27 *Let $m_1, m_2, \ldots, m_r$ be r integers > 1 which are pairwise coprime, and let $m = m_1 m_2 \cdots m_r$. Then*

$$\phi(m) = \phi(m_1)\phi(m_2)\cdots\phi(m_r). \tag{2.10}$$

Proof. In Example 2.4 we knew that $\mathbb{Z}_m$ is the set of residue classes $\overline{n} = n + m\mathbb{Z}$, $n = 0, 1, 2, \ldots, m-1$. Then $\mathbb{Z}_m^*$ is the set of residue classes $\overline{n}$ such that $\gcd(n, m) = 1$. In Example 2.13 we identified the residue class $\overline{n} = n + m\mathbb{Z}$ ($n = 0, 1, 2, \ldots, m-1$) with the number $n \in \{0, 1, 2, \ldots, m-1\}$. Under this identification, the residue class $\overline{n} = n + m\mathbb{Z}$ ($0 \leq n < m$), which is coprime with m, is identified with the number n ($0 \leq n < m$), which is coprime with m. Similarly, for each $i = 1, 2, \ldots, r$ the residue class

$\overline{n}_i = n_i + m_i\mathbb{Z}$ $(0 \le n_i < m_i)$, which is coprime with m_i is identified with
the number n_i $(0 \le n_i < m_i)$, which is coprime with m_i. By Theorem 2.26,
the map

$$\mathbb{Z}^*_{m_1} \times \mathbb{Z}^*_{m_2} \times \cdots \times \mathbb{Z}^*_{m_r} \to \mathbb{Z}^*_m$$
$$(n_1, \, n_2, \, \ldots, \, n_r) \quad \mapsto n,$$

where $0 \le n_i < m_i$ $(i = 1, 2, \ldots, r)$, $\gcd(n_i, m_i) = 1$, and $0 \le n < m$,
$\gcd(n, m) = 1$, $n \equiv n_i \pmod{m_i}$, is a bijection. Hence (2.10) follows. $\square$

From Corollaries 2.25 and 2.27 we deduce immediately

Theorem 2.28 *Let m be a positive integer and*

$$m = p_1^{e_1} p_2^{e_2} \cdots p_r^{e_r}$$

be the prime factorization of m into a product of powers of distinct primes
$p_1, p_2, \ldots, p_r$. *Then*

$$\phi(m) = \prod_{i=1}^{r} \phi(p_i^{e_i}) = \prod_{i=1}^{r} p_i^{e_i - 1}(p_i - 1) = m \prod_{i=1}^{r}(1 - 1/p_i).$$

Now let us return to the study of cyclic groups.

Theorem 2.29   (i) *Any subgroup of a cyclic group is cyclic.*

(ii) *If G is a finite cyclic group of order n, then the order of any subgroup
of G is a divisor of n, and for any positive divisor d of n, G has a
unique subgroup of order d.*

(iii) *If G is an infinite cyclic group, then any subgroup $\ne \{e\}$ of G is
infinite cyclic and is of finite index in G, and for any positive integer
s, G has a unique subgroup of index s in G.*

Proof. (i) Let $G = \langle a \rangle$ be a cyclic group and H a subgroup of G. If $H = \{e\}$,
then $H = \langle e \rangle$. Now let $H \ne \{e\}$, then there exists an $n \ne 0$ in $\mathbb{Z}$ such that
$a^n \in H$. If $n < 0$, then $-n > 0$ and $a^{-n} = (a^n)^{-1} \in H$. Therefore we can
assume that $a^n \in H$ with $n > 0$. Let s be the smallest positive integer such
that $a^s \in H$. We assert that $H = \langle a^s \rangle$. For any element $a^m \in H$, by the
division algorithm we can write $m = qs + r$, where $q, r \in \mathbb{Z}$ and $0 \le r < s$.
Then $a^r = a^{m-qs} = a^m (a^s)^{-q} \in H$. Since s is the smallest positive integer
such that $a^s \in H$, we must have $r = 0$. Then $a^m = (a^s)^q \in \langle a^s \rangle$. Hence
$H = \langle a^s \rangle$.

(ii) Suppose $G = \langle a \rangle$ is of finite order n, then $G = \{e, a, a^2, \ldots, a^{n-1}\}$.
Let H be a subgroup of G and $H \ne \{e\}$. By the proof of (i), $H = \langle a^s \rangle$,
where s is the smallest positive integer such that $a^s \in H$. We claim that

$s \mid n$. By the division algorithm we write $n = qs + r$, where $q, r \in \mathbb{Z}$ and $0 \leq r < s$. Then $e = a^n = (a^s)^q a^r$, so $a^r = (a^s)^{-q} \in H$. The minimality of s forces $r = 0$, so $n = qs$, $s \mid n$, $H = \{e, a^s, a^{2s}, \ldots, a^{(n/s-1)s}\}$ is of order n/s.

Let d be any positive divisor of n. Clearly, $\langle a^{n/d} \rangle$ is a subgroup of order d. Let H be any subgroup of order d. By the preceding paragraph, we can assume that $H = \langle a^{s'} \rangle$ where s' is the smallest positive integer such that $a^{s'} \in H$, $s' \mid n$ and H is of order n/s'. Then $d = n/s'$ and $H = \langle a^{n/d} \rangle$. Hence G has a unique subgroup of order d.

(iii) Suppose $G = \langle a \rangle$ is infinite cyclic. Let H be a subgroup of G and $H \neq \{e\}$. By the proof of (i) $H = \langle a^s \rangle$, where s is the smallest positive integer such that $a^s \in H$. Since $H \neq \{e\}$, $s \neq 0$. Clearly, $\text{ord}(a^s) = \infty$. Therefore H is infinite cyclic. The cosets

$$\langle a^s \rangle, \ a\langle a^s \rangle, \ a^2\langle a^s \rangle, \ldots, a^{s-1}\langle a^s \rangle$$

are all the different cosets of G relative to $\langle a^s \rangle$, so $G : \langle a^s \rangle = s$.

For distinct integer m and n, $m - n \neq 0$, so $a^{m-n} \neq e$ and $a^m \neq a^n$. Hence for any positive integer s, the elements a^{ms}, $m = 0, \pm 1, \pm 2, \ldots$, are distinct, so $\langle a^s \rangle$ is an infinite cyclic subgroup of G. By the above proof, $G : \langle a^s \rangle = s$. Moreover, let H be a subgroup of index s in G. Assume $H = \langle a^{s'} \rangle$, then $G : H = s'$. Thus $s = s'$. Hence G has a unique subgroup of index s in G. $\qquad\square$

2.4 Exercises

2.1 Let m be a positive integer and $a, b, c, d \in \mathbb{Z}$. Prove that

   (i) $a \equiv b \ (\text{mod } m)$, $c \equiv d \ (\text{mod } m) \Rightarrow a \pm c \equiv b \pm d \ (\text{mod } m), ac \equiv bd \ (\text{mod } m)$.

   (ii) $ac \equiv bc \ (\text{mod } m) \Rightarrow a \equiv b \ (\text{mod } m/\gcd(c, m))$.

2.2 Let m be a positive integer and $a \in \mathbb{Z}$. Prove that there exists $b \in \mathbb{Z}$ such that $ab \equiv 1 \ (\text{mod } m)$ if and only if $\gcd(a, m) = 1$. Moreover, for any two $b, b' \in \mathbb{Z}$ such that $ab \equiv 1 \ (\text{mod } m)$ and $ab' \equiv 1 \ (\text{mod } m)$ we must have $b \equiv b' \ (\text{mod } m)$.

2.3 Let $a_1, a_2, \ldots, a_r$ be r elements of a group G. Prove that

$$(a_1 a_2 \cdots a_r)^{-1} = a_r^{-1} \cdots a_2^{-1} a_1^{-1}.$$

2.4 Prove that S_n is not abelian, when $n \geq 3$.

2.5 Let

$$\sigma = \begin{pmatrix} 1 & 2 & 3 & 4 & 5 & 6 \\ 5 & 6 & 3 & 1 & 4 & 2 \end{pmatrix}, \quad \tau = \begin{pmatrix} 1 & 2 & 3 & 4 & 5 & 6 \\ 3 & 4 & 6 & 2 & 5 & 1 \end{pmatrix}$$

be two elements of S_6. Compute $\sigma\tau, \tau\sigma, \sigma^{-1}, \sigma^2, \sigma^3$ and $\tau^{-1}\sigma\tau$.

2.6 Let

$$\sigma = \begin{pmatrix} 1 & 2 & 3 & 4 \\ 3 & 4 & 1 & 2 \end{pmatrix}, \quad \tau = \begin{pmatrix} 1 & 2 & 3 & 4 \\ 4 & 1 & 2 & 3 \end{pmatrix}$$

be two elements of S_4. Find permutations x and $y \in S_4$ such that $\sigma x = \tau$ and $y\sigma = \tau$.

2.7 Prove Theorem 2.3.

2.8 If the square of every element of a group is equal to the identity, prove that the group is abelian.

2.9 Prove that the intersection of two subgroups of a group is also a subgroup.

2.10 Let G be a group and define

$$C = \{c \in G : cx = xc \quad \text{for all } x \in G\}.$$

Prove that C is an abelian subgroup of G. C is called the *center* of G.

2.11 If $n \geq 3$, then the center of S_n consists of the identity permutation alone.

2.12 Let G be a group and $a \in G$. Define

$$C_G(a) = \{x \in G : x^{-1}ax = a\}.$$

Prove that $C_G(a)$ is a subgroup of G and contains a and C. $C_G(a)$ is called the *centralizer* of a in G.

2.13 Determine the centralizer of

$$\begin{pmatrix} 1 & 2 & 3 & 4 & 5 \\ 2 & 3 & 1 & 4 & 5 \end{pmatrix}$$

in S_5.

2.14 Let a and b be two elements of a group G. They are said to be *conjugate* if there is an element $x \in G$ such that $x^{-1}ax = b$. Prove that

(i) For a given $a \in G$ any two elements of the set

$$\{x^{-1}ax \: : \: x \in G\}$$

are conjugate and any element conjugate to one of the elements of the set is contained also in the set. This set is called a *conjugate class* of G.

(ii) For a given $a \in G$ and any $x, y \in G$, $x^{-1}ax = y^{-1}ay$ if and only if x and y belong to the same coset of G relative to $C_G(a)$.

(iii) If G is finite, then the number of elements of G conjugate to a given element a is equal to the index $G : C_G(a)$.

(iv) Two conjugate classes of G either have no elements in common or coincide.

2.15 List all the conjugate classes of S_4.

2.16 Let G be a group and H be a subgroup of G. Define

$$N_G(H) = \{x \in G \: : \: x^{-1}Hx = H\}.$$

Prove that $N_G(H)$ is a subgroup of G and contains H. $N_G(H)$ is called the *normalizer* of H in G.

2.17 Let H_1 and H_2 be two subgroups of a group G. They are said to be *conjugate* if there is an element $x \in G$ such that $x^{-1}H_1x = H_2$. Prove that

(i) For a given subgroup H of G any two subgroups of the set

$$\{x^{-1}Hx \: : \: x \in G\}$$

are conjugate and any subgroup conjugate to one of the subgroups of the set is contained also in the set.

(ii) For a given subgroup $H \subset G$ and any $x, y \in G$, $x^{-1}Hx = y^{-1}Hy$ if and only if x and y belong to the same coset of G relative to $N_G(H)$.

(iii) If G is finite, then the number of subgroups of G conjugate to a given subgroup H is equal to the index $G : N_G(H)$.

2.18 Prove that the centralizers of two conjugate elements of a group are conjugate.

2.19 Prove that the normalizers of two conjugate subgroups of a group are conjugate.

2.20 If a finite group G has only two conjugate classes, then $|G| = 2$.

2.21 Let G be a group, H a subgroup of G, and $a \in G$. Prove that

$$Ha^{-1} = \{x^{-1} \ : \ x \in aH\}.$$

2.22 Let a and b be any two elements of G. Prove that

   (i) a, a^{-1} and $b^{-1}ab$ have the same order.

   (ii) ab and ba have the same order.

2.23 Let a and b be any two elements of orders m and n, respectively, in an abelian group and assume $\gcd(m, n) = 1$. Prove that $\mathrm{ord}(ab) = mn$.

2.24 Let a and b be any two elements of orders m and n, respectively, in a group. If $ab = ba$ then $\mathrm{ord}(ab)$ divides the least common multiple $\mathrm{lcm}[m, n]$ of m and n. Give an example showing that $\mathrm{ord}(ab) \neq \mathrm{lcm}[m, n]$.

2.25 Prove the case when $\mathrm{ord}(a) = \infty$ in Theorem 1.17.

2.26 Prove that any two cyclic groups of the same order are isomorphic.

Chapter 3

Fields and Rings

3.1 Fields

Many algebraic structures, such as $\mathbb{Q}$, $\mathbb{R}$, and $\mathbb{C}$, have two binary operations, the addition and multiplication. Thus it is natural to introduce the concept of a field with two binary operations, which covers $\mathbb{Q}$, $\mathbb{R}$, and $\mathbb{C}$.

Definition 3.1 *Let F be a set with two binary operations, one is called the addition and denoted by $+$, and the other is called the multiplication and denoted by $\cdot$. F is called a field with respect to the addition and multiplication, if the following manipulation rules are fulfilled:*

F1 $(F, +)$ is an abelian group.

F2 $(F^*, \cdot)$ is an abelian group, where $F^* = F \setminus \{0\}$ and 0 is the zero of the group $(F, +)$.

F3 $a(b + c) = ab + ac$ for all $a, b, c \in F$.

We also say that $(F; +, \cdot)$ is a field. $(F, +)$ is called the additive group of the field F and $(F^*, \cdot)$ is called the multiplicative group of F.

We call $F1, F2$, and $F3$ *the axioms of fields.*

Definition 3.2 *Let F be any field. For any a, $b \in F$ $a + b$ is called the sum of a and b, and ab the product of a and b. The zero of the group $(F, +)$ is denoted by 0 and is called the zero of the field F. For any $a \in F$, the inverse of a in the group $(F, +)$ is denoted by $-a$ and is called the negative of a in the field F. The identity of $(F^*, \cdot)$ is denoted by e and is called the identity of F. For any $a \in F^*$, the inverse of a in the group $(F^*, \cdot)$ is denoted by a^{-1} and is called the inverse of a.*

Notice that, by Theorem 2.1 both 0 and e are uniquely determined in F and $-0 = 0$, and for any $a \in F^*$, both $-a$ and a^{-1} are also uniquely determined in F and $1^{-1} = 1$.

Example 3.1 $\mathbb{Q}$ is a field with respect to the addition and multiplication of rational numbers and is called the field of rational numbers. Similarly, $\mathbb{R}$ and $\mathbb{C}$ are fields and they are called the field of real numbers and the field of complex numbers, respectively.

We know that $\mathbb{R}$ is a subset of $\mathbb{C}$ and that the addition and multiplication of elements in $\mathbb{R}$ are the same as considering them as being elements in $\mathbb{C}$. In this way, we call $\mathbb{R}$ a subfield of $\mathbb{C}$. In general, we have the following definition.

Definition 3.3 *Let F be a field and F_0 a subset of F. If F_0 is a field with respect to the addition and multiplication of F, i.e., for any two elements in F_0, when the addition and multiplication according to F yield the sum and product which are elements of F_0, (in other words, F_0 is closed with respect to the addition and multiplication of F,) and the field axioms F1, F2, and F3 hold in F_0, then we say that F_0 is a subfield of F. If F_0 is a subfield of F we also say that F is an extension field of F_0 or an extension of F_0.*

By Definition 3.3, we can say that $\mathbb{R}$ is a subfield of $\mathbb{C}$, $\mathbb{Q}$ is a subfield of $\mathbb{C}$ and also a subfield of $\mathbb{R}$. Moreover, $\mathbb{R}$ is an extension field of $\mathbb{Q}$, and $\mathbb{C}$ is an extension field of $\mathbb{Q}$ and also of $\mathbb{R}$.

In order to clarify the concept of a field, we give more examples.

Example 3.2 Consider the set of real numbers of the form

$$a + b\sqrt{2}, \qquad a, b \in \mathbb{Q},$$

and denote the set by $\mathbb{Q}[\sqrt{2}]$. We define the sum and product of two elements in $\mathbb{Q}[\sqrt{2}]$ to be those by considering them as real numbers. Let us prove that $\mathbb{Q}[\sqrt{2}]$ is a field with respect to the addition and multiplication of real numbers, and hence is a subfield of $\mathbb{R}$. First, let $a + b\sqrt{2},\ c + d\sqrt{2} \in \mathbb{Q}[\sqrt{2}]$, then

$$(a + b\sqrt{2}) + (c + d\sqrt{2}) = (a + c) + (b + d)\sqrt{2},$$

$$(a + b\sqrt{2})(c + d\sqrt{2}) = (ac + 2bd) + (ad + bc)\sqrt{2}.$$

Clearly, $a + c,\ b + d,\ ac + 2bd,\ ad + bc \in \mathbb{Q}$. Hence $\mathbb{Q}[\sqrt{2}]$ is closed with respect to the addition and multiplication. Next, one needs to prove that field axioms $F1$, $F2$, and $F3$ holds in $\mathbb{Q}[\sqrt{2}]$. Since $F3$ holds in $\mathbb{R}$, it also holds in $\mathbb{Q}[\sqrt{2}]$. The verification of $F1$ is easy and is omitted. For $F2$, the associative law holds also automatically in $\mathbb{Q}[\sqrt{2}]$ and $1 = 1 + 0\sqrt{2} \in \mathbb{Q}[\sqrt{2}]$ is the identity of $\mathbb{Q}[\sqrt{2}]$. For $a + b\sqrt{2} \in \mathbb{Q}[\sqrt{2}]$ and $a + b\sqrt{2} \neq 0$, we have $a^2 - 2b^2 \neq 0$ so that

$$\frac{a}{a^2 - 2b^2} - \frac{b}{a^2 - 2b^2}\sqrt{2} \in \mathbb{Q}[\sqrt{2}]$$

and

$$(a + b\sqrt{2})(\frac{a}{a^2 - 2b^2} - \frac{b}{a^2 - 2b^2}\sqrt{2}) = 1.$$

Therefore $F2$ is verified. This completes the proof that $\mathbb{Q}[\sqrt{2}]$ is a subfield of $\mathbb{R}$.

Example 3.3 Consider the set of real numbers of the form

$$a + b\sqrt[3]{2}, \qquad a, b \in \mathbb{Q}$$

and denote this set by S. If we define the sum and product of two elements in S to be those by considering the elements as real numbers, S will not be a field, since S is not closed with respect to the multiplication. For example,

$$\sqrt[3]{2}\sqrt[3]{2} = \sqrt[3]{4}$$

which is not a number of the form $a + b\sqrt[3]{2}$ $(a, b \in \mathbb{Q})$.

But if we consider the set of real numbers of the form

$$a + b\sqrt[3]{2} + c\sqrt[3]{4}, \qquad a, b, c \in \mathbb{Q},$$

denote this set by $\mathbb{Q}[\sqrt[3]{2}]$, and define the sum and product of two elements in $\mathbb{Q}[\sqrt[3]{2}]$ to be those by considering them as being real numbers, then it can be proved that $\mathbb{Q}[\sqrt[3]{2}]$ is a field. This is left as an exercise.

In any of $\mathbb{C}$, $\mathbb{R}$, $\mathbb{Q}$, $\mathbb{Q}[\sqrt{2}]$, or $\mathbb{Q}[\sqrt[3]{2}]$, the number of elements is infinite. But there are fields with a finite number of elements. The following are some examples.

Example 3.4 Let $F_2 = \{0, 1\}$ and define the addition and multiplication in F_2 by the following tables:

+	0	1		·	0	1
0	0	1		0	0	0
1	1	0		1	0	1

The left table reads $0 + 0 = 0$, $0 + 1 = 1$, $1 + 0 = 1$, $1 + 1 = 0$ and the right table reads $0 \cdot 0 = 0$, $0 \cdot 1 = 0$, $1 \cdot 0 = 0$, $1 \cdot 1 = 1$. It is easy to verify that F_2 is a field.

Example 3.5 Let $F_3 = \{0, 1, 2\}$ and define the addition and multiplication in F_3 by the following tables:

+	0	1	2		·	0	1	2
0	0	1	2		0	0	0	0
1	1	2	0		1	0	1	2
2	2	0	1		2	0	2	1

It can be verified that F_3 is a field.

Example 3.6 Let $F_4 = \{0, 1, x, 1 + x\}$ and define the addition and multiplication in F_4 by the following tables:

+	0	1	x	$1+x$		·	0	1	x	$1+x$
0	0	1	x	$1+x$		0	0	0	0	0
1	1	0	$1+x$	x		1	0	1	x	$1+x$
x	x	$1+x$	0	1		x	0	x	$1+x$	1
$1+x$	$1+x$	x	1	0		$1+x$	0	$1+x$	1	x

It can be verified that F_4 is a field. Moreover, let $F_0 = \{0, 1\}$, then F_0 is a subfield of F_4.

Example 3.7 Let p be a prime. By Example 2.4, $\mathbb{Z}_p$ consists of p residue classes

$$\bar{a} = a + p\mathbb{Z} = \{a + pk : k \in \mathbb{Z}\}, \ a = 0, 1, 2, \ldots, p - 1.$$

In Example 2.4 we defined both the addition and multiplication in $\mathbb{Z}_p$, i.e., for $\bar{a}, \bar{b} \in \mathbb{Z}_p$, we defined

$$\bar{a} + \bar{b} = \overline{a + b}$$

and

$$\bar{a} \cdot \bar{b} = \overline{ab}.$$

We already proved that both $(\mathbb{Z}_p, +)$ and $(\mathbb{Z}_p^*, \cdot)$ are abelian groups, where $\mathbb{Z}_p^* = \{\bar{1}, \bar{2}, \ldots, \overline{p-1}\} = \mathbb{Z}_p \setminus \{\bar{0}\}$. Therefore both $F1$ and $F2$ hold in $\mathbb{Z}_p$. For any $\bar{a}, \bar{b}, \bar{c} \in \mathbb{Z}_p$, we have

$$\bar{a}(\bar{b} + \bar{c}) = \bar{a}(\overline{b + c}) = \overline{a(b + c)} = \overline{ab + ac}$$
$$= \overline{ab} + \overline{ac} = \bar{a}\bar{b} + \bar{a}\bar{c},$$

thus $F3$ also holds in $\mathbb{Z}_p$. Hence $\mathbb{Z}_p$ is a field. Clearly $\mathbb{Z}_p$ is a field with p elements. In $\mathbb{Z}_p$ we have

$$\underbrace{\bar{1} + \bar{1} + \cdots + \bar{1}}_{p \ \bar{1}'s} = \bar{0}.$$

Sometimes, for simplicity, we also identify $\bar{a} \in \mathbb{Z}_p$ ($a = 0, 1, 2, \ldots, p-1$) with $a \in \{0, 1, 2, \ldots, p-1\}$ as we did in Example 2.13. Then we have $\mathbb{Z}_p = \{0, 1, 2, \ldots, p-1\}$.

When m is a positive composite number, we also defined the addition and multiplication in $\mathbb{Z}_m$ in Example 2.4. Clearly, the axioms $F1$ and $F3$ hold in $\mathbb{Z}_m$, but $F2$ does not hold. In fact, $\bar{1}$ is the only element in $\mathbb{Z}_m \setminus \{0\}$ such that $\bar{a} \cdot \bar{1} = \bar{a}$ for all $\bar{a} \in \mathbb{Z}_m$. Assume $m = m_1 m_2$, $0 < m_1, m_2 < m$. Then $\overline{m_1} \neq \bar{0}$, $\overline{m_2} \neq \bar{0}$, and $\overline{m_1} \overline{m_2} = \bar{0}$. $\overline{m_1}$ does not have an inverse in $\mathbb{Z}_m \setminus \{0\}$. Suppose $\overline{m_1}$ has an inverse $\overline{m_1}^{-1} \in \mathbb{Z}_m \setminus \{\bar{0}\}$, then $\bar{0} = \overline{m_1}^{-1} \cdot \bar{0} = \overline{m_1}^{-1}(\overline{m_1}\overline{m_2}) = (\overline{m_1}^{-1}\overline{m_1})\overline{m_2} = \bar{1}\overline{m_2} = \overline{m_2}$, which is a contradiction. Therefore when m is composite, $\mathbb{Z}_m$ is not a field.

Definition 3.4 *Let F be any field. If the number of elements in F is infinite, F is called an infinite field. If the number of elements in F is finite, F is called a finite field or a Galois field.*

From the axioms of fields we can deduce other manipulation rules, which are summarized in the following two theorems.

Theorem 3.1 *Let F be any field, then the following manipulation rules hold in F.*

(i) *The cancellation law of addition. Let a, b, and c be any three elements of F. If $a + c = b + c$, then $a = b$.*

(ii) *$a0 = 0$ for all $a \in F$.*

(iii) *The cancellation law of multiplication. Let a, b, and c be any three elements of F and $c \neq 0$. If $ac = bc$, then $a = b$.*

(iv) *Let $a, b \in F$. If $ab = 0$, then either $a = 0$ or $b = 0$.*

Proof. (i) follows from Theorem 2.2(i).

(ii) For any $a \in F$ we have $a0 + 0 = a0 = a(0 + 0) = a0 + a0$. By (i) we have $a0 = 0$.

(iii) Let $ac = bc$ and $c \neq 0$. If both $a \neq 0$ and $b \neq 0$, by Theorem 2.2(i) we have $a = b$. If $a = 0$ but $b \neq 0$, by (ii) $bc = ac = 0c = 0$ and then $0 = b^{-1}0 = b^{-1}(bc) = (b^{-1}b)c = ec = c$, which is a contradiction. Therefore if $a = 0$ we must have $b = 0$.

(iv) Let $ab = 0$ and $a \neq 0$. Then $ab = 0 = a0$. Since $a \neq 0$, by (iii) we have $b = 0$. □

Theorem 3.2 *Let F be a field. Then the following manipulation rules hold in F.*

(i) $-(-a) = a$ for any $a \in F$.

(ii) $-(a + b) = (-a) + (-b)$ for any $a, b \in F$.

(iii) $a(-b) = (-a)b = -(ab)$ for any $a, b \in F$.

(iv) $(-a)(-b) = ab$ for any $a, b \in F$.

(v) $(a^{-1})^{-1} = a$ for any $a \in F$ and $a \neq 0$.

(vi) $(ab)^{-1} = a^{-1}b^{-1}$ for any $a, b \in F$ and $a \neq 0$, $b \neq 0$.

(vii) $(-a)^{-1} = -a^{-1}$ for any $a \in F$ and $a \neq 0$.

Proof.   (i) and (v) follow from Theorem 2.2(iii).

(ii) and (vi) follow from Theorem 2.2(ii).

(iii) We have

$$0 = a0 = a(b + (-b)) = ab + a(-b).$$

On the other hand, it is clear that

$$0 = ab + (-(ab)),$$

so

$$ab + a(-b) = ab + (-(ab)),$$

which, by the cancellation law of addition, gives $a(-b) = -(ab)$. Similarly, one may prove that $(-a)b = -(ab)$.

(iv) With the help of (iii) and (i), we derive

$$(-a)(-b) = -(a(-b)) = -(-(ab)) = ab.$$

(vii) For any $a \in F$ and $a \neq 0$, by (iv),

$$(-a)(-a^{-1}) = aa^{-1} = e.$$

Hence $(-a)^{-1} = -a^{-1}$.   □

In any field F, we can also use the negative and inverse to introduce the substraction and division. That is, if $a, b \in F$, then define

$$a - b = a + (-b).$$

Next let $b \neq 0$ and define

$$a \div b = ab^{-1}.$$

From the manipulation rules satisfied by the addition and multiplication in fields, i.e., the axioms of fields and the rules contained in the above two theorems, one can also derive some other manipulation rules of fields, in which the addition, substraction, multiplication, and division appear. For example, for any three elements a, b, c in a field F, we have

$$a - (b + c) = a - b - c,$$

$$a - (b - c) = a - b + c,$$

$$a(b - c) = ab - ac,$$

$$a \div (bc) = ab^{-1}c^{-1}, \quad \text{when } b \neq 0, c \neq 0,$$

$$a \div (b \div c) = ab^{-1}c, \quad \text{when } b \neq 0, c \neq 0,$$

and so on.

Let a be an arbitrary element of a field F and n be any positive integer. We denote the sum of n a's by na for simplicity, that is,

$$na = \underbrace{a + a + \cdots + a}_{n \ a's}$$

and define the integer 0 times a to be the zero of F, i.e.,

$$0a = 0.$$

Note that the 0 on the left-hand side of the last equation is the integer 0 while the one on the right-hand side is the zero of field F. Now let n be a positive integer, then $-n$ is a negative integer and we define

$$(-n)a = -(na).$$

Then we have

Theorem 3.3 *Let F be a field. Then the following manipulation rules hold in F:*

(i) *For any $a \in F$, any integers m and n, we have*

$$(m + n)a = ma + na, \quad (mn)a = m(na).$$

(ii) *For any $a, b \in F$ and any integer n, we have*

$$n(a + b) = na + nb.$$

(iii) *For any $a, b \in F$ and any integers m and n, we have*

$$(ma)(nb) = (mn)(ab).$$

Let a be an arbitrary element of a field F and n be any positive integer. We denote the product of n a's by a^n for simplicity, i.e.,

$$a^n = \underbrace{a\, a \, \cdots \, a}_{n \ a's}.$$

When $a \neq 0$, we define

$$a^0 = e.$$

We also define

$$a^{-n} = (a^{-1})^n.$$

Then we have

Theorem 3.4 *Let F be a field. Then the following manipulation rules hold in F.*

(i) *For any $a \in F$ and any integers m and n, we have*

$$a^{m+n} = a^m a^n, \quad a^{mn} = (a^m)^n,$$

and when $a = 0$, we also require that $m > 0$ and $n > 0$.

(ii) *For any $a, b \in F$ and any integer m, we have*

$$(ab)^m = a^m b^m,$$

and when $a = 0$ or $b = 0$, we also require that $m > 0$.

(iii) *The binomial theorem. For any $a, b \in F$ and any positive integer n,*

$$(a + b)^n = \sum_{i=0}^{n} \binom{n}{i} a^{n-i} b^i, \qquad (3.1)$$

where $\binom{n}{i}$ is the number of ways of choosing i objects from a total of n objects and is also called the number of combinations of choosing i items from n items, i.e.,

$$\binom{n}{i} = \frac{n!}{i!(n-i)!}.$$

The proofs of the above two theorems are rather easy except that for the binomial theorem. We shall prove it by mathematical induction.

Proof of the binomial theorem. When $n = 1$, (3.1) trivially holds. Now let $n \geq 1$, and assume (3.1) holds for n. Then

$$\begin{aligned}
(a+b)^{n+1} &= (a+b)^n(a+b) \\
&= \left(\sum_{i=0}^{n} \binom{n}{i} a^{n-i}b^i \right)(a+b) \\
&= \sum_{i=0}^{n} \binom{n}{i} a^{n+1-i}b^i + \sum_{i=0}^{n} \binom{n}{i} a^{n-i}b^{i+1} \\
&= \sum_{i=0}^{n} \binom{n}{i} a^{n+1-i}b^i + \sum_{i=1}^{n+1} \binom{n}{i-1} a^{n+1-i}b^i \\
&= a^{n+1} + \sum_{i=1}^{n} \left(\binom{n}{i} + \binom{n}{i-1} \right) a^{n+1-i}b^i + b^{n+1}.
\end{aligned}$$

But

$$\binom{n}{i} + \binom{n}{i-1} = \binom{n+1}{i}.$$

Therefore (3.1) holds also for $n+1$. □

We have the following properties of subfields.

Theorem 3.5 *Let F be a field and F_0 be a subfield of F. Then the zero and identity of F both are elements of F_0 and moreover, they are, respectively, the zero and identity of F_0. If $a \in F_0$ then the negative of a in F is also the negative of a in F_0; if $a \in F_0$ and $a \neq 0$, then the inverse of a in F is also the inverse of a in F_0.*

Proof. By Theorem 2.6. □

Theorem 3.6 *Let F be a field and F_0 be a non-empty subset of F. If for any $a, b \in F_0$ we have $a - b \in F_0$ and we also have $ab^{-1} \in F_0$, provided $b \neq 0$, then F_0 is a subfield of F.*

Proof. By Theorem 2.7. □

Finally, we introduce the isomorphism of fields.

Definition 3.5 *Let F and F' be two fields. Assume that a bijective map from F to F'*

$$\sigma : a \mapsto \sigma(a) \quad (a \in F, \ \sigma(a) \in F')$$

can be established such that it preserves the addition and multiplication of fields. That is, for any $a, b \in F$,

$$\sigma(a+b) = \sigma(a) + \sigma(b)$$

and
$$\sigma(ab) = \sigma(a)\sigma(b).$$
Then we say that F and F' are isomorphic, which is denoted by $F \simeq F'$, and that σ is an isomorphic map from F to F', or, in short, an isomorphism. An isomorphism from a field F to itself is called an automorphism of F.

For example, it is easy to verify that the map

$$
\begin{array}{ccc}
F_2 & \to & \mathbb{Z}_2 \\
0 & \mapsto & \overline{0} \\
1 & \mapsto & \overline{1}
\end{array}
$$

is an isomorphism of fields and the map

$$
\begin{array}{ccc}
F_3 & \to & \mathbb{Z}_3 \\
0 & \mapsto & \overline{0} \\
1 & \mapsto & \overline{1} \\
2 & \mapsto & \overline{2}
\end{array}
$$

is also an isomorphism of fields.

In isomorphic fields, only the symbols of their corresponding elements are different. Hence we will often regard isomorphic fields as the same field.

Theorem 3.7 *Let σ be an isomorphism from a field F to a field F'. Denote the zeroes of F and F' by 0 and $0'$, respectively, and their identities by e and e', respectively. Then $\sigma(0) = 0'$ and $\sigma(e) = e'$. Moreover, for any $a \in F$ $\sigma(-a) = -\sigma(a)$, and for any $a \in F^*$ $\sigma(a^{-1}) = \sigma(a)^{-1}$.*

Proof. By Theorem 2.4.   □

Parallel to Theorem 2.5 we have

Theorem 3.8 *Let F be a field and F' be a set in which two binary operations called the addition and multiplication are defined and denoted by $+$ and $\cdot$, respectively. Let $\sigma\colon F \to F'$ be a bijective map which preserves the operations, i.e., $\sigma(a+b) = \sigma(a)+\sigma(b)$ and $\sigma(ab) = \sigma(a)\sigma(b)$ for all $a, b \in F$. Then F' is a field and σ is an isomorphism from F to F'.*

The proof is similar to that of Theorem 2.5.

3.2   The Characteristic of a Field

In the previous section we have seen that fields such as $\mathbb{Q}$, $\mathbb{R}$, and $\mathbb{C}$ are quite different from $\mathbb{Z}_p$ in one respect. That is, for the former, any sum of a finite number of 1's is nonzero and for the latter, the sum of p 1's is equal to 0. Thus, in order to distinguish these two types of fields, it is useful to consider the sequence of elements

$$e, 2e, 3e, \ldots,$$

where e is the identity of the field.

Definition 3.6 *Let F be a field and e be its identity. If for any positive integer m, we have $me \neq 0$, then we say that the characteristic of F is 0 or that F is a field of characteristic 0. If there exists a positive integer m such that $me = 0$, then the smallest positive integer p satisfying $pe = 0$ is called the characteristic of F and F is called a field of characteristic p.*

Note that all of $\mathbb{Q}$, $\mathbb{R}$, and $\mathbb{C}$ are fields of characteristic 0, and $\mathbb{Z}_p$ is a field of characteristic p.

Theorem 3.9 *Let F be any field, then the characteristic of F is either 0 or a prime p.*

Proof. Let the characteristic of F be p and $p \neq 0$. We need to prove that p must be a prime. If p is not a prime, it has a factorization $p = p_1 p_2$, where $1 < p_1, p_2 < p$. Hence

$$pe = (p_1 p_2)e = 0.$$

Since $e^2 = e$, we have

$$(p_1 e)(p_2 e) = (p_1 p_2)e.$$

Thus

$$p_1 e = 0 \quad \text{or} \quad p_2 e = 0.$$

But $1 < p_1, p_2 < p$, which contradicts that p is the smallest. Therefore p must be a prime. $\qquad \square$

Theorem 3.10 *Let F and F' be two isomorphic fields, then their characteristics must be equal.*

Proof. Let

$$\sigma : a \mapsto \sigma(a)$$

be an isomorphism from F to F'. By Theorem 3.7 the image $\sigma(e)$ of the identity e of F under σ is the identity e' of F', and the image $\sigma(0)$ of the zero 0 of F is the zero $0'$ of F'. F is of characteristic p, where p is a prime, if and only if $pe = 0$. But $pe = 0$ if and only if $pe' = p\sigma(e) = \sigma(pe) = \sigma(0) = 0'$. And $pe' = 0'$ if and only if F' is of characteristic p. $\qquad \square$

From Theorem 3.10 it can be seen that the characteristic of a field is indeed a distinguished property.

Theorem 3.11 *Let F be any field. If F is of characteristic 0, then for any nonzero element a in F and any positive integer m, $ma \neq 0$ and the elements*

$$0, \pm a, \pm 2a, \pm 3a, \ldots \qquad (3.1)$$

are all distinct. If F is a field of characteristic p, then for any nonzero element a in F we have $pa = 0$, p is the smallest positive integer satisfying $pa = 0$, and the following p elements

$$0, a, 2a, 3a, \ldots, (p-1)a \qquad (3.2)$$

are all distinct; further, for any integer m, $ma = 0$ if and only if $p \,|\, m$.

Proof. First let F be a field of characteristic 0. If for a nonzero element a in F there exists a positive integer m such that $ma = 0$, then

$$a(me) = m(ae) = ma = 0.$$

But $a \neq 0$, so we must have $me = 0$, which contradicts that F is of characteristic 0. Hence for any nonzero element a in F and any positive integer m, there must be $ma \neq 0$. That is, $\mathrm{ord}(a) = \infty$ in the additive group of F. It follows from Theorem 2.17 that the elements in (3.1) are all distinct.

Next let the characteristic of F be p, then $pe = 0$. Let a be any nonzero element in F, then

$$pa = p(ea) = (pe)a = 0a = 0.$$

Again if m is a positive integer such that $ma = 0$ then

$$(me)a = m(ea) = ma = 0,$$

but $a \neq 0$ so there must be $me = 0$. By the definition of the characteristic of a field, p is the smallest positive integer satisfying $pe = 0$, so $p \leq m$. Hence p is the smallest positive integer such that $pa = 0$. That is, $\mathrm{ord}(a) = p$ in the additive group of F. It follows from Theorem 2.17 that the p elements in (3.2) are all distinct and it follows from Theorem 2.20 that for any integer m, $ma = 0$ if and only if $p \,|\, m$. $\qquad \square$

Now let us study fields of characteristic p. Let F be any field of characteristic p and e be its identity. Let

$$\Pi = \{0, e, 2e, \ldots, (p-1)e\}.$$

We assert that Π is a subfield of F. Let $ke, le \in \Pi$. From $le + (p-l)e = pe = 0$ we deduce $-le = (p-l)e$. Thus $ke - le = ke + (p-l)e = (k + (p-l))e \in \Pi$, by Theorem 2.17. Assume that $le \neq 0$, then $\gcd(p, l) = 1$. There are c and

$d \in \mathbb{Z}$ such that $cp + dl = 1$, which implies $(de)(le) = e$, i.e., $(le)^{-1} = de$. Then $(ke)(le)^{-1} = (ke)(de) = (kd)e \in \Pi$, by Theorem 2.17. Therefore by Theorem 3.6 Π is a subfield of F.

Furthermore, by Theorem 3.5 any subfield of F contains the identity e of F. Hence it must contain Π. That is, Π is the smallest subfield of F. We call Π the *prime field* of F.

By comparing Π with $\mathbb{Z}_p$ in Example 3.7, it can be seen that the map from Π to $\mathbb{Z}_p$

$$ke \mapsto \bar{k} \quad (0 \le k < p)$$

is an isomorphism. We usually identify Π with $\mathbb{Z}_p$.

Next we study fields of characteristic 0. Let F be any field of characteristic 0 and e be its identity. Then by Theorem 3.11 the following elements

$$\ldots, -2e, -e, 0, e, 2e, \ldots \tag{3.3}$$

are distinct. When $n \ne 0$, we denote the inverse of ne by $(ne)^{-1}$. Let

$$\Pi = \{(me)(ne)^{-1} : m, n \in \mathbb{Z}, n \ne 0\}.$$

Clearly, $(0e)(ne)^{-1} = 0$ and $(ne)(ne)^{-1} = e$. Thus the zero 0 and identity e of F are all in Π. Moreover,

$$(me)(ne)^{-1} = (m'e)(n'e)^{-1},$$

if and only if

$$mn' = nm'.$$

It is not difficult to prove that

$$(me)(ne)^{-1} + (m'e)(n'e)^{-1} = [(mn' + nm')e][(nn')e]^{-1},$$

$$((me)(ne)^{-1})((m'e)(n'e)^{-1}) = [(mm')e][(nn')e]^{-1}.$$

Hence Π is closed with respect to the addition and multiplication of F. Let $(me)(ne)^{-1}$ be any element in Π, then $(-me)(ne)^{-1}$ is also in Π and

$$(me)(ne)^{-1} + (-me)(ne)^{-1} = 0.$$

When $(me)(ne)^{-1} \ne 0$, then $me \ne 0$ and $m \ne 0$, so $(ne)(me)^{-1}$ is also in Π and

$$((me)(ne)^{-1}((ne)(me)^{-1}) = e.$$

This proves that Π is a subfield of F.

Further, by Theorem 3.5, any subfield of F contains the identity e of F, it contains all the elements in (3.3) and also all the elements of the form

$$(me)(ne)^{-1}, \quad (m, n \in \mathbb{Z}, n \neq 0).$$

That is, F must contain Π. In other words, Π is the smallest subfield of F. We call Π the *prime field* of F.

To make a comparison with the rational number field we know that every rational number can be expressed in the form

$$m/n \quad (m, n \in \mathbb{Z}, n \neq 0),$$

and

$$m/n = m'/n',$$

if and only if

$$mn' = nm'.$$

It can be seen that the map from Π to $\mathbb{Q}$

$$(me)(ne)^{-1} \mapsto m/n \quad (m, n \in \mathbb{Z}, \ n \neq 0)$$

is an isomorphism. We usually identity Π with $\mathbb{Q}$.

Summarizing the above discussions we have

Theorem 3.12 *Let F be any field and denote by Π the prime field of F, then Π and $\mathbb{Z}_p$ are isomorphic when the characteristic of F is a prime p, and Π and $\mathbb{Q}$ are isomorphic when the characteristic of F is 0.*

Corollary 3.13 *If F is a finite field, then the characteristic of F is not equal to 0.*

Proof. If the characteristic of F is 0, then the prime field Π of F and the rational number field $\mathbb{Q}$ are isomorphic. But $\mathbb{Q}$ is an infinite field, so is F. A contradiction is obtained. □

Finally we prove the following manipulation rules which hold only in fields of characteristic p.

Theorem 3.14 *Let F be a field of characteristic p, $p \neq 0$, and a, b be any two elements of F, then*

$$(a + b)^p = a^p + b^p.$$

Proof. By the binomial theorem, we have

$$(a + b)^p = \sum_{i=0}^{p} \binom{p}{i} a^{p-i} b^i.$$

Since

$$\binom{p}{i} = \frac{p!}{i!(p-i)!}$$

is the number of combinations when choosing i items amongst p items, it must be an integer. Clearly, $p \mid p!$. Moreover, since p is a prime, when $0 < i < p$, there must be $p \nmid i!$ and $p \nmid (p-i)!$ so that $p \nmid i!(p-i)!$. Hence, when $0 < i < p$,

$$p \mid \binom{p}{i}.$$

Then by Theorem 3.11, when $0 < i < p$,

$$\binom{p}{i} a^{p-i} b^i = 0.$$

Hence

$$(a + b)^p = a^p + b^p.$$

$\square$

Corollary 3.15 *Let F be a field of characteristic p, $p \neq 0$, and a, b be any two elements of F, then*

$$(a - b)^p = a^p - b^p.$$

Proof. By Theorem 3.14, we have

$$(a - b)^p = (a + (-b))^p = a^p + (-b)^p$$
$$= a^p + ((-1)b)^p = a^p + (-1)^p b^p.$$

When $p > 2$, p is odd, we have $(-1)^p = -1$. Hence

$$(a - b)^p = a^p - b^p.$$

When $p = 2$, $2a = 0$ for any $a \in F$, so $a = -a$ for any $a \in F$. Therefore we also have

$$(a - b)^2 = a^2 + b^2 = a^2 - b^2.$$

$\square$

Corollary 3.16 *Let F be a field of characteristic p, $p \neq 0$ and a_1, a_2, ..., a_m be any m elements of F, then*

$$(a_1 + a_2 + \cdots + a_m)^p = a_1^p + a_2^p + \cdots + a_m^p.$$

Corollary 3.17 *Let F be a field of characteristic p, $p \neq 0$, a, b be any two elements of F, and n be any nonnegative integer, then*

$$(a \pm b)^{p^n} = a^{p^n} \pm b^{p^n}.$$

The above two corollaries can be proved by using induction on m and n, respectively, and the details are left to the reader.

Corollary 3.18 *Let F be a finite field of characteristic p, $p \neq 0$, and n be any nonnegative integer, then the map from F to itself*

$$\sigma_n : \ a \mapsto a^{p^n} \ \ (a \in F)$$

is an automorphism of F.

Proof.   First we prove that σ_n is a bijection. Let $\sigma_n(a) = \sigma_n(b)$, i.e., $a^{p^n} = b^{p^n}$. Then by Corollary 3.17 we have $(a - b)^{p^n} = a^{p^n} - b^{p^n} = 0$. From Theorem 3.1(iv) we deduce $a - b = 0$. Hence $a = b$. Therefore σ_n is an injection. Since the number of elements in F is finite, σ_n is also a surjection. Hence σ_n is a bijection from F to F'.

Next, by Corollary 3.17, we have $(a + b)^{p^n} = a^{p^n} + b^{p^n}$. So $\sigma_n(a + b) = \sigma_n(a) + \sigma_n(b)$. By the commutative law of multiplication we have $(ab)^{p^n} = a^{p^n} b^{p^n}$. So $\sigma_n(ab) = \sigma_n(a)\sigma_n(b)$. This proves that σ_n is an automorphism of F. $\qquad\qquad\square$

For the field F_4 in Example 3.6 the automorphism $a \mapsto a^2$ of F_4 is the map

$$0 \mapsto 0, \ 1 \mapsto 1, \ x \mapsto 1 + x, \ 1 + x \mapsto x.$$

3.3   Rings and Integral Domains

In Section 3.1 we introduced the concept of a field. It has been seen that it generalizes the usual fields such that the field of rational numbers $\mathbb{Q}$, the field of real number $\mathbb{R}$, the field of complex numbers $\mathbb{C}$, and the finite field $\mathbb{Z}_p$ with p elements, where p is a prime. But it can not cover some other usual algebraic structures such as the set of integers $\mathbb{Z}$. Also when m is a composite positive number, $\mathbb{Z}_m$ is not a field with respect to the modulo m addition and multiplication. In order to cover these algebraic structures and some more general ones, we have to introduce the concept of a ring.

Definition 3.7 *Let R be a set with two binary operations, called the addition and multiplication and denoted by $+$ and $\cdot$, respectively. R is called a ring with respect to the addition and multiplication, if the following axioms hold:*

$R1$ $(R, +)$ *is an abelian group.*

$R2.1$ $(ab)c = a(bc)$ *for all $a, b, c \in R$.*

$R2.2$ *There is an element, denoted by e, in R such that $ae = a$ for all $a \in R$.*

$R2.3$ $ab = ba$ *for all $a, b \in R$.*

$R3$ $a(b + c) = ab + ac$ *for all $a, b, c \in R$.*

Clearly any field is a ring. It is also clear that both $\mathbb{Z}$ and $\mathbb{Z}_m$ are rings and they are called the *ring of integers* and the *ring of integers modulo m*, respectively. We call $R1$, $R2.1$, $R2.2$, $R2.3$, and $R3$ in Definition 3.7 the *axioms of rings*. $(R, +)$ is called the *additive group of the ring R*. The element $e \in R$ satisfying $R2.2$ is unique. In fact, if $e' \in R$ satisfies also $ae' = a$ for all $a \in R$, then $e = ee' = e'e = e'$.

Actually, the ring defined in Definition 3.7 is called a *commutative ring with identity* in the literature, while "commutative" means that $R2.3$ holds and "with identity" means that $R2.2$ holds. Since we study only commutative rings with identity in this book, we call them simply rings.

Definition 3.8 *Let R be a ring. The identity of the additive group of R is called the zero of R and denoted by 0. The inverse of any element a of R in the additive group of R is called the negative of a and denoted by $-a$. The unique element e in R such that $ae = a$ for all $a \in R$ is called the identity of R.*

From now on we denote by 1 the identity of a ring. In particular, the identity of a field is also denoted by 1.

Theorem 3.19 *Let R be a ring, then the following manipulation rules hold in R.*

(i) *The cancellation law of addition. Let a, b, c be any three elements in R. If $a + c = b + c$, then $a = b$.*

(ii) $a0 = 0$ *for all $a \in R$.*

Proof. (i) follows from Theorem 2.2(i). (ii) can be proved in the same way as Theorem 3.1(ii). □

Naturally one can define the substraction in R with the help of the negative elements. That is, if $a, b \in R$, define

$$a - b = a + (-b).$$

It can also be proved that the addition, substraction, and multiplication in a ring R satisfy some of the manipulation rules listed on page 55 and we will not repeat them here.

As in the case of fields we have

Definition 3.9 *Let R and R' be two rings. Assume that there is a bijective map from R to R'*

$$\sigma : \ a \mapsto \sigma(a) \ \ (a \in R, \ \sigma(a) \in R')$$

which preserves the addition and multiplication of rings. That is, for any $a, b \in R$

$$\sigma(a + b) = \sigma(a) + \sigma(b),$$

$$\sigma(ab) = \sigma(a)\sigma(b).$$

Then we say that R and R' are isomorphic, which is denoted by $R \simeq R'$, and σ is called an isomorphic map from R to R', or, in short, an isomorphism. An isomorphism from a ring R to itself is called an automorphism of R.

Similar to Theorem 3.7 we have

Theorem 3.20 *Let σ be an isomorphism from a ring R to a ring R'. Denote the zeros of R and R' by 0 and $0'$, respectively, and their identities by 1 and $1'$, respectively. Then $\sigma(0) = 0'$ and $\sigma(1) = 1'$. Moreover, for any $a \in R$, $\sigma(-a) = -\sigma(a)$.*

Parallel to Theorem 3.8 we have

Theorem 3.21 *Let R be a ring and R' be a set in which two binary operations called the addition and multiplication are defined and denoted by $+$ and $\cdot$, respectively. Let $\sigma : R \to R'$ be a bijective map which preserves the operations, i.e., $\sigma(a + b) = \sigma(a) + \sigma(b)$ and $\sigma(ab) = \sigma(a)\sigma(b)$ for all $a, b \in R$. Then R' is a ring and σ is an isomorphism from R to R'.*

Moreover, we have

Theorem 3.22 *Let R be a ring and assume that $|R| > 1$. Then $1 \neq 0$.*

Proof. Suppose that $1 = 0$. Then for any $a \in R$, $a \cdot 1 = a \cdot 0$. By $R2.2$ $a \cdot 1 = a$ and by Theorem 3.19(ii), $a \cdot 0 = 0$. Therefore $a = 0$. Hence $R = \{0\}$. $\square$

In rings, the cancellation law of multiplication does not necessarily hold. For example, in $\mathbb{Z}_4$, $\bar{2} \cdot \bar{2} = \bar{0} = \bar{2} \cdot \bar{0}$, but $\bar{2} \neq \bar{0}$. More generally, let m be a composite positive integer and assume that $m = m_1 m_2$, where $0 < m_1, m_2 < m$. Then $\overline{m}_1 \overline{m}_2 = \bar{0} = \overline{m}_1 \cdot \bar{0}$, but $\overline{m}_2 \neq 0$.

Definition 3.10 *Let R be a ring. If the cancellation law of multiplication holds in R, that is, if for any three elements $a, b, c \in R$ with $c \neq 0$, from $ac = bc$ one can deduce $a = b$, then we say that R is an integral domain.*

It is easy to prove that $\mathbb{Z}$ is an integral domain. But when m is a composite positive number, $\mathbb{Z}_m$ is not a integral domain.

Clearly, if R is an integral domain and $\sigma \colon R \to R'$ is an isomorphism of rings, R' is also an integral domain (Exercise 3.14).

Definition 3.11 *Let R be a ring and a, b be two nonzero elements of R. If $ab = 0$, we say that a, b are zero divisors of R. If R does not have any zero divisor, then R is called a ring without zero divisors.*

Theorem 3.23 *Let R be a ring, then R is an integral domain if and only if R does not have zero divisors.*

Proof. First assume that R is an integral domain. Let $a, b \in R$ and $ab = 0$. Assume that $a \neq 0$. By Theorem 3.19(ii), $a0 = 0$. Then by the cancellation law of multiplication, from $ab = a0$ we deduce $b = 0$. Hence R cannot have zero divisors.

Conversely, assume that R is a ring without zero divisors. Let $a, b, c \in R$, $c \neq 0$, and $ac = bc$, then $(a - b)c = ac - bc = 0$. Since R does not have zero divisors, $a - b = 0$ and $a = b$. This proves that the cancellation law of multiplication holds in R. Hence R is an integral domain. $\square$

Parallel to subfields of a field and subgroups of a group subrings of a ring can be defined.

Definition 3.12 *Let R be a ring and R_0 a subset of R. If R_0 is a ring with respect to the addition and multiplication in R, then R_0 is called a subring of R. If R_0 is a subring of R, R is called an extension ring of R_0.*

For example, $\mathbb{Z}$ is a subring of the field $\mathbb{Q}$.

3.4   Field of Fractions of an Integral Domain

From the ring of integers $\mathbb{Z}$ we can form the fractions a/b, where $a, b \in \mathbb{Z}$ and $b \neq 0$. Defining the sum and product of two fractions as

$$a/b + c/d = (ad + bc)/bd,$$

$$(a/b)(c/d) = ac/bd,$$

we know that the set of fractions forms the rational field $\mathbb{Q}$. This procedure can be generalized to any integral domain.

Let D be an arbitrary integral domain and assume that D has more than one element. Denote by D^* the set of nonzero elements of D. Consider the Cartesian product set

$$D \times D^* = \{(a, b) : a \in D, b \in D^*\}$$

and we introduce a relation $\sim$ on $D \times D^*$ by defining

$$(a, b) \sim (c, d) \text{ if and only if } ad = bc.$$

Then $(a, b) \sim (a, b)$, since $ab = ba$. If $(a, b) \sim (c, d)$, then $ad = bc$; hence $cb = da$ and so $(c, d) \sim (a, b)$. Finally, if $(a, b) \sim (c, d)$ and $(c, d) \sim (e, f)$, then $ad = bc$ and $cf = de$. Hence $adf = bcf = bde$. Since $d \in D^*$ and D is an integral domain, d may be canceled to give $af = be$, which is the condition that $(a, b) \sim (e, f)$. We have therefore proved that $\sim$ is an equivalence relation. We shall call the equivalence class determined by (a, b) the fraction a/b. Thus we have $a/b = c/d$ if and only if $ad = bc$. Let F be the set of equivalence classes determined by the equivalence relation $\sim$ on $D \times D^*$.

We introduce an addition and a multiplication in F as follows:

$$a/b + c/d = (ad + bc)/bd,$$

$$(a/b)(c/d) = ac/bd.$$

Since $b \neq 0$ and $d \neq 0$, $bd \neq 0$. It follows that both $(ad + bc)/bd$ and ac/bd are fractions. If $a/b = a'/b'$ and $c/d = c'/d'$, then $ab' = ba'$ and $cd' = dc'$. Hence

$$ab'dd' = ba'dd' \text{ and } cd'bb' = dc'bb',$$

so that

$$ab'dd' + cd'bb' = ba'dd' + dc'bb',$$

or

$$(ad + bc)b'd' = (a'd' + b'c')bd,$$

which implies

$$a/b + c/d = (ad + bc)/bd = (a'd' + b'c')b'd' = a'/b' + c'/d'.$$

Thus the addition of fractions is well-defined. Moreover, if $a/b = a'/b'$, $c/d = c'/d'$, then $ab' = ba'$, $cd' = dc'$, so $ab'cd' = ba'dc'$. Hence $ac/bd = a'c'/b'd'$. Thus the multiplication is also well-defined.

Let us show that F is a field with respect to the above defined addition and multiplication. First, we verify $F1$. The associative law of addition can be proved as follows.

$$
\begin{aligned}
(a/b + c/d) + e/f &= (ad + bc)/(bd) + e/f \\
&= ((ad + bc)f + bde)/bdf \\
&= (adf + b(cf + de))/bdf \\
&= a/b + (cf + de)/df \\
&= a/b + (c/d + e/f).
\end{aligned}
$$

If we put $0 = 0/1$, then

$$a/b + 0 = a/b + 0/1 = (a1 + b0)/b1 = a/b$$

and

$$a/b + (-a)/b = (ab + (-a)b)/b^2 = 0/b^2 = 0/1 = 0.$$

The commutative law of addition is easy. So, $F1$ is verified.

Then, we verify $F2$. The associative law and the commutative law of multiplication are easy. If we put $1 = 1/1$, then

$$(a/b)1 = (a/b)(1/1) = a/b$$

and for $a/b \neq 0$, i.e., $a \neq 0$, we have $b/a \in F$ and

$$(a/b)(b/a) = ab/ab = 1/1 = 1.$$

Hence $F2$ also holds in F.

Finally, the distributive law is verified as follows.

$$
\begin{aligned}
(a/b)(c/d + e/f) &= (a/b)((cf + de)/df) \\
&= a(cf + de)/bdf \\
&= (acf + ade))/bdf
\end{aligned}
$$

and
$$(a/b)(c/d) + (a/b)(e/f) = ac/bd + ae/bf$$
$$= (acbf + bdae)/b^2 df$$
$$= (acf + ade)/bdf.$$

Therefore
$$(a/b)(c/d + e/f) = (a/b)(c/d) + (a/b)(e/f).$$

Hence F is a field. F is called the *field of fractions* of D.

Moreover, consider the map

$$\sigma : D \to F$$
$$a \mapsto a/1$$

If $a/1 = b/1$, then $a = b$. Hence the map is an injection. We also have $a/1 + b/1 = (a + b)/1$ and $(a/1)(b/1) = ab/1$. Therefore the map preserves the addition and multiplication. By Theorem 3.21 the image set $D' = \{a/1 : a \in D\}$ is a ring and $\sigma : D \to D'$ is an isomorphism of rings. Then D' is also an integral domain. We can identity D with its image set $D' = \{a/1 : a \in D\}$ by the above map σ, and we say that D is a subring of its field of fractions or D is imbedded in its field of fractions.

3.5   Divisibility in a Ring

The divisibility in $\mathbb{Z}$ can be generalized to an arbitrary ring R in which $1 \neq 0$ as follows.

Let $a, b \in R$ and $a \neq 0$. If there is an element $c \in R$ such that $b = ac$, we say that a *divides* b and write $a \mid b$. In this case, a is called a *divisor* of b or a *factor* of b, and b is called a *multiple* of a.

Let $a, b_1, b_2, \ldots, b_r \in R$ and $a \neq 0$. If $a \mid b_i$ for all $i = 1, 2, \ldots, r$, a is called a *common divisor* of $b_1, b_2, \ldots, b_r$. If $b_1 = b_2 = \cdots = b_r = 0$, then any element of R is a common divisor of them. Now we assume that $b_1, b_2, \ldots, b_r$ are not all 0; a common divisor d of $b_1, b_2, \ldots, b_r$ is called a *greatest common divisor* of them if any common divisor d' of them divides d. We denote *any* greatest common divisor of $b_1, b_2, \ldots, b_r$ by $\gcd(b_1, b_2, \ldots, b_r)$ if it exists.

The above definitions are generalizations of the concepts of divisors, common divisors, and the greatest common divisors in $\mathbb{Z}$. But it should be remarked that for $b_1, b_2, \ldots, b_r \in R$, which are not all 0, if the greatest common divisor of $b_1, b_2, \ldots, b_r$ exists, it may not be unique. For instance, even in $\mathbb{Z}$, both 3 and -3 are greatest common divisors of 6, 9, 12 according to the above definition. But we also use the notation $\gcd(b_1, b_2, \ldots, b_r)$ to denote any greatest common divisor of $b_1, b_2, \ldots, b_r$ if it exists. Clearly, if both d and d' are greatest common divisors of $b_1, b_2, \ldots, b_r$, then $d \mid d'$ and $d' \mid d$. We recall that in Section 1.2 we defined the greatest common divisor of a set of integers, not all 0, to be the largest common divisor of them, then it is uniquely determined.

An element $u \in R$ is called a *unit* if there is an element $v \in R$ such that $uv = 1$. For example, in $\mathbb{Z}$ only ± 1 are units, in any field F any nonzero element is a unit, and in $\mathbb{Z}_m$ the elements in $\mathbb{Z}_m^*$ are units. The set of units in a ring R forms a group with respect to the multiplication (Exercise 3.17), which is called the *group of units* of the ring R and is denoted by R^*.

Let $a, b \in R$, a and b are said to be *associates*, if $a = ub$ for some unit $u \in R$. In this case we also say that a is associate with b and b is associate with a. For example, in $\mathbb{Z}$, 3 and -3 are associates; and any two units in a ring R are associates. Clearly, if a and b are associates and one of them is not a zero divisor, so is the other.

Moreover, we have

Lemma 3.24 *Let R be any ring in which $1 \neq 0$ and let $a, b \in R$. If a and b are associates then $a \mid b$ and $b \mid a$. Conversely, assume that one of a and b is not a zero divisor, then $a \mid b$ and $b \mid a$ imply that a and b are associates and that the other one of them is also not a zero divisor.*

The proof is left as an exercise.

Let $a, b \in R$, $a \neq 0$ and $b \neq 0$. a is called a *proper divisor* of b if a is a divisor of b, a is not a unit, and a and b are not associates. For example, in $\mathbb{Z}$ if $a \neq 0$ is a proper divisor of b, then $|a| < |b|$.

Let $a \in R$, $a \neq 0$, and a is not a unit. a is called a *prime element* in R if a has no proper divisors. For example, in $\mathbb{Z}$ the prime elements are the prime numbers and their negatives.

The following lemmas will be used later.

Lemma 3.25 *Let R be any ring in which $1 \neq 0$. Let $a, b, c \in R$, and $a \neq 0$. Assume that both $\gcd(a, b)$ and $\gcd(a, b - ca)$ exist and that one of them is not a zero divisor. Then*

$$\gcd(a, b) = u \gcd(a, b - ca),$$

where u is a unit in R.

Proof. Let $d =\gcd(a,b)$ and $d' =\gcd(a, b - ca)$. Then $d\,|\,a$ and $d\,|\,b$. $d\,|\,a$ implies $d\,|\,ca$. From $d\,|\,b$ and $d\,|\,ca$ we deduce $d\,|\,(b - ca)$. Then, from $d\,|\,a$ and $d\,|\,(b - ca)$ we deduce $d\,|\,d'$. Similarly, $d'\,|\,d$. By Lemma 3.24, d and d' are associates, i.e., $d = ud'$, where u is a unit in R.

$\square$

Lemma 3.26 *Let R be any ring in which $1 \neq 0$. Assume further that the greatest common divisor of any two elements of R, which are not both 0, exists and is not a zero divisor. Let m and n be positive integers, $t \in R$ and $t \neq 1$. Then*

$$\gcd(t^m - 1, t^n - 1) = u\,(t^{\gcd(m,n)} - 1),$$

where u is a unit in R.

Proof. When $m = n$, the lemma holds trivially. Now assume that $m \neq n$. We apply induction on $\max\{m, n\}$. When $\max\{m, n\} = 1$, the lemma is trivial. Now let $\max\{m, n\} > 1$ and without loss of generality we can assume that $m < n$. By Lemma 3.25,

$$\gcd(t^m - 1, t^n - 1) = u_1\gcd(t^m - 1, t^n - 1 - t^{n-m}(t^m - 1))$$
$$= u_1\gcd(t^m - 1, t^{n-m} - 1),$$

where u_1 is a unit in R. Then, by induction hypothesis,

$$u_1\gcd(t^m - 1, t^{n-m} - 1) = u_1 u_2(t^{\gcd(m,n-m)} - 1)$$
$$= u\,(t^{\gcd(m,n)} - 1),$$

where u_2 is a unit in R and $u = u_1 u_2$. $\square$

Finally, the coprimeness of two integers which are not both zero can be generalized as follows.

For $a, b \in R$, if there exist $x, y \in R$ such that $xa + yb = 1$, then a, b are said to be *coprime* in R. Clearly, a and b are coprime in R, if and only if $Ra + Rb = R$, where $Ra = \{ra : r \in R\}$ and $Ra + Rb = \{ra + r'b : r, r' \in R\}$.

3.6   Exercises

3.1 Prove that a field has at least two elements.

3.2 Let $\mathbb{Q}[i]$ denote the set of complex numbers $\{a + bi : a, b \in \mathbb{Q}\}$. Prove that $\mathbb{Q}[i]$ is a field with respect to the usual addition and multiplication of complex numbers.

3.3 Let $\mathbb{Q}[\sqrt[3]{2}]$ denote the set of real numbers $\{a+b\sqrt[3]{2}+c\sqrt[3]{4} : a, b, c \in \mathbb{Q}\}$. Prove that $\mathbb{Q}[\sqrt[3]{2}]$ is a field with respect to the usual addition and multiplication of real numbers.

3.4 Prove Theorems 3.3 and 3.4.

3.5 Prove that any field containing only two elements is isomorphic to the field F_2 of Example 3.4.

3.6 Prove that any field containing only three elements is isomorphic to the field F_3 of Example 3.5.

3.7 Prove that all nonzero elements of a field F have the same order in the additive group $(F, +)$.

3.8 Show that any automorphism of a field maps every element of its prime field to itself. Then deduce that the identity automorphism is the only automorphism of the field of rational numbers $\mathbb{Q}$ and the field $\mathbb{Z}_p$, where p is any prime.

3.9 Prove that the field of real numbers $\mathbb{R}$ has only one automorphism.

3.10 Prove Corollaries 3.16 and 3.17.

3.11 Write down the proof that $\mathbb{Z}_m$ is a ring in detail.

3.12 Determine all rings with two or three elements.

3.13 Determine all rings with four elements whose additive groups are cyclic of order four.

3.14 Let $\sigma : R \rightarrow R'$ be an isomorphism of rings. Prove that if R is an integral domain, so is R'.

3.15 Prove that a finite integral domain is a field.

3.16 Let R be a finite ring, $a \in R$, and $a \neq 0$. Then a is either a zero divisor or a unit.

3.17 Prove that the set of units in a ring forms an abelian group with respect to the multiplication of the ring.

3.18 Prove Lemma 3.24.

3.19 Prove that in $\mathbb{Z}$ the only units are ± 1 and the only prime elements are the prime numbers and their negatives.

3.20 Let $\mathbb{Z}[i]$ denote the set of complex numbers $\{a + bi : a, b \in \mathbb{Z}\}$. For any $a + bi \in \mathbb{Z}[i]$ define $N(a + bi) = a^2 + b^2$. Prove that

   (i) $\mathbb{Z}[i]$ is an integral domain with respect to the addition and multiplication of complex numbers.

   (ii) $N(\alpha\beta) = N(\alpha)N(\beta)$ for any $\alpha, \beta \in \mathbb{Z}[i]$.

   (iii) $\alpha \in \mathbb{Z}[i]$ is a unit if and only if $N(\alpha) = 1$. Hence deduce that $1, -1, i, -i$ are the only units in $\mathbb{Z}[i]$.

   (iv) For $\alpha \in \mathbb{Z}[i]$, if $N(\alpha)$ is a prime number, then α is a prime element of $\mathbb{Z}[i]$.

   (v) $1 + i, 2 + i, 3 + 2i, 3$, and 7 are prime elements in $\mathbb{Z}[i]$, and $2, 5$, and 13 are not prime elements.

3.21 Let $\mathbb{Z}[\sqrt{-3}] = \{a + b\sqrt{-3} : a, b \in \mathbb{Z}\}$. For $a + b\sqrt{-3} \in \mathbb{Z}[\sqrt{-3}]$ define $N(a + b\sqrt{-3}) = a^2 + 3b^2$. Prove that

   (i) $\mathbb{Z}[\sqrt{-3}]$ is an integral domain with respect to the addition and multiplication of complex numbers.

   (ii) $N(\alpha\beta) = N(\alpha)N(\beta)$ for any $\alpha, \beta \in \mathbb{Z}[\sqrt{-3}]$.

   (iii) The number 4 has two factorizations into products of prime elements in $\mathbb{Z}[\sqrt{-3}]$:

$$4 = 2 \cdot 2 = (1 + \sqrt{-3})(1 - \sqrt{-3}).$$

   $2, 1 + \sqrt{-3}$, and $1 - \sqrt{-3}$ are prime elements in $\mathbb{Z}[\sqrt{-3}]$ and 2 is not associate with both $1 + \sqrt{-3}$ and $1 - \sqrt{-3}$.

3.22 Let $\mathbb{Z}[\sqrt{-5}] = \{a + b\sqrt{-5} : a, b \in \mathbb{Z}\}$ and $b_1 = 9, b_2 = 6 + 3\sqrt{-5}, d_1 = 3, d_2 = 2 + \sqrt{-5}$. Prove that

   (i) $\mathbb{Z}[\sqrt{-5}]$ is an integral domain with respect to the addition and multiplication of complex numbers.

   (ii) $d_i \,|\, b_j$ for $i, j = 1, 2$. Thus both d_1 and d_2 are common divisors of b_1 and b_2.

   (iii) There is no number d such that $d \,|\, b_j, j = 1, 2$ and $d_i \,|\, d, i = 1, 2$. Thus b_1 and b_2 have no greatest common divisor.

Chapter 4

Polynomials

4.1 Polynomial Rings

Let F be a given field and x be an indeterminate. Assume that i is a nonnegative integer, then expressions of the form

$$a_i x^i, \quad a_i \in F,$$

are called *monomials* of *degree* i in the *indeterminate* x with coefficients in F. A formal sum

$$a_0 x^0 + a_1 x^1 + a_2 x^2 + \cdots + a_n x^n \tag{4.1}$$

of a finite number of monomials

$$a_0 x^0, \ a_1 x^1, \ a_2 x^2, \ \ldots, a_n x^n$$

with coefficients $a_0, a_1, a_2, \ldots, a_n$ in F, where n is any nonnegative integer, is called a *polynomial* in the *indeterminate* x with coefficients in F or simply, a polynomial in x over the field F. In polynomial (4.1), $a_i x^i$ is called its *i-th degree term* and a_i is called the *coefficient* of the term. When $a_i = 1$ we write $1 x^i = x^i$. When $a_i = 0$, we usually omit the term $a_i x^i$ in the expression (4.1). In particular, terms $0 x^{n+1}, 0 x^{n+2}, \ldots$ can be regarded as omitted in (4.1). In what follows, we define $x^0 = 1$, $x^1 = x$ and write $a_0 x^0$ as a_0, x^1 as x, then the polynomial (4.1) becomes

$$a_0 + a_1 x + a_2 x^2 + \cdots + a_n x^n.$$

Often the simple notations $f(x), g(x), h(x), f_i(x), \ldots$ are used for polynomials.

Let $f(x)$ and $g(x)$ be two polynomials in x over F. If all their coefficients of terms of the same degree are equal, we say that $f(x)$ and $g(x)$ are *equal* or, in other words, that $f(x)$ and $g(x)$ are the same polynomial, written as

$$f(x) = g(x).$$

In particular,

$$a_0 + a_1 x + a_2 x^2 + \cdots + a_n x^n, \quad a_i \in F$$

and

$$a_0 + a_1 x + a_2 x^2 + \cdots + a_n x^n + 0 x^{n+1} + 0 x^{n+2} + \cdots + 0 x^{n+m}$$

are equal polynomials.

Later on we will frequently use the summation sign $\sum$ to simplify the notation of the polynomial

$$f(x) = a_0 + a_1 x + a_2 x^2 + \cdots + a_n x^n$$

as

$$f(x) = \sum_{i=0}^{n} a_i x^i.$$

If $a_n \neq 0$, $f(x)$ is called a polynomial of *degree n*, which is denoted by $\deg f(x) = n$, and a_n is called its *leading coefficient*. If $a_n = 1$, $f(x)$ is called a *monic polynomial*. Nonzero elements of F are polynomials of degree 0 in $F[x]$. A polynomial of degree 1, 2, 3, or 4 is called a *linear, quadratic, cubic,* or *biquadratic polynomial*, respectively. When all the coefficients of $f(x)$ are 0, $f(x)$ is called the *zero polynomial*, denoted by 0 and we define $\deg 0 = -\infty$.

Denote by $F[x]$ the set of all polynomials in x over F. Let $f(x)$ and $g(x)$ be any two elements of $F[x]$ and

$$f(x) = \sum_{i=0}^{n} a_i x^i, \quad g(x) = \sum_{i=0}^{m} b_i x^i. \tag{4.2}$$

Let $M = \max\{n, m\}$, i.e., when $n \neq m$, M is the larger one in n and m, while when $n = m, M = n = m$. Set

$$a_{n+1} = a_{n+2} = \cdots = a_M = 0, \quad \text{if } n < M,$$

$$b_{m+1} = b_{m+2} = \cdots = b_M = 0, \quad \text{if } m < M,$$

then $f(x)$ and $g(x)$ can be written as

$$f(x) = \sum_{i=0}^{M} a_i x^i, \quad g(x) = \sum_{i=0}^{M} b_i x^i.$$

The *sum* of $f(x)$ and $g(x)$, denoted by $f(x) + g(x)$, is defined as

$$f(x) + g(x) = \sum_{i=0}^{M} (a_i + b_i) x^i.$$

Clearly $f(x) + g(x) \in F[x]$, i.e., $F[x]$ is closed with respect to the above-defined addition. Next set

$$a_{n+1} = a_{n+2} = \cdots = a_{n+m} = 0, \quad \text{if } m \geq 1,$$

$$b_{m+1} = b_{m+2} = \cdots = b_{m+n} = 0, \quad \text{if } n \geq 1,$$

then $f(x)$ and $g(x)$ can be written as

$$f(x) = \sum_{i=0}^{n+m} a_i x^i, \quad g(x) = \sum_{i=0}^{n+m} b_i x^i.$$

We define the *product* of $f(x)$ and $g(x)$, denoted by $f(x)g(x)$, as follows:

$$f(x)g(x) = \sum_{i=0}^{n+m} \left(\sum_{j=0}^{i} a_j b_{i-j} \right) x^i.$$

Clearly $f(x)g(x) \in F[x]$, i.e., $F[x]$ is closed with respect to the above-defined multiplication. Further, it is routine to verify that $F[x]$ is a ring with respect to the above-defined addition and multiplication in $F[x]$. $F[x]$ is called *the ring of polynomials in an indeterminate x over the field F* or *the polynomial ring in x over F*. The zero 0 of F is the zero of $F[x]$ and the identity 1 of F is the identity of $F[x]$. Moreover, we have

Theorem 4.1 *Let $f(x), g(x) \in F[x]$. Then*

$$\deg\left(f(x)g(x)\right) = \deg f(x) + \deg g(x), \tag{4.3}$$

$$\deg\left(f(x) + g(x)\right) \leq \max\{\deg f(x), \deg g(x)\}. \tag{4.4}$$

Proof. If one or both of $f(x)$ and $g(x)$ is 0, then $f(x)g(x) = 0$ and both sides of (4.3) are $-\infty$. Let both $f(x)$ and $g(x)$ be not 0, assume that they are given by (4.2) and that $\deg f(x) = n$, and $\deg g(x) = m$. Then

$a_n \neq 0$ and $b_m \neq 0$. By the definition of multiplication of polynomials, the coefficient of x^{n+m} in $f(x)g(x)$ is $a_n b_m \neq 0$ and the coefficients of terms of degree $> n + m$ are all 0's. Therefore $\deg(f(x)g(x)) = n + m = \deg f(x) + \deg g(x)$.

(4.4) is obvious. $\square$

From (4.3) we deduce immediately that $F[x]$ has no zero divisors. Therefore we have

Theorem 4.2 *Let F be any field and x an indeterminate. Then $F[x]$ is an integral domain.*

It follows also from (4.3) that polynomials of degree ≥ 1 do not have inverses. Therefore $F[x]$ is not a field.

By the construction given in Section 3.4, $F[x]$ has a field of fractions, which is denoted by $F(x)$, and elements in $F(x)$ are of the form $f(x)/g(x)$, where $f(x)$, $g(x) \in F[x]$ and $g(x) \neq 0$, and are called *rational functions* in x over F. The field $F(x)$ is called the *field of rational functions in x over F*. We also regard $F[x]$ to be embedded in $F(x)$ by the map $f(x) \mapsto f(x)/1$.

Let R be a ring with $1 \neq 0$. The above construction of polynomials over a field F can be carried over to polynomials over R without any difficulty. Then we obtain a *polynomial ring $R[x]$*. Moreover, if R is an integral domain D, then both Theorem 4.1 and 4.2 hold also in $D[x]$, and, in particular, $D[x]$ is also an integral domain. The details will not be given in the text.

Now we discuss briefly polynomials in n indeterminates $x_1, x_2, \ldots, x_n$ over a field F (or an integral domain D). An expression of the form

$$ax_1^{k_1}x_2^{k_2}\cdots x_n^{k_n}$$

where $a \in F$ (or D) and $k_1, k_2, \ldots, k_n$ are all nonnegative integers, is called a *monomial* in n *indeterminates* $x_1, x_2, \ldots, x_n$ with coefficient in F (or D). In short, we call it an *n-ary monomial* over F (or D) and call $k_1 + k_2 + \cdots + k_n$ its *degree* if $a \neq 0$. If two monomials have the same powers for the same indeterminate, then they are called *similar terms*. A formal sum of a finite number of non-similar n-ary monomials over F (or D),

$$\sum_{k_1, k_2, \ldots, k_n} a_{k_1, k_2, \ldots, k_n} x_1^{k_1} x_2^{k_2} \cdots x_n^{k_n},$$

is called a *polynomial* in n *indeterminates* $x_1, x_2, \ldots, x_n$ with coefficients in F (or D) and in short, an *n-ary polynomial* over F (or D). The highest degree of the monomials with nonzero coefficients in a polynomial is called the *degree* of the polynomial. The polynomial whose coefficients are all 0 is called the *zero polynomial* and denoted by 0, and the degree of the zero polynomial 0 is defined to be $-\infty$. Two n-ary polynomials are said to be *equal* or they are the same polynomial if their coefficients of similar terms are all equal. We often denote n-ary polynomials over F (or D) by $f(x_1, \ldots, x_n)$, $g(x_1, \ldots, x_n)$, etc. The set of all the n-ary polynomials over F (or D) is denoted by $F[x_1, x_2, \ldots, x_n]$ (or $D[x_1, x_2, \ldots, x_n]$). Just as in elementary algebra, one can define the *sum* and *product* of two n-ary polynomials

$$f(x_1, x_2, \ldots, x_n) = \sum_{k_1, k_2, \ldots, k_n} a_{k_1, k_2, \ldots, k_n} x_1^{k_1} x_2^{k_2} \cdots x_n^{k_n},$$

$$g(x_1, x_2, \ldots, x_n) = \sum_{k_1, k_2, \ldots, k_n} b_{k_1, k_2, \ldots, k_n} x_1^{k_1} x_2^{k_2} \cdots x_n^{k_n},$$

as

$$f(x_1, x_2, \ldots, x_n) + g(x_1, x_2, \ldots, x_n)$$
$$= \sum_{k_1, k_2, \ldots, k_n} (a_{k_1, k_2, \ldots, k_n} + b_{k_1, k_2, \ldots, k_n}) x_1^{k_1} x_2^{k_2} \cdots x_n^{k_n}$$

and

$$f(x_1, x_2, \ldots, x_n) g(x_1, x_2, \ldots, x_n)$$
$$= \sum_{k_1, k_2, \ldots, k_n} \sum_{l_1, l_2, \ldots, l_n} a_{k_1, k_2, \ldots, k_n} b_{l_1, l_2, \ldots, l_n} x_1^{k_1 + l_1} x_2^{k_2 + l_2} \cdots x_n^{k_n + l_n},$$

respectively. Then it can be easily proved that $F[x_1, x_2, \ldots, x_n]$ (and also $D[x_1, x_2, \ldots, x_n]$) is an integral domain with respect to the above definitions of addition and multiplication. Similar to the case of one indeterminate, $F[x_1, x_2, \ldots, x_n]$ has a field of fractions $F(x_1, x_2, \ldots, x_n)$, the elements of which are *rational functions*

$$f(x_1, x_2, \ldots, x_n) / g(x_1, x_2, \ldots, x_n),$$

where

$$f(x_1, x_2, \ldots, x_n), \ g(x_1, x_2, \ldots, x_n) \in F[x_1, x_2, \ldots, x_n]$$

and $g(x_1, x_2, \ldots, x_n) \neq 0$.

It is interesting to see that $F[x, y] \simeq F[x][y]$, where $F[x][y]$ is the polynomial ring in the indeterminate y with coefficients in the polynomial ring $F[x]$ (Exercise 4.17). More generally,

$$F[x_1, x_2, \ldots, x_n] \simeq F[x_1][x_2] \ldots [x_n].$$

Finally, we call an n-ary polynomial a *homogeneous polynomial* of degree r or a *form* of degree r, if it is a sum of a finite number of monomials of the same degree r. For example, $x_1^2 + x_2 x_3$ is a homogeneous polynomial of degree 2 and $x_1^2 x_2 + x_2 x_3^2 + x_4^3$ is a homogeneous polynomial of degree 3.

4.2   Division Algorithm

Because $F[x]$ contains nonzero elements which do not have inverses, one can not proceed, as in Section 3.1, to introduce division by means of inverse and multiplication. But similar to the case of $\mathbb{Z}$, there is a division algorithm in $F[x]$.

Theorem 4.3 (*Division Algorithm*): Let $a(x)$ and $b(x) \in F[x]$ and $b(x) \neq 0$. Then there exists a unique pair of polynomials $q(x)$ and $r(x)$ such that

$$a(x) = q(x)b(x) + r(x), \quad \deg r(x) < \deg b(x). \qquad (4.5)$$

Proof. First we prove that there exists a pair of polynomials $q(x)$ and $r(x)$ satisfying (4.5). We apply induction on the degree of $a(x)$.

When $\deg a(x) < \deg b(x)$, let $q(x) = 0$ and $r(x) = a(x)$, then we have (4.5). When $\deg a(x) \geq \deg b(x)$, let $\deg a(x) = n$, $\deg b(x) = m$, and

$$a(x) = a_0 + a_1 x + a_2 x^2 + \cdots + a_n x^n,$$
$$b(x) = b_0 + b_1 x + b_2 x^2 + \cdots + b_m x^m,$$

where $a_n \neq 0$ and $b_m \neq 0$. Then $n \geq m$. Let

$$a_1(x) = a(x) - a_n b_m^{-1} x^{n-m} b(x),$$

then $\deg a_1(x) \leq n - 1$. By induction hypothesis there exists a pair of polynomials $q_1(x)$ and $r_1(x)$ such that

$$a_1(x) = q_1(x)b(x) + r_1(x), \quad \deg r_1(x) < \deg b(x).$$

So,

$$a(x) = (a_n b_m^{-1} x^{n-m} + q_1(x))b(x) + r_1(x).$$

Let

$$q(x) = a_n b_m^{-1} x^{n-m} + q_1(x), \quad r(x) = r_1(x),$$

then (4.5) follows.

Next we prove that the polynomials $q(x)$ and $r(x)$ satisfying (4.5) are unique. Assume that there is another pair of polynomials $q_1(x)$ and $r_1(x)$ such that

$$a(x) = q_1(x)b(x) + r_1(x), \quad \deg r_1(x) < \deg b(x).$$

Then

$$(q(x) - q_1(x))b(x) = -r(x) + r_1(x).$$

So,

$$\deg\left((q(x) - q_1(x))b(x)\right) = \deg\left(-r(x) + r_1(x)\right).$$

But

$$\deg\left(-r(x) + r_1(x)\right) \le \max\{\deg r(x), \deg r_1(x)\} < \deg b(x).$$

If $q(x) - q_1(x) \ne 0$, by Theorem 4.1,

$$\deg\left((q(x) - q_1(x))b(x)\right) = \deg\left(q(x) - q_1(x)\right) + \deg b(x) \ge \deg b(x).$$

Therefore it is necessary that $q(x) - q_1(x) = 0$, i.e., $q(x) = q_1(x)$. Also $r(x) = r_1(x)$. $\square$

The inductive proof of Theorem 4.3 actually gives a method of obtaining $q(x)$ and $r(x)$. We illustrate this with an example.

Example 4.1 Let $F = \mathbb{Z}_2$,

$$a(x) = x^5 + x^4 + x^2 + 1,$$

and

$$b(x) = x^3 + x + 1.$$

The division algorithm can be performed with the help of the following scheme.

$x^3 + x + 1$	x^5	$+x^4$		$+x^2$		$+1$	$x^2 + x + 1$
	x^5		$+x^3$	$+x^2$			
		x^4	$+x^3$			$+1$	
		x^4		$+x^2$	$+x$		
			x^3	$+x^2$	$+x$	$+1$	
			x^3		$+x$	$+1$	
			x^2				

So we obtain the quotient $q(x) = x^2 + x + 1$ and the remainder $r(x) = x^2$ and the result can be written as

$$x^5 + x^4 + x^2 + 1 = (x^2 + x + 1)(x^3 + x + 1) + x^2.$$

For the convenience of writing and computing we denote the polynomial

$$a_n x^n + a_{n-1} x^{n-1} + \cdots + a_1 x + a_0$$

by its simplified form

$$a_n a_{n-1} \ldots a_1 a_0$$

For example, one can write the polynomials

$$x^5 + x^4 + x^2 + 1$$

and

$$x^3 + x + 1$$

as

$$1\,1\,0\,1\,0\,1$$

and

$$1\,0\,1\,1$$

respectively. Doing this, the above scheme can be simplified as

$$
\begin{array}{r|l|l}
1011 & 1\,1\,0\,1\,0\,1 & 111 \\
     & \underline{1\,0\,1\,1} & \\
     & 1\,1\,0\,0\,1 & \\
     & \underline{1\,0\,1\,1} & \\
     & 1\,1\,1\,1 & \\
     & \underline{1\,0\,1\,1} & \\
     & 1 & \\
\end{array}
$$

Let

$$f(x) = a_0 + a_1 x + a_2 x^2 + \cdots + a_{n-1} x^{n-1} + a_n x^n$$

be a polynomial in $F[x]$ and $\alpha \in F$. Substituting x by α in $f(x)$, we obtain an element in F, i.e.,

$$a_0 + a_1 \alpha + a_2 \alpha^2 + \cdots + a_{n-1} \alpha^{n-1} + a_n \alpha^n$$

which is called the *value* of $f(x)$ at $x = \alpha$, and denoted by $f(\alpha)$. From the division algorithm, we deduce

Theorem 4.4 (*Remainder Theorem*). *Let $f(x)$ be a polynomial in $F[x]$ and $\alpha \in F$. Then the remainder of dividing $f(x)$ by the linear polynomial $x - \alpha$ is the element $f(\alpha)$ in F.*

Proof. Divide $f(x)$ by $x - \alpha$ and let the quotient be $q(x)$ and the remainder be the element c in F, i.e.,

$$f(x) = q(x)(x - \alpha) + c.$$

Substitute x by α, then $f(\alpha) = c$. □

If the value of the polynomial $f(x)$ in $F[x]$ at $x = \alpha$ is 0, i.e., $f(\alpha) = 0$, then α is called a *root* of $f(x)$.

Corollary 4.5 *Let $f(x)$ be a polynomial in $F[x]$ and $\alpha \in F$. Then α is a root of $f(x)$, if and only if $(x - \alpha) \mid f(x)$.*

From Corollary 4.5 we deduce

Theorem 4.6 *Let $f(x)$ be a polynomial of degree n in $F[x]$. Then $f(x)$ has at most n distinct roots in F.*

Proof. We apply induction on the degree n of $f(x)$. When $n = 1$, let $f(x) = ax + b$, where $a, b \in F$ and $a \neq 0$, then $-b/a$ is the only root of $f(x)$. Now consider the case $n > 1$. If $f(x)$ has no root in F, the theorem is trivially true. Let $\alpha \in F$ be a root of $f(x)$. By Corollary 4.5, $(x - \alpha) \mid f(x)$. Let $f(x) = (x - \alpha)g(x)$. Let $\beta \in F$ be a root of $f(x)$ and $\beta \neq \alpha$, then $f(\beta) = (\beta - \alpha)g(\beta)$, which implies $g(\beta) = 0$, i.e., β is also a root of $g(x)$. By Theorem 4.1, $\deg g(x) = n - 1$. By induction hypothesis, $g(x)$ has at most $n - 1$ distinct roots in F. Hence $f(x)$ has at most n distinct roots in F. □

Theorem 4.6 does not always hold for polynomials over rings.

Example 4.2 Consider the polynomial $x^2 - 1$ over the ring $\mathbb{Z}_8$. It is easy to check that $x^2 - 1$ has four roots in $\mathbb{Z}_8$, and they are $1, 3, 5, 7$.

4.3 Euclidean Algorithm

Let $a(x)$ and $b(x)$ be nonzero polynomials in $F[x]$. Parallel to the case $\mathbb{Z}$ in Section 1.2, by repeated uses of the division algorithm we can obtain $\gcd\left((a(x), b(x)\right)$. First, dividing $a(x)$ by $b(x)$ yields the quotient $q_1(x)$ and remainder $r_1(x)$ with $\deg r_1(x) < \deg b(x)$. If $r_1(x) \neq 0$, dividing $b(x)$ by $r_1(x)$ yields the quotient $q_2(x)$ and remainder $r_2(x)$ with $\deg r_2(x) < \deg r_1(x)$. If $r_2(x) \neq 0$, dividing $r_1(x)$ by $r_2(x)$ yields the quotient $q_3(x)$ and remainder $r_3(x)$ with $\deg r_3(x) < \deg r_2(x)$. As this process continues, the degrees of the remainders are monotonically reduced, i.e.,

$$\deg b(x) > \deg r_1(x) > \deg r_2(x) > \cdots,$$

hence after a finite number of divisions, the remainder must be 0. Thus we have the following sequence of formulas:

$$
\begin{aligned}
a(x) &= q_1(x)b(x) + r_1(x), & 0 &\leq \deg r_1(x) < \deg b(x), \\
b(x) &= q_2(x)r_1(x) + r_2(x), & 0 &\leq \deg r_2(x) < \deg r_1(x), \\
r_1(x) &= q_3(x)r_2(x) + r_3(x), & 0 &\leq \deg r_3(x) < \deg r_2(x), \\
&\cdots & &\cdots \\
r_{n-3}(x) &= q_{n-1}(x)r_{n-2}(x) + r_{n-1}(x), & 0 &\leq \deg r_{n-1}(x) < \deg r_{n-2}(x), \\
r_{n-2}(x) &= q_n(x)r_{n-1}(x) + r_n(x), & 0 &\leq \deg r_n(x) < \deg r_{n-1}(x), \\
r_{n-1}(x) &= q_{n+1}(x)r_n(x).
\end{aligned}
$$

This is the *Euclidean algorithm*.

As in the case of integers in Section 1.2 we can prove that $\gcd(a(x), b(x))$ exists,

$$\gcd(a(x), b(x)) = r_n(x),$$

and $r_n(x)$ can be expressed as a linear combination of $a(x)$ and $b(x)$ with polynomial coefficients, i.e.,

$$\gcd((a(x), b(x)) = c(x)a(x) + d(x)b(x), \tag{4.6}$$

where $c(x)$ and $d(x)$ are polynomials in $F[x]$. But the details are omitted. Let the leading coefficient of $r_n(x)$ be c, then usually we take

$$\gcd(a(x), b(x)) = c^{-1}r_n(x),$$

which is a monic polynomial in $F[x]$.

Theorem 4.7 *Let F be any field and $a(x)$ and $b(x)$ be nonzero polynomials in $F[x]$. Then $\gcd(a(x), b(x))$ exists, can be computed by the Euclidean algorithm and can be expressed as a linear combination of $a(x)$ and $b(x)$ with polynomials in $F[x]$ as coefficients, i.e., (4.6)*

$$\gcd(a(x), b(x)) = c(x)a(x) + d(x)b(x),$$

where $c(x)$ and $d(x) \in F[x]$. Furthermore, if we assume that $\deg a(x) > 0$, $\deg b(x) > 0$ and $a(x) \neq c\,b(x)$ for any $c \in F$, then we can require that $c(x)$ and $d(x)$ in (4.6) satisfy the following condition

$$\deg c(x) < \deg b(x) - \deg \gcd(a(x), b(x)),$$
$$\deg d(x) < \deg a(x) - \deg \gcd(a(x), b(x)), \tag{4.7}$$

and moreover, $c(x)$ and $d(x)$ which satisfy (4.7) are unique.

Proof. Clearly we have (4.6), but $c(x)$ and $d(x)$ do not necessarily satisfy condition (4.7). For simplicity, let $g(x) = \gcd(a(x), b(x))$, then (4.6) becomes

$$g(x) = c(x)a(x) + d(x)b(x).$$

Let

$$a(x) = a_1(x)g(x) \text{ and } b(x) = b_1(x)g(x),$$

then

$$\deg a_1(x) = \deg a(x) - \deg g(x) \text{ and } \deg b_1(x) = \deg b(x) - \deg g(x).$$

By the division algorithm, we have

$$c(x) = q_1(x)b_1(x) + c_1(x), \quad \deg c_1(x) < \deg b_1(x),$$
$$d(x) = q_2(x)a_1(x) + d_1(x), \quad \deg d_1(x) < \deg a_1(x).$$

Then

$$g(x) = (q_1(x) + q_2(x))a_1(x)b_1(x)g(x) + c_1(x)a(x) + d_1(x)b(x). \tag{4.8}$$

Let $n = \deg a(x) + \deg b(x) - \deg g(x)$, then

$$n = \deg a_1(x) + \deg b_1(x) + \deg g(x). \tag{4.9}$$

Clearly,

$$\deg(c_1(x)a(x)) < \deg b_1(x) + \deg a(x)$$
$$= \deg b_1(x) + \deg a_1(x) + \deg g(x) = n.$$

Similarly,

$$\deg(d_1(x)b(x)) < n.$$

So,

$$\deg(c_1(x)a(x) + d_1(x)b(x)) < n. \tag{4.10}$$

Since $a(x) \neq cb(x)$ for any $c \in F$,

$$\deg a_1(x) + \deg b_1(x) > 0.$$

Thus

$$\deg g(x) < n. \qquad (4.11)$$

From (4.8), (4.9), (4.10), and (4.11) we deduce

$$q_1(x) + q_2(x) = 0.$$

Then (4.8) becomes

$$\gcd(a(x), b(x)) = c_1(x)a(x) + d_1(x)b(x),$$

where

$$\deg c_1(x) < \deg b_1(x) = \deg b(x) - \deg \gcd(a(x), b(x)),$$
$$\deg d_1(x) < \deg a_1(x) = \deg a(x) - \deg \gcd(a(x), b(x)).$$

Let $c_1(x)$ and $d_1(x)$ be $c(x)$ and $d(x)$, respectively, then (4.7) is satisfied.

To prove the uniqueness of $c(x)$ and $d(x)$ we need the following lemma.

Lemma 4.8 *Let $f(x), a(x), b(x) \in F[x]$ and assume that $f(x) \neq 0$. If $f(x) \mid a(x)b(x)$ and $\gcd(f(x), a(x)) = 1$, then $f(x) \mid b(x)$.*

Proof. Since $\gcd(f(x), a(x)) = 1$, there exists $c(x)$ and $d(x) \in F[x]$ such that

$$1 = c(x)f(x) + d(x)a(x).$$

Multiplying both sides by $b(x)$, we have

$$b(x) = b(x)c(x)f(x) + d(x)a(x)b(x).$$

By hypothesis $f(x) \mid a(x)b(x)$, so $f(x) \mid b(x)$. $\square$

Proof of Theorem 4.7(continued). Suppose we also have

$$\gcd(a(x), b(x)) = c_1(x)a(x) + d_1(x)b(x), \qquad (4.12)$$

where $c_1(x), d_1(x) \in F[x]$ and

$$\deg c_1(x) < \deg b(x) - \deg \gcd(a(x), b(x)),$$
$$\deg d_1(x) < \deg a(x) - \deg \gcd(a(x), b(x)). \qquad (4.13)$$

Then subtracting (4.12) from (4.6), we obtain

$$(c(x) - c_1(x))a(x) + (d(x) - d_1(x))b(x) = 0.$$

Canceling $\gcd(a(x), b(x))$ from the above equation, we have

$$(c(x) - c_1(x))a_1(x) + (d(x) - d_1(x))b_1(x) = 0.$$

Since $\gcd(a_1(x), b_1(x)) = 1$, by Lemma 4.8 we have

$$a_1(x) \mid (d(x) - d_1(x)).$$

By (4.7) and (4.13)

$$\deg(d(x) - d_1(x)) < \deg a(x) - \deg \gcd(a(x), b(x)) = \deg a_1(x).$$

Hence $d(x) - d_1(x) = 0$ and $d(x) = d_1(x)$. Similarly $c(x) = c_1(x)$. $\square$

As in the case of integers, from Theorem 4.7 we deduce

Corollary 4.9 *Let F be any field and $a(x)$, $b(x) \in F[x]$. Then $a(x)$ and $b(x)$ are coprime, i.e., there exists polynomials $c(x)$ and $d(x)$ in $F[x]$ such that*

$$1 = c(x)a(x) + d(x)b(x),$$

if and only if $\gcd(a(x), b(x)) = 1$. Furthermore if we assume $a(x)$ and $b(x)$ are coprime polynomials both of degree ≥ 1, then we can require that

$$\deg c(x) < \deg b(x), \ \deg d(x) < \deg a(x),$$

and that $c(x)$ and $d(x)$ satisfying these conditions are unique.

We use an example to show how to obtain the greatest common divisor of two polynomials by the Euclidean algorithm and how to express it as a linear combination of the two polynomials.

Example 4.3 Let us compute the greatest common divisor of the polynomials

$$x^5 + x^4 + x^3 + x^2 + x + 1 \text{ and } x^4 + x^2 + x + 1$$

over F_2. We can use the following scheme for the Euclidean algorithm:

```
          x^4       +x^2 +x +1  | x^5 +x^4 +x^3 +x^2 +x +1 | x+1
x^2+x     x^4 +x^3               | x^5      +x^3 +x^2 +x
          -----------------------|-------------------------
              x^3 +x^2 +x +1     |     x^4                +1
              x^3 +x^2           |     x^4      +x^2 +x +1
              -------------------|-------------------------
                       x     +1  |              x^2 +x      | x
                                 |              x^2 +x
                                 |              -----------
                                 |                  0
```

Hence

$$x + 1 = \gcd(x^5 + x^4 + x^3 + x^2 + x + 1, \; x^4 + x^2 + x + 1).$$

The above scheme can also be replaced by the following simplified one:

```
1  1  0 | 1  0  1  1  1 | 1  1  1  1  1  1 | 1  1
        | 1  1          | 1  0  1  1  1    |
        +---------------+------------------+
        |    1  1  1  1 |    1  0  0  0  1  |
        |    1  1       |    1  0  1  1  1  |
        +---------------+------------------+
        |          1  1 |             1  1 | 1  0
        |               |             1  1 |
        |               |             -----+
        |               |                0 |
```

To express $x + 1$ as a linear combination of $x^5 + x^4 + x^3 + x^2 + x + 1$ and $x^4 + x^2 + x + 1$, it is necessary to rewrite the above scheme as

$$x^5 + x^4 + x^3 + x^2 + x + 1 = (x+1)(x^4 + x^2 + x + 1) + (x^2 + x),$$
$$x^4 + x^2 + x + 1 = (x^2 + x)(x^2 + x) + (x + 1),$$
$$x^2 + x = x(x + 1).$$

Then

$$
\begin{aligned}
x + 1 &= (x^4 + x^2 + x + 1) + (x^2 + x)(x^2 + x) \\
&= (x^4 + x^2 + x + 1) + (x^2 + x)[(x^5 + x^4 + x^3 + x^2 + x + 1) \\
&\quad + (x+1)(x^4 + x^2 + x + 1)] \\
&= (x^2 + x)(x^5 + x^4 + x^3 + x^2 + x + 1) \\
&\quad + (x^3 + x + 1)(x^4 + x^2 + x + 1).
\end{aligned}
$$

Theorem 4.10 (*Modified Euclidean Algorithm*) *Let $a(x)$ and $b(x)$ be two nonzero polynomials in $F[x]$. Denote*

$$r_{-1}(x) = a(x), \; r_0(x) = b(x).$$

Define

$$c_{-1}(x) = 1, \; c_0(x) = 0,$$
$$d_{-1}(x) = 0, \; d_0(x) = 1.$$

Calculate $q_k(x)$, $r_k(x)$, $c_k(x)$, $d_k(x)$, $k = 1, 2, 3, \ldots$, respectively, by the following rules

$$r_{k-2}(x) = q_k(x)r_{k-1}(x) + r_k(x), \quad \deg r_k(x) < \deg r_{k-1}(x), \quad (4.14)$$
$$c_k(x) = -q_k(x)c_{k-1}(x) + c_{k-2}(x), \quad (4.15)$$
$$d_k(x) = -q_k(x)d_{k-1}(x) + d_{k-2}(x) \quad (4.16)$$

until $r_{n+1}(x) = 0$. *Then*

$$r_n(x) = c_n(x)a(x) + d_n(x)b(x). \qquad (4.17)$$

Moreover, $\gcd(a(x), b(x)) = c^{-1}r_n(x)$, *where* c *is the leading coefficient of* $r_n(x)$.

Proof. We shall prove more generally

$$r_k(x) = c_k(x)r_{-1}(x) + d_k(x)r_0(x), \ \ k = 1, 2, \dots, n. \qquad (4.18)$$

Then substituting $k = n$ in (4.18), we obtain (4.17).

We apply induction on k to prove (4.18). When $k = -1$, we have

$$c_{-1}(x)r_{-1}(x) + d_{-1}(x)r_0(x) = r_{-1}(x),$$

and when $k = 0$, we have

$$c_0(x)r_{-1}(x) + d_0(x)r_0(x) = r_0(x).$$

Therefore (4.18) holds when $k = -1$ and 0. Now we assume that $k > 0$ and that (4.18) holds for $k - 2$ and $k - 1$. Then by induction hypothesis, (4.14), (4.15), and (4.16)

$$\begin{aligned}
r_k(x) &= r_{k-2}(x) - q_k(x)r_{k-1}(x) \\
&= c_{k-2}(x)r_{-1}(x) + d_{k-2}(x)r_0(x) \\
&\quad -q_k(x)(c_{k-1}(x)r_{-1}(x) + d_{k-1}(x)r_0(x)) \\
&= (c_{k-2}(x) - q_k(x)c_{k-1}(x))r_{-1}(x) \\
&\quad +(d_{k-2}(x) - q_k(x)d_{k-1}(x))r_0(x) \\
&= c_k(x)r_{-1}(x) + d_k(x)r_0(x).
\end{aligned}$$

Thus (4.18) holds also for k.

As we mentioned already in the paragraph before Theorem 4.7 that the last statement can be proved as in the case of integers. $\qquad \square$

Corollary 4.11 *Under the assumption of Theorem 4.10, assume further that* $\deg a(x) > 0$, $\deg b(x) > 0$, *and* $a(x) \neq c\,b(x)$ *for all* $c \in F^*$, *then*

$$\deg c_n(x) < \deg b(x) - \deg \gcd(a(x), b(x)),$$
$$\deg d_n(x) < \deg a(x) - \deg \gcd(a(x), b(x)).$$

Proof. It is clear that

$$\deg q_k(x) > 0, \ k = 1, 2, \ldots, n+1.$$

From (4.15) and by induction on k we deduce

$$\deg c_{k-1}(x) < \deg c_k(x), \ k = 1, 2, \ldots, n+1.$$

In particular,

$$\deg c_n(x) < \deg c_{n+1}(x). \tag{4.19}$$

Now apply induction on k to prove

$$c_k(x)r_{k-1}(x) - c_{k-1}(x)r_k(x) = (-1)^{k-1}r_0(x), \ k = 1, 2, \ldots, n+1. \tag{4.20}$$

For $k = 0$ we have

$$c_0(x)r_{-1}(x) - c_{-1}(x)r_0(x) = -r_0(x).$$

Therefore (4.20) holds for $k = 0$. Now assume $k > 0$ and (4.20) is true for $k - 1$. By (4.14), (4.15) and induction hypothesis

$$\begin{aligned}
c_k(x)&r_{k-1}(x) - c_{k-1}(x)r_k(x) \\
&= (-q_k(x)c_{k-1}(x) + c_{k-2}(x))r_{k-1}(x) - c_{k-1}(x)(r_{k-2}(x) - q_k(x)r_{k-1}(x)) \\
&= c_{k-2}(x)r_{k-1}(x) - c_{k-1}(x)r_{k-2}(x) \\
&= (-1)^{k-1}r_0(x).
\end{aligned}$$

Therefore (4.20) is true for k.

Taking $k = n + 1$ in (4.20), we have

$$c_{n+1}(x)r_n(x) - c_n(x)r_{n+1}(x) = (-1)^n r_0(x).$$

Since $r_{n+1}(x) = 0$, $r_n(x) = c \gcd(a(x), b(x))$, and $r_0(x) = b(x)$, we have

$$cc_{n+1}(x) \gcd(a(x), b(x)) = (-1)^n b(x).$$

Thus

$$\deg c_{n+1}(x) = \deg b(x) - \deg \gcd(a(x), b(x)). \tag{4.21}$$

From (4.19) and (4.21) we deduce

$$\deg c_n(x) < \deg b(x) - \deg \gcd(a(x), b(x)).$$

Similarly,

$$\deg d_n(x) < \deg a(x) - \deg \gcd(a(x), b(x)).$$

□

Example 4.4 Let

$$a(x) = x^5 + x^4 + x^3 + x^2 + x + 1,$$
$$b(x) = x^4 + x^2 + x + 1$$

be polynomials over $\mathbb{Z}_2$. Now let us use the modified Euclidean algorithm to determinate their greatest common divisor $\gcd(a(x), b(x))$ and express it as a linear combination of $a(x)$ and $b(x)$.

Let

$$r_{-1}(x) = a(x) = x^5 + x^4 + x^3 + x^2 + x + 1,$$
$$r_0(x) = b(x) = x^4 + x^2 + x + 1,$$
$$c_{-1}(x) = 1, \ c_0(x) = 0,$$
$$d_{-1}(x) = 0, \ d_0(x) = 1.$$

The first step is to divide $r_{-1}(x)$ by $r_0(x)$. We have

$$r_{-1}(x) = q_1(x)r_0(x) + r_1(x),$$
$$q_1(x) = x + 1, \ r_1(x) = x^2 + x.$$

Then calculate

$$c_1(x) = q_1(x)c_0(x) + c_{-1}(x) = 1,$$
$$d_1(x) = q_1(x)d_0(x) + d_{-1}(x) = x + 1.$$

The second step is to divide $r_0(x)$ by $r_1(x)$. We have

$$r_0(x) = q_2(x)r_1(x) + r_2(x),$$
$$q_2(x) = x^2 + x, \ r_2(x) = x + 1.$$

Then calculate

$$c_2(x) = q_2(x)c_1(x) + c_0(x) = x^2 + x,$$
$$d_2(x) = q_2(x)d_1(x) + d_0(x) = x^3 + x + 1.$$

The third step is to divide $r_1(x)$ by $r_2(x)$. We have

$$r_1(x) = q_3(x)r_2(x), \ q_3(x) = x, \ r_3(x) = 0.$$

Since $r_3(x) = 0$, the calculation is terminated here. We have

$$\gcd(a(x), b(x)) = r_2(x) = x + 1.$$

Further

$$\gcd(a(x), b(x)) = c_2(x)a(x) + d_2(x)b(x)$$
$$= (x^2 + x)(x^5 + x^4 + x^3 + x^2 + x + 1)$$
$$+ (x^3 + x + 1)(x^4 + x^2 + x + 1).$$

The above computation can be tabulated as follows:

i	c_i	d_i	r_i	q_i
-1	1	0	$x^5 + x^4 + x^3 + x^2 + x + 1$	
0	0	1	$x^4 + x^2 + x + 1$	
1	1	$x + 1$	$x^2 + x$	$x + 1$
2	$x^2 + x$	$x^3 + x + 1$	$x + 1$	$x^2 + x$
3			0	x

The modified Euclidean algorithm for the case of integers can be found in Qin Jiushao, *Nine Chapters in Arithmetic* (in Chinese), 1274AD, where it was called the *Dayanqiuyi Method*. It has the advantage that only six polynomials are needed to be recorded in each step of the computation of the greatest common divisor of two polynomials.

Let $a(x)$, $b(x)$, and $c(x)$ be polynomials in $F[x]$ and $a(x) \neq 0$, $b(x) \neq 0$. If $c(x)$ is a multiple of both $a(x)$ and $b(x)$, we say that $c(x)$ is a *common multiple* of $a(x)$ and $b(x)$. Clearly, $a(x)b(x)$ is a common multiple of $a(x)$ and $b(x)$. Hence amongst the common multiples of $a(x)$ and $b(x)$, there exists one which has the lowest degree and is monic. We call this common multiple the *least common multiple* of $a(x)$ and $b(x)$ and denote it by

$$\text{lcm}\,[a(x), b(x)].$$

Similar to the case of integers, one can also define the greatest common divisor of n polynomials $a_1(x), a_2(x), \ldots, a_n(x)$ in $F[x]$, which are not all zero. This is denoted by

$$\gcd\,(a_1(x), a_2(x), \ldots, a_n(x)).$$

When none of the $a_i(x)\,(1 \leq i \leq n)$ are zero, their least common multiple, denoted by

$$\text{lcm}\,[a_1(x), a_2(x), \ldots, a_n(x)],$$

can also be defined. As in the case of integers, parallel to Theorem 1.4 we have

Theorem 4.12 *Let $f_1(x), f_2(x), \ldots, f_n(x)$ be polynomials in $F[x]$, which are not all zero. Then there exist polynomials $c_1(x), c_2(x), \ldots, c_n(x)$ such that*

$$\gcd\,(c_1(x), c_2(x), \ldots, c_n(x)) = c_1(x)f_1(x) + c_2(x)f_2(x) + \cdots + c_n(x)f_n(x).$$

We will not discuss such details.

Analogous to the Chinese Remainder Theorem for integers (Theorem 2.26) we have the *Chinese Remainder Theorem for polynomials over a field*.

Theorem 4.13 *Let F be any field and $f_1(x), f_2(x), \ldots, f_r(x)$ be r polynomials of degree ≥ 1 in $F[x]$, which are pairwise coprime. Let $f(x) = f_1(x)f_2(x) \cdots f_r(x)$. Then for any given $g_i(x) \in F[x]$, $i = 1, 2, \ldots, r$, there exists a polynomial $g(x) \in F[x]$ such that the following r congruences*

$$g(x) \equiv g_i(x) \pmod{f_i(x)}, \quad i = 1, 2, \ldots, r \qquad (4.22)$$

hold simultaneously. Moreover, any $h(x) \in F[x]$ is a solution of (4.22) if and only if

$$g(x) \equiv h(x) \pmod{f(x)}.$$

In particular, there is a unique $g(x) \in F[x]$ with $\deg g(x) < \deg f(x)$ satisfying (4.22). Finally, if $\gcd(g_i(x), f_i(x)) = 1$ for all $i = 1, 2, \ldots, r$, then $\gcd(g(x), f(x)) = 1$, and conversely.

The proof is completely parallel to that of Theorem 2.26 and is omitted.

4.4 Unique Factorization of Polynomials

Let $p(x)$ be a polynomial of degree ≥ 1 in $F[x]$. If $p(x)$ is a prime element in $F[x]$, we say that $p(x)$ is an *irreducible* polynomial in $F[x]$. Otherwise, $p(x)$ is said to be *reducible*.

Clearly, a polynomial of degree ≥ 1 in $F[x]$ is irreducible if and only if $p(x)$ can not be expressed as (or factored into) a product of two polynomials of degree > 0 in $F[x]$. The concept of reducibility of polynomials are closely related to fields.

Example 4.5 $x^2 - 2$ is an irreducible polynomial in $\mathbb{Q}[x]$; but in $\mathbb{R}[x]$, $x^2 - 2$ is reducible, since in $\mathbb{R}[x]$

$$x^2 - 2 = (x + \sqrt{2})(x - \sqrt{2}),$$

and both $x + \sqrt{2}$ and $x - \sqrt{2}$ are proper factors of $x^2 - 2$.

Parallel to the fundamental theorem of arithmetic, we have the following

Theorem 4.14 (*Unique Factorization Theorem*) *Any polynomial $f(x)$ of degree ≥ 1 over a field F can be expressed as a product of a finite number of irreducible polynomials in $F[x]$. Furthermore, if*

$$f(x) = p_1(x)p_2(x) \cdots p_r(x) = q_1(x)p_2(x) \cdots q_s(x)$$

represents two ways of factorizing $f(x)$ into products of irreducible polynomials, then we must have $r = s$ and after a rearrangement of the factors we have

$$p_i(x) = c_i q_i(x), \quad i = 1, 2, \ldots, r,$$

where c_i $(i = 1, 2, \ldots, r)$ are some nonzero elements of F.

The proof of this theorem is completely parallel to that of the fundamental theorem of arithmetic in Section 1.2. For the proof of the first part of the theorem, one can use induction on the degree of $f(x)$. For the proof of the second part, induction is to be used on r and Lemma 4.8 will be needed.

Let $f(x)$ and $g(x)$ be polynomials in $F[x]$, $f(x) \neq 0$ and $\deg g(x) \geq 1$. If $g(x)^2 \mid f(x)$, then we say that $g(x)$ is a *multiple factor* of $f(x)$. Moreover, if $g(x)^m \mid f(x)$, but $g(x)^{m+1} \nmid f(x)$, then m is called the *multiplicity* of $g(x)$ as a factor of $f(x)$. In particular, let α be a root of $f(x)$. If $(x-\alpha)^m \mid f(x)$, where $m > 1$, then α is called a *multiple root* of $f(x)$. Moreover, if $(x - \alpha)^m \mid f(x)$ and $(x - \alpha)^{m+1} \nmid f(x)$, then m is called the *multiplicity* of the root α of $f(x)$. A root of $f(x)$ with multiplicity 1 is called a *simple root* .

We will discuss the condition under which a nonzero polynomial $f(x)$ in $F[x]$ has multiple factors. To do this, we introduce the formal differentiation of a polynomial.

Let

$$f(x) = a_0 + a_1 x + a_2 x^2 + \cdots + a_{n-1} x^{n-1} + a_n x^n$$

be a polynomial in $F[x]$. We define the *formal derivative* $f'(x)$ of $f(x)$ as

$$f'(x) = a_1 + 2a_2 x + \cdots + (n - 1)a_{n-1} x^{n-2} + na_n x^{n-1}.$$

This definition is naturally adopted from the rule of differentiation of a polynomial in calculus. But in this book, we only consider it as a formal definition. It can be proved that the formal differentiation of polynomials satisfies the following basic rules:

$$(f(x) + g(x))' = f(x)' + g(x)',$$
$$(cf(x))' = cf(x)',$$
$$(f(x)g(x))' = f'(x)g(x) + f(x)g'(x),$$
$$(f(x)^m)' = mf(x)^{m-1}f'(x),$$

where $f(x)$, $g(x) \in F[x]$, $c \in F$, and m is any positive integer.

Theorem 4.15 *Let $f(x)$ be a nonzero polynomial of positive degree in $F[x]$. Then $f(x)$ does not have multiple factors if and only if $f(x)$ and $f'(x)$ are coprime.*

Proof. Assume that $f(x)$ has multiple factors and let $g(x)$ be one of them. Then $\deg g(x) \geq 1$. We may write

$$f(x) = g(x)^2 f_1(x),$$

so

$$f'(x) = 2g(x)g'(x)f_1(x) + g(x)^2 f_1'(x).$$

Hence $g(x)$ is a common factor of $f(x)$ and $f'(x)$. $f(x)$ and $f'(x)$ are not coprime.

Conversely, assume $f(x)$ and $f'(x)$ are not coprime. Let

$$d(x) = \gcd(f(x), f'(x)),$$

then $\deg d(x) \geq 1$. We may write

$$f(x) = d(x)f_1(x).$$

Then

$$f'(x) = d'(x)f_1(x) + d(x)f_1'(x).$$

From $d(x)|f'(x)$ and the above equality we deduce $d(x)|d'(x)f_1(x)$. Since $\deg d'(x) < \deg d(x), d(x) \nmid d'(x)$. Then by the Unique Factorization Theorem there is a polynomial in $F[x]$, say $g(x)$, of degree ≥ 1 such that $g(x)|d(x)$ and $g(x)|f_1(x)$. Hence $g(x)$ is a multiple factor of $f(x)$. $\qquad\square$

Inspired by Theorem 4.14 we have the following definition.

Definition 4.1 *Let D be an integral domain and assume that $|D| > 1$. D is called a unique factorization domain, if*

(i) *every nonzero and nonunit element a can be factored into a product of prime elements.*

(ii) *If $a = p_1 p_2 \cdots p_r = q_1 q_2 \cdots q_s$ are two factorizations of a into products of prime elements, then $r = s$ and after a rearrangement of $q_1, q_2, \ldots, q_s$, p_i and q_i are associates, $i = 1, 2, \ldots, r$.*

By Theorem 4.14 $F[x]$ is a unique factorization domain. We know also from Theorem 1.5 that $\mathbb{Z}$ is a unique factorization domain.

In a unique factorization domain D any nonzero element a can be expressed in the form

$$a = u p_1^{e_1} p_2^{e_2} \ldots p_r^{e_r}, \tag{4.23}$$

where u is a unit, $p_1, p_2, \cdots, p_r$ are prime elements in D, no two of them are associates, and $e_1, e_2, \ldots, e_r$ are positive integers. The expression (4.23) is called a *prime factorization* of a and also a *complete factorization* of a.

Theorem 4.16 *Let D be a unique factorization domain. Then for any finite set of elements $a_1, a_2, \ldots, a_n \in D$, which are not all zero, their greatest common divisor* $\gcd(a_1, a_2, \ldots, a_n)$ *exists. If all $a_1, a_2, \ldots, a_n$ are nonzero, their least common multiple* $\mathrm{lcm}[a_1, a_2, \ldots, a_n]$ *also exists.*

Proof. Without loss of generality we can assume all $a_1, a_2, \ldots, a_n$ are nonzero. We represent $a_1, a_2, \ldots, a_n$ as

$$a_1 = u_1 p_1^{e_{11}} p_2^{e_{12}} \cdots p_r^{e_{1r}},$$
$$a_2 = u_2 p_1^{e_{21}} p_2^{e_{22}} \cdots p_r^{e_{2r}},$$
$$\cdots$$
$$a_n = u_n p_1^{e_{n1}} p_2^{e_{n2}} \cdots p_r^{e_{nr}},$$

where u_i, $1 \le i \le n$, are units in D, $p_1, p_2, \ldots, p_r$ are prime elements of D such that no two of them are associates, and e_{ij} $(1 \le i \le n, 1 \le j \le r)$ are nonnegative integers. Then

$$\gcd(a_1, a_2, \ldots, a_n) = p_1^{\min\{e_{11}, e_{21}, \ldots, e_{n1}\}} p_2^{\min\{e_{12}, e_{22}, \ldots, e_{n2}\}}$$
$$\cdots p_r^{\min\{e_{1r}, e_{2r}, \ldots, e_{nr}\}}$$

and

$$\mathrm{lcm}[a_1, a_2, \cdots, a_n] = p_1^{\max\{e_{11}, e_{21}, \cdots, e_{n1}\}} p_2^{\max\{e_{12}, e_{22}, \cdots, e_{n2}\}}$$
$$\cdots p_r^{\max\{e_{1r}, e_{2r}, \cdots, e_{nr}\}}.$$

$\square$

Lemma 4.17 *Let D be a unique factorization domain, and $a, b, c \in D$, all of which are nonzero. If $c | ab$ and $\gcd(c, a) = 1$, then $c | b$.*

Proof. This is an immediate consequence of Definition 4.1. $\square$

In the following we shall prove an important theorem which states that if D is a unique factorization domain so is the polynomial ring $D[x]$ in an indeterminate x over D.

Let D be a unique factorization domain. It is clear that the units of $D[x]$ are the units in D. By Theorem 4.16 any finite set of elements of D, which are not all 0, have a greatest common divisor, which is uniquely determined up to a unit factor. Let $f(x) = a_0 + a_1x + \cdots + a_nx^n \in D[x]$ and $\neq 0$, we define the *content* $c(f)$ of $f(x)$ to be $c(f) = \gcd(a_0, a_1, \ldots, a_n)$. Then $c(f)$ is also uniquely determined up to a unit factor. If $c(f) = 1$ (or a unit), $f(x)$ is called a *"primitive" polynomial.*

We are going to study the factorization of nonzero elements of $D[x]$. Given a nonzero polynomial $f(x) \in D[x]$, it can be factored into $f(x) = cg(x)$, where $c \in D$, $c \neq 0$ and $g(x)$ is "primitive". For example, let $c = c(f)$, we can write $a_i = ca'_i$, $0 \leq i \leq n$, and $f(x) = cg(x)$ where $g(x) = a'_0 + a'_1x + \cdots + a'_nx^n$. Clearly, $c(g) = 1$. Suppose that $f(x)$ has two factorizations of this form, say $f(x) = cg(x) = c_1g_1(x)$ where $c, c_1 \in D$, $c \neq 0$, $c_1 \neq 0$, and $g(x)$ and $g_1(x)$ are both "primitive". From $f(x) = cg(x)$ we deduce $c(f) = cc(g) = c$. Similarly, $c(f) = c_1$. Therefore c and c_1 are associates. It follows that $g(x)$ and $g_1(x)$ are associates. Therefore we have proved

Lemma 4.18 *Let $f(x) \in D[x]$ be nonzero. Then $f(x) = cg(x)$, where $c \in D$, $c \neq 0$ and $g(x)$ is "primitive". Moreover, if we have also $f(x) = c_1g_1(x)$, where $c_1 \in D$, $c_1 \neq 0$ and $g_1(x)$ is "primitive", then c and c_1 are associates, and $g(x)$ and $g_1(x)$ are associates.*

By Lemma 4.18 the factorization of nonzero polynomials of $D[x]$ reduces to the factorization of nonzero elements of D and the factorization of "primitive" polynomials of $D[x]$. So we concentrate on the factorization of "primitive" polynomials of $D[x]$.

Lemma 4.19 *(Gauss' Lemma) The product of two "primitive" polynomials is "primitive".*

Proof. Suppose $f(x)$ and $g(x)$ are both "primitive" but $h(x) = f(x)g(x)$ is not. Then there is a prime element p in D such that $p \nmid f(x)$, $p \nmid g(x)$, but $p \mid h(x)$. Let

$$f(x) = a_0 + a_1x + \cdots + a_nx^n, \ a_i \in D, \ 0 \leq i \leq n, \text{ and } a_n \neq 0,$$

$$g(x) = b_0 + b_1x + \cdots + b_mx^m, \ b_i \in D, \ 0 \leq i \leq m, \text{ and } b_m \neq 0,$$

and

$$h(x) = c_0 + c_1x + \cdots + c_{n+m}x^{n+m}, \ c_i \in D, \ 0 \leq i \leq n+m, \text{ and } c_{n+m} \neq 0.$$

We can assume that $p \mid a_0, a_1, \ldots, a_{k-1}$, but $p \nmid a_k$, where $0 \le k \le n$, and that $p \mid b_0, b_1, \ldots, b_{l-1}$, but $p \nmid b_l$, where $0 \le l \le m$. By the definition of multiplication of polynomials

$$c_{k+l} = a_0 b_{k+l} + a_1 b_{k+l-1} + \cdots + a_k b_l + \cdots + a_{k+l-1} b_1 + a_{k+l} b_0, \quad (4.24)$$

in which we assume $a_i = 0$ if $i > n$ and $b_j = 0$ if $j > m$. It is clear p divides all terms on the right-hand side of (4.24) except $a_k b_l$. Therefore $p \nmid c_{k+l}$, which contradicts $p \mid h(x)$. □

Corollary 4.20 *Let $f(x)$, $g(x)$ be nonzero polynomials of $D[x]$ and $h(x) = f(x)g(x)$. Then $c(h) = c(f)c(g)$.*

Lemma 4.21 *Let F be the field of fractions of D, and $f(x)$ be a "primitive" polynomial of degree > 0 of $D[x]$. If $f(x)$ is a prime element in $D[x]$, then it is an irreducible polynomial in $F[x]$.*

Proof. Assume that $f(x) = f_1(x)f_2(x)$, where $f_1(x), f_2(x)$ are polynomials of degree > 0 of $F[x]$. There are non-zero elements c_1, c_2 of F such that both $g_1(x) = c_1 f_1(x)$ and $g_2(x) = c_2 f_2(x)$ are "primitive" polynomials of $D[x]$. Then $c_1 c_2 f(x) = g_1(x)g_2(x)$. By Lemma 4.19 $g_1(x)g_2(x)$ is "primitive" in $D[x]$. But $f(x)$ is assumed to be "primitive". Therefore $c_1 c_2$ is a unit in D. It follows that $f(x)$ and $g_1(x)g_2(x)$ differ by a unit multiplier in D, which contradicts the primeness of $f(x)$. □

Theorem 4.22 *Let D be a unique factorization domain, then so is $D[x]$.*

Proof. Let $f(x) \in D[x]$ be non-zero and not a unit. By Lemma 4.18 $f(x) = cf_1(x)$, where $c \in D$ and $f_1(x)$ is "primitive". If $\deg f_1(x) = 0$ then $f_1(x)$ is a unit and thus c is not a unit, by the unique factorization in D we have $c = p_1 p_2 \cdots p_r$, where p_i are prime elements in D. If $\deg f_1(x) > 0$ then $f_1(x)$ is not a unit. If $f_1(x)$ is a prime element, let $q_1(x) = f_1(x)$. If $f_1(x)$ is not a prime element, we have $f_1(x) = g_1(x)g_2(x)$ where $g_1(x), g_2(x) \in D[x]$ and $0 < \deg g_1(x), \deg g_2(x) < \deg f(x)$. By Corollary 4.20 both $g_1(x)$ and $g_2(x)$ are "primitive". Using induction on the degrees we obtain that $f_1(x) = q_1(x)q_2(x) \cdots q_s(x)$, where $q_i(x)$ are prime elements of degree > 0 in $D[x]$ and hence are "primitive". If c is not a unit, by the unique factorization in D we have $c = p_1 p_2 \cdots p_r$, where p_i are prime elements in D. In all cases we obtain

$$f(x) = p_1 p_2 \cdots p_r g_1(x)g_2(x) \cdots g_s(x).$$

Assume that $f(x)$ has another factorization

$$f(x) = p_1'p_2'\cdots p_{r'}'g_1'(x)g_2'(x)\cdots g_{s'}'(x),$$

where $p_1', p_2', \cdots, p_r'$ are prime elements in D and $g_1'(x)g_2'(x)\cdots g_s'(x)$ are prime elements of degree > 0 in $D[x]$. Then $g_j'(x)$ are "primitive". By Lemma 4.19

$$c(f) = p_1p_2\cdots p_r = up_1'p_2'\cdots p_{r'}',$$

where u is a unit in D. Since D is a unique factorization domain, $r = r'$ and after a rearrangement, for each $i = 1, 2, \cdots, r$, p_i and p_i' are associates. It follows that

$$ug_1(x)g_2(x)\cdots g_s(x) = g_1'(x)g_2'(x)\cdots g_{s'}'(x).$$

By Lemma 4.21, all $g_i(x)$ and $g_j'(x)$ are irreducible in $F[x]$. By the unique factorization theorem in $F[x]$, $s = s'$ and after a rearrangement, for each $i = 1, 2, \cdots, s$, $g_i(x)$ and $g_i'(x)$ differ by an element of F^*. Since both $g_i(x)$ and $g_i'(x)$ are primitive, they differ only by a unit multiplier of D.

Hence $D[x]$ is a unique factorization domain. $\qquad\square$

Corollary 4.23 $\mathbb{Z}[x]$ *is a unique factorization domain.*

Corollary 4.24 *Let F be any field and $x_1, x_2, \ldots, x_n$ be n indeterminates over F, where $n \geq 1$. Then $F[x_1, x_2 \ldots, x_n]$ is a unique factorization domain.*

4.5 Exercises

4.1 Let F be a field. Prove that the units of $F[x]$ are the non-zero elements of F.

4.2 Let $a(x) = x^9 + x^8 + x^5 + x^3 + 1$, $b(x) = x^4 + x + 1 \in \mathbb{Z}_2[x]$. Is $a(x)$ divisible by $b(x)$?

4.3 Let $a(x) = x^6 + x^5 + x^2 + x + 1$, $b(x) = x^2 + x + 1 \in \mathbb{Z}_2[x]$. Determine $q(x), r(x) \in \mathbb{Z}_2[x]$ such that $a(x) = q(x)b(x) + r(x)$, $\deg r(x) < \deg b(x)$.

4.4 Let $a(x) = x^6 + 3x^5 + 4x^2 + x + 1$, $b(x) = x^2 + x + 1 \in \mathbb{Z}_5[x]$. Determine $q(x), r(x) \in \mathbb{Z}_5[x]$ such that $a(x) = q(x)b(x) + r(x)$, $\deg r(x) < \deg b(x)$.

4.5 Let $a(x) = x^7 + x^4 + x^2 + x + 1$, $b(x) = x^5 + x^2 + x + 1 \in \mathbb{Z}_2[x]$. Use Euclidean algorithm to compute $\gcd(a(x), b(x))$, express it as a linear combination of $a(x)$ and $b(x)$ with polynomials in $\mathbb{Z}_2[x]$ as coefficients, and try to reduce the degree of the coefficients so that they satisfy (4.7).

4.6 Do the same for the polynomials $a(x) = x^8 + 3x^7 + 2x^6 + 3x^5 + 3x^4 + 2x^3 + x$, $b(x) = x^3 + 2x^2 + x \in \mathbb{Z}_5[x]$, but use the modified Euclidean algorithm.

4.7 Write down the proofs of Theorems 4.12, 4.13, and 4.14 in detail.

4.8 Decompose $x^6 + x^4 + x + 1 \in \mathbb{Z}_2[x]$ into a product of irreducible polynomials in $\mathbb{Z}_2[x]$.

4.9 Let $f(x)$ be a quadratic or cubic polynomial over a field F. If $f(\alpha) \neq 0$ for every $\alpha \in F$, then $f(x)$ is irreducible over F.

4.10 Prove that $x^7 + x^6 + x^5 + x^4 + x^3 + x^2 + x + 1 \in \mathbb{Z}_2[x]$ has a multiple factor. Determine this multiple factor and its multiplicity.

4.11 Do the same for the polynomial $x^5 + x^4 + x^3 + x^2 + x + 1 \in \mathbb{Z}_2[x]$.

4.12 Determine those $a \in \mathbb{Z}_5[x]$ such that $x^2 - a$ is reducible over $\mathbb{Z}_5$ and factorize them into product of linear factors.

4.13 Prove that the rules of formal differentiation of polynomials listed on page 94.

4.14 Let F be a field of finite characteristic p and $f(x) \in F[x]$. Prove that $f'(x) = 0$ if and only if there is a polynomial $g(x) \in F[x]$ such that $f(x) = g(x)^p$.

4.15 We saw in Example 4.2 that $x^2 - 1$ has four roots in the ring $\mathbb{Z}_8$. How many roots does $x^2 - 1$ have in the ring $\mathbb{Z}_4$ and also in the ring $\mathbb{Z}_{2^n}$?

4.16 Prove that any two distinct monic irreducible polynomials over a field F can not have a common root in any extension field of F.

4.17 Let F be any field and x, y are indeterminates. Prove that $F[x, y] \simeq F[x][y]$.

4.18 Let D be a unique factorization domain with field of fractions F. Prove that if $f(x), g(x) \in D[x]$ have no common factors in $D[x]$, then they have no common factors in $F[x]$.

Chapter 5

Residue Class Rings

5.1 Residue Class Rings

Let m be a fixed positive integer. In Example 1.8 we introduced the set $\mathbb{Z}_m$ of residue classes modulo m, in Example 2.4 we defined the addition and multiplication in $\mathbb{Z}_m$, and in Section 3.3 we mentioned that $\mathbb{Z}_m$ is a ring and is called the ring of integers modulo m. This construction of $\mathbb{Z}_m$ can be generalized as follows.

Let R be any ring and m be a fixed element of R. For any two elements $a, b \in R$, define

$$a \equiv b \,(\text{mod } m) \text{ if and only if } m \mid (a - b). \tag{5.1}$$

We read $a \equiv b \,(\text{mod } m)$ as a *is congruent to* b modulo m. It is easy to verify that this is an equivalence relation on R (Exercise 5.1). Each equivalence class is called a *residue class* modulo m in R. The residue class $\text{mod}\, m$ containing $a \in R$ will be denoted by $\overline{a}$. Denote the set of residue classes $\text{mod}\, m$ in R by $R/(m)$. Clearly, when $m = 0, R/(m) = R$.

Let $a \in R$, then $\overline{a} \in R/(m)$. Clearly, for any $b \in R, b \in \overline{a}$ if and only if $b \equiv a \,(\text{mod } m)$, i.e., $b = a + rm$ for some $r \in R$. Thus

$$\overline{a} = \{a + rm : r \in R\},$$

and for $\overline{a}, \overline{a'} \in R/(m), \overline{a} = \overline{a'}$ if and only if $a \equiv a' \,(\text{mod } m)$.

Let R be a ring and $m \in R$. The set of all multiples of m forms a subgroup of the additive group of R and denoted by (m). That is,

$$(m) = \{rm : r \in R\}.$$

101

Let $a \in R$, then $\bar{a} = a + (m)$, i.e., the residue class mod m containing a is the coset of a relative to the subgroup (m). A complete system of coset representatives of the additive group of R relative to (m) is called *a complete system of representatives of the residue classes* mod m in R.

Let $\bar{a}, \bar{b}$ be two residue classes mod m. Define the *addition* and *multiplication* of $\bar{a}$ and $\bar{b}$ by

$$\bar{a} + \bar{b} = \overline{a + b} \tag{5.2}$$

and

$$\bar{a}\bar{b} = \overline{ab}; \tag{5.3}$$

As in Example 2.4 we can prove that these two definitions are independent of the particular choices of the elements a and b in the residue classes $\bar{a}$ and $\bar{b}$, respectively (Exercise 5.2), hence they are well-defined. We call $\overline{a+b}$ and $\overline{ab}$ the *sum* and *product* of $\bar{a}$ and $\bar{b}$, respectively.

Theorem 5.1 *Let R be a ring and m be a fixed element of R. Then $R/(m)$ is a ring with respect to the addition and multiplication defined by* (5.2) *and* (5.3), *respectively.*

Proof. First, as in Example 2.4 we can show that $(R/(m), +)$ is an abelian group. That is, $R1$ holds in $R/(m)$. Next, for any $\bar{a}, \bar{b}, \bar{c} \in R/(m)$, we have

$$(\bar{a}\bar{b})\bar{c} = \overline{ab}\,\bar{c} = \overline{(ab)c} = \overline{a(bc)} = \bar{a}\overline{bc} = \bar{a}(\bar{b}\bar{c}).$$

The third equality follows from the associativity of multiplication in R and all other equalities follow from the definition (5.3). Similarly, for any $\bar{a}, \bar{b} \in R/(m)$, we have

$$\bar{a}\bar{b} = \overline{ab} = \overline{ba} = \bar{b}\bar{a},$$

and for any $\bar{a} \in R/(m)$, we have

$$\bar{a}\bar{1} = \overline{a1} = \bar{a}.$$

Therefore $R2.1, R2.2$ and $R2.3$ hold in $R/(m)$. Finally, for any $\bar{a}, \bar{b}, \bar{c} \in R/(m)$, we have

$$\bar{a}(\bar{b} + \bar{c}) = \bar{a}(\overline{b + c}) = \overline{a(b + c)} = \overline{ab + ac} = \overline{ab} + \overline{ac} = \bar{a}\bar{b} + \bar{a}\,\bar{c}.$$

Therefore $R3$ also holds in $R/(m)$. Hence $R/(m)$ is a ring. $\square$

Definition 5.1 *Let R be a ring and m be a fixed element of R. The ring $R/(m)$ is called the residue class ring of R modulo m.*

Clearly, for any prime number p, $\mathbb{Z}/(p) = \mathbb{Z}_p$ is a field with p elements, and for any fixed positive integer m, $\mathbb{Z}/(m) = \mathbb{Z}_m$. When m is a negative integer, $\mathbb{Z}/(m) = \mathbb{Z}/(-m) = \mathbb{Z}_{-m}$. More generally, we have

Theorem 5.2 *Let R be any ring and $m, m' \in R$. If m and m' are associates, then $R/(m) = R/(m')$.*

Proof. Assume that m and m' are associates, i.e., there is a unit $u \in R$ such that $m' = um$. For any $a \in R$, let $\bar{a}$ and $\bar{a}'$ be the residue classes mod m and mod m' respectively. Then

$$\bar{a} = \{a + rm : r \in R\} = \{a + ru^{-1}m' : r \in R\} = \{a + r'm' : r' \in R\} = \bar{a}'.$$

Consequently, $R/(m) = R/(m')$. $\square$

Even when R is a ring with zero divisors, $R/(m)$ may be an integral domain or even a field. For example, we have

Theorem 5.3 *Let p be a prime number and $s \geq 1$ be an integer. If $s > 1$, $\mathbb{Z}_{p^s}$ is a ring with zero divisors and for any $s \geq 1$, $\mathbb{Z}_{p^s}/(p)$ is a field which is isomorphic to $\mathbb{Z}_p$.*

Proof. Following the identification introduced in Example 2.13, we have

$$\mathbb{Z}_{p^s} = \{a \in \mathbb{Z} : 0 \leq a < p^s\}.$$

Suppose $s > 1$, then $p \neq 0$ and $p^{s-1} \neq 0$ in $\mathbb{Z}_{p^s}$, but $pp^{s-1} = p^s = 0$. Therefore $\mathbb{Z}_{p^s}$ is a ring with zero divisors.

Let $s \geq 1$. For any $a \in \mathbb{Z}_{p^s}$, dividing a by p as they are integers, we obtain

$$a = qp + r, \qquad 0 \leq r < p,$$

which can also be regarded as an equation in $\mathbb{Z}_{p^s}$. Therefore $\bar{a} = \bar{r}$ in $\mathbb{Z}_{p^s}/(p)$, in other words, we can choose the representative of a residue class of $\mathbb{Z}_{p^s}$ mod p to be a number r with $0 \leq r < p$. Then

$$\mathbb{Z}_{p^s}/(p) = \{\bar{0}, \bar{1}, \bar{2}, \ldots, \overline{p-1}\}.$$

Let $\bar{r} \in \mathbb{Z}_{p^s}/(p)$ and $\bar{r} \neq \bar{0}$, i.e., $0 < r < p$. Then $\gcd(r, p) = 1$, thus there exist integers c and d such that

$$cr + dp = 1.$$

We may regard the above equation as an equation in $\mathbb{Z}_{p^s}$. Taking mod p, the above equation gives $\bar{c}\bar{r} = \bar{1}$. Therefore $\bar{r}$ has an inverse in $\mathbb{Z}_{p^s}/(p)$. Hence $\mathbb{Z}_{p^s}/(p)$ is a field.

Since $\mathbb{Z}_{p^s}/(p)$ is a field with p elements, it coincides with its prime field which is isomorphic to $\mathbb{Z}_p$.   □

On the other hand, even we assume that R is an integral domain, $R/(m)$ is not necessarily an integral domain for some $m \in R$. For example, when $R = \mathbb{Z}$, $\mathbb{Z}/(4) = \mathbb{Z}_4$ has zero divisors: $\overline{2} \cdot \overline{2} = \overline{4} = \overline{0}$. More generally, when m is a composite number, $\mathbb{Z}/(m)$ has also zero divisors: if $m = m_1 m_2, 1 < |m_1|, |m_2| < |m|$, then $\overline{m_1} \neq 0, \overline{m_2} \neq 0$, but $\overline{m_1 m_2} = \overline{m} = \overline{0}$. As another example, let $D = F[x]$, where F is any field, and $m(x)$ be a polynomial of degree ≥ 2 in $F[x]$, which is not irreducible. We can assume that $m(x) = m_1(x)m_2(x)$, where $0 < \deg m_1(x), \deg m_2(x) < \deg m(x)$. Then $F[x]/(m(x))$ has also zero divisors. In fact, $\overline{m_1(x)} \neq \overline{0}, \overline{m_2(x)} \neq \overline{0}$, and $\overline{m_1(x)}\,\overline{m_2(x)} = \overline{m_1(x)m_2(x)} = \overline{m(x)} = \overline{0}$. But we have

Theorem 5.4 *Let D be a unique factorization domain and p a prime element of D. Then $D/(p)$ is an integral domain.*

Proof. Assume $\overline{a}\overline{b} = \overline{0}$ for $\overline{a}, \overline{b} \in D/(p)$. Then $\overline{ab} = \overline{a}\overline{b} = \overline{0}$. Thus $p \mid ab$. Since D is a unique factorization domain and p is a prime element, from $p \mid ab$ we deduce $p \mid a$ or $p \mid b$. If $p \mid a$ then $\overline{a} = \overline{0}$, and if $p \mid b$ then $\overline{b} = \overline{0}$. Therefore $D/(p)$ has no zero divisors and, hence, $D/(p)$ is an integral domain.   □

But when p is a prime element of D, $D/(p)$ is not necessarily a field. For example, let $D = F[x, y]$, then x is a prime element of D. It can be shown that $D/(x) \simeq F[y]$ (Exercise 5.6), but $F[y]$ is not a field. However, we have

Theorem 5.5 *Let F be a field, $F[x]$ be the polynomial ring in an indeterminate x over F, and $p(x)$ be an irreducible polynomial in $F[x]$. Then $F[x]/(p(x))$ is a field.*

Proof. Let $\overline{a(x)} \in F[x]/(p(x))$ and $\overline{a(x)} \neq \overline{0}$. Then $p(x) \nmid a(x)$. Thus $\gcd(p(x), a(x)) = 1$. By Corollary 4.9 there are polynomials $c(x), d(x) \in F[x]$ such that $c(x)p(x)+d(x)a(x) = 1$. Then $\overline{c(x)}\,\overline{p(x)}+\overline{d(x)}\,\overline{a(x)} = \overline{1}$. But $\overline{p(x)} = \overline{0}$. So, $\overline{d(x)}\,\overline{a(x)} = \overline{1}$. Hence $\overline{a(x)}$ has an inverse $\overline{d(x)}$ in $F[x]/(p(x))$. Therefore $F[x]/(p(x))$ is a field.   □

Definition 5.2 *Let F be a field, $F[x]$ be the polynomial ring in an indeterminate x over F, and $f(x)$ be a polynomial in $F[x]$. Then the ring $F[x]/(f(x))$ is called the residue class ring of the polynomial ring $F[x]$ modulo the polynomial $f(x)$. Moreover, if $f(x)$ is an irreducible polynomial over F, the field $F[x]/(f(x))$ is called the residue class field of the polynomial ring $F[x]$ modulo the irreducible polynomial $f(x)$.*

Theorem 5.6 *Let F be a field and $f(x)$ be a polynomial of degree $n > 0$ of $F[x]$. Then the set of elements*

$$S = \{a_0 + a_1 x + a_2 x^2 + \cdots + a_{n-1} x^{n-1} : a_0, a_1, a_2, ..., a_{n-1} \in F\}$$

is a complete system of representatives of the residue classes $\mod f(x)$ *in $F[x]$.*

Proof. Let $g(x) \in F[x]$. By the division algorithm there are polynomials $q(x)$ and $r(x) \in F[x]$ such that

$$g(x) = q(x)f(x) + r(x) \text{ and } \deg r(x) < \deg f(x).$$

Then $r(x) \in S$ and $\overline{g(x)} = \overline{r(x)}$. This proves that any polynomial in $F[x]$ lies in a residue class modulo $f(x)$ containing a polynomial in S.

Let

$$g_1(x) = a_0 + a_1 x + a_2 x^2 + \cdots + a_{n-1} x^{n-1}$$

and

$$g_2(x) = b_0 + b_1 x + b_2 x^2 + \cdots + b_{n-1} x^{n-1}$$

be two elements in S. If $\overline{g_1(x)} = \overline{g_2(x)}$, then $f(x) \,|\, (g_1(x) - g_2(x))$. But $\deg f(x) = n$, and $\deg (g_1(x) - g_2(x)) \leq n - 1$. By Theorem 4.1 we must have $g_1(x) = g_2(x)$. Hence S is a complete system of representatives of the residue classes modulo $f(x)$ in $F[x]$. $\square$

Let F be a field and $f(x)$ be a polynomial of degree $n > 0$. For each $a \in F$ there is a residue class

$$\overline{a} = \{a + q(x)f(x) : q(x) \in F[x]\}.$$

Let

$$\overline{F} = \{\overline{a} : a \in F\}.$$

Clearly, the map

$$- : F \to \overline{F}$$

$$a \mapsto \overline{a}$$

is a bijection. Moreover, for all $a, b \in F$, we have

$$\overline{a} + \overline{b} = \overline{a + b}$$

and

$$\overline{a}\,\overline{b} = \overline{ab}.$$

By Theorem 3.8 $\overline{F}$ is a field and the map $- : F \to \overline{F}$ is an isomorphism of fields. Thus there will not arise any confusion in computation in the residue class ring $F[x]/(f(x))$ if we identify $a \in F$ with $\bar{a} \in \bar{F}$. In other words, the residue class $\bar{a}$ of $a \in F$ mod $f(x)$ will also be denoted by a and, more generally, the residue class of $a_0 + a_1 x + a_2 x^2 + \cdots + a_{n-1} x^{n-1}$ mod $f(x)$, where $a_0, a_1, a_2, \ldots, a_{n-1} \in F$, i.e., $\overline{a_0 + a_1 x + a_2 x^2 + \cdots + a_{n-1} x^{n-1}} = \overline{a_0} + \overline{a_1}\,\bar{x} + \overline{a_2}\,\bar{x}^2 + \cdots + \overline{a_{n-1}}\,\bar{x}^{n-1}$ will also be denoted by $a_0 + a_1 \bar{x} + a_2 \bar{x}^2 + \cdots + a_{n-1}\bar{x}^{n-1}$. Then F is regarded as a subset of $F[x]/(f(x))$ and

$$F[x]/(f(x)) = \{a_0 + a_1 \bar{x} + a_2 \bar{x}^2 + \cdots + a_{n-1}\bar{x}^{n-1} : a_0, a_1, a_2, \ldots, a_{n-1} \in F\}.$$

There are more notations for the residue class ring $F[x]/(f(x))$, which will be adopted in this book. For example, we usually denote the residue class of x mod $f(x)$ by α, then we have

$$\overline{a_0 + a_1 x + a_2 x^2 + \cdots + a_{n-1} x^{n-1}} = a_0 + a_1 \alpha + a_2 \alpha^2 + \cdots + a_{n-1} \alpha^{n-1}.$$

Thus

$$F[x]/(f(x)) = \{a_0 + a_1 \alpha + a_2 \alpha^2 + \cdots + a_{n-1} \alpha^{n-1} : a_0, a_1, a_2, \ldots, a_{n-1} \in F\}.$$

Sometimes, for simplicity, we delete the bar and write the polynomial $a_0 + a_1 x + a_2 x^2 + \cdots + a_{n-1} x^{n-1}$ for $\overline{a_0 + a_1 x + a_2 x^2 + \cdots + a_{n-1} x^{n-1}}$. Then we also write

$$F[x]/(f(x)) = \{a_0 + a_1 x + a_2 x^2 + \cdots + a_{n-1} x^{n-1} : a_0, a_1, a_2, \ldots, a_{n-1} \in F\}.$$

For any $g(x) \in F[x]$ we also use the notation

$$g(x) \bmod f(x)$$

to denote the unique polynomial of degree $< n$, which is congruent to $g(x)$ modulo $f(x)$.

5.2 Examples

We illustrate the construction of residue class rings of Section 5.1 with the following examples.

Example 5.1 Let $p(x) = x^2 + 1 \in Q[x]$. Clearly, $x^2 + 1$ has no root in Q and it is irreducible in $Q[x]$. By Theorem 5.5, $Q[x]/(x^2 + 1)$ is a field. By our convention
$$Q[x]/(x^2 + 1) = \{a + b\bar{x} : a, b \in Q\}.$$
We prove that $Q[x]/(x^2 + 1)$ is isomorphic to the field

$$Q[i] = \{a + bi : a, b \in Q\}$$

introduced in Exercise 3.2.

Consider the map

$$\sigma : Q[x]/(x^2 + 1) \to Q[i]$$
$$a + b\overline{x} \mapsto a + bi.$$

Clearly, σ is a bijection and preserves the addition. In $Q[x]/(x^2 + 1)$ the multiplication is performed as follows

$$\overline{(a + bx)} \; \overline{(c + dx)} = \overline{ac + (ad + bc)x + bdx^2}$$

$$= \overline{ac + (ad + bc)x + bd(x^2 + 1) - bd}$$

$$= \overline{(ac - bd) + (ad + bc)x}.$$

$Q[i]$ is a subfield of the complex field $\mathbb{C}$ and its multiplication is performed as follows
$$(a + bi)(c + di) = (ac - bd) + (ad + bc)i.$$

Therefore, σ preserves also the multiplication. Hence

$$Q[x]/(x^2 + 1) \simeq Q[i].$$

Example 5.2 We may regard $p(x) = x^2 + 1 \in \mathbb{Z}[x]$. $p(x)$ is a prime element of $\mathbb{Z}[x]$. By Theorem 5.4, $\mathbb{Z}[x]/(x^2 + 1)$ is an integral domain. Similarly, we write
$$\mathbb{Z}[x]/(x^2 + 1) = \{a + b\overline{x} : a, b \in \mathbb{Z}\}.$$
It can be proved in the same way as in Example 5.1 that $\mathbb{Z}[x]/(x^2 + 1)$ is isomorphic to the integral domain

$$\mathbb{Z}[i] = \{a + bi : a, b \in \mathbb{Z}\}$$

introduced in Exercise 3.20.

Example 5.3 We may also regard $p(x) = x^2 + 1 \in \mathbb{R}[x]$. $x^2 + 1$ has no real roots, hence it is irreducible in $\mathbb{R}[x]$. By Theorem 5.5 $\mathbb{R}[x]/(x^2 + 1)$ is a field. Similarly, we can prove that $\mathbb{R}[x]/(x^2 + 1)$ is isomorphic to the complex field $\mathbb{C}$ and that the following is an isomorphic map.

$$\mathbb{R}[x]/(x^2 + 1) \to \mathbb{C}$$
$$a + b\overline{x} \mapsto a + bi.$$

Example 5.4 Let $p(x) = Ax^2 + Bx + C$ be an irreducible quadratic polynomial in $\mathbb{R}[x]$. Then $A \neq 0$ and $B^2 - 4AC < 0$. We call $\triangle = B^2 - 4AC$ the *discriminant* of $p(x)$. By Theorem 5.5 $\mathbb{R}[x]/(Ax^2 + Bx + C)$ is a field. Using the notation introduced in Section 5.1, we have

$$\mathbb{R}[x]/(Ax^2 + Bx + C) = \{a + b\alpha : a, b \in \mathbb{R}\},$$

where α denotes the residue class of $x \bmod p(x)$. The multiplication in $\mathbb{R}[x]/(Ax^2 + Bx + C)$ is performed as follows

$$(a + b\alpha)(c + d\alpha) = (ac - bdC/A) + (ad + bc - bdB/A)\alpha.$$

We can prove that $\mathbb{R}[x]/(Ax^2 + Bx + C)$ is isomorphic to the complex field and that the map

$$\mathbb{R}[x]/(Ax^2 + Bx + C) \to \mathbb{C}$$
$$a + b\alpha \mapsto (a - bB/(2A)) + (b\sqrt{-\Delta}/(2A))i \qquad (5.4)$$

is an isomorphism (Exercise 5.7).

However, if $p(x) = Ax^2 + Bx + C$ is an irreducible quadratic polynomial in $\mathbb{Q}[x]$, $\mathbb{Q}[x]/(Ax^2 + Bx + C)$ is not necessarily isomorphic to $\mathbb{Q}[i]$. For example, let $p(x) = x^2 - 2$, then $\mathbb{Q}[x]/(x^2 - 2) \not\simeq \mathbb{Q}[i]$. In fact, we have $\mathbb{Q}[x]/(x^2 - 2) \simeq \mathbb{Q}[\sqrt{2}]$ and $(\sqrt{2})^2 = 2$, but $(a + bi)^2 \neq 2$ for all $a + bi \in \mathbb{Q}[i]$.

Example 5.5 Let $f(x) = x^2 + 1$ be a polynomial in $\mathbb{C}[x]$. Consider the residue class ring $\mathbb{C}[x]/(x^2 + 1)$. Denote the residue class of $x \bmod x^2 + 1$ by α, then we have

$$\mathbb{C}[x]/(x^2 + 1) = \{a + b\alpha : a, b \in \mathbb{C}\},$$

where $\alpha^2 + 1 = 0$. Now, $\mathbb{C}[x]/(x^2 + 1)$ is a ring with zero divisors. In fact, $i + \alpha \neq 0$ and $i - \alpha \neq 0$, but

$$(i + \alpha)(i - \alpha) = i^2 - \alpha^2 = -1 - (-1) = 0.$$

Example 5.6 Let F be a field, $n > 1$, and $x^n - 1$ be a polynomial in $F[x]$. The residue class ring $F[x]/(x^n - 1)$ is also a ring with zero divisors. Using the notation introduced at the end of Section 5.1, we have

$$F[x]/(x^n - 1) = \{a_0 + a_1x + a_2x^2 + \cdots + a_{n-1}x^{n-1} : a_0, a_1, \ldots, a_{n-1} \in F\}.$$

Then

$$(x - 1)(x^{n-1} + x^{n-2} + \cdots + 1) = x^n - 1 = 0 \text{ in } F[x]/(x^n - 1),$$

while $x - 1 \neq 0$ and $x^{n-1} + x^{n-2} + \cdots + 1 \neq 0$ in $F[x]/(x^n - 1)$.

5.3   Residue Class Fields

Let E be a field and F be a subfield of E. Let α be any fixed element of E. For any $f(x) = a_0 + a_1x + \cdots + a_nx^n \in F[x]$, the value of $f(x)$ at $x = \alpha$ is defined as

$$f(\alpha) = a_0 + a_1\alpha + \cdots + a_n\alpha^n.$$

Define
$$F[\alpha] = \{f(\alpha) : f(x) \in F[x]\}$$
and
$$F(\alpha) = \{f(\alpha)/g(\alpha) : f(x), g(x) \in F[x] \text{ and } g(\alpha) \neq 0\}.$$
Then $F[\alpha] \subseteq E$ and $F(\alpha) \subseteq E$. Moreover, we have

Theorem 5.7 *Let E be a field, F be a subfield of E, and α be any element of E. Then $F[\alpha]$ is a subring of E and is an integral domain; moreover, $F(\alpha)$ is a subfield of E.*

Proof. Let $f(x), g(x) \in F[x]$, then $f(x) - g(x), f(x) \cdot g(x) \in F[x]$. It follows that $f(\alpha) - g(\alpha) \in F[\alpha]$ and $f(\alpha)g(\alpha) \in F[\alpha]$. Clearly, $1 \in F[\alpha]$. Therefore $F[\alpha]$ is a subring of E. Being a subring of the field E, $F[\alpha]$ is of course an integral domain.

Similarly we can show that $F(\alpha)$ is a subfield of E. $\square$

Theorem 5.8 *Let E be a field, F be a subfield of E, and $\alpha \in E$. Let $p(x)$ be an irreducible polynomial of degree n over F and assume that $p(\alpha) = 0$. Then $F[\alpha]$ is a subfield of E,*

$$F[\alpha] = \{a_0 + a_1\alpha + a_2\alpha^2 + \cdots + a_{n-1}\alpha^{n-1} : a_i \in F\}, \qquad (5.5)$$

and every element of $F[\alpha]$ can be expressed uniquely in the form

$$a_0 + a_1\alpha + a_2\alpha^2 + \cdots + a_{n-1}\alpha^{n-1}, \text{ where } a_0, a_1, \ldots, a_{n-1} \in F.$$

Moreover, $F[\alpha]$ is isomorphic to the residue class field $F[x]/(p(x))$ and $F(\alpha) = F[\alpha]$. If F is a finite field with q elements, then $|F[\alpha]| = q^n$.

Proof. Let
$$p(x) = p_0 + p_1 x + p_2 x^2 + \cdots + p_n x^n,$$
where $p_i \in F$, $i = 0, 1, \ldots, n$ and $p_n \neq 0$. Then
$$p(\alpha) = p_0 + p_1\alpha + p_2\alpha^2 + \cdots + p_n\alpha^n = 0,$$
from which we deduce
$$
\begin{aligned}
\alpha^n &= -p_n^{-1}p_0 - p_n^{-1}p_1\alpha - p_n^{-1}p_2\alpha^2 - \cdots - p_n^{-1}p_{n-1}\alpha^{n-1}, \\
\alpha^{n+1} &= p_n^{-2}p_0 p_{n-1} + (p_n^{-2}p_1 p_{n-1} - p_n^{-1}p_0)\alpha \\
&\quad + (p_n^{-2}p_2 p_{n-1} - p_n^{-1}p_1)\alpha^2 + \cdots + (p_n^{-2}p_{n-1}^2 - p_n^{-1}p_{n-2})\alpha^{n-1}, \\
&\quad \cdots .
\end{aligned}
$$

Using these relations, any element of $F[\alpha]$ can be expressed as a linear combination of 1, α, α^2, $\ldots$, α^{n-1} with coefficients in F or as a polynomial in α of degree $\leq n - 1$ with coefficients in F. Therefore we have (5.5). Let us prove the uniqueness of the expression. Assume that

$$a_0 + a_1\alpha + \cdots + a_{n-1}\alpha^{n-1} = 0, \text{ where } a_0, \cdots, a_{n-1} \in F.$$

Let

$$h(x) = a_0 + a_1 x + \cdots + a_{n-1}x^{n-1},$$

then $h(\alpha) = 0$. By the Remainder Theorem (Theorem 4.4) $(x - \alpha) \,|\, h(x)$. We also have $(x - \alpha) \,|\, p(x)$. Therefore $\gcd(p(x), h(x)) \neq 1$. Since $p(x)$ is irreducible over F, $\gcd(p(x), h(x)) = p_n^{-1}p(x)$ and $p(x) \,|\, h(x)$. But $\deg h(x) < n = \deg p(x)$, thus $h(x) = 0$ and, hence, $a_0 = a_1 = \cdots = a_{n-1} = 0$. Then we deduce easily that every element of $F[\alpha]$ can be expressed uniquely as a polynomial in α of degree $\leq n - 1$ with coefficients in F.

Consider the map

$$F[x]/(p(x)) \rightarrow F[\alpha]$$
$$a_0 + a_1 x + \cdots + a_{n-1}x^{n-1} + (p(x)) \mapsto a_0 + a_1\alpha + \cdots + a_{n-1}\alpha^{n-1}.$$
$$(5.6)$$

Clearly, it is bijective. It is routine to verify that the map (5.6) preserves the addition and multiplication. By Theorem 5.5, $F[x]/(p(x))$ is a field. Thus by Theorem 3.8, $F[\alpha]$ is also a field and (5.5) is an isomorphism of fields.

Now let us prove that $F(\alpha) = F[\alpha]$. Let $f(\alpha)/g(\alpha) \in F(\alpha)$, where $f(x), g(x) \in F[x]$ and $g(\alpha) \neq 0$. Since $p(\alpha) = 0$, $p(x) \nmid g(x)$. But $p(x)$ is irreducible over F, so $\gcd(p(x), g(x))=1$. Then there exist polynomials $c(x)$ and $d(x) \in F[x]$ such that

$$c(x)p(x) + d(x)g(x) = 1.$$

Substituting $x = \alpha$ into the above equation, we obtain $d(\alpha)g(\alpha) = 1$. Then $f(\alpha)/g(\alpha) = d(\alpha)f(\alpha)/d(\alpha)g(\alpha) = d(\alpha)f(\alpha) \in F[\alpha]$. Therefore $F(\alpha) \subset F[\alpha]$. Clearly, $F[\alpha] \subset F(\alpha)$, hence $F(\alpha) = F[\alpha]$.

The last assertion is clear.   □

Theorem 5.9 *Let F be a field and $p(x)$ be an irreducible polynomial of degree n over F. Denote the residue class of x mod $p(x)$ by α. Then*

$$F[x]/(p(x)) \simeq F[\alpha] = F(\alpha). \tag{5.7}$$

Moreover, if F is a finite field with q elements then $|F[x]/(p(x))| = q^n$.

Proof. Let $E = F[x]/(p(x))$, then by Theorem 5.5 E is a field. Regard F as a subfield of E through the identification

$$- : F \to \overline{F} \subseteq E$$
$$a \mapsto \overline{a}$$

given in Section 5.1 and let α be the residue class of x mod $p(x)$, then $p(\alpha) = 0$. By Theorem 5.8 we get (5.7).

If F is a finite field with q elements, $|F[x]/(p(x))| = |F[\alpha]| = q^n$ by Theorem 5.8. $\qquad\square$

Corollary 5.10 *Let $\mathbb{Z}_p$ be the prime field of characteristic p and $p(x)$ be an irreducible polynomial of degree n in $\mathbb{Z}_p[x]$. Then $\mathbb{Z}_p[x]/(p(x))$ is a finite field with p^n elements.*

5.4 More Examples

To be more concrete, consider the following examples.

Example 5.7 Let $f(x) = x^2 + x + 1$ be a quadratic polynomial in $\mathbb{Z}_2[x]$. Clearly, $f(0) = f(1) = 1$. Therefore $f(x)$ has no roots in $\mathbb{Z}_2$ and $f(x)$ is irreducible over $\mathbb{Z}_2$. The finite field $\mathbb{Z}_2[x]/(x^2 + x + 1)$ has 2^2 elements and

$$\mathbb{Z}_2[x]/(x^2 + x + 1) = \{0, 1, \alpha, 1 + \alpha\},$$

where α is the residue class of x mod $x^2 + x + 1$.

We proceed to show that $\mathbb{Z}_2[x]/(x^2 + x + 1)$ is isomorphic to the field F_4 given in Example 3.6. Clearly, $\mathbb{Z}_2[x]/(x^2 + x + 1)$ is of characteristic 2 and its addition table is

$+$	0	1	α	$1 + \alpha$
0	0	1	α	$1 + \alpha$
1	1	0	$1 + \alpha$	α
α	α	$1 + \alpha$	0	1
$1 + \alpha$	$1 + \alpha$	α	1	0

Since $x^2 \equiv x + 1 \,(\text{mod } x^2 + x + 1)$, we have

$$\overline{(a + bx)}\,\overline{(c + dx)} = \overline{(a + bx)(c + dx)}$$

$$= \overline{ac + (ad + bc)x + bdx^2}$$

$$= \overline{(ac + bd) + (ad + bc + bd)x}.$$

Therefore

$$(a + b\alpha)(c + d\alpha) = (ac + bd) + (ad + bc + bd)\alpha \quad in \ \mathbb{Z}_2[x]/(x^2 + x + 1).$$

From this multiplication rule we write down immediately the multiplication table of $\mathbb{Z}_2[x]/(x^2 + x + 1)$ as follows:

$\cdot$	0	1	α	$1 + \alpha$
0	0	0	0	0
1	0	1	α	$1 + \alpha$
α	0	α	$1 + \alpha$	1
$1 + \alpha$	0	$1 + \alpha$	1	α

Clearly, the map

$$0 \to 0, 1 \to 1, x \to \alpha, 1 + x \to 1 + \alpha$$

is an isomorphism from F_4 to $\mathbb{Z}_2[x]/(x^2 + x + 1)$.

Now we prove that any two finite fields with 4 elements are isomorphic.

Let F be any field with four elements. By Corollary 3.13 the characteristic of F must be a prime p. Let 1 be the identity of F. By Theorem 3.12 the prime field of F

$$\Pi = \{0, 1, 2, ..., p - 1\}$$

is a field containing p elements and is isomorphic to $\mathbb{Z}_p$. But F contains only four elements, hence $p \leq 3$.

We first prove that $p \neq 3$. If $p = 3$ then Π is a field containing three elements $0, 1, 2$. Hence there must be an element of F which is not in Π. Denote this element by a. Then $F = \{0, 1, 2, a\}$. F is closed with respect to the addition, so $a + 1 \in F$. If $a + 1 = 0$ then $a = 2$; if $a + 1 = 1$ then $a = 0$; if $a + 1 = 2$ then $a = 1$; if $a + 1 = a$ then $1 = 0$; none of these is possible. This proves $p \neq 3$. Hence $p = 2$ and

$$\Pi = \{0, 1\}.$$

By Theorem 3.12, Π and $\mathbb{Z}_2$ are isomorphic.

Now let $y \in F \setminus \Pi$, then $1 + y \neq 0, 1, y$. Hence

$$F = \{0, 1, y, 1 + y\}.$$

Since F is closed with respect to the multiplication, $y^2 \in F$. It is clear that $y^2 \neq 0$. If $y^2 = 1$, then $(y - 1)^2 = 0$, so $y = 1$, which is impossible; if $y^2 = y$ then the cancellation law gives $y = 1$, which is also impossible. Hence we must have $y^2 = 1 + y$ and $y^2 + y + 1 = 0$. Then the addition and multiplication tables of F are as follows:

$+$	0	1	y	$1 + y$		$\cdot$	0	1	y	$1 + y$
0	0	1	y	$1 + y$		0	0	0	0	0
1	1	0	$1 + y$	y		1	0	1	y	$1 + y$
y	y	$1 + y$	0	1		y	0	y	$1 + y$	1
$1 + y$	$1 + y$	y	1	0		$1 + y$	0	$1 + y$	1	y

Clearly, F and F_4 have similar addition and multiplication tables. In other words, the map

$$0 \mapsto 0, 1 \mapsto 1, y \mapsto x, 1 + y \mapsto 1 + x$$

is an isomorphism. Hence there is only one field with four element within isomorphism.

From the multiplication table of F_4 we see that $x^1 = x, x^2 = 1 + x, x^3 = x \cdot x^2 = x(1 + x) = 1$, Thus $F_4^* = <x>$ is a cyclic group of order 3.

Example 5.8 Let $f(x) = x^3 + x + 1$ be a cubic polynomial in $\mathbb{Z}_2[x]$. Clearly, $f(0) = f(1) = 1$. Therefore $f(x)$ is irreducible. The finite field $\mathbb{Z}_2[x]/(x^3 + x + 1)$ has 2^3 elements. Denote $\mathbb{Z}_2[x]/(x^3 + x + 1)$ by F_8. Let us see how the multiplication is performed in F_8. We have

$$(a_0 + a_1 x + a_2 x^2)(b_0 + b_1 x + b_2 x^2)$$
$$= a_0 b_0 + (a_0 b_1 + a_1 b_0)x + (a_0 b_2 + a_1 b_1 + a_2 b_0)x^2$$
$$+ (a_1 b_2 + a_2 b_1)x^3 + a_2 b_2 x^4$$

and

$$\overline{x^3} = \overline{x + 1}, \quad \overline{x^4} = \overline{x^2 + x}.$$

Denote by α the residue class of x modulo $x^3 + x + 1$. Then

$$(a_0 + a_1 \alpha + a_2 \alpha^2)(b_0 + b_1 \alpha + b_2 \alpha^2)$$
$$= (a_0 b_0 + a_1 b_2 + a_2 b_1) + (a_0 b_1 + a_1 b_0 + a_1 b_2 + a_2 b_1 + a_2 b_2)\alpha \qquad (5.8)$$
$$+ (a_0 b_2 + a_1 b_1 + a_2 b_0 + a_2 b_2)\alpha^2.$$

This is the multiplication rule in $\mathbb{Z}_2[x]/(x^3 + x + 1)$. The addition and multiplication tables of F_8 can be listed easily, which is left as an exercise.

Now let us compute the power of α.

$$\alpha^0 = 1,$$
$$\alpha^1 = \alpha,$$
$$\alpha^2 = \alpha^2,$$
$$\alpha^3 = 1 + \alpha,$$
$$\alpha^4 = \alpha + \alpha^2,$$
$$\alpha^5 = 1 + \alpha + \alpha^2,$$
$$\alpha^6 = 1 + \alpha^2,$$
$$\alpha^7 = 1 = \alpha^0.$$

Thus $\{\alpha^0, \alpha^1, \alpha^2, \alpha^3, \alpha^4, \alpha^5, \alpha^6\}$ are all the nonzero elements of F_8 and α is an element of order 7. Therefore the multiplicative group F_8^* is cyclic and α is a generator.

5.5   Exercises

5.1 Let R be a ring and m be a fixed element of R. Prove that the congruence $a \equiv b \pmod{m}$ defined by (5.1) is an equivalence relation.

5.2 Let R be a ring and $m \in R$. Prove that the definitions (5.2) and (5.3) are well-defined.

5.3 Let R be a ring and m be a unit of R. Describe the residue class ring $R/(m)$.

5.4 Let R be a ring and $m = 0 \in R$. Describe the residue class ring $R/(0)$.

5.5 Let D be an integral domain and $m, m' \in D$. If $D/(m) = D/(m')$, then m and m' are associates.

5.6 Let F be any field and x, y be indeterminates. Prove that $F[x,y]/(x) \simeq F[y]$.

5.7 Let $Ax^2 + Bx + C$ be an irreducible quadratic polynomial in $\mathbb{R}[x]$. Prove that the map (5.4) is an isomorphism from $\mathbb{R}[x]/(Ax^2 + Bx + C)$ onto $\mathbb{C}$.

5.8 Write down the addition and multiplication tables of F_8 in Example 5.8.

5.9 Prove that $x^4 + x + 1$ is an irreducible polynomial over $\mathbb{Z}_2$. Then give a rule analogue to the multiplication rule of $\mathbb{Z}_2[x]/(x^3 + x + 1)$ in Example 5.8 for multiplying elements in the field $\mathbb{Z}_2[x]/(x^4 + x + 1)$ and write down its multiplication table.

5.10 Prove that $x^4 + x^3 + x^2 + x + 1$ is an irreducible polynomial over $\mathbb{Z}_2$. Then give a rule analogue to (5.8) for multiplying elements in the field $\mathbb{Z}_2[x]/(x^4 + x^3 + x^2 + x + 1)$ and write down its multiplication table.

5.11 Prove that the fields $\mathbb{Z}_2[x]/(x^4 + x + 1)$ and $\mathbb{Z}_2[x]/(x^4 + x^3 + x^2 + x + 1)$ are isomorphic.

5.12 Prove that $x^2 - x - 1$ is an irreducible polynomial over $\mathbb{Z}_3$. Give a rule analogue to (5.8) for multiplying elements in the field $\mathbb{Z}_3[x]/(x^2 - x - 1)$ and write down its addition and multiplication tables.

Chapter 6

Structure of Finite Fields

6.1 The Multiplicative Group of a Finite Field

Let F_q be a finite field with q elements. The multiplicative group F_q^* of F_q is of order $q - 1$. By Corollary 2.18, every element of F_q^* is of finite order and its order is a divisor of $q - 1$. Therefore we have

Theorem 6.1 *Let F_q be a finite field with q elements. Then $a^{q-1} = 1$ for all $a \in F_q^*$.*

This theorem can be regarded as a generalization of Fermat's small theorem, which asserts that for any prime number p and $a \in \mathbb{Z}$ with $p \nmid a$, $a^{p-1} \equiv 1 \pmod{p}$.

Corollary 6.2 *Let F_q be a finite field with q elements and E be a field which contains F_q as a subfield. Then $a^q = a$ for all $a \in F_q$ and, moreover, for any $\alpha \in E$, $\alpha^q = \alpha$ implies $\alpha \in F_q$.*

Proof. It follows from Theorem 6.1 that $a^q = a$ for all $a \in F_q$. By Theorem 4.6 $x^q - x$ has at most q distinct roots in E. Thus the q elements of F_q are all the roots of $x^q - x$ in E. Now if $\alpha^q = \alpha$ for $\alpha \in E$, α is a root of $x^q - x$. Therefore α must be one of the elements of F_q, i.e., $\alpha \in F_q$. $\square$

In Example 5.7 we saw that the multiplicative group F_4^* of the finite field F_4 is cyclic of order 3 and in Example 5.8 we saw that the multiplicative group F_8^* of the finite field F_8 is cyclic of order 7. In general, we have

Theorem 6.3 *The multiplicative group of any finite field is cyclic.*

Proof. Let F_q be a finite field with q elements and F_q^* be its multiplicative group. F_q^* is of order $q - 1$. By Corollary 2.18, the order of every element of F_q^* is a divisor of $q - 1$. Let d be a positive divisor of $q - 1$. Denote by $\psi(d)$ the number of elements of order d in F_q^*. Clearly,

$$\sum_{d|(q-1),d>0} \psi(d) = q - 1, \tag{6.1}$$

where the d under the summation sign runs through all positive divisors of $q - 1$.

Assume that $\psi(d) > 0$, then there is an element of order d in F_q^*. Let a be such an element, then the cyclic group $\langle a \rangle$ generated by a is of order d and every element a^i, $0 \le i \le d - 1$, in $\langle a \rangle$ satisfies the polynomial $x^d - 1$. Let b be any element of order d in F_q^*, then b satisfies also $x^d - 1$. By Theorem 4.6 the number of distinct roots of $x^d - 1$ in F_q is at most d. Then we must have $b = a^i$ for some i, $1 \le i \le d - 1$. Thus $b \in \langle a \rangle$. By Theorem 2.23 the number of elements of order d in $\langle a \rangle$ is $\phi(d)$. Therefore we have proved that

$$\psi(d) = 0 \text{ or } \phi(d). \tag{6.2}$$

By Theorem 2.24,

$$\sum_{d|(q-1),d>0} \phi(d) = q - 1. \tag{6.3}$$

From (6.1), (6.2) and (6.3) we deduce $\psi(d) = \phi(d)$ for all positive divisors d of $q-1$. In particular, $\psi(q-1) = \phi(q-1) > 0$. That is, there is an element of order $q - 1$ in F_q^*. Hence F_q^* is cyclic. $\square$

Let F_q be a finite field with q elements. The generators of the cyclic group F_q^* are called *primitive elements* (or *primitive roots*) of F_q. By Theorem 2.23 the number of primitive elements of F_q is $\phi(q - 1)$.

Let α be a fixed generator of the cyclic group F_q^*. We can use α as a base to define a "logarithm" in F_q^*, which is called the *discrete logarithm* in F_q. That is, we define

$$\log_\alpha(\beta) = k \text{ if and only if } \alpha^k = \beta,$$

where $0 \le k \le q - 2$.

Example 6.1 Consider the field F_8 constructed in Example 5.8 as

$$F_8 = \mathbb{Z}_2[x]/(x^3 + x + 1) = \{a\alpha^2 + b\alpha + c : a, b, c \in F_2\},$$

where $\alpha = \overline{x}$ is the residue class of x modulo $x^3 + x + 1$. α is an element of order 7 in F_8^*, and we have the multiplication rule (5.8) for elements of F_8. We can use the 3-tuple (abc) to represent the element $a\alpha^2 + b\alpha + c \in F_8$. Using α as a base, we have Table 6.1, the table of logarithms and anti-logarithms in F_8. Moreover, use Table 6.1, it is easier to compute the product in F_8. For example, let $a(\alpha) = \alpha + 1$ and $b(\alpha) = \alpha^2$, then $a(\alpha)$ is represented by the 3-tuple (011) and $b(\alpha)$ by (100). From the table of logarithms we have $\log_\alpha a(\alpha) = 3$, $\log_\alpha b(\alpha) = 2$. So $a(\alpha)b(\alpha) = \alpha^3\alpha^2 = \alpha^5$. By the table of anti-logarithms $\log_\alpha(\alpha^2 + \alpha + 1) = 5$. Therefore $(\alpha+1)\alpha^2 = \alpha^2 + \alpha + 1$.

Table 6.1

β	$\log_\alpha \beta$		k	α^k
000	*		*	000
001	0		0	001
010	1		1	010
011	3		2	100
100	2		3	011
101	6		4	110
110	4		5	111
111	5		6	101

Example 6.2 Consider the field F_{16}, which is obtained from $\mathbb{Z}_2[x]$ modulo the irreducible polynomial $x^4 + x + 1$ (Exercise 5.9). That is,

$$F_{16} = \mathbb{Z}_2[x]/(x^4 + x + 1)$$
$$= \{a_3\alpha^3 + a_2\alpha^2 + a_1\alpha + a_0 : a_3, a_2, a_1, a_0 \in \mathbb{Z}_2\},$$

where $\alpha = \overline{x}$ is the residue class of x modulo $x^4 + x + 1$. We have $\alpha^4 = \alpha + 1$, $\alpha^5 = \alpha^2 + \alpha$, $\alpha^6 = \alpha^3 + \alpha^2$, and the multiplication rule in F_{16}

$$(a_3\alpha^3 + a_2\alpha^2 + a_1\alpha + a_0)(b_3\alpha^3 + b_2\alpha^2 + b_1\alpha + b_0)$$
$$= (a_3b_3 + a_3b_0 + a_2b_1 + a_1b_2 + a_0b_3)\alpha^3$$
$$+(a_3b_3 + a_3b_2 + a_2b_3 + a_2b_0 + a_1b_1 + a_0b_2)\alpha^2$$
$$+(a_3b_2 + a_2b_3 + a_3b_1 + a_2b_2 + a_1b_3 + a_1b_0 + a_0b_1)\alpha$$
$$+(a_3b_1 + a_2b_2 + a_1b_3 + a_0b_0).$$

We use the 4-tuple $(a_3a_2a_1a_0)$ to represent the element $a_3\alpha^3 + a_2\alpha^2 + a_1\alpha + a_0$ of F_{16}. It is easy to show that α is a primitive element. We can use α as a base to build Table 6.2, the table of logarithms and anti-logarithms in F_{16}.

Table 6.2

β	$\log_\alpha \beta$		k	α^k
0000	*		*	0000
0001	0		0	0001
0010	1		1	0010
0011	4		2	0100
0100	2		3	1000
0101	8		4	0011
0110	5		5	0110
0111	10		6	1100
1000	3		7	1011
1001	14		8	0101
1010	9		9	1010
1011	7		10	0111
1100	6		11	1110
1101	13		12	1111
1110	11		13	1101
1111	12		14	1001

Example 6.3 Consider the field $\mathbb{Z}_7 = \{0, 1, \ldots, 6\}$. The powers of 2 are $2^0 = 1$, $2^1 = 2$, $2^2 = 4$, $2^3 = 1$. Thus neither 2 nor any of its powers is a primitive element in $\mathbb{Z}_7$. Then we try 3. The powers of 3 are $3^0 = 1$, $3^1 = 3$, $3^2 = 2$, $3^3 = 6$, $3^4 = 4$, $3^5 = 5$, $3^6 = 1$. Thus 3 is a primitive element in $\mathbb{Z}_7$. $\phi(7 - 1) = \phi(6) = 2$. Therefore besides 3 there is another primitive element, which is $3^5 = 5$. Table 6.3 is the table of base 3 logarithm and antilogarithm of $\mathbb{Z}_7$.

Table 6.3

i	$\log_3 i$		k	3^k	$\mathrm{ord}\,(3^k)$
0	*		*	0	*
1	0		0	1	1
2	2		1	3	6
3	1		2	2	3
4	4		3	6	2
5	5		4	4	3
6	3		5	5	6

The last column in Table 6.3 shows the orders of the elements of $\mathbb{Z}_7^*$. We see that there is $\phi(1) = 1$ element of order 1, $\phi(2) = 1$ of order 2, $\phi(3) = 2$

of order 3, and $\phi(6) = 2$ of order 6. We have

$$\phi(1) + \phi(2) + \phi(3) + \phi(6) = 6.$$

Example 6.4 $x^2 - 2$ is an irreducible polynomial over the field $\mathbb{Z}_5$, since it has no root in $\mathbb{Z}_5$. Then $\mathbb{Z}_5[x]/(x^2 - 2)$ is a field with 5^2 elements and we denote it by F_{5^2}. We try to find a primitive element of F_{5^2}. Let $\alpha = \bar{x}$ be the residue class of x modulo $x^2 - 2$, then $\alpha^2 = 2$ in F_{5^2}. We compute

$$\alpha^0 = 1, \ \alpha^1 = \alpha, \ \alpha^2 = 2, \ \alpha^3 = 2\alpha, \ \alpha^4 = 4, \ \alpha^5 = 4\alpha,$$
$$\alpha^6 = 3, \ \alpha^7 = 3\alpha, \ \alpha^8 = 1.$$

That is, α is an element of order 8 in F_{5^2} and not a primitive element. We try $\alpha + 1$. We compute

$$(\alpha + 1)^0 = 1, \quad (\alpha + 1)^1 = \alpha + 1, \quad (\alpha + 1)^2 = \alpha^2 + 2\alpha + 1 = 2\alpha + 3,$$
$$(\alpha + 1)^3 = (\alpha + 1)^2(\alpha + 1) = (2\alpha + 3)(\alpha + 1) = 2\alpha^2 + 3 = 2,$$
$$(\alpha + 1)^4 = 2\alpha + 2, \quad (\alpha + 1)^5 = 4\alpha + 1, \quad (\alpha + 1)^6 = 4,$$
$$\cdots, (\alpha + 1)^{12} = 1.$$

That is, $\alpha + 1$ is an element of order 12 in F_{5^2}. We take $d = 8$, $e = 3$. Then $d \mid 8$, $e \mid 12$, $\gcd(d, e) = 1$ and $de = \text{lcm}[8, 12]$. Let $\beta = \alpha^{8/d}(\alpha + 1)^{12/e}$, then $\text{ord}(\beta) = 24$ and

$$\beta = \alpha(\alpha + 1)^4 = 2\alpha^2 + 2\alpha = 2\alpha + 4,$$

is a primitive element of F_{5^2}.

The algorithm we adopt in Example 6.4 to find a primitive element of a finite field was suggested by Gauss. It can be formulated as follows.

Gauss's Algorithm: Let F_q be a finite field with q elements.

G_1 Set $i = 1$, and let α_1 be an arbitrary element of F_q^*. Let $\text{ord}(\alpha_1) = t_1$.

G_2 If $t_i = q - 1$, stop; α_i is a primitive element.

G_3 Otherwise, choose an element $\beta \in F_q^* \backslash \langle \alpha_i \rangle$. Let $\text{ord}(\beta) = s$. If $s = q - 1$, set $\alpha_{i+1} = \beta$ and stop.

G_4 Otherwise, find $d \mid t_i$ and $e \mid s$ with $\gcd(d, e) = 1$ and $d \cdot e = \text{lcm}[t_i, s]$. Let $\alpha_{i+1} = \alpha_i^{t_i/d} \beta^{s/e}$, $t_{i+1} = de$. Increase i by 1 and go to G_2.

Theorem 6.4 *The Gauss's Algorithm applied to a finite field F_q with q elements produces a primitive element of F_q.*

Proof. It is enough to prove that the Gauss's Algorithm produces a sequence of elements $\alpha_1, \alpha_2, \ldots \in F_q$ with $\operatorname{ord}(\alpha_1) < \operatorname{ord}(\alpha_2) < \cdots \leq q - 1$. Then there is a $k > 0$ such that $\operatorname{ord}(\alpha_k) = q - 1$.

First, we observe that in Step G_3 the order s of the element $\beta \in F_j^* \backslash \langle \alpha_i \rangle$ cannot be a divisor of $t_i = \operatorname{ord}(\alpha_i)$, since all solutions of $x^{t_i} = 1$ must be powers of α. Hence $\operatorname{lcm}[t_i, s]$ will be a multiple of t_i and strictly greater than t_i.

Secondly, we remark that in Step G_4 we can find $d \mid t_i$ and $e \mid s$ with $\gcd(d, e) = 1$ and $de = \operatorname{lcm}[t_i, s]$. In fact, we can assume

$$t_i = p_1^{e_1} \cdots p_r^{e_r} p_{r+1}^{e_{r+1}} \cdots p_{r+r'}^{e_{r+r'}},$$

$$s = p_1^{f_1} \cdots p_r^{f_r} p_{r+1}^{f_{r+1}} \cdots p_{r+r'}^{f_{r+r'}},$$

where $p_1, \ldots, p_{r+r'}$ are distinct primes, $e_i \geq f_i \geq 0$ for $1 \leq i \leq r$, and $0 \leq e_{r+j} < f_{r+j}$ for $1 \leq j \leq r'$. Let $d = p_1^{e_1} \cdots p_r^{e_r}$, $e = p_{r+1}^{f_{r+1}} \cdots p_{r+r'}^{f_{r+r'}}$. Then $d \mid t_i$, $e \mid s$, $\gcd(d, e) = 1$ and $de = \operatorname{lcm}[t_i, s]$. $\square$

6.2   The Number of Elements in a Finite Field

In this section we will prove the following

Theorem 6.5 *Let F be a finite field of characteristic p, then the number of elements of F must be a power of p.*

We can prove a generalization of Theorem 6.5.

Theorem 6.6 *Let F be a finite field which contains a subfield F_q with q elements, then the number of elements of F must be a power of q.*

Proof. Rewrite F_q as F_1. If $F = F_1$, then F is a finite field containing exactly q elements. Hence our theorem holds. If $F \neq F_1$, then F will contain an element e_2 and $e_2 \notin F_1$. Let

$$F_2 = \{a_1 + a_2 e_2 : a_1, a_2 \in F_1\}.$$

We will prove: if

$$a_1 + a_2 e_2 = b_1 + b_2 e_2, \quad a_1, a_2, b_1, b_2 \in F_1,$$

then $a_1 = b_1, a_2 = b_2$. In fact, from this equation it follows that

$$(a_2 - b_2)e_2 = b_1 - a_1.$$

If $a_2 \neq b_2$ then

$$e_2 = (a_2 - b_2)^{-1}(b_1 - a_1) \in F_1.$$

A contradiction follows. Hence we must have $a_2 = b_2$ and $a_1 = b_1$. Therefore F_2 contains exactly q^2 distinct elements.

If $F = F_2$, then F is a finite field with exactly q^2 elements. So our theorem holds. If $F \neq F_2$, then F will contain an element e_3 and $e_3 \notin F_2$. Let

$$F_3 = \{a_1 + a_2e_2 + a_3e_3 : a_1, a_2, a_3 \in F_1\}.$$

Assume that $a_1 + a_2e_2 + a_3e_3 = b_1 + b_2e_2 + b_3e_3$, $a_i, b_i \in F_1$. Then

$$(a_3 - b_3)e_3 = (b_1 - a_1) + (b_2 - a_2)e_2.$$

If $a_3 \neq b_3$, then

$$e_3 = (a_3 - b_3)^{-1}(b_1 - a_1) + (a_3 - b_3)^{-1}(b_2 - a_2)e_2 \in F_2.$$

This is a contradiction. Hence we must have $a_3 = b_3$. Then we have

$$a_1 + a_2e_2 = b_1 + b_2e_2,$$

which, by the above proof, yields $a_1 = b_1, a_2 = b_2$. Therefore we conclude that F_3 contains exactly q^3 distinct elements.

If $F = F_3$, then F is a finite field with exactly q^3 elements. Thus our theorem holds. If $F \neq F_3$, then F will contain an element e_4 and $e_4 \notin F_3$. Let

$$F_4 = \{a_1 + a_2e_2 + a_3e_3 + a_4e_4 : a_1, a_2, a_3, a_4 \in F_1\}.$$

Similarly it can be proved that F_4 contains exactly q^4 distinct elements. Continue in this way. If the number of elements in F is N and $q^n \leq N < q^{n+1}$, then a sequence of subsets of F

$$F_1, F_2, F_3, \ldots, F_n$$

is obtained, where

$$F_i = \{a_1 + a_2e_2 + \cdots + a_ie_i : a_1, a_2, \ldots, a_i \in F_1\},$$

$$e_2 \notin F_1, e_3 \notin F_2, \cdots, e_n \notin F_{n-1},$$

and $F_i(1 \leq i \leq n)$ contains exactly q^i distinct elements. If $F \neq F_n$, then F will contain an element e_{n+1} and $e_{n+1} \notin F_n$. Let

$$F_{n+1} = \{a_1 + a_2e_2 + \cdots + a_ne_n + a_{n+1}e_{n+1} : a_1, a_2, \ldots, a_n, a_{n+1} \in F_1\}.$$

In a similar way one can prove that F_{n+1} contains exactly q^{n+1} distinct elements. But F_{n+1} is a subset of F, the number of elements in F is N, and $N < q^{n+1}$. This is impossible. Hence $F = F_n$. So F is a finite field which contains exactly q^n elements. $\square$

Remark: We can view F as a vector space over F_q, if we define the scalar multiplication as follows

$$F_q \times F \to F$$
$$(a, \alpha) \mapsto a\alpha.$$

Then we can give a short proof of Theorem 6.6 as follows.

Since F is finite, the *dimension* of F over $\mathbb{F}_q$, which is denoted by $\dim_{F_q} F$, is also finite. Let $\dim_{F_q} F = n$, and $\alpha_1, \alpha_2, \ldots, \alpha_n$ be a basis of F over F_q. Then every element $\alpha \in F$ can be written uniquely as

$$\alpha = a_1\alpha_1 + a_2\alpha_2 + \cdots + a_n\alpha_n, \ a_1, a_2, \ldots, a_n \in F_q.$$

Then $|F| = q^n$. $\square$

Proof of Theorem 6.5: Let F be a finite field of characteristic p, then p must be a prime and the prime field Π of F is a finite field with exactly p elements. Letting $F_q = \Pi$ in Theorem 6.6 yields Theorem 6.5. $\square$

6.3   Existence of Finite Field with p^n Elements

Let p be a prime and n be any integer. We will prove in this section that there exists an irreducible polynomial of degree n over the prime field $\mathbb{Z}_p$. Then it will follow from Corollary 5.10 that there exists a finite field with p^n elements.

We need the Möbius function and the Möbius inversion formula.

We define the Möbius function $\mu(n)$ as a function defined on the set of positive integers with values in the set of integers as follows.

$$\mu(n) = \begin{cases} 1, & \text{if } n = 1, \\ 0, & \text{if there is a prime number } p \text{ such that } p^2 | n, \\ (-1)^r, & \text{if } n = p_1 p_2 \cdots p_r, \text{where } p_1, \ldots, p_r \text{ are distinct.} \end{cases}$$

Lemma 6.7

$$\sum_{d|n} \mu(d) = \begin{cases} 1, & \text{if } n = 1, \\ 0, & \text{if } n > 1, \end{cases}$$

where the d under summation sign runs through all positive divisors of n.

Proof. When $n = 1$, $\mu(1) = 1$ by definition. Assume now that $n > 1$ and $n = p_1^{e_1} \cdots p_r^{e_r}$ is the prime factorization of n. Then

$$\sum_{d|n} \mu(d) = \mu(1) + \sum_{i=1}^{r} \mu(p_i) + \sum_{i,j=1, i \neq j}^{r} \mu(p_i p_j) + \cdots + \mu(p_1 p_2 \cdots p_r)$$

$$= 1 + \binom{r}{1}(-1) + \binom{r}{2}(-1)^2 \cdots + \binom{r}{r}(-1)^r$$

$$= (1 - 1)^r = 0.$$

$\square$

We have the following Möbius inversion formula.

Lemma 6.8 *Let $a(n)$ and $b(n)$ both be functions defined on the set of positive integers with values in an abelian group whose operation is written as addition. Then*

$$b(n) = \sum_{d|n} a(d) \tag{6.4}$$

if and only if

$$a(n) = \sum_{d|n} \mu\left(\frac{n}{d}\right) b(d), \tag{6.5}$$

where under the summation signs of (6.4) and (6.5) d runs through all positive divisors of n.

Proof. Assume that (6.4) holds. Then

$$\sum_{d|n} \mu\left(\frac{n}{d}\right) b(d) = \sum_{d|n} \mu\left(\frac{n}{d}\right) \sum_{e|d} a(e)$$

$$= \sum_{e|n} a(e) \sum_{e|d|n} \mu\left(\frac{n}{d}\right)$$

$$= \sum_{e|n} a(e) \sum_{\frac{n}{d}|\frac{n}{e}} \mu\left(\frac{n}{d}\right)$$

$$= a(n). \qquad \text{(By Lemma 6.7)}$$

Therefore (6.5) holds. Conversely, (6.5)⇒(6.4) can be proved in the same way. $\square$

Using Möbius inversion formula, we can give another proof of Theorem 2.28. By Theorem 2.24, for any $n > 0$

$$\sum_{d|n} \phi(d) = n.$$

By Lemma 6.8,

$$\phi(n) = \sum_{d|n} \mu(\frac{n}{d})d = \sum_{d|n} \mu(d)\frac{n}{d}.$$

Let $n = p_1^{e_1} \cdots p_r^{e_r}$ be the prime factorization of n, then

$$\sum_{d|n} \mu(d)\frac{n}{d} = n \sum_{d|n} \frac{\mu(d)}{d}$$

$$= n \left(1 + \sum_{i=1}^{r} \frac{\mu(p_i)}{p_i} + \sum_{i,j=1}^{r} \frac{\mu(p_i p_j)}{p_i p_j} + \cdots + \frac{\mu(p_1 \cdots p_r)}{p_1 \cdots p_r} \right)$$

$$= n \left(1 + \sum_{i=1}^{r} \frac{-1}{p_i} + \sum_{i,j=1}^{r} \frac{1}{p_i p_j} + \cdots + \frac{(-1)^r}{p_1 p_2 \cdots p_r} \right)$$

$$= n \prod_{i=1}^{r} \left(1 - \frac{1}{p_i} \right).$$

$\square$

We shall now prove more generally in the following that for any finite field F_q with q elements there exist irreducible polynomials of any degree in $F_q[x]$. In the following series of lemmas, let F_q be a finite field with q elements.

Lemma 6.9 *Let $f(x)$ be an irreducible polynomial of degree n in $F_q[x]$. Then $f(x) \,|\, (x^{q^n} - x)$.*

Proof. Let $F = F_q[x]/(f(x))$. By Theorem 5.9 F is a finite field with q^n elements. By Corollary 6.2 all elements of F satisfy the polynomial $X^{q^n} - X$. In particular, for $\overline{x} \in F$ we have $\overline{x}^{q^n} - \overline{x} = 0$. Then $x^{q^n} - x \equiv 0 \pmod{f(x)}$. That is, $f(x) \,|\, (x^{q^n} - x)$. $\square$

Lemma 6.10 *Let $f(x)$ be an irreducible polynomial of degree d in $F_q[x]$. Then $f(x) \,|\, (x^{q^n} - x)$ if and only if $d \,|\, n$.*

Proof. Assume that $d \,|\, n$. Then $\gcd(d, n) = d$. By Lemma 3.26

$$\gcd(q^n - 1, q^d - 1) = q^{\gcd(n,d)} - 1 = q^d - 1$$

and

$$\gcd(x^{q^n-1} - 1, x^{q^d-1} - 1) = x^{\gcd(q^n-1, q^d-1)} - 1 = x^{q^d-1} - 1.$$

Thus

$$\gcd(x^{q^n} - x, x^{q^d} - x) = \gcd(x(x^{q^n-1} - 1), x(x^{q^d-1} - 1))$$
$$= x\gcd(x^{q^n-1} - 1, x^{q^d-1} - 1)$$
$$= x(x^{q^d-1} - 1) = x^{q^d} - x.$$

Therefore $(x^{q^d} - x) \mid (x^{q^n} - x)$. By Lemma 6.9 $f(x) \mid (x^{q^d} - x)$. Hence $f(x) \mid (x^{q^n} - x)$.

Conversely, assume that $f(x) \mid (x^{q^n} - x)$. By Lemma 6.9 $f(x) \mid (x^{q^d} - x)$. Then $f(x) \mid \gcd(x^{q^n} - x, x^{q^d} - x)$. By Lemma 3.26

$$\gcd(x^{q^n} - x, x^{q^d} - x) = x\gcd(x^{q^n-1} - 1, x^{q^d-1} - 1)$$
$$= x(x^{\gcd(q^n-1, q^d-1)} - 1)$$
$$= x(x^{q^{\gcd(n,d)}-1} - 1)$$
$$= x^{q^{\gcd(n,d)}} - x.$$

Let $d' = \gcd(n, d)$. Then $f(x) \mid (x^{q^{d'}} - x)$. We identity $a \in F_q$ with its residue class mod $f(x)$ and denote the residue class of x mod $f(x)$ by α as we did in Section 5.1. Then

$$F_q[x]/(f(x)) = \{a_0 + a_1\alpha + \cdots + a_{d-1}\alpha^{d-1} : a_0, a_1, \ldots, a_{d-1} \in F_q\}.$$

From $f(x) \mid (x^{q^{d'}} - x)$ we deduce $\alpha^{q^{d'}} = \alpha$. By Corollary 6.2 $a^q = a$ for all $a \in F_q$, and, hence, $a^{q^{d'}} = a$. By Theorem 6.5 q is a power of the characteristic p of F_q. For any element $a_0 + a_1\alpha + \cdots + a_{d-1}\alpha^{d-1} \in F_q[x]/(f(x))$, by Corollaries 3.16 and 3.17

$$(a_0 + a_1\alpha + \cdots + a_{d-1}\alpha^{d-1})^{q^{d'}} = a_0 + a_1\alpha^{q^{d'}} + \cdots + a_{d-1}(\alpha^{q^{d'}})^{d-1}$$
$$= a_0 + a_1\alpha + \cdots + a_{d-1}\alpha^{d-1}.$$

Thus all elements of $F_q[x]/(f(x))$ satisfy the polynomial $X^{q^{d'}} - X$. But $|F_q[x]/(f(x))| = q^d$. Then by Theorem 4.6 we must have $d \leq d'$. But $d' = \gcd(n, d) \leq d$. Hence $d = d'$ and $d \mid n$. $\qquad\square$

Lemma 6.11 *For any positive integer* n, $x^{q^n} - x$ *does not possess multiple factors in* $F_q[x]$.

Proof. By Theorem 6.5 q is a power of the characteristic of F_q. Thus $(x^{q^n} - x)' = -1$. Therefore $\gcd(x^{q^n} - x, (x^{q^n} - x)') = 1$. Then our lemma follows from Theorem 4.15. $\qquad\square$

Denote by $\Phi_{q,n}(x)$ the product of all monic irreducible polynomials of degree n in $F_q[x]$ and by $\Phi_{q,n}$ the number of monic irreducible polynomials of degree n in $\mathbb{F}_q[x]$.

Theorem 6.12 *Let F_q be a finite field with q elements and n be a positive integer. Then*

$$x^{q^n} - x = \prod_{d|n} \Phi_{q,d}(x), \tag{6.6}$$

$$q^n = \sum_{d|n} d\Phi_{q,d}, \tag{6.7}$$

and

$$n\Phi_{q,n} = q^n - \sum_{i=1}^{r} q^{n/p_i} + \sum_{1 \le i < j \le r} q^{n/p_i p_j} + \cdots + (-1)^r q^{n/p_1 \cdots p_r}, \tag{6.8}$$

where $p_1, \ldots, p_r$ are all the distinct positive prime divisors of n.

Proof. (6.6) follows from Lemmas 6.10 and 6.11. Comparing the degrees of both sides of (6.6), we obtain (6.7). By Möbius inversion formula we derive (6.8) from (6.7).   □

Corollary 6.13 $\Phi_{q,n} > 0$.

Proof. Clearly $\Phi_{q,n} \ge 0$. Hence in order to prove this corollary, it is sufficient to prove $\Phi_{q,n} \ne 0$. On the right-hand side of (6.8), except the last term, all the terms are multiples of

$$q^{(n/p_1 p_2 \cdots p_r)+1}.$$

Hence $n\Phi_{q,n}$ is not divisible by $q^{(n/p_1 \cdots p_r)+1}$, then $n\Phi_{q,n} \ne 0$. Therefore $\Phi_{q,n} > 0$.   □

Theorem 6.14 *Let p be any prime number and n be any positive integer, then there exists a finite field which contains exactly p^n elements.*

Proof. When p is a prime and n is any positive integer, by Corollary 6.13 we have $\Phi_{p,n} > 0$. Thus there always exist monic irreducible polynomials of degree n over the prime field $\mathbb{Z}_p$ with p elements. Let $p(x)$ be one of them. Then by Corollary 5.10 $\mathbb{Z}_p[x]/(p(x))$ is a finite field with p^n elements.   □

6.4   Uniqueness of Finite Field with p^n Elements

In this section we shall prove

Theorem 6.15 *Any two finite fields containing the same number of elements are isomorphic.*

First we prove the following lemma.

Lemma 6.16 *Let F_q and F_{q^n} be finite fields with q and q^n elements, respectively, where q is a prime power and n is a positive integer. Assume that F_q is a subfield of F_{q^n}. Let $f(x)$ be a monic irreducible polynomial of degree n over F_q. Then $f(x)$ has n distinct roots in F_{q^n}; moreover, let α be any one of them, then $f(x)$ is also the monic polynomial of lowest degree over F_q having α as one of its roots.*

Proof. By Corollary 6.2 all q^n elements of F_{q^n} are roots of $x^{q^n} - x$. It follows from Corollary 4.5 that $x^{q^n} - x$ factors into a product of linear factors in $F_{q^n}[x]$, i.e.,

$$x^{q^n} - x = \prod_{\alpha \in F_{q^n}} (x - \alpha).$$

By Lemma 6.9 $f(x) \mid (x^{q^n} - x)$. Thus by the unique factorization theorem,

$$f(x) = (x - \alpha_1)(x - \alpha_2) \cdots (x - \alpha_n),$$

where $\alpha_1, \alpha_2, \ldots, \alpha_n$ are distinct elements of F_{q^n}. Therefore $f(x)$ has n distinct roots in F_{q^n}.

Let $\alpha \in F_{q^n}$ be a root of $f(x)$, i.e., $f(\alpha) = 0$, and let $g(x)$ be the monic polynomial of lowest degree over F_q such that $g(\alpha) = 0$. Dividing $f(x)$ by $g(x)$, we obtain

$$f(x) = q(x)g(x) + r(x),$$

where $q(x), r(x) \in F_q[x]$ and $\deg r(x) < \deg g(x)$. Substituting α into the above identity, we obtain $r(\alpha) = 0$. By the definition of $g(x)$, we must have $r(x) = 0$. Thus $g(x) \mid f(x)$. Both $g(x)$ and $f(x)$ are monic and $f(x)$ is irreducible, therefore $g(x) = f(x)$. $\qquad\square$

Proof of Theorem 6.15: By Corollary 6.13 for any prime number p and positive integer n there exists a monic irreducible polynomial of degree n over $\mathbb{Z}_p$. Let $f(x)$ be such a polynomial, by Corollary 5.10 $\mathbb{Z}_p[x]/(f(x))$ is a field with p^n elements. To prove that any two finite fields both with p^n

elements are isomorphic, it is enough to prove that any finite field with p^n elements is isomorphic to $\mathbb{Z}_p[x]/(f(x))$.

Let F_{p^n} be a finite field with p^n elements. The additive group of F_{p^n} is of order p^n. Let 1 be the identity of F_{p^n}, then $p^n 1 = 0$. But $p^n 1 = (p1)^n$, therefore $p1 = 0$. Thus F_{p^n} is of characteristic p and the prime field of F_{p^n} is isomorphic to $\mathbb{Z}_p$. We identity the element $m1$ ($0 \leq m \leq p - 1$) of the prime field with $m \in \mathbb{Z}_p$ ($0 \leq m \leq p - 1$), then we can assume that the prime field of F_{p^n} is $\mathbb{Z}_p$. By Lemma 6.16 $f(x)$ has a root $\alpha \in F_{p^n}$ and $f(x)$ is the monic polynomial of lowest degree over $\mathbb{Z}_p$ having α as one of its roots. Then the following elements of F_{p^n}

$$a_0 + a_1\alpha + \cdots + a_{n-1}\alpha^{n-1},$$

where $a_0, a_1, \ldots, a_{n-1} \in \mathbb{Z}_p$, are distinct. But they are p^n in number and $|F_{p^n}| = p^n$. Therefore

$$F_{p^n} = \{a_0 + a_1\alpha + \cdots + a_{n-1}\alpha^{n-1} : a_0, \ldots, a_{n-1} \in \mathbb{Z}_p\}.$$

On the other hand, by Theorem 5.8

$$\mathbb{Z}_p[\alpha] = \{a_0 + a_1\alpha + \cdots + a_{n-1}\alpha^{n-1} : a_0, a_1, \ldots, a_{n-1} \in \mathbb{Z}_p\}$$

and $\mathbb{Z}_p[\alpha]$ is isomorphic to $\mathbb{Z}_p[x]/(f(x))$. Clearly, $F_{p^n} = \mathbb{Z}_p[\alpha]$. Therefore F_{p^n} is isomorphic to $\mathbb{Z}_p[x]/(f(x))$. $\qquad\square$

Theorems 6.5, 6.14, and 6.15 are the three structure theorems of finite fields. From these theorems we know that for any prime number p and positive integer n, there exists a finite field with p^n elements within isomorphism. Isomorphic fields are usually regarded as the same field. From now on we will use $\mathbb{F}_{p^n}$ to denote the finite field with p^n elements. In particular, $\mathbb{F}_p = \mathbb{Z}_p$.

6.5   Subfields of Finite Fields

As an application of the structure theorems of finite fields, we prove

Theorem 6.17 *Let $\mathbb{F}_{p^n}$ be a finite field with p^n elements and $\mathbb{F}_p$ be its prime field. If F_0 is a subfield of $\mathbb{F}_{p^n}$, assume that F_0 contains p^m elements, then m must be a factor of n. Conversely, if m is a factor of n then $\mathbb{F}_{p^n}$ has a unique subfield of p^m elements. Further, let $\mathbb{F}_{p^m}$ and $\mathbb{F}_{p^{m'}}$ both be subfields of $\mathbb{F}_{p^n}$, then $\mathbb{F}_{p^m} \subset \mathbb{F}_{p^{m'}}$ if and only if m is a factor of m'.*

Proof. Let F_0 be a subfield of $\mathbb{F}_{p^n}$. Since F_0 contains the identity 1 of $\mathbb{F}_{p^n}$, it contains the prime field $\mathbb{F}_p$ of $\mathbb{F}_{p^n}$ as its prime field. By Theorem 6.5, we can assume that F_0 contains p^m elements. Since $\mathbb{F}_{p^n}$ contains the field F_0 with p^m elements as its subfield, by Theorem 6.6, $p^n = (p^m)^k$ for some positive integer k. Hence $n = mk$, that is, m is a factor of n.

Conversely, assume that m is a factor of n. Since the p^n elements of $\mathbb{F}_{p^n}$ are just the p^n roots of the polynomial $x^{p^n} - x$ and

$$\left(x^{p^m} - x\right) \mid \left(x^{p^n} - x\right),$$

there exist p^m elements in $\mathbb{F}_{p^n}$ which are roots of the polynomial $x^{p^m} - x$. Denote by F_0 the set of the p^m roots of $x^{p^m} - x$ in $\mathbb{F}_{p^n}$. Let $\alpha, \beta \in F_0$, i.e.,

$$\alpha^{p^m} = \alpha, \beta^{p^m} = \beta.$$

Then

$$(\alpha + \beta)^{p^m} = \alpha^{p^m} + \beta^{p^m} = \alpha + \beta,$$
$$(\alpha\beta)^{p^m} = \alpha^{p^m}\beta^{p^m} = \alpha\beta,$$

hence $\alpha + \beta, \alpha\beta \in F_0$, i.e., F_0 is closed with respect to the addition and multiplication in $\mathbb{F}_{p^n}$. Clearly $0, 1 \in F_0$. Let $\alpha \in F_0$, then

$$(-\alpha)^{p^m} = (-1)^{p^m}\alpha^{p^m} = -\alpha,$$

and if $\alpha \neq 0$, then

$$(\alpha^{-1})^{p^m} = (\alpha^{p^m})^{-1} = \alpha^{-1}.$$

Hence $-\alpha \in F_0$ and $\alpha^{-1} \in F_0$, when $\alpha \neq 0$. This proves that F_0 is a subfield of $\mathbb{F}_{p^n}$ and by the definition of F_0 we know that it contains p^m elements.

If F_1 is another subfield of $\mathbb{F}_{p^n}$ which contains p^m elements. Then the p^m elements in F_1 are also the p^m roots of the polynomial $x^{p^m} - x$. But $x^{p^m} - x$ has at most p^m roots in F_{p^n}, hence $F_1 = F_0$.

Further, let $\mathbb{F}_{p^m}$ and $\mathbb{F}_{p^{m'}}$ both be subfields of $\mathbb{F}_{p^n}$. If $\mathbb{F}_{p^m} \subset \mathbb{F}_{p^{m'}}$, then by Theorem 6.6, $p^{m'} = (p^m)^l$ for some positive integer l, hence m is a factor of m'. Conversely, if m is a factor of m', then $\mathbb{F}_{p^{m'}}$ has a subfield F_0 which contains p^m elements and which is also a subfield of $\mathbb{F}_{p^n}$. But $\mathbb{F}_{p^n}$ has only a unique subfield $\mathbb{F}_{p^m}$ containing p^m elements. Hence $F_0 = \mathbb{F}_{p^m}$ and $\mathbb{F}_{p^m} \subset \mathbb{F}_{p^{m'}}$. □

From Theorem 6.17 we deduce the following generalization immediately.

Theorem 6.18 *Let q be a prime power, $\mathbb{F}_q$ and $\mathbb{F}_{q^n}$ be finite fields with q and q^n elements, respectively. Assume $\mathbb{F}_q \subset \mathbb{F}_{q^n}$. Then $\mathbb{F}_{q^m}$ is a subfield* *of $\mathbb{F}_{q^n}$ if and only if $m \mid n$. When $m \mid n$, $\mathbb{F}_{q^m}$ is the unique subfield of $\mathbb{F}_{q^n}$ containing q^m elements and containing $\mathbb{F}_q$. Furthermore, let $\mathbb{F}_{q^m}$ and $\mathbb{F}_{q^{m'}}$ both be subfields of $\mathbb{F}_{q^n}$, then $\mathbb{F}_{q^m} \subset \mathbb{F}_{q^{m'}}$ if and only if m is a factor of m'.*

Moreover, we have

Theorem 6.19 *Let q be a prime power and m, n be positive integers. Let $d = \gcd(m, n)$ and $l = \mathrm{lcm}\,[m, n]$. Assume both $\mathbb{F}_{q^m}$ and $\mathbb{F}_{q^n}$ contains $\mathbb{F}_q$ and are contained in $\mathbb{F}_{q^{mn}}$. Then the largest subfield contained in both $\mathbb{F}_{q^m}$ and $\mathbb{F}_{q^n}$ is $\mathbb{F}_{q^d}$ and the smallest subfield of $\mathbb{F}_{q^{mn}}$ containing both $\mathbb{F}_{q^m}$ and $\mathbb{F}_{q^n}$ is $\mathbb{F}_{q^l}$.*

Example 6.5 The lattice of divisors of the integer 12 is

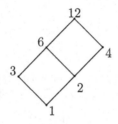

It follows from Theorem 6.17 that the lattice of subfields of $\mathbb{F}_{2^{12}}$ is

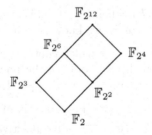

6.6   A Distinction between Finite Fields of Characteristic 2 and Not 2

The foregoing discussion on the structure of finite fields holds for finite fields of characteristic 2 and not 2 simultaneously. But there is a remarkable distinction between the structures of them.

Let $\mathbb{F}_q$ be a finite field with q elements, where q is a prime power. Let

$$\mathbb{F}_q^2 = \{x^2 : x \in \mathbb{F}_q\}$$

and

$$\mathbb{F}_q^{*2} = \{x^2 : x \in \mathbb{F}_q^*\}.$$

Theorem 6.20 *Let $\mathbb{F}_q$ be a finite field with q elements, where q is a prime power.*

(i) *If $q = 2^n$, every element of $\mathbb{F}_{2^n}$ is a square element; in other words, $\mathbb{F}_{2^n} = \mathbb{F}_{2^n}^2$ and $\mathbb{F}_{2^n}^* = \mathbb{F}_{2^n}^{*2}$. Moreover, every element of $\mathbb{F}_{2^n}$ has a unique square root in $\mathbb{F}_{2^n}$.*

(ii) *If q is odd, then $\mathbb{F}_q^{*2}$ is a subgroup of index 2 in $\mathbb{F}_q^*$; let $z \in \mathbb{F}_q^* \setminus \mathbb{F}_q^{*2}$, then $\mathbb{F}_q^* = \mathbb{F}_q^{*2} \bigcup z\mathbb{F}_q^{*2}$ is a disjoint union. Moreover, every element of $\mathbb{F}_q^{*2}$ has two square roots in $\mathbb{F}_q^*$ and every element of $z\mathbb{F}_q^{*2}$ has no square root in $\mathbb{F}_q^*$.*

Proof. (i) Consider the map

$$\sigma_2 : \quad \mathbb{F}_{2^n} \to \mathbb{F}_{2^n}$$
$$x \mapsto x^2.$$

Suppose $\sigma_2(x) = \sigma_2(y)$, i.e., $x^2 = y^2$. Then $(x - y)^2 = x^2 - y^2 = 0$. But $\mathbb{F}_{2^n}$ has no zero divisors, therefore $x = y$. This proves that σ_2 is injective. Since $\mathbb{F}_{2^n}$ is finite, σ_2 is bijective. Therefore $\mathbb{F}_{2^n}^2 = \mathbb{F}_{2^n}$. Since $0^2 = 0$, we also have $\mathbb{F}_{2^n}^{*2} = \mathbb{F}_{2^n}^*$.

Moreover, for any $x \in \mathbb{F}_{2^n}$, since $\mathbb{F}_{2^n} = \mathbb{F}_{2^n}^2$, there is an element $x_0 \in \mathbb{F}_{2^n}$ such that $x = x_0^2$. Suppose $x = x_0^2 = y_0^2$, where $x_0, y_0 \in \mathbb{F}_{2^n}$, then $(x_0 - y_0)^2 = 0$, and, hence, $x_0 = y_0$. Therefore, every element of $\mathbb{F}_{2^n}$ has a unique square root in $\mathbb{F}_{2^n}$.

(ii) By Theorem 6.1 $\mathbb{F}_q^*$ is a cyclic group. If q is odd, $|\mathbb{F}_q^*| = q - 1$ is even. Let ξ be a primitive element of $\mathbb{F}_q$, i.e., $\mathbb{F}_q^* = \langle \xi \rangle$ and ξ is of order $q - 1$. Clearly, $\mathbb{F}_q^{*2} = \langle \xi^2 \rangle$, ξ^2 is of order $(q - 1)/2$, and

$$\mathbb{F}_q^* = \mathbb{F}_q^{*2} \bigcup \xi \mathbb{F}_q^{*2},$$

which is a disjoint union. That is, $\mathbb{F}_q^{*2}$ is a subgroup of index 2 in $\mathbb{F}_q^*$. For any $z \in \mathbb{F}_q^* \setminus \mathbb{F}_q^{*2}$, we also have a disjoint union

$$\mathbb{F}_q^* = \mathbb{F}_q^{*2} \bigcup z\mathbb{F}_q^{*2}.$$

Moreover, for any $x \in \mathbb{F}_q^{*2}$, clearly there is an element $x_0 \in \mathbb{F}_q^*$ such that $x = x_0^2$. We also have $(-x_0)^2 = x$ and $-x_0 \neq x_0$. Suppose $x = x_0^2 = y_0^2$, where $x_0, y_0 \in \mathbb{F}_q^*$, then $(x_0^{-1} y_0)^2 = 1$, and hence $x_0^{-1} y_0 = \pm 1$, $y_0 = \pm x_0$. Therefore every element of $\mathbb{F}_q^{*2}$ has two square roots in $\mathbb{F}_q^*$. For any element $y \in z\mathbb{F}_q^{*2}$, we have $y = zu^2$ where $u \in \mathbb{F}_q^*$. Suppose there is an element $y_0 \in \mathbb{F}_q^{*2}$ such that $y = y_0^2$, then $zu^2 = y_0^2$ and, hence, $z = (y_0 u^{-1})^2 \in \mathbb{F}_q^{*2}$. This is a contradiction. Therefore every element of $z\mathbb{F}_q^{*2}$ has no square root in $\mathbb{F}_q^*$.   □

When q is odd, elements of $\mathbb{F}_q^{*2}$ are called *square elements* of $\mathbb{F}_q^*$ and elements of $\mathbb{F}_q^* \setminus \mathbb{F}_q^{*2}$ are called *non-square elements* of $\mathbb{F}_q^*$. It is obvious that the product of any two square elements or any two non-square elements is a square element and that the product of a square element with a non-square element is a non-square element. Moreover, each $x \in \mathbb{F}_q^{*2}$ has two square roots in $\mathbb{F}_q$, which are denoted by $x^{1/2}$ and $-x^{1/2}$.

When q is even, every element $x \in \mathbb{F}_q$ has a unique square root in $\mathbb{F}_q$, which is denoted by $x^{1/2}$.

Both statements (i) and (ii) of Theorem 6.20 can be generalized.

Let $q = p^n$, where p is a prime. Define

$$\mathbb{F}_q^p = \{ x^p : x \in \mathbb{F}_q \}$$

and

$$\mathbb{F}_q^{*p} = \{ x^p : x \in \mathbb{F}_q^* \}.$$

Then we have the following generalization of statement (i) of Theorem 6.20.

Theorem 6.21 *Let $\mathbb{F}_q$ be a finite field with q elements, where $q = p^n$ and p is a prime, then every element of $\mathbb{F}_q^*$ is a p-th power; in other words, $\mathbb{F}_q = \mathbb{F}_q^p$ and $\mathbb{F}_q^* = \mathbb{F}_q^{*p}$. Moreover, every element of $\mathbb{F}_q$ has a unique p-th root in $\mathbb{F}_q$.*

Proof. The proof is the same as that of Theorem 6.20(i), but the map

$$\sigma_p : \quad \mathbb{F}_q \quad \to \quad \mathbb{F}_q$$
$$x \quad \mapsto \quad x^p$$

is used instead of σ_2.   □

When $q = p^n$, for any $x \in \mathbb{F}_q$ the unique p-th root of x in $\mathbb{F}_q$ is usually denoted by $x^{1/p}$. Then $(x^{1/p})^p = x$.

For any positive integer r define

$$\mathbb{F}_q^{*r} = \{ x^r : x \in \mathbb{F}_q^* \}.$$

Then statement (ii) of Theorem 6.20 is generalized as follows.

Theorem 6.22 *Let $\mathbb{F}_q$ be a finite field with q elements and let r be a positive integer. Then $\mathbb{F}_q^{*r}$ is a subgroup of $\mathbb{F}_q^*$. Let $d = \gcd(q-1, r)$. Then $\mathbb{F}_q^{*r} = \mathbb{F}_q^{*d}$, $\mathbb{F}_q^* : \mathbb{F}_q^{*d} = d$, and*

$$\mathbb{F}_q^* = \mathbb{F}_q^{*d} \bigcup \xi \mathbb{F}_q^{*d} \bigcup \xi^2 \mathbb{F}_q^{*d} \bigcup \cdots \bigcup \xi^{d-1} \mathbb{F}_q^{*d}, \qquad (6.9)$$

where ξ is a primitive element of $\mathbb{F}_q$, is a disjoint union.

Proof. It is obvious that $\mathbb{F}_q^{*r}$ is a subgroup of $\mathbb{F}_q^*$.

Let ξ be a primitive element of $\mathbb{F}_q$, then $\mathbb{F}_q^* = \langle \xi \rangle$, $\mathbb{F}_q^{*r} = \langle \xi^r \rangle$, and $\mathbb{F}_q^{*d} = \langle \xi^d \rangle$. Since $d = \gcd(q-1, r)$, $d \mid r$ and $\mathbb{F}_q^{*r} \subseteq \mathbb{F}_q^{*d}$. There are integers a and b such that $a(q-1) + br = d$. Then $\xi^d = \xi^{a(q-1)+br} = (\xi^r)^b$ and, hence, $\xi^d \in \langle \xi^r \rangle$. Thus $\mathbb{F}_q^{*d} \subseteq \mathbb{F}_q^{*r}$. Therefore $\mathbb{F}_q^{*r} = \mathbb{F}_q^{*d}$.

(6.9) is clear, from which we deduce $\mathbb{F}_q^* : \mathbb{F}_q^{*d} = d$.

6.7 Exercises

6.1 Prove that both 2 and 3 are primitive elements in $\mathbb{F}_5$.

6.2 Prove that both 3 and 5 are primitive elements in $\mathbb{F}_7$, but 2 is not. Find x and $y \in \{1, 2, \ldots, 6\}$ such that $2 = 3^x = 5^y$ in $\mathbb{F}_7$.

6.3 What is the smallest prime number p such that neither 2 nor 3 is a primitive element of $\mathbb{F}_p$.

6.4 Consider the field $\mathbb{F}_{41}$.

   (i) Compute $\mathrm{ord}\,(2)$ and $\mathrm{ord}\,(3)$ in the multiplicative group of $\mathbb{F}_{41}$.

   (ii) Use Gauss algorithm to obtain a primitive element of $\mathbb{F}_{41}$.

   (iii) Find the least primitive element g of $\mathbb{F}_{41}$, i.e., g is a primitive element, $0 < g < 40$, and $g \leq h$ for any primitive element h with $0 < h < 40$.

6.5 Do the same for the field $\mathbb{F}_{13}$.

6.6 Prove that every element of $\mathbb{F}_q$ is the k-th power of some element of $\mathbb{F}_q$ if and only if $(k, q-1) = 1$.

6.7 Show that the sum of all elements of a finite field is 0, except for $\mathbb{F}_2$.

6.8 Show that the product of all non-zero elements of a finite field is -1.

6.9 Let n be a positive integer. Prove that $(n-1)! \equiv -1 \,(\text{mod } n)$ if and only if n is a prime.

6.10 Prove that for any positive integer m,

$$\sum_{a \in \mathbb{F}_q} a^m = \begin{cases} -1, & \text{if } (q-1)\,|\,m, \\ 0, & \text{otherwise.} \end{cases}$$

6.11 Prove that for any field F, every finite subgroup of the multiplicative group F^* is cyclic.

6.12 Prove that if the multiplicative group F^* of a field F is cyclic then F is finite.

6.13 Let $a(n)$ and $b(n)$ be functions defined on the set of positive integers with values in a multiplicative group G, and assume that G is abelian. Then

$$b(n) = \prod_{d|n} a(d)$$

if and only if

$$a(n) = \prod_{d|n} b(d)^{\mu(\frac{n}{d})}.$$

This is the multiplicative version of Möbius inversion formula.

6.14 Let n be an integer > 0. Prove that $\Phi_{q,n} = (q^n - \sum d\Phi_{q,n})/n$, where the summation extends over all positive divisors d of n and $\neq n$.

6.15 Derive from (6.6) the following formula

$$n\Phi_{q,d}(x) = \prod_{d|n}(x^{q^d} - x)^{\mu(\frac{n}{d})}.$$

Let $n = p_1^{e_1} \dots p_r^{e_r}$ be the prime factorization of n, then

$$n\Phi_{q,d}(x) = (x^{q^n} - x)\prod_{i=1}^{r}(x^{q^{n/r_i}} - x)^{-1} \prod_{1 \leq i < j \leq r}(x^{q^{n/p_i p_j}} - x)$$

$$\cdots (x^{q^{n/p_1 p_2 \cdots p_r}} - x)^{(-1)^r}.$$

Deduce (6.8) from the above equation.

6.16 Draw the lattice of subfields for $\mathbb{F}_{2^{24}}$ and also for $\mathbb{F}_{3^{30}}$.

6.17 Let a, b be elements of $\mathbb{F}_{2^n}$, n odd. Prove that $a^2 + ab + b^2 = 0$ implies $a = b = 0$.

6.18 Let F be a field and $\phi : F \to F$ be the map defined by $\phi(a) = a^{-1}$ if $a \neq 0$ and $\phi(a) = 0$ if $a = 0$. Prove that ϕ is an automorphism of F if and only if F has at most four elements.

6.19 Assume that $\mathbb{F}_{16}$ was built as $\mathbb{F}_2[x]/(x^4 + x + 1)$ and let α be the residue class of x modulo $x^4 + x + 1$.

  (i) Prove that α is a primitive element of $\mathbb{F}_{16}$.

  (ii) Using the table of logarithms and antilogarithms to the base α to compute $(\alpha^3 + \alpha^2 + \alpha + 1)(\alpha^3 + \alpha + 1)$, $(\alpha + 1)^7$, and $(\alpha^2 + \alpha + 1)/(\alpha^3 + 1)$.

6.20 (i) Prove that $x^3 + 2x + 1$ is irreducible over $\mathbb{F}_3$.

  (ii) Determine a primitive element of $\mathbb{F}_{27} = \mathbb{F}_3[x]/(x^3 + 2x + 1)$.

  (iii) Express all nonzero elements of $\mathbb{F}_{27}$ as powers of the primitive element.

6.21 Do the same thing for the polynomial $x^2 - 3$ over the field $\mathbb{F}_5$.

6.22 Let q be a power of an odd prime. Prove that an element $\alpha \in \mathbb{F}_q^*$ is a square element of $\mathbb{F}_q^*$ if and only if $\alpha^{(q-1)/2} = 1$. In particular, $-1 \in \mathbb{F}_q^{*2}$ if and only if $q \equiv 1 \pmod 4$.

6.23 Let q be a power of an odd prime. Prove that

$$x^{(q-1)/2} - 1 = \prod_{\alpha \in \mathbb{F}_q^{*2}} (x - \alpha), \quad x^{(q-1)/2} + 1 = \prod_{\alpha \in \mathbb{F}_q^* \setminus \mathbb{F}_q^{*2}} (x - \alpha).$$

Then deduce that for any $f(x) \in \mathbb{F}_q[x]$ without multiple roots, let

$$g(x) = \gcd(f(x), x^{(q-1)/2} - 1),$$

then $g(x)$ is a proper divisor of $f(x)$ if and only if $f(x)$ has at least one root in $\mathbb{F}_q^{*2}$ and at least one root in $\mathbb{F}_q^* \setminus \mathbb{F}_q^{*2}$.

6.24 Let q be a prime power and r be a prime divisor of $q - 1$. Let $a \in \mathbb{F}_q^*$ and $\mathrm{ord}(a) = m > 1$ in $\mathbb{F}_q^*$. Prove that $r \mid (q-1)/m$ if and only if $a \in \mathbb{F}_q^{*r}$.

6.25 Let $a, b \in \mathbb{F}_q^*$. Prove that for any $c \in \mathbb{F}_q$, the equation $ax^2 + by^2 = c$ has always a solution (x, y) in $\mathbb{F}_q$.

6.26 Prove that any irreducible quadratic polynomial over $\mathbb{F}_q$ can be factorized into a product of two linear polynomials over $\mathbb{F}_{q^2}$.

6.27 Let q be a power of a prime, $\mathbb{F}_q$ be the finite field with q elements and $GL_n(\mathbb{F}_q)$ be the *general lineal group of degree n over $\mathbb{F}_q$*, i.e. the group consisting of all $n \times n$ nonsingular matrices over $\mathbb{F}_q$. Prove that the order of $GL_n(\mathbb{F}_q)$ is

$$q^{n(n-1)/2} \prod_{i=1}^{n} (q^i - 1).$$

Moreover, prove that the number of elements of order $q^n - 1$ in $GL_n(\mathbb{F}_q)$ is

$$q^{n(n-1)/2} \frac{\phi(q^n - 1)}{n} \prod_{i=1}^{n} (q^i - 1).$$

Chapter 7

Further Properties of Finite Fields

7.1 Automorphisms

Let q be a prime power and n a positive integer. We may view $\mathbb{F}_q$ as a subfield of $\mathbb{F}_{q^n}$. By Corollary 3.18 the map

$$\sigma : \alpha \mapsto \alpha^q$$

from $\mathbb{F}_{q^n}$ to itself is an automorphism of $\mathbb{F}_{q^n}$. By Corollary 6.2 $\sigma(a) = a$ if and only if $a \in \mathbb{F}_q$. An automorphism of $\mathbb{F}_{q^n}$ which leaves every element of $\mathbb{F}_q$ fixed is called an *automorphism* of $\mathbb{F}_{q^n}$ *over* $\mathbb{F}_q$. In particular, σ is an automorphism of $\mathbb{F}_{q^n}$ over $\mathbb{F}_q$. We call σ the *Frobenius automorphism* of $\mathbb{F}_{q^n}$ *over* $\mathbb{F}_q$. Let $\sigma^2 = \sigma \circ \sigma$ be the composite of σ with σ. Then σ^2 is also an automorphism of $\mathbb{F}_{q^n}$ over $\mathbb{F}_q$. More generally, let

$$\sigma^{i+1} = \sigma \circ \sigma^i, \; i = 1, 2, \ldots$$

and define

$$\sigma^0 = 1,$$

where 1 denotes the identity map of $\mathbb{F}_{q^n}$, then all σ^i, $i = 0, 1, \ldots$, are automorphisms of $\mathbb{F}_{q^n}$ over $\mathbb{F}_q$. By Corollary 6.2 $\sigma^n(\alpha) = \alpha^{q^n} = \alpha$ for all $\alpha \in \mathbb{F}_{q^n}$, hence $\sigma^n = 1$. Let ξ be a primitive element of $\mathbb{F}_{q^n}$, i.e., $\xi \in \mathbb{F}_{q^n}$, $\xi^i \neq 1$ for $1 \leq i < q^n - 1$ and $\xi^{q^n-1} = 1$. For $1 \leq k < n$, $q^k < q^n$ and $q^k - 1 < q^n - 1$, thus $\sigma^k(\xi) = \xi^{q^k} \neq \xi$. Therefore, $\sigma^k \neq 1$. It follows that $\sigma^0 = 1$, $\sigma^1 = \sigma, \ldots, \sigma^{n-1}$ are n distinct automorphisms of $\mathbb{F}_{q^n}$ over $\mathbb{F}_q$.

We have proved the following

Theorem 7.1 *The map*

$$\sigma : \alpha \mapsto \alpha^q$$

from $\mathbb{F}_{q^n}$ to itself is an automorphism of $\mathbb{F}_{q^n}$, which leaves every element of $\mathbb{F}_q$ fixed, i.e., $\sigma(a) = a$ for every $a \in \mathbb{F}_q$. Moreover, $\sigma^n = 1$, $\sigma^k \neq 1$ for $1 \leq k \leq n$, and $\sigma^0 = 1$, $\sigma^1 = \sigma, \ldots, \sigma^{n-1}$ are n distinct automorphisms of $\mathbb{F}_{q^n}$ over $\mathbb{F}_q$.

Clearly, the set of automorphisms of $\mathbb{F}_{q^n}$ over $\mathbb{F}_q$ forms a group with respect to the composition of maps, which is called the *Galois group* of $\mathbb{F}_{q^n}$ over $\mathbb{F}_q$ and denoted by $\mathrm{Gal}(\mathbb{F}_{q^n}/\mathbb{F}_q)$. We will prove $\mathrm{Gal}(\mathbb{F}_{q^n}/\mathbb{F}_q) = \langle \sigma \rangle$. We need some preparation.

Let τ be an automorphism of $\mathbb{F}_{q^n}$. We extend τ to a map $\tilde{\tau}$ from $\mathbb{F}_{q^n}[x]$ to itself in the following way. Let

$$f(x) = \alpha_0 + \alpha_1 x + \alpha_2 x^2 + \cdots + \alpha_m x^m, \text{ where } \alpha_0, \alpha_1, \alpha_2, \ldots, \alpha_m \in \mathbb{F}_{q^n}. \quad (7.1)$$

Define

$$\tilde{\tau}(f(x)) = \tau(\alpha_0) + \tau(\alpha_1)x + \tau(\alpha_2)x^2 + \cdots + \tau(\alpha_m)x^m.$$

It is routine to verify that $\tilde{\tau} : \mathbb{F}_{q^n}[x] \to \mathbb{F}_{q^n}[x]$ is an automorphism of the polynomial ring $\mathbb{F}_{q^n}[x]$. $\tilde{\tau}$ is called the *extended automorphism* of τ to $\mathbb{F}_{q^n}[x]$.

Lemma 7.2 *Let σ be the Frobenius automorphism of $\mathbb{F}_{q^n}$ over $\mathbb{F}_q$ and $\tilde{\sigma}$ be the extended automorphism of σ to the polynomial ring $\mathbb{F}_{q^n}[x]$. Let $f(x) \in \mathbb{F}_{q^n}[x]$. If $\tilde{\sigma}(f(x)) = f(x)$, then $f(x) \in \mathbb{F}_q[x]$.*

Proof. Let $f(x)$ be the polynomial (7.1). Then

$$\tilde{\sigma}(f(x)) = \sigma(\alpha_0) + \sigma(\alpha_1)x + \sigma(\alpha_2)x^2 + \cdots + \sigma(\alpha_m)x^m.$$

From $\tilde{\sigma}(f(x)) = f(x)$ we deduce $\sigma(\alpha_i) = \alpha_i$ for $i = 0, 1, 2, \ldots, m$. But we also have $\sigma(\alpha_i) = \alpha_i^q$. Therefore $\alpha_i^q = \alpha_i$. By Corollary 6.2 $\alpha_i \in \mathbb{F}_q$ for $i = 0, 1, 2, \ldots, m$. Hence $f(x) \in \mathbb{F}_q[x]$. $\square$

Theorem 7.3 $\mathrm{Gal}(\mathbb{F}_{q^n}/\mathbb{F}_q) = \langle \sigma \rangle$. *More precisely, $\sigma^0 = 1$, $\sigma^1 = \sigma, \ldots,$ σ^{n-1} are all the automorphisms of $\mathbb{F}_{q^n}$ over $\mathbb{F}_q$.*

Proof. Let ξ be a primitive element of $\mathbb{F}_{q^n}$ and let

$$f(x) = (x - \xi)(x - \sigma(\xi)) \cdots (x - \sigma^{n-1}(\xi))$$
$$= x^n + a_1 x^{n-1} + \cdots + a_{n-1} x + a_n.$$

Clearly, $\tilde{\sigma}(x - \sigma^i(\xi)) = x - \sigma^{i+1}(\xi)$ for $i = 0, 1, 2, \ldots, n-2$ and $\tilde{\sigma}(x - \sigma^{n-1}(\xi)) = x - \sigma^n(\xi) = x - \xi$. Thus $\tilde{\sigma}(f(x)) = f(x)$. By Lemma 7.2, $f(x) \in \mathbb{F}_q[x]$, i.e., $a_k \in \mathbb{F}_q$ for $k = 1, 2, \ldots, n$.

Let τ be an automorphism of $\mathbb{F}_{q^n}$ over $\mathbb{F}_q$ and suppose that $\tau(\xi) = \xi^i$ where $1 \leq i \leq q^n - 2$. Clearly, $f(\xi) = 0$, i.e.,

$$\xi^n + a_1 \xi^{n-1} + \cdots + a_{n-1} \xi + a_n = 0.$$

Applying the automorphism τ to the above equation we obtain

$$(\xi^i)^n + a_1(\xi^i)^{n-1} + \cdots + a_{n-1}\xi^i + a_n = 0,$$

i.e., $f(\xi^i) = 0$. But $f(x)$ has n roots $\xi, \sigma(\xi), \ldots, \sigma^{n-1}(\xi)$ in $\mathbb{F}_{q^n}$. Thus ξ^i must be one of them. Say $\xi^i = \sigma^j(\xi)$, $0 \leq j \leq n-1$, then $\tau(\xi) = \xi^i = \sigma^j(\xi)$, which implies $\tau(\alpha) = \sigma^j(\alpha)$ for all $\alpha \in \mathbb{F}_{q^n}$, hence $\tau = \sigma^j$. $\qquad\square$

Corollary 7.4 *Let σ be the Frobenius automorphism of $\mathbb{F}_{q^n}$ over $\mathbb{F}_q$, and let d be a positive divisor of n. Then*

(i) *σ^d is the Frobenius automorphism of $\mathbb{F}_{q^n}$ over $\mathbb{F}_{q^d}$ and $\mathrm{Gal}(\mathbb{F}_{q^n}/\mathbb{F}_{q^d}) = \langle \sigma^d \rangle$ is a cyclic group of order n/d.*

(ii) *For any $\alpha \in \mathbb{F}_{q^d}$, $\sigma(\alpha) \in \mathbb{F}_{q^d}$. Denote the restriction of σ to $\mathbb{F}_{q^d}$ by $\sigma|_{\mathbb{F}_{q^d}}$, then the map*

$$\sigma|_{\mathbb{F}_{q^d}} : \quad \mathbb{F}_{q^d} \to \mathbb{F}_{q^d}$$
$$\alpha \mapsto \sigma(\alpha)$$

is well-defined, $\sigma|_{\mathbb{F}_{q^d}}$ is the Frobenius automorphism of $\mathbb{F}_{q^d}$ over $\mathbb{F}_q$ and $\mathrm{Gal}(\mathbb{F}_{q^d}/\mathbb{F}_q) = \langle \sigma|_{\mathbb{F}_{q^d}} \rangle$ is a cyclic group of order d.

Proof. (i) For any $\alpha \in \mathbb{F}_{q^n}$, $\sigma^d(\alpha) = \alpha^{q^d}$. By definition, σ^d is the Frobenius automorphism of $\mathbb{F}_{q^n}$ over $\mathbb{F}_{q^d}$. By Theorem 7.3 $\mathrm{Gal}(\mathbb{F}_{q^n}/\mathbb{F}_{q^d}) = \langle \sigma^d \rangle$ and, hence, is a cyclic group of order n/d.

(ii) For any $\alpha \in \mathbb{F}_{q^d}$, $\sigma(\alpha) = \alpha^q \in \mathbb{F}_{q^d}$. Thus the map $\sigma|_{\mathbb{F}_{q^d}} : \mathbb{F}_{q^d} \to \mathbb{F}_{q^d}$ is well-defined. Moreover, $\sigma|_{\mathbb{F}_{q^d}}(\alpha) = \sigma(\alpha) = \alpha^q$ for all $\alpha \in \mathbb{F}_{q^d}$. By definition $\sigma|_{\mathbb{F}_{q^d}}$ is the Frobenius automorphism of $\mathbb{F}_{q^d}$ over $\mathbb{F}_q$. By Theorem 7.3 $\mathrm{Gal}(\mathbb{F}_{q^d}/\mathbb{F}_q) = \langle \sigma|_{\mathbb{F}_{q^d}} \rangle$ and is cyclic of order d. $\qquad\square$

In the following we denote $\sigma|_{\mathbb{F}_{q^d}}$ simply also by σ.

7.2 Characteristic Polynomials and Minimal Polynomials

As an application of the foregoing discussion on the automorphisms of $\mathbb{F}_{q^n}$ over $\mathbb{F}_q$ let us study the characteristic polynomial and the minimal polynomial of an arbitrary element $\alpha \in \mathbb{F}_{q^n}$ over $\mathbb{F}_q$.

Definition 7.1 *Let* $\alpha \in \mathbb{F}_{q^n}$, σ *be the Frobenius automorphism of* $\mathbb{F}_{q^n}$ *over* $\mathbb{F}_q$, *and*

$$g(x) = (x - \alpha)(x - \sigma(\alpha)) \cdots (x - \sigma^{n-1}(\alpha))$$
$$= (x - \alpha)(x - \alpha^q) \cdots (x - \alpha^{q^{n-1}}),$$

then $\tilde{\sigma}(g(x)) = g(x)$. *By Lemma 7.2* $g(x)$ *is a monic polynomial of* $\mathbb{F}_q[x]$. $g(x)$ *is called the characteristic polynomial of* $\alpha \in \mathbb{F}_{q^n}$ *over* $\mathbb{F}_q$. *The monic polynomial of least degree over* $\mathbb{F}_q$ *having* α *as a root is called the minimal polynomial of* $\alpha \in \mathbb{F}_{q^n}$ *over* $\mathbb{F}_q$.

The minimal polynomial already appeared in the proof of Lemma 6.16 but we did not give it the name minimal polynomial there.

Theorem 7.5 *Let* α *be an arbitrary element of* $\mathbb{F}_{q^n}$. *Then*

(i) *The minimal polynomial of* α *over* $\mathbb{F}_q$ *exists and is unique; moreover, it is irreducible over* $\mathbb{F}_q$.

(ii) *Let* $m(x)$ *be the minimal polynomial of* α *over* $\mathbb{F}_q$. *If* $f(x) \in \mathbb{F}_q[x]$ *and* $f(\alpha) = 0$, *then* $m(x) \mid f(x)$.

(iii) *Let* d *be the least positive integer such that* $\sigma^d(\alpha) = \alpha$. *Then* $d \mid n$, $d = \deg m(x)$, *and*

$$m(x) = (x - \alpha)(x - \sigma(\alpha)) \cdots (x - \sigma^{d-1}(\alpha)).$$

(iv) *Let* $g(x)$ *be the characteristic polynomial of* α *over* $\mathbb{F}_q$, *then*

$$g(x) = m(x)^{n/d}.$$

Proof. (i) Let $\alpha \in \mathbb{F}_{q^n}$. Clearly, α is a root of the characteristic polynomial

$$g(x) = (x - \alpha)(x - \sigma(\alpha)) \cdots (x - \sigma^{n-1}(\alpha))$$

of α. Thus there exists a monic polynomial $m(x)$ of least degree over $\mathbb{F}_q$ having α as a root, which is by definition the minimal polynomial of α over $\mathbb{F}_q$. Assume that $m_1(x)$ is also a monic polynomial of least degree over $\mathbb{F}_q$ having α as a root, then $\deg m(x) = \deg m_1(x)$. If $m(x) \neq m_1(x)$, let c be the leading coefficient of $m(x) - m_1(x)$, then $c^{-1}(m(x) - m_1(x))$ is a monic polynomial over $\mathbb{F}_q$ having α as a root and $\deg c^{-1}(m(x) - m_1(x)) < \deg m(x)$, this is a contradiction. Therefore $m(x) = m_1(x)$ and $m(x)$ is unique.

Assume that $m(x) = f_1(x)f_2(x)$ is a factorization of $m(x)$ in $\mathbb{F}_q[x]$ and $0 < \deg f_1(x)$, $\deg f_2(x) < \deg m(x)$. We can assume that both $f_1(x)$ and $f_2(x)$ are monic. Substituting $x = \alpha$ into $m(x) = f_1(x)f_2(x)$, we obtain $0 = m(\alpha) = f_1(\alpha)f_2(\alpha)$. Since $\mathbb{F}_q$ has no zero-divisors, $f_1(\alpha) = 0$ or $f_2(\alpha) = 0$. But both $f_1(x)$ and $f_2(x)$ are monic and has degrees $< \deg m(x)$. This is a contradiction. Therefore $m(x)$ is irreducible over $\mathbb{F}_q$.

(ii) is trivial.

(iii) Let d be the least positive integer such that $\sigma^d(\alpha) = \alpha$. We know $\sigma^n = 1$, thus $\sigma^n(\alpha) = \alpha$. Dividing n by d, we obtain $n = ld + r$, where $l, r \in \mathbb{Z}$, $0 \leq l$ and $0 \leq r < d$; then $\alpha = \sigma^n(\alpha) = \sigma^{ld+r}(\alpha) = (\sigma^r \circ \sigma^{ld})(\alpha) = \sigma^r(\sigma^{ld}(\alpha)) = \sigma^r(\alpha)$, which implies $r = 0$. Thus $d \mid n$.

We assert that $\alpha, \sigma(\alpha), \ldots, \sigma^{d-1}(\alpha)$ are d distinct elements of $\mathbb{F}_{q^n}$. Suppose that $\sigma^i(\alpha) = \sigma^j(\alpha)$, where $1 \leq i, j \leq d-1$, and $i > j$, then $\sigma^{i-j}(\alpha) = (\sigma^{-j} \circ \sigma^i)(\alpha) = (\sigma^{-j} \circ \sigma^j)(\alpha) = \sigma^0(\alpha) = \alpha$, a contradiction. Let $m_1(x) = (x - \alpha)(x - \sigma(\alpha)) \cdots (x - \sigma^{d-1}(\alpha))$. Then $\bar{\sigma}(m_1(x)) = m_1(x)$. By Lemma 7.2 $m_1(x) \in \mathbb{F}_q[x]$. Clearly $m_1(\alpha) = 0$. By (ii) $m(x) \mid m_1(x)$. By the definition of the minimal polynomial, $m(\alpha) = 0$. Applying σ^i, $1 \leq i \leq d-1$, to $m(\alpha) = 0$ we obtain $m(\sigma^i(\alpha)) = 0$, thus $\alpha, \sigma(\alpha), \ldots, \sigma^{d-1}(\alpha)$ are d distinct roots of $m(x)$. It follows that $\deg m(x) \geq d$. But $\deg m_1(x) = d$ and $m(x) \mid m_1(x)$, therefore $m(x) = m_1(x)$ and $\deg m(x) = d$.

(iv) We have

$$g(x) = (x - \alpha)(x - \sigma(\alpha)) \cdots (x - \sigma^{n-1}(\alpha))$$

$$= \prod_{k=0}^{(n/d)-1} (x - \sigma^{kd}(\alpha)) \cdots (x - \sigma^{kd+(d-1)}(\alpha))$$

$$= \prod_{k=0}^{(n/d)-1} (x - \alpha)(x - \sigma(\alpha)) \cdots (x - \sigma^{d-1}(\alpha))$$

$$= m(x)^{n/d}.$$

$\square$

The elements $\alpha, \sigma(\alpha), \ldots, \sigma^{d-1}(\alpha)$, which are all the roots of the minimal polynomial $m(x)$ of α, are called the *conjugates* of α over $\mathbb{F}_q$. The number of conjugates of α is called the *degree* of α over $\mathbb{F}_q$ and denoted by $\deg(\alpha)$. By Theorem 7.5 $\deg(\alpha) = \deg m(x)$.

Theorem 7.6 *Let $m(x)$ be a monic irreducible polynomial of degree d over $\mathbb{F}_q$ and $d \mid n$. Then*

(i) *$m(x)$ has d distinct roots in $\mathbb{F}_{q^n}$.*

(ii) *Let α be any root of $m(x)$ in $\mathbb{F}_{q^n}$, then $m(x)$ is the minimal polynomial of α over $\mathbb{F}_q$, and $\mathbb{F}_q[\alpha] = \mathbb{F}_{q^d}$.*

(iii) *Let σ be the Frobenius automorphism of $\mathbb{F}_{q^n}$ over $\mathbb{F}_q$, then d is the least positive integer such that $\sigma^d(\alpha) = \alpha$, $\deg(\alpha) = d$, and $\alpha, \sigma(\alpha), \ldots, \sigma^{d-1}(\alpha)$ are all the distinct conjugates of α over $\mathbb{F}_q$; moreover, they have the same order in $\mathbb{F}_{q^n}^*$.*

Proof. (i) By Lemma 6.16 $m(x)$ has d distinct roots in $\mathbb{F}_{q^d}$. Since $d \mid n$, by Theorem 6.18 $\mathbb{F}_{q^d}$ is a subfield of $\mathbb{F}_{q^n}$. Thus $m(x)$ has its d distinct roots in $\mathbb{F}_{q^n}$.

(ii) Let α be any root of $m(x)$ in $\mathbb{F}_{q^n}$. It has already been proved in Lemma 6.16 that $m(x)$ is the minimum polynomial of α over $\mathbb{F}_q$. Clearly, $\mathbb{F}_q[\alpha] \subseteq \mathbb{F}_{q^d}$. By Theorem 5.8 $|\mathbb{F}_q[\alpha]| = q^d$. Therefore $\mathbb{F}_q[\alpha] = \mathbb{F}_{q^d}$.

(iii) Let d' be the least positive integer such that $\sigma^{d'}(\alpha) = \alpha$, then by Theorem 7.5(iii) $d' = \deg m(x)$. Therefore $d' = d$. Also by Theorem 7.5(iii),

$$m(x) = (x - \alpha)(x - \sigma(\alpha)) \cdots (x - \sigma^{d-1}(\alpha)).$$

Thus $\sigma(\alpha), \ldots, \sigma^{d-1}(\alpha)$ are all the distinct conjugates of α over $\mathbb{F}_q$ and $d = \deg(\alpha)$. Since σ is an automorphism of $\mathbb{F}_{q^n}, \alpha, \sigma(\alpha), \ldots, \sigma^{d-1}(\alpha)$ have the same order. □

Example 7.1 Let us compute the minimal polynomials of the elements of $\mathbb{F}_{2^3}$. By Example 5.8 $\mathbb{F}_{2^3} = \mathbb{F}_2[x]/(x^3 + x + 1)$. Denote by α the residue class containing x, then $\alpha^3 = \alpha + 1$, α is a primitive element, and

$$\mathbb{F}_{2^3} = \{0, 1, \alpha, \ldots, \alpha^6\}.$$

Clearly, the minimal polynomials of 0 and 1 are x and $x + 1$, respectively. Denote the minimal polynomials of α^i by $f_i(x)$, $i = 1, 2, \ldots, 6$. By Theorem 7.5,

$$\begin{aligned}
f_1(x) &= (x - \alpha)(x - \alpha^2)(x - \alpha^4) \\
&= x^3 - (\alpha + \alpha^2 + \alpha^4)x^2 + (\alpha\alpha^2 + \alpha\alpha^4 + \alpha^2\alpha^4)x - (\alpha\alpha^2\alpha^4).
\end{aligned}$$

By the computation in Example 5.8,

$$\alpha + \alpha^2 + \alpha^4 = 0,$$
$$\alpha\alpha^2 + \alpha\alpha^4 + \alpha^2\alpha^4 = \alpha^3 + \alpha^5 + \alpha^6$$
$$= (1 + \alpha) + (1 + \alpha + \alpha^2) + (1 + \alpha^2)$$
$$= 1,$$
$$\alpha \cdot \alpha^2 \cdot \alpha^4 = \alpha^7 = 1.$$

Therefore

$$f_1(x) = x^3 + x + 1.$$

We also have

$$f_1(x) = f_2(x) = f_4(x).$$

By Theorem 7.5,

$$f_3(x) = (x - \alpha^3)(x - \alpha^{3 \cdot 2})(x - \alpha^{3 \cdot 2^2}).$$

We can compute $f_3(x)$ as we computed $f_1(x)$. But we can also do it as follows. By Example 6.1 α^0 is represented by the 3-tuple (001), α^3 by (011). $\alpha^{3 \cdot 2} = \alpha^6$ by (101), and $\alpha^{3 \cdot 3} = \alpha^9 = \alpha^2$ by (100). Clearly, (001), (011), and (101) are linearly independent. But we have the linear relation

$$(001) + (101) + (100) = 0,$$

that is, $1 + \alpha^{3 \cdot 2} + \alpha^{3 \cdot 3} = 0$. Therefore

$$f_3(x) = x^3 + x^2 + 1.$$

We also have

$$f_3(x) = f_6(x) = f_5(x).$$

We can tabulate as follows:

Table 7.1

i	α^i	ord (α^i)	deg(α^i)	*minimal polynomial of* α^i
*	0	*	1	x
0	1	1	1	$x + 1$
1	α	7	3	$x^3 + x + 1$
2	α^2	7	3	$x^3 + x + 1$
3	$\alpha + 1$	7	3	$x^3 + x^2 + 1$
4	$\alpha^2 + \alpha$	7	3	$x^3 + x + 1$
5	$\alpha^2 + \alpha + 1$	7	3	$x^3 + x^2 + 1$
6	$\alpha^2 + 1$	7	3	$x^3 + x^2 + 1$

We have the complete factorization of $x^8 - x$:

$$x^8 - x = x(x - 1)(x^3 + x + 1)(x^3 + x^2 + 1).$$

Example 7.2 Let $\mathbb{F}_{2^4} = \mathbb{F}_2[x]/(x^4 + x + 1)$ and α denote the residue class containing x in $\mathbb{F}_2[x]/(x^4 + x + 1)$. Let us compute the minimal polynomial $f(x)$ of α over $\mathbb{F}_2$. By Theorem 7.5,

$$f(x) = (x - \alpha)(x - \alpha^2)(x - \alpha^4)(x - \alpha^8).$$

We have $\alpha^4 = \alpha + 1$. Therefore

$$f(x) = x^4 + x + 1.$$

We can also use the table in Example 6.2 to compute $f(x)$ as follows. We know that α^0 is represented by the 4-tuple (0001), α by (0010), α^2 by (0100), α^3 by (1000), and α^4 by (0011). Clearly, (0001), (0010), (0100), and (1000) are linearly independent. But

$$(0001) + (0010) + (0011) = 0,$$

that is, $1 + \alpha + \alpha^4 = 0$. Therefore

$$f(x) = x^4 + x + 1.$$

The minimal polynomials of elements α^i, $i = 0, 1, \ldots, 14$, are tabulated as follows:

Table 7.2

i	α^i	ord(α^i)	deg(α^i)	*minimal polynomial*	*conjugate to*
0	0001	1	1	$x + 1$	
1	0010	15	4	$x^4 + x + 1$	
2	0100	15	4	$x^4 + x + 1$	α
3	1000	5	4	$x^4 + x^3 + x^2 + x + 1$	
4	0011	15	4	$x^4 + x + 1$	α
5	0110	3	2	$x^2 + x + 1$	
6	1100	5	4	$x^4 + x^3 + x^2 + x + 1$	α^3
7	1011	15	4	$x^4 + x^3 + 1$	
8	0101	15	4	$x^4 + x + 1$	α
9	1010	5	4	$x^4 + x^3 + x^2 + x + 1$	α^3
10	0111	3	2	$x^2 + x + 1$	α^5
11	1110	15	4	$x^4 + x^3 + 1$	α^7
12	1111	5	2	$x^4 + x^3 + x^2 + x + 1$	α^3
13	1101	15	4	$x^4 + x^3 + 1$	α^7
14	1001	15	4	$x^4 + x^3 + 1$	α^7

(Exercise 7.1))

7.3 Primitive Polynomials

Definition 7.2 *Let $f(x)$ be a monic polynomial of degree n over $\mathbb{F}_q$. If $f(x)$ has a primitive element of $\mathbb{F}_{q^n}$ as one of its roots, $f(x)$ is called a primitive polynomial of degree n over $\mathbb{F}_q$.*

Theorem 7.7 *For any positive integer n there always exist primitive polynomials of degree n over $\mathbb{F}_q$. All the n roots of a primitive polynomial of degree n over $\mathbb{F}_q$ are primitive elements of $\mathbb{F}_{q^n}$. All primitive polynomials over $\mathbb{F}_q$ are irreducible over $\mathbb{F}_q$. The number of primitive polynomials of degree n over $\mathbb{F}_q$ is equal to $\phi(q^n - 1)/n$.*

Proof. Let ξ be a primitive element of $\mathbb{F}_{q^n}$ and σ be the Frobenius automorphism of $\mathbb{F}_{q^n}$ over $\mathbb{F}_q$. Let

$$f(x) = (x - \xi)(x - \sigma(\xi)) \cdots (x - \sigma^{n-1}(\xi)).$$

By the proof of Theorem 7.3, $\tilde{\sigma}(f(x)) = f(x)$. Therefore $f(x) \in \mathbb{F}_q[x]$. Clearly, $f(x)$ is a monic polynomial of degree n and has the primitive element ξ as one of its roots. Hence $f(x)$ is a primitive polynomial of degree n over $\mathbb{F}_q$.

Let $f(x)$ be a primitive polynomial of degree n over $\mathbb{F}_q$. Then $f(x)$ has a primitive element, say ξ, as one of its roots. Thus $f(\xi) = 0$. Applying $\sigma, \sigma^2, \ldots, \sigma^{n-1}$ successively to $f(\xi) = 0$, we obtain $f(\sigma^i(\xi)) = 0$ for $i = 1, 2, \ldots, n - 1$. Clearly, $\sigma^n(\xi) = \xi$. We assert that n is the least positive integer such that $\sigma^n(\xi) = \xi$. Assume d is the least positive integer such that $\sigma^d(\xi) = \xi$ and $d < n$. Then $d \mid n$ and $\xi \in \mathbb{F}_{q^d}$. It follows that the order of ξ is a divisor of $q^d - 1$, contradicting that ξ being a primitive element of $\mathbb{F}_{q^n}$ is of order $q^n - 1$. Therefore $\xi, \sigma(\xi), \sigma^2(\xi), \ldots, \sigma^{n-1}(\xi)$ are n distinct roots of $f(x)$ and

$$f(x) = (x - \xi)(x - \sigma(\xi))(x - \sigma^2(\xi)) \cdots (x - \sigma^{n-1}(\xi)).$$

By Theorem 7.6(iii) $\sigma(\xi), \sigma^2(\xi), \ldots, \sigma^{n-1}(\xi)$ are of the same order and, hence, they are all primitive elements of $\mathbb{F}_{q^n}$. Therefore all the n roots of a primitive polynomial of degree n over $\mathbb{F}_q$ are primitive elements of $\mathbb{F}_{q^n}$.

By Theorem 7.5 $f(x)$ is the minimum polynomial of ξ and is irreducible. Notice that distinct monic irreducible polynomials and, in particular, primitive polynomials can not have a common root (Exercise 4.19). There are altogether $\phi(q^n - 1)$ primitive elements of $\mathbb{F}_{q^n}$. It follows that the number of primitive polynomials of degree n over $\mathbb{F}_q$ is equal to $\phi(q^n - 1)/n$. $\square$

The proof of Theorem 7.7 proves also the following:

Corollary 7.8 *Let ξ be a primitive element of $\mathbb{F}_{q^n}$. Then the minimal polynomial of ξ over $\mathbb{F}_q$ is a primitive polynomial of degree n over $\mathbb{F}_q$.*

Next we shall give an equivalent definition of a primitive polynomial, which is independent of its roots (cf. Corollary 7.10 below).

Definition 7.3 *Let $f(x)$ be a monic irreducible polynomial of degree n over $\mathbb{F}_q$ and assume that $f(x) \neq x$. By Theorem 5.8, $\mathbb{F}_q[x]/(f(x))$ is a field with q^n elements. By Theorem 6.15 and our convention $\mathbb{F}_q[x]/(f(x)) = \mathbb{F}_{q^n}$. The order of the residue class containing x in the group $\mathbb{F}_{q^n}^*$ is called the period of $f(x)$.*

Theorem 7.9 *Let $f(x)$ be a monic irreducible polynomial of degree n over $\mathbb{F}_q$ and assume that $f(x) \neq x$. Then the common order of the n roots of $f(x)$ is equal to the period of $f(x)$.*

Proof. By Theorem 7.6(iii) the roots of $f(x)$ in $\mathbb{F}_{q^n}$ have the same order in $\mathbb{F}_{q^n}^*$. Denote the common order of the n roots of $f(x)$ by l and denote the period of $f(x)$ by l'. Let α be a root of $f(x)$. Then $\alpha^l = 1$. Since $f(\alpha) = 0$ and $\alpha^l - 1 = 0$, $\gcd(f(x), x^l - 1) \neq 1$. But $f(x)$ is irreducible, so $\gcd(f(x), x^l - 1) = f(x)$. Thus $f(x) \mid (x^l - 1)$ and $x^l \equiv 1 \pmod{f(x)}$. Therefore $l' \mid l$.

From $x^{l'} \equiv 1 \pmod{f(x)}$ we deduce $\alpha^{l'} = 1$. Thus $l \mid l'$.

From $l' \mid l$ and $l \mid l'$ we conclude $l = l'$. $\qquad\qquad\square$

Corollary 7.10 *Let $f(x)$ be a monic irreducible polynomial of degree n over $\mathbb{F}_q$ and assume that $f(x) \neq x$. Then $f(x)$ is a primitive polynomial of degree n over $\mathbb{F}_q$ if and only if the period of $f(x)$ is $q^n - 1$.*

Definition 7.4 *Let*

$$f(x) = a_0 + a_1 x + a_2 x^2 + \cdots + a_n x^n$$

be a polynomial of degree n over $\mathbb{F}_q$ and $a_0 a_n \neq 0$. Define

$$\tilde{f}(x) = x^n f(1/x) = a_0 x^n + a_1 x^{n-1} + a_2 x^{n-2} + \cdots + a_n.$$

$f(x)$ and $\tilde{f}(x)$ are called reciprocal polynomials, and $\tilde{f}(x)$ (or $f(x)$) is called the reciprocal polynomial of $f(x)$ (or $\tilde{f}(x)$, respectively).

Clearly,

(i) $\tilde{\tilde{f}}(x) = f(x)$,

(ii) $f(x)$ and $\tilde{f}(x)$ are either reducible or irreducible polynomials simultaneously,

(iii) For irreducible polynomials $f(x)$ and $\tilde{f}(x)$, the period of $f(x)$ and $\tilde{f}(x)$ are equal,

(iv) $f(x)$ and $\tilde{f}(x)$ are either both primitive polynomials or not simultaneously.

Generally speaking, irreducible polynomials are not necessarily primitive polynomials.

Example 7.3 The irreducible polynomial $x^2 + 1$ over $\mathbb{F}_3$ has the residue classes $\overline{x}$ and $-\overline{x}$ in $\mathbb{F}_3[x]/(x^2+1)$ as its roots. We have $(\pm\overline{x})^4 = (-1)^2 = 1$. But $4 \neq 3^2 - 1$. Therefore $x^2 + 1$ is not a primitive polynomial over $\mathbb{F}_3$.

Example 7.4 Let

$$f(x) = x^4 + x^3 + x^2 + x + 1 \in \mathbb{F}_2[x],$$

x and $x + 1$ are the only polynomials of degree 1 over $\mathbb{F}_2$, and $x^2 + x + 1$ is the only irreducible polynomial of degree 2 over $\mathbb{F}_2$. It is easy to show that $f(x)$ is not divisible by $x, x+1$, and x^2+x+1. Therefore $f(x)$ is irreducible over $\mathbb{F}_2$. Let $\overline{x}$ be the residue class containing x in the residue class field $\mathbb{F}_2[x]/(f(x))$, then $f(\overline{x}) = 0$. We have

$$\overline{x}^5 - 1 = (\overline{x} - 1)(\overline{x}^4 + \overline{x}^3 + \overline{x}^2 + \overline{x} + 1) = 0$$

and $5 \neq 2^4 - 1$. So $f(x)$ is not a primitive polynomial over $\mathbb{F}_2$.

But we have

Theorem 7.11 *Let q be a power of a prime, n be a positive integer, and assume that $q^n - 1$ is a prime. If q is odd, then $q = 3$ and $n = 1$. If $q = 2$, then n must be a prime. In particular, when $2^n - 1$ is a prime, the irreducible polynomials of degree n over $\mathbb{F}_2$ must be primitive polynomials of degree n over $\mathbb{F}_2$.*

Proof. First, let q be a power of an odd prime, then $q-1$ is an even number. But clearly

$$(q - 1) \mid (q^n - 1),$$

hence 2 is a factor of $q^n - 1$. When $q > 3$, or $q = 3$ and $n > 1$, we have $2 < q^n - 1$. Hence $q^n - 1$ is not a prime. We have proved that if q is a power of an odd prime and $q^n - 1$ is a prime, then $q = 3$ and $n = 1$.

Next let $q = 2$. If n is not a prime and $n > 1$, assume that m is a factor of n and $1 < m < n$, then

$$(2^m - 1) \mid (2^n - 1),$$

hence $2^m - 1$ is a factor of $2^n - 1$ and $1 < 2^m - 1 < 2^n - 1$. So $2^n - 1$ is not a prime. If $n = 1, 2^n - 1 = 2^1 - 1 = 1$ is not a prime either. This proves that when $2^n - 1$ is a prime, n must be a prime.

When $2^n - 1$ is a prime, the orders of the elements in $\mathbb{F}_{2^n}^*$ which are $\neq 1$, are all equal to $2^n - 1$, i.e., the elements which are $\neq 1$ in $\mathbb{F}_{2^n}^*$ are all primitive elements of $\mathbb{F}_{2^n}$. Hence the irreducible factors of $x^{2^n-1} - 1$ over $\mathbb{F}_2$, which are $\neq x - 1$, are just the minimal polynomials of primitive elements in $\mathbb{F}_{2^n}$, thus they are all primitive polynomials of degree n over $\mathbb{F}_2$. But when $2^n - 1$ is a prime, n must be a prime, hence $n > 1$. Consequently, the irreducible polynomials of degree n over $\mathbb{F}_2$ are all irreducible factors of $x^{2^n-1} - 1$, which are $\neq x - 1$, hence they are all primitive polynomials. $\square$

From Theorem 7.11, it can be seen that to discuss when $2^n - 1$ is a prime is meaningful. However, even if n is a prime, $2^n - 1$ can still be composite. For example, $2^{11} - 1 = 2047 = 23 \cdot 89$. When p is a prime and $2^p - 1$ is a prime as well, $2^p - 1$ is called a *Mersenne prime*, denoted by

$$M_p = 2^p - 1.$$

The number of Mersenne primes which were known up to 2009 is altogether 47, i.e., when

$$
\begin{aligned}
p = \ & 2, 3, 5, 7, 13, 17, 19, 31, 61, 89, 107, 127, \\
& 521, 607, 1279, 2203, 2281, 3217, 4253, \\
& 4423, 9689, 9941, 11213, 19937, 21701, \\
& 23209, 44497, 86243, 110503, 132049, \\
& 216091, 756839, 859433, 1257787, 1398269, \\
& 2976221, 3021377, 6972593, 13466917, \\
& 20996011, 24036583, 25964951, 30402457, \\
& 32582657, 37156667, 42643801, 43112609
\end{aligned}
$$

M_p is a prime. (cf. Caldwell (2009).)

7.4 Trace and Norm

Definition 7.5 *Let q be a prime power and n be a positive integer. Assume that $\mathbb{F}_q$ is a subfield of $\mathbb{F}_{q^n}$. Let σ be the Frobenius automorphism of $\mathbb{F}_{q^n}$ over $\mathbb{F}_q$. If α is an element of $\mathbb{F}_{q^n}$, its trace and norm relative to $\mathbb{F}_q$ are defined by:*

$$\mathrm{Tr}_{\mathbb{F}_{q^n}/\mathbb{F}_q}(\alpha) = \alpha + \sigma(\alpha) + \sigma^2(\alpha) + \cdots + \sigma^{n-1}(\alpha) = \alpha + \alpha^q + \alpha^{q^2} + \cdots + \alpha^{q^{n-1}}$$

and

$$\mathrm{N}_{\mathbb{F}_{q^n}/\mathbb{F}_q}(\alpha) = \alpha\sigma(\alpha)\sigma^2(\alpha)\cdots\sigma^{n-1}(\alpha) = \alpha\alpha^q\alpha^{q^2}\cdots\alpha^{q^{n-1}} = \alpha^{(q^n-1)/(q-1)},$$

respectively, which can also be written simply as

$$\mathrm{Tr}_{\mathbb{F}_{q^n}/\mathbb{F}_q}(\alpha) = \sum_{i=0}^{n-1} \sigma^i(\alpha) = \sum_{i=0}^{n-1} \alpha^{q^i}$$

and

$$\mathrm{N}_{\mathbb{F}_{q^n}/\mathbb{F}_q}(\alpha) = \prod_{i=0}^{n-1} \sigma^i(\alpha) = \prod_{i=0}^{n-1} \alpha^{q^i}.$$

If $\mathbb{F}_{q^n}$ and $\mathbb{F}_q$ are clear from the context, we simply write $\mathrm{Tr}(\alpha)$ and $\mathrm{N}(\alpha)$ for $\mathrm{Tr}_{\mathbb{F}_{q^n}/\mathbb{F}_q}(\alpha)$ and $\mathrm{N}_{\mathbb{F}_{q^n}/\mathbb{F}_q}(\alpha)$,respectively.

Theorem 7.12 *For $\alpha, \beta \in \mathbb{F}_{q^n}$ and $a \in \mathbb{F}_q$ we have*

(i)	$\mathrm{Tr}(\alpha) \in \mathbb{F}_q$;	(i')	$\mathrm{N}(\alpha) \in \mathbb{F}_q$;
(ii)	$\mathrm{Tr}(\alpha + \beta) = \mathrm{Tr}(\alpha) + \mathrm{Tr}(\beta)$;	(ii')	$\mathrm{N}(\alpha\beta) = \mathrm{N}(\alpha)\mathrm{N}(\beta)$;
(iii)	$\mathrm{Tr}(a\alpha) = a\mathrm{Tr}(\alpha)$ and,	(iii')	$\mathrm{N}(a\alpha) = a^n \mathrm{N}(\alpha)$ and,
	in particular, $\mathrm{Tr}(a) = na$;		*in particular,* $\mathrm{N}(a) = a^n$;
(iv)	$\mathrm{Tr}(\alpha^q) = \mathrm{Tr}(\alpha)$;	(iv')	$\mathrm{N}(\alpha^q) = \mathrm{N}(\alpha)$.

Proof. We prove only the statements about the trace; the proofs for the norm are similar and are left as an exercise.

(i) Let σ be the Frobenius automorphism of $\mathbb{F}_{q^n}$ over $\mathbb{F}_q$. Then $\sigma(\alpha) = \alpha^q$ for all $\alpha \in \mathbb{F}_{q^n}$ and $\mathrm{ord}(\sigma)=n$. Thus

$$\mathrm{Tr}(\alpha)^q = \sigma(\mathrm{Tr}(\alpha)) = \sigma(\alpha + \sigma(\alpha) + \sigma^2(\alpha) + \cdots + \sigma^{n-1}(\alpha))$$
$$= \sigma(\alpha) + \sigma^2(\alpha) + \cdots + \sigma^{n-1}(\alpha) + \alpha = \mathrm{Tr}(\alpha).$$

By Corollary 6.3, $\mathrm{Tr}(\alpha) \in \mathbb{F}_q$.

(ii) follows from that σ^i $(i = 0, 1, 2, \cdots)$ are automorphisms of $\mathbb{F}_{q^n}$.

(iii) follows from $\sigma^i(a) = a$ for all $a \in \mathbb{F}_q$ and $\mathrm{Tr}(1) = n$.

(iv) Since $\sigma^n(\alpha) = \alpha$ for all $\alpha \in \mathbb{F}_{q^n}$, we have

$$\mathrm{Tr}(\alpha^q) = \mathrm{Tr}(\sigma(\alpha)) = \sigma(\alpha) + \sigma(\sigma(\alpha)) + \sigma^2(\sigma(\alpha)) + \cdots + \sigma^{n-1}(\sigma(\alpha))$$
$$= \sigma(\alpha) + \sigma^2(\alpha) + \cdots + \sigma^{n-1}(\alpha) + \alpha$$
$$= \mathrm{Tr}(\alpha).$$

$\square$

By Theorem 7.12(i) and (i$'$), the maps

$$\alpha \mapsto \mathrm{Tr}(\alpha)$$

and

$$\alpha \mapsto \mathrm{N}(\alpha)$$

are maps from $\mathbb{F}_{q^n}$ to $\mathbb{F}_q$, which are called the *trace map* and the *norm map* from $\mathbb{F}_{q^n}$ to $\mathbb{F}_q$, respectively. By Theorem 7.12(ii) and (iii), Tr is a linear map. Clearly, the map $\alpha \mapsto \mathrm{N}(\alpha)$ maps $\mathbb{F}_{q^n}^*$ to $\mathbb{F}_q^*$ and by Theorem 7.12(ii$'$) N is a homomorphism from $\mathbb{F}_{q^n}^*$ to $\mathbb{F}_q^*$.

Example 7.5 Let $q=2$ and $n=3$ .We regard $\mathbb{F}_{2^3}$ as the field $\mathbb{F}_2[x]/(x^3+x+1)$. Represent the residue class of x mod x^3+x+1 by α, then $\alpha^3+\alpha+1 = 0$ and every element $\beta \in \mathbb{F}_{2^3}$ can be expressed uniquely in the form $\beta = a_0 + a_1\alpha + a_2\alpha^2$, where $a_0, a_1, a_2 \in \mathbb{F}_2$. By Theorem 7.12(ii) and (iii), to compute $\mathrm{Tr}(\beta)$ we need to compute $\mathrm{Tr}(1), \mathrm{Tr}(\alpha)$, and $\mathrm{Tr}(\alpha^2)$. We have

$$\begin{aligned}
\mathrm{Tr}(1) &= 1 + 1 + 1 = 1, \\
\mathrm{Tr}(\alpha) &= \alpha + \alpha^2 + \alpha^4 = \alpha(1 + \alpha + \alpha^3) = 0, \\
\mathrm{Tr}(\alpha^2) &= \mathrm{Tr}(\alpha) = 0 \ (\text{by Theorem 7.12(iv)}).
\end{aligned}$$

Therefore $\mathrm{Tr}(\beta) = a_0$.

For the norm, from Theorem 7.12(i$'$), $\mathrm{N}(\alpha) \in \mathbb{F}_2$ for all $\alpha \in \mathbb{F}_{2^3}$. Clearly, $\mathrm{N}(0) = 0$ and $\mathrm{N}(\alpha) \neq 0$ for all $\alpha \in \mathbb{F}_{2^3}^*$. Thus $\mathrm{N}(\alpha) = 1$ for $\alpha \in \mathbb{F}_{2^3}^*$.

Example 7.6 Let $q=2$ and $n=4$. Clearly, $x^4 + x + 1$ has no root in $\mathbb{F}_2$ and is not divisible by the unique irreducible quadratic polynomial $x^2 + x + 1$ over $\mathbb{F}_2$. Therefore $x^4 + x + 1$ is irreducible over $\mathbb{F}_2$ and we can assume that $\mathbb{F}_{2^4} = \mathbb{F}_2[x]/(x^4 + x + 1)$. As in Example 7.5 $\mathrm{N}(0) = 0$ and $\mathrm{N}(\alpha) = 1$ for $\alpha \in \mathbb{F}_{2^4}^*$.

Represent the residue class of x mod x^4+x+1 by α, then $\alpha^4+\alpha+1 = 0$ and every element β can be expressed uniquely in the form $\beta = a_0 + a_1\alpha +$

$a_2\alpha^2 + a_3\alpha^3$, where $a_0, a_1, a_2, a_3 \in \mathbb{F}_2$. We compute

$$
\begin{aligned}
\mathrm{Tr}(1) &= 1 + 1 + 1 + 1 = 0, \\
\mathrm{Tr}(\alpha) &= \alpha + \alpha^2 + \alpha^4 + \alpha^8 = \alpha + \alpha^2 + (\alpha + 1) + (\alpha + 1)^2 = 0, \\
\mathrm{Tr}(\alpha^2) &= \mathrm{Tr}(\alpha) = 0.
\end{aligned}
$$

To compute $\mathrm{Tr}(\alpha^3)$ we proceed as follows.

$$
\mathrm{Tr}(\alpha^3) = \alpha^3 + \alpha^6 + \alpha^{12} + \alpha^{24}.
$$

From $\alpha^4 = \alpha + 1$ we deduce $\alpha^6 = \alpha^3 + \alpha^2$. Thus $\alpha^3 + \alpha^6 = \alpha^2$ and $\alpha^{12} + \alpha^{24} = (\alpha^3 + \alpha^6)^4 = (\alpha^2)^4 = (\alpha^4)^2 = (\alpha + 1)^2 = \alpha^2 + 1$. Therefore $\mathrm{Tr}(\alpha^3) = 1$. Hence $\mathrm{Tr}(\beta) = a_3$.

Let d be a positive divisor of n, then $\mathbb{F}_{q^d}$ can be viewed as a subfield of $\mathbb{F}_{q^n}$ and $\mathbb{F}_q$ as a subfield of $\mathbb{F}_{q^d}$.

Theorem 7.13 *Let d be a positive divisor of n. Then for all $\alpha \in \mathbb{F}_{q^n}$*

$$
\mathrm{Tr}_{\mathbb{F}_{q^n}/\mathbb{F}_q}(\alpha) = \mathrm{Tr}_{\mathbb{F}_{q^d}/\mathbb{F}_q}(\mathrm{Tr}_{\mathbb{F}_{q^n}/\mathbb{F}_{q^d}}(\alpha)); \quad \mathrm{N}_{\mathbb{F}_{q^n}/\mathbb{F}_q}(\alpha) = \mathrm{N}_{\mathbb{F}_{q^d}/\mathbb{F}_q}(\mathrm{N}_{\mathbb{F}_{q^n}/\mathbb{F}_{q^d}}(\alpha)).
$$

Proof. We give only the proof of the first formula, since the proof of the second one is similar. Let σ be the Frobenius automorphism of $\mathbb{F}_{q^n}$ over $\mathbb{F}_q$, i.e., $\sigma(\alpha) = \alpha^q$ for all $\alpha \in \mathbb{F}_{q^n}$. By Theorem 7.1 $\mathrm{ord}(\sigma) = n$ and by Theorem 7.3 $\mathrm{Gal}(\mathbb{F}_{q^n}/\mathbb{F}_q) = \langle\sigma\rangle$. We have $\sigma^d(\alpha) = \alpha^{q^d}$ for all $\alpha \in \mathbb{F}_{q^n}$, $\mathrm{ord}(\sigma^d) = n/d$ and $\mathrm{Gal}(\mathbb{F}_{q^n}/\mathbb{F}_{q^d}) = \langle\sigma^d\rangle$. Let σ' be the restriction of σ to $\mathbb{F}_{q^d}$. Then $\sigma'(\alpha) = \alpha^q$ for all $\alpha \in \mathbb{F}_{q^d}$, $\mathrm{ord}(\sigma') = d$ and $\mathrm{Gal}(\mathbb{F}_{q^d}/\mathbb{F}_q) = \langle\sigma'\rangle$. By Theorem 7.12(i), $\mathrm{Tr}_{\mathbb{F}_{q^n}/\mathbb{F}_{q^d}}(\alpha) \in \mathbb{F}_{q^d}$ for all $\alpha \in \mathbb{F}_{q^n}$. Therefore $\mathrm{Tr}_{\mathbb{F}_{q^d}/\mathbb{F}_q}(\mathrm{Tr}_{\mathbb{F}_{q^n}/\mathbb{F}_{q^d}}(\alpha))$ is well-defined and

$$
\begin{aligned}
\mathrm{Tr}_{\mathbb{F}_{q^d}/\mathbb{F}_q}(\mathrm{Tr}_{\mathbb{F}_{q^n}/\mathbb{F}_{q^d}}(\alpha)) &= \sum_{i=0}^{d-1} \sigma'^i(\mathrm{Tr}_{\mathbb{F}_{q^n}/\mathbb{F}_{q^d}}(\alpha)) \\
&= \sum_{i=0}^{d-1} \sigma^i \Big(\sum_{j=0}^{(n/d)-1} \sigma^{jd}(\alpha) \Big) \\
&= \sum_{i=0}^{n-1} \sigma^i(\alpha) \\
&= \mathrm{Tr}_{\mathbb{F}_{q^n}/\mathbb{F}_q}(\alpha).
\end{aligned}
$$

$\square$

Theorem 7.14 *Let $\alpha \in \mathbb{F}_{q^n}$ and the minimal polynomial of α relative to $\mathbb{F}_q$ be*

$$m(x) = x^d + c_1 x^{d-1} + c_2 x^{d-2} + \cdots + c_d. \tag{7.2}$$

Then

$$\mathrm{Tr}_{\mathbb{F}_{q^n}/\mathbb{F}_q}(\alpha) = -(n/d)c_1, \quad \mathrm{N}_{\mathbb{F}_{q^n}/\mathbb{F}_q}(\alpha) = (-1)^n c_d^{n/d}.$$

Proof. By Theorem 7.5, $d \mid n$. By Theorem 7.6(ii), we have $\mathbb{F}_q[\alpha] = \mathbb{F}_{q^d}$. In particular, $\alpha \in \mathbb{F}_{q^d}$. Let σ be the Frobenius automorphism of $\mathbb{F}_{q^n}$ over $\mathbb{F}_q$. Then $\sigma^d(\alpha) = \alpha^{q^d} = \alpha$. For any m, $0 \le m \le n-1$, dividing m by d, we obtain $m = ld + r$, where $l, r \in \mathbb{Z}$ and $0 \le r < d$. Then $\sigma^m(\alpha) = \sigma^{ld+r}(\alpha) = \sigma^r(\sigma^{ld}(\alpha)) = \sigma^r(\alpha)$. Therefore

$$\begin{aligned}
\mathrm{Tr}_{\mathbb{F}_{q^n}/\mathbb{F}_q}(\alpha) &= \alpha + \sigma(\alpha) + \sigma^2(\alpha) + \cdots + \sigma^{n-1}(\alpha) \\
&= \frac{n}{d}(\alpha + \sigma(\alpha) + \sigma^2(\alpha) + \cdots + \sigma^{d-1}(\alpha)) \\
&= -(n/d)c_1.
\end{aligned}$$

Similarly,

$$\begin{aligned}
\mathrm{N}_{\mathbb{F}_{q^n}/\mathbb{F}_q}(\alpha) &= \alpha\sigma(\alpha)\sigma^2(\alpha)\cdots\sigma^{n-1}(\alpha) \\
&= (\alpha\sigma(\alpha)\sigma^2(\alpha)\cdots\sigma^{d-1}(\alpha))^{n/d} \\
&= ((-1)^d c_d)^{n/d} \\
&= (-1)^n c_d^{n/d}.
\end{aligned}$$

Example 7.7 We know that $x^4 + x^3 + x^2 + x + 1$ is irreducible over $\mathbb{F}_2$, hence $\mathbb{F}_{2^4}$ can also be regarded as the residue class field $\mathbb{F}_2[x]/(x^4 + x^3 + x^2 + x + 1)$. Let the residue class of $x \bmod x^4 + x^3 + x^2 + x + 1$ be represented by α, then $\alpha^4 + \alpha^3 + \alpha^2 + \alpha + 1 = 0$ and every element $\beta \in \mathbb{F}_{2^4}$ can be expressed uniquely as $\beta = a_0 + a_1\alpha + a_2\alpha^2 + a_3\alpha^3$. We also have $\mathrm{Tr}(1) = 0$. The minimal polynomial of α relative to $\mathbb{F}_2$ is $x^4 + x^3 + x^2 + x + 1$. By Theorem 7.14, $\mathrm{Tr}(\alpha) = 1$. By Theorem 7.12(iv), $\mathrm{Tr}(\alpha^2) = \mathrm{Tr}(\alpha) = 1$. Clearly, $\alpha^5 - 1 = (\alpha - 1)(\alpha^4 + \alpha^3 + \alpha^2 + \alpha + 1) = 0$, thus $\alpha^5 = 1$, which implies $\alpha^8 = \alpha^3$ and $\mathrm{Tr}(\alpha^3) = \mathrm{Tr}(\alpha^8) = \mathrm{Tr}(\alpha^2) = 1$. Therefore for $\beta = a_0 + a_1\alpha + a_2\alpha^2 + a_3\alpha^3$, $\mathrm{Tr}(\beta) = a_1 + a_2 + a_3$.

Let $\{v_1, v_2, \ldots, v_n\}$ be a basis of $\mathbb{F}_{q^n}$ over $\mathbb{F}_q$. For any $\alpha \in \mathbb{F}_{q^n}$, define

$$\alpha(v_1, v_2, \ldots, v_n) = (\alpha v_1, \alpha v_2, \ldots, \alpha v_n),$$

then there is an $n \times n$ matrix $M(\alpha)$ over $\mathbb{F}_q$ such that

$$(\alpha v_1, \alpha v_2, \ldots, \alpha v_n) = (v_1, v_2, \ldots, v_n)M(\alpha).$$

Denote the sum of the diagonal elements of $M(\alpha)$ by $\operatorname{Tr} M(\alpha)$. Then we have

Theorem 7.15 *Let $\alpha \in \mathbb{F}_{q^n}$, then*

$$\operatorname{Tr}_{\mathbb{F}_{q^n}/\mathbb{F}_q}(\alpha) = \operatorname{Tr} M(\alpha), \quad \operatorname{N}_{\mathbb{F}_{q^n}/\mathbb{F}_q}(\alpha) = \det M(\alpha).$$

Proof. By linear algebra, $\operatorname{Tr} M(\alpha)$ and $\det M(\alpha)$ are independent of the particular choice of the basis of $\mathbb{F}_{q^n}$ over $\mathbb{F}_q$. We can choose a basis of $\mathbb{F}_{q^n}$ over $\mathbb{F}_q$ in the following way. Let $\deg \alpha = d$, then $\mathbb{F}_q[\alpha] = \mathbb{F}_{q^d}$, $\mathbb{F}_{q^n} \supset \mathbb{F}_{q^d} \supset \mathbb{F}_q$, $n = md$, where m is a positive integer, and $\mathbb{F}_{q^n} = \mathbb{F}_{(q^d)^m}$. Clearly, $\{1, \alpha, \alpha^2, \ldots, \alpha^{d-1}\}$ is a basis of $\mathbb{F}_{q^d}$ over $\mathbb{F}_q$. Let $\{u_1, u_2, \ldots, u_m\}$ be a basis of $\mathbb{F}_{q^n}$ over $\mathbb{F}_{q^d}$. Then $\{\alpha^i u_j : 0 \le i \le d-1, 1 \le j \le m\}$ is a basis of $\mathbb{F}_{q^n}$ over $\mathbb{F}_q$ (Exercise 7.16). As in Theorem 7.14 let $m(x)$ be the minimal polynomial of α relative to $\mathbb{F}_q$ and $m(x)$ be given by (7.2). Then

$$\alpha(1, \alpha, \ldots, \alpha^{d-1}) = (1, \alpha, \ldots, \alpha^{d-1})M,$$

where

$$M = \begin{pmatrix} 0 & & & & -c_d \\ 1 & 0 & & & -c_{d-1} \\ & 1 & \ddots & & \vdots \\ & & \ddots & 0 & -c_2 \\ & & & 1 & -c_1 \end{pmatrix},$$

in which blanks denote omitted zeroes. Therefore $\operatorname{Tr} M = -c_1$ and $\det M = (-1)^d c_d$. Moreover, for each $j = 1, 2, \ldots, m$, we have

$$\alpha(u_j, \alpha u_j, \ldots, \alpha^{d-1}u_j) = (u_j, \alpha u_j, \ldots, \alpha^{d-1}u_j)M.$$

Let $M(\alpha)$ be the $n \times n$ matrix over $\mathbb{F}_q$ such that

$$\alpha(u_1, \alpha u_1, \ldots, \alpha^{d-1}u_1, \ldots, u_m, \alpha u_m, \ldots, \alpha^{d-1}u_m)$$
$$= (u_1, \alpha u_1, \ldots, \alpha^{d-1}u_1, \ldots, u_m, \alpha u_m, \ldots, \alpha^{d-1}u_m)M(\alpha),$$

then

$$M(\alpha) = \begin{pmatrix} M & & \\ & \ddots & \\ & & M \end{pmatrix},$$

where the $d \times d$ block $M's$ are m in number. Consequently,

$$\operatorname{Tr} M(\alpha) = m\operatorname{Tr} M = (n/d)(-c_1) = -(n/d)c_1 = \operatorname{Tr}_{\mathbb{F}_{q^n}/\mathbb{F}_q}(\alpha)$$

and

$$\det M(\alpha) = (\det M)^m = ((-1)^d c_d)^m = (-1)^n c_d^m = N_{\mathbb{F}_{q^n}/\mathbb{F}_q}(\alpha).$$

$\square$

Now let us come to Hilbert's Theorem 90.

Theorem 7.16   (i) *The map* $\mathrm{Tr} : \mathbb{F}_{q^n} \to \mathbb{F}_q$ *is surjective. For* $\alpha \in \mathbb{F}_{q^n}$,
$\mathrm{Tr}(\alpha) = 0$ *if and only if there exists an element* $\beta \in \mathbb{F}_{q^n}$ *such that*
$\alpha = \beta - \beta^q$.

  (ii) *The map* $\mathrm{N} : \mathbb{F}_{q^n}^* \to \mathbb{F}_q^*$ *is surjective. For* $\alpha \in \mathbb{F}_{q^n}^*$, $\mathrm{N}(\alpha) = 1$ *if and
only if there exists an element* $\beta \in \mathbb{F}_{q^n}^*$ *such that* $\alpha = \beta^{1-q}$.

Proof. We give only the proof of (i). First we prove that for $\alpha \in \mathbb{F}_{q^n}$,
$\mathrm{Tr}(\alpha)=0$ if and only if there exists an element $\beta \in \mathbb{F}_{q^n}$ such that $\alpha = \beta - \beta^q$.

Let $\alpha \in \mathbb{F}_{q^n}$ be of the form $\alpha = \beta - \beta^q$, where $\beta \in \mathbb{F}_{q^n}$. Then

$$\begin{aligned}
\mathrm{Tr}(\alpha) &= \mathrm{Tr}(\beta - \beta^q) \\
&= (\beta - \beta^q) + (\beta - \beta^q)^q + \cdots + (\beta - \beta^q)^{q^{n-1}} \\
&= 0.
\end{aligned}$$

Conversely, let $\alpha \in \mathbb{F}_{q^n}$ be such that $\mathrm{Tr}(\alpha) = 0$. Consider the polynomial
$x^q - x + \alpha \in \mathbb{F}_{q^n}[x]$. Let $p(x)$ be an irreducible factor of $x^q - x + \alpha$ over
$\mathbb{F}_{q^n}$. By Theorem 5.5 we have the residue class field $\mathbb{F}_{q^n}[x]/(p(x))$. Denote
the residue class of $x \bmod p(x)$ by β, then $\beta - \beta^q = \alpha$. Thus

$$\begin{aligned}
0 = \mathrm{Tr}(\alpha) &= \alpha + \alpha^q + \alpha^{q^2} + \cdots + \alpha^{q^{n-1}} \\
&= (\beta - \beta^q) + (\beta - \beta^q)^q + (\beta - \beta^q)^{q^2} + \cdots + (\beta - \beta^q)^{q^{n-1}} \\
&= \beta - \beta^q + \beta^q - \beta^{q^2} + \beta^{q^2} - \beta^{q^3} + \cdots + \beta^{q^{n-1}} - \beta^{q^n} \\
&= \beta - \beta^{q^n}.
\end{aligned}$$

By Corollary 6.2, $\beta \in \mathbb{F}_{q^n}$.

Then we prove the surjectivity of the map $\mathrm{Tr}: \mathbb{F}_{q^n} \to \mathbb{F}_q$. By Theorem
7.12(iii), it is enough to show that there is an element $\gamma \in \mathbb{F}_{q^n}$ such that
$\mathrm{Tr}(\gamma) \neq 0$. Then for any $a \in \mathbb{F}_q$ we have $\mathrm{Tr}(a\mathrm{Tr}(\gamma)^{-1}\gamma) = a\mathrm{Tr}(\gamma)^{-1}\mathrm{Tr}(\gamma) =$
a. Let K be the set of elements $\alpha \in \mathbb{F}_{q^n}$ such that $\mathrm{Tr}(\alpha) = 0$. By what we
have proved above, $K = \{\beta - \beta^q : \beta \in \mathbb{F}_{q^n}\}$. Clearly, $\beta - \beta^q = \beta' - \beta'^q$
for $\beta, \beta' \in \mathbb{F}_{q^n}$ if and only if $\beta' - \beta \in \mathbb{F}_q$, i.e., β and β' belong to the same
coset relative to $\mathbb{F}_q$ in the additive group $\mathbb{F}_{q^n}$. It follows that $|K| = \mathbb{F}_{q^n} : \mathbb{F}_q$
$= q^{n-1}$. Therefore there are elements $\gamma \in \mathbb{F}_{q^n}$ such that $\mathrm{Tr}(\gamma) \neq 0$. $\square$

Corollary 7.17 (i) *For any $a \in \mathbb{F}_q$ the number of elements $\alpha \in \mathbb{F}_{q^n}$ such that $\mathrm{Tr}(\alpha) = a$ is q^{n-1}.*

(ii) *For any $a \in \mathbb{F}_q^*$, the number of elements $\alpha \in \mathbb{F}_{q^n}^*$ such that $\mathrm{N}(\alpha) = a$ is $(q^{n-1} - 1)/(q - 1)$.*

Proof. We give only the proof of (i). Let K be the set of elements $\alpha \in \mathbb{F}_{q^n}$ such that $\mathrm{Tr}(\alpha) = 0$. We proved $|K| = q^{n-1}$. If $\alpha \in \mathbb{F}_{q^n}$ and $\mathrm{Tr}(\alpha) = a$, then $\mathrm{Tr}(\alpha + \beta) = a$ for all $\beta \in K$. If $\alpha' \in \mathbb{F}_q$ such that $\mathrm{Tr}(\alpha') = a$, then $\mathrm{Tr}(\alpha' - \alpha) = 0$, thus $\alpha' - \alpha = \beta \in K$ and $\alpha' = \alpha + \beta$. Therefore $\alpha + \beta$ ($\beta \in K$) are all the elements of $\mathbb{F}_{q^n}$ such that $\mathrm{Tr}(\alpha + \beta) = a$ and they are q^{n-1} in number. $\qquad\square$

The proof of the second statement of Theorem 7.16(i) above is not constructive in the sense that it gives no effective procedure for solving the equation $\alpha = \beta - \beta^q$ for β. Hillbert gave a constructive proof based on the following lemma.

Lemma 7.18 *Let $\alpha, \theta \in \mathbb{F}_{q^n}$ and β be defined by*

$$\beta = \alpha\theta^q + (\alpha + \alpha^q)\theta^{q^2} + \cdots + (\alpha + \alpha^q + \alpha^{q^2} + \cdots + \alpha^{q^{n-2}})\theta^{q^{n-1}}. \quad (7.3)$$

Then

$$\beta - \beta^q = \alpha(\mathrm{Tr}(\theta) - \theta) - \theta(\mathrm{Tr}(\alpha) - \alpha).$$

Proof. From (7.3) we deduce

$$\beta^q = \alpha^q\theta^{q^2} + (\alpha^q + \alpha^{q^2})\theta^{q^3} + \cdots + (\alpha^q + \alpha^{q^2} + \cdots + \alpha^{q^{n-1}})\theta^{q^n}.$$

By substraction,

$$\beta - \beta^q = \alpha(\theta^q + \theta^{q^2} + \cdots + \theta^{q^{n-1}}) - (\alpha^q + \alpha^{q^2} + \cdots + \alpha^{q^{n-1}})\theta^{q^n}.$$

But $\theta^{q^n} = \theta$,

$$\theta^q + \theta^{q^2} + \cdots + \theta^{q^{n-1}} = \mathrm{Tr}(\theta) - \theta,$$

and

$$\alpha^q + \alpha^{q^2} + \cdots + \alpha^{q^{n-1}} = \mathrm{Tr}(\alpha) - \alpha.$$

$\qquad\square$

Now we can give Hilbert's constructive proof of the second statement of Theorem 7.16(i). By the first statement of Theorem 7.16(i), we can choose θ to be a fixed element of $\mathbb{F}_{q^n}$ with $\mathrm{Tr}(\theta) = 1$. Then if $\mathrm{Tr}(\alpha) = 0$, the element β defined by (7.3) satisfies $\beta - \beta^q = \alpha$.

7.5   Quadratic Equations

Consider the quadratic equation

$$ax^2 + bx + c = 0$$

over a field F, where $a, b, c \in F$ and $a \neq 0$. If F is not of characteristic 2, we have the well-known formula

$$x = \frac{-b \pm \sqrt{(b^2 - 4ac)}}{2a}$$

for its solutions. If F is of characteristic 2, the above formula breaks down. In the present section we shall study the solution of the quadratic equation over the finite field $\mathbb{F}_{2^n}$.

Let us start with a special case. Consider the quadratic equation

$$x^2 + x + \alpha = 0 \tag{7.4}$$

over $\mathbb{F}_{2^n}$, where $\alpha \in \mathbb{F}_{2^n}$. By Theorem 7.16(i), (7.4) is solvable in $\mathbb{F}_{2^n}$ if and only if $\mathrm{Tr}(\alpha) = 0$. Here and after we use Tr to denote $\mathrm{Tr}_{\mathbb{F}_{2^n}/\mathbb{F}_2}$. If $\mathrm{Tr}(\alpha) = 0$, we choose $\theta \in \mathbb{F}_{2^n}$ such that $\mathrm{Tr}(\theta) = 1$. By Lemma 7.18,

$$\beta = \alpha \theta^2 + (\alpha + \alpha^2)\theta^{2^2} + \cdots + (\alpha + \alpha^2 + \cdots + \alpha^{2^{n-2}})\theta^{2^{n-1}}$$

is a solution of (7.4). Clearly, $\beta + 1$ is another solution of (7.4). We have proved

Lemma 7.19 *Let $\alpha, \theta \in \mathbb{F}_{2^n}$ be such that $\mathrm{Tr}(\alpha) = 0$ and $\mathrm{Tr}(\theta) = 1$. Then the quadratic equation*
$$x^2 + x + \alpha = 0$$
has solutions

$$\beta = \alpha \theta^2 + (\alpha + \alpha^2)\theta^{2^2} + \cdots + (\alpha + \alpha^2 + \cdots + \alpha^{2^{n-2}})\theta^{2^{n-1}}$$

and $\beta + 1$.

Example 7.8 In Example 7.5 we viewed $\mathbb{F}_{2^3} = \mathbb{F}_2[x]/(x^3 + x + 1)$ and denoted the residue class of x mod $x^3 + x + 1$ by α. Then $\alpha^3 + \alpha + 1 = 0$ and we have computed $\mathrm{Tr}(1) = 1, \mathrm{Tr}(\alpha) = \mathrm{Tr}(\alpha^2) = 0$. For any $\beta = a_0 + a_1\alpha + a_2\alpha^2$, where $a_0, a_1, a_2 \in \mathbb{F}_2$, we have $\mathrm{Tr}(\beta) = a_0$. Therefore by Theorem 7.16(i) the quadratic equation

$$x^2 + x + \beta = 0$$

is solvable in $\mathbb{F}_{2^3}$ if and only if $\text{Tr}(\beta) = 0$, and, hence, if and only if $\beta = a_1\alpha + a_2\alpha^2$. We can take simply $\theta = 1$. By Lemma 7.19 $x^2 + x + \beta = 0$ has β^2 as one of its solution.

More generally, in $\mathbb{F}_{2^n}$ with odd n we always have $\text{Tr}(1) = 1$ and we can simply take $\theta = 1$. For any $\alpha \in \mathbb{F}_{2^n}$ with $\text{Tr}(\alpha) = 0$, by Lemma 7.19 the equation $x^2 + x + \alpha = 0$ has

$$\alpha^2 + \alpha^{2^3} + \alpha^{2^5} + \cdots + \alpha^{2^{n-2}}$$

as one of its solutions.

Example 7.9 In Example 7.6 we viewed $\mathbb{F}_{2^4} = \mathbb{F}_2[x]/(x^4 + x + 1)$ and denoted the residue class of $x \bmod x^4 + x + 1$ by α. Then $\alpha^4 + \alpha + 1 = 0$ and we have computed

$$\text{Tr}(1) = \text{Tr}(\alpha) = \text{Tr}(\alpha^2) = 0, \ \text{Tr}(\alpha^3) = 1.$$

For any $\beta = a_0 + a_1\alpha + a_2\alpha^2 + a_3\alpha^3$, where $a_0, a_1, a_2, a_3 \in \mathbb{F}_2$, we have $\text{Tr}(\beta) = a_3$. Therefore by Theorem 7.16(i) the quadratic equation

$$x^2 + x + \beta = 0$$

is solvable in $\mathbb{F}_{2^4}$ if and only if $\text{Tr}(\beta) = 0$, and, hence, if and only if $\beta = a_0 + a_1\alpha + a_2\alpha^2$. In Lemma 7.19 we take $\theta = \alpha^3$. Then the equation $x^2 + x + 1 = 0$ has a solution $\alpha^2 + \alpha$, the equation $x^2 + x + \alpha = 0$ has a solution $\alpha^3 + \alpha + 1$, and the equation $x^2 + x + \alpha^2 = 0$ has a solution $\alpha^3 + 1$. Hence, the equation $x^2 + x + \beta = 0$ where $\beta = a_0 + a_1\alpha + a_2\alpha^2$ and $a_0, a_1, a_2 \in \mathbb{F}_2$, has a solution $a_0(\alpha^2 + \alpha) + a_1(\alpha^3 + \alpha^2 + 1) + a_2(\alpha^3 + 1)$, and the other solution is $1 + a_0(\alpha^2 + \alpha) + a_1(\alpha^3 + \alpha^2 + 1) + a_2(\alpha^3 + 1)$.

For the general case, we have

Theorem 7.20 *Let*

$$ax^2 + bx + c = 0 \tag{7.5}$$

be a quadratic equation over $\mathbb{F}_{2^n}$, *where* $a, b, c \in \mathbb{F}_{2^n}$ *and* $a \neq 0$. *When* $b = 0$, *(7.5) has a unique solution* $x = (c/a)^{2^{n-1}}$. *When* $b \neq 0$, *(7.5) has*

$$\begin{cases} \text{no solution,} & \text{if } \text{Tr}(ac/b^2) = 1, \\ \text{two solutions,} & \text{if } \text{Tr}(ac/b^2) = 0. \end{cases}$$

Proof. When $b = 0$, (7.5) becomes

$$x^2 = c/a. \tag{7.6}$$

Clearly, $(c/a)^{2^{n-1}}$ is a solution of (7.6). Let x_0 be any solution of (7.6), then $x_0^2 = c/a$. Raising both sides to 2^{n-1}-th power, we obtain $x_0 = (c/a)^{2^{n-1}}$. Therefore (7.6) has a unique solution, so does (7.5).

When $b \neq 0$, if we set $y = (a/b)x$, (7.5) becomes

$$y^2 + y = ac/b^2. \tag{7.7}$$

By Lemma 7.19, (7.7) is solvable and has two solutions if and only if $\text{Tr}(ac/b^2) = 0$, so is (7.5). $\qquad\square$

7.6 Exercises

7.1 Regard $\mathbb{F}_{16} = \mathbb{F}_2[x]/(x^4 + x + 1)$. Compute the minimal polynomials over $\mathbb{F}_2$ of the elements of $\mathbb{F}_{16}$ and point out the primitive ones.

7.2 Do the same thing for the field $\mathbb{F}_{27} = \mathbb{F}_3[x]/(x^3 + 2x + 1)$.

7.3 Do the same thing for the field $\mathbb{F}_{25} = \mathbb{F}_5[x]/(x^2 - 3)$.

7.4 Compute the minimal polynomials over $\mathbb{F}_4$ of the elements of $\mathbb{F}_{16}$.

7.5 Let $f(x)$ be a monic polynomial of degree $n \geq 1$ over $\mathbb{F}_q$ and assume that $f(0) \neq 0$. Prove that $f(x)$ is a primitive polynomial over $\mathbb{F}_q$ if and only if $q^n - 1$ is the smallest integer such that $f(x) \mid (x^{q^n-1} - 1)$.

7.6 Let $\{v_1, v_2, \ldots, v_n\}$ be a basis of $\mathbb{F}_{q^n}$ over $\mathbb{F}_q$. For any $\alpha \in \mathbb{F}_{q^n}$, there are elements $a_{ij} \in \mathbb{F}_q$ such that

$$\alpha v_j = \sum_{i=1}^{n} v_i a_{ij}.$$

Let

$$M(\alpha) = (a_{ij})_{1 \leq i,j \leq n},$$

which is an $n \times n$ matrix over $\mathbb{F}_q$. Prove that the characteristic (and minimal) polynomial of α over $\mathbb{F}_q$ is equal to the characteristic (and minimal) polynomial of the matrix $M(\alpha)$, (respectively).

7.7 Give the proofs for the norm in Theorem 7.12.

7.8 Prove Theorem 7.13 for the norm.

7.9 Prove that $x^5 + x^2 + 1$ is irreducible over $\mathbb{F}_2$. Let $\mathbb{F}_{2^5} = \mathbb{F}_2[x]/(x^5 + x^2 + 1)$, then give a formula for the trace in $\mathbb{F}_{2^5}$.

7.10 Prove that every function $f : \mathbb{F}_q \to \mathbb{F}_q$ may be represented uniquely by a polynomial of degree $\leq q - 1$ and conversely, if R is a commutative ring with identity such that every function $R \to R$ is given by a polynomial then R is a finite ring.

7.11 Let $\mathbb{F}_q$ and $\mathbb{F}_{q^n}$ be finite fields with q and q^n elements, respectively, and assume that $\mathbb{F}_q \subset \mathbb{F}_{q^n}$. Let $\{\alpha_1, \alpha_2, \ldots, \alpha_n\}$ be a basis of $\mathbb{F}_{q^n}$ over $\mathbb{F}_q$, then any $\beta \in \mathbb{F}_{q^n}$ can be expressed uniquely as $\beta = b_1\alpha_1 + b_2\alpha_2 + \cdots + b_n\alpha_n$, where $b_1, b_2, \ldots, b_n \in \mathbb{F}_q$. Let $(c_1, c_2, \ldots, c_n)$ be a nonzero vector in $\mathbb{F}_q^n$. Define $T(\beta) = b_1c_1 + b_2c_2 + \cdots + b_nc_n$, then T satisfies properties (i), (ii), and (iii). Prove that $T(\beta^q) = T(\beta)$ for all $\beta \in \mathbb{F}_{q^n}$ if and only if $T(\beta) = \lambda\mathrm{Tr}(\beta)$ for some $\lambda \in \mathbb{F}_q^*$.

7.12 Assume $\mathbb{F}_q$ is a subfield of $\mathbb{F}_{q^n}$. Prove that every linear map from $\mathbb{F}_{q^n}$ to $\mathbb{F}_q$ is of the form $\alpha \mapsto \mathrm{Tr}(\beta\alpha)$ for some $\beta \in \mathbb{F}_{q^n}$.

7.13 In the field $\mathbb{F}_{3^3} = \mathbb{F}_3[x]/(x^3 + 2x + 1)$, let α be the residue class of $x \bmod x^3 + 2x + 1$. For any $\beta \in \mathbb{F}_{3^3}$, let $\beta = \beta_0 + \beta_1\alpha + \beta_2\alpha^2$, where $\beta_0, \beta_1, \beta_2 \in \mathbb{F}_3$. Find a vector $(c_0, c_1, c_2) \in \mathbb{F}_3^3$ such that $\mathrm{Tr}(\beta) = \beta_0c_0 + \beta_1c_1 + \beta_2c_2$ for all $\beta \in \mathbb{F}_{3^3}$.

7.14 Prove Theorem 7.16 and Corollary 7.17 for the norm.

7.15 Let $\alpha \in \mathbb{F}_{q^n}$ and the minimal polynomial of α relative to $\mathbb{F}_q$ be (7.2). Prove that

$$\mathrm{Tr}_{\mathbb{F}_{q^n}/\mathbb{F}_q}(\alpha^{-1}) = -\frac{n}{d}c_d^{-1}c_{d-1}, \quad \mathrm{N}_{\mathbb{F}_{q^n}/\mathbb{F}_q}(\alpha^{-1}) = (-1)^n c_d^{-n/d}.$$

7.16 Let $\mathbb{F}_q$ be a subfield of $\mathbb{F}_{q^d}$ and $\mathbb{F}_{q^d}$ be a subfield of $\mathbb{F}_{q^n}$. Then $n = md$, where m is an integer, $\mathbb{F}_{q^n}$ is a vector space of dimension m over $\mathbb{F}_{q^d}$, and $\mathbb{F}_{q^d}$ is a vector space of dimension d over $\mathbb{F}_q$. Let $\{u_i : 1 \leq i \leq m\}$ be a basis of $\mathbb{F}_{q^n}$ over $\mathbb{F}_{q^d}$ and $\{v_j : \leq j \leq d\}$ be a basis of $\mathbb{F}_{q^d}$ over $\mathbb{F}_q$. Prove that $\{u_iv_j : 1 \leq i \leq m, 1 \leq j \leq d\}$ is a basis of $\mathbb{F}_{q^n}$ over $\mathbb{F}_q$.

7.17 Let $\mathbb{F}_{2^5} = \mathbb{F}_2[x]/(x^5+x^2+1)$ and denote the residue class of $x \bmod x^5 + x^2 + 1$ by α. For any $\gamma \in \mathbb{F}_{2^5}$, let $\gamma = \gamma_0 + \gamma_1\alpha + \gamma_2\alpha^2 + \gamma_3\alpha^3 + \gamma_4\alpha^4$, where $\gamma_i \in \mathbb{F}_2$. Find necessary and sufficient conditions on the element γ in terms of γ_i such that the equation $y^2 + y = \gamma$ has two solutions in $\mathbb{F}_{2^5}$ and give a general solution of the equation when it is solvable.

7.18 Do the same for $\mathbb{F}_{2^6} = \mathbb{F}_2[x]/(x^6 + x + 1)$ and the equation $y^2 + y = \gamma$ for $\gamma \in \mathbb{F}_{2^6}$.

7.19 Given elements $\alpha, \theta \in \mathbb{F}_{q^n}$ and define

$$\beta = \theta + \alpha\theta^q + \alpha^{1+q}\theta^{q^2} + \cdots + \alpha^{1+q+\cdots+q^{n-2}}\theta^{q^{n-1}}.$$

   (i) Prove that $\alpha\beta^q + \theta = \beta + \mathrm{N}(\alpha)\theta$.

   (ii) Prove that for any given $\alpha \in \mathbb{F}_{q^n}$, there exists $\theta \in \mathbb{F}_{q^n}$ such that $\beta \neq 0$.

   (iii) Show that if $\mathrm{N}(\alpha) = 1$, there exists $\alpha, \beta \in \mathbb{F}_{q^n}$ such that $\alpha = \beta^{1-q}$.

This is the Hilbert's constructive proof of the second statement of Theorem 7.16(ii) for the norm.

7.20 Let q be a prime power and $n \geq 1$. Prove that the map $\mathrm{N}_{\mathbb{F}_{q^n}/\mathbb{F}_q} : \mathbb{F}_{q^n}^* \to \mathbb{F}_q^*$ carries $\mathbb{F}_{q^n}^{*2}$ onto $\mathbb{F}_q^{*2}$.

Chapter 8

Bases

8.1 Bases and Polynomial Bases

Let $\mathbb{F}_{q^n}$ and $\mathbb{F}_q$ be finite fields with q^n and q elements, respectively, where q is a prime power and n is a positive integer. We assume that $\mathbb{F}_q$ is a subfield of $\mathbb{F}_{q^n}$. We can view $\mathbb{F}_{q^n}$ as a vector space over $\mathbb{F}_q$, if we define the scalar multiplication of an element $\alpha \in \mathbb{F}_{q^n}$ by a scalar $a \in \mathbb{F}_q$ as follows:

$$\begin{aligned} \mathbb{F}_q \times \mathbb{F}_{q^n} &\longrightarrow \mathbb{F}_{q^n} \\ (a, \alpha) &\longmapsto a\alpha. \end{aligned}$$

Assume $\dim_{\mathbb{F}_q} \mathbb{F}_{q^n} = m$ and let $\{\alpha_1, \alpha_2, \ldots, \alpha_m\}$ be a *basis* of the vector space $\mathbb{F}_{q^n}$ over $\mathbb{F}_q$, i.e., a maximal set of linearly independent elements of $\mathbb{F}_{q^n}$ over $\mathbb{F}_q$. Then any element $\beta \in \mathbb{F}_{q^n}$ can be expressed uniquely as a linear combination of $\alpha_1, \alpha_2, \ldots, \alpha_m$ with coefficients in $\mathbb{F}_q$:

$$\beta = b_1\alpha_1 + b_2\alpha_2 + \cdots + b_m\alpha_m, \text{ where } b_i \in \mathbb{F}_q \text{ for } i = 1, 2, \cdots, m.$$

Thus $|\mathbb{F}_{q^n}| = q^m$, which implies $m = n$.

Theorem 8.1 $\mathbb{F}_{q^n}$ *is a vector space over* $\mathbb{F}_q$ *and* $\dim_{\mathbb{F}_q} \mathbb{F}_{q^n} = n$.

Therefore a basis of $\mathbb{F}_{q^n}$ over $\mathbb{F}_q$ has cardinality n and n is called the *degree* of $\mathbb{F}_{q^n}$ over $\mathbb{F}_q$, which is denoted by $[\mathbb{F}_{q^n} : \mathbb{F}_q] = n$.

Theorem 8.2 *Let* $\{\alpha_1, \alpha_2, \ldots, \alpha_n\}$ *be a basis of* $\mathbb{F}_{q^n}$ *over* $\mathbb{F}_q$, $(a_{ij})_{1 \leq i,j \leq n}$ *be an* $n \times n$ *matrix over* $\mathbb{F}_q$, *and* $\beta_j = \sum_{i=1}^n a_{ij}\alpha_i$, $j = 1, 2, \ldots, n$. *Then* $\{\beta_1, \beta_2, \ldots, \beta_n\}$ *is also a basis of* $\mathbb{F}_{q^n}$ *over* $\mathbb{F}_q$ *if and only if* $(a_{ij})_{1 \leq i,j \leq n}$ *is nonsingular.*

Proof. We prove only the "if" part. Assume $(a_{ij})_{1 \le i,j \le n}$ is nonsingular. Let $c_1\beta_1 + c_2\beta_2 + \cdots + c_n\beta_n = 0$, where $c_1, c_2, \ldots, c_n \in \mathbb{F}_q$. Then

$$0 = \sum_{j=1}^{n} c_j\beta_j = \sum_{j=1}^{n} c_j\left(\sum_{i=1}^{n} a_{ij}\alpha_i\right) = \sum_{i=1}^{n}\left(\sum_{j=1}^{n} c_j a_{ij}\right)\alpha_i.$$

Since $\{\alpha_1, \alpha_2, \ldots, \alpha_n\}$ is a basis of $\mathbb{F}_{q^n}$ over $\mathbb{F}_q$, $\sum_{j=1}^{n} c_j a_{ij} = 0$, $i = 1, 2, \ldots, n$. But $(a_{ij})_{1 \le i,j \le n}$ is nonsingular, $c_1 = c_2 = \cdots = c_n = 0$. This proves that $\beta_1, \beta_2, \ldots, \beta_n$ are linearly independent over $\mathbb{F}_q$. Since $\{\beta_1, \beta_2, \ldots, \beta_n\}$ has cardinality n, it is a basis of $\mathbb{F}_{q^n}$ over $\mathbb{F}_q$. $\square$

Let $\{\alpha_1, \alpha_2, \ldots, \alpha_n\}$ be a basis of $\mathbb{F}_{q^n}$ over $\mathbb{F}_q$. Then any element $\beta \in \mathbb{F}_{q^n}$ can be expressed uniquely as a linear combination of $\alpha_1, \alpha_2, \ldots, \alpha_n$ with coefficients in $\mathbb{F}_q$:

$$\beta = b_1\alpha_1 + b_2\alpha_2 + \cdots + b_n\alpha_n, \text{ where } b_i \in \mathbb{F}_q \text{ for } i = 1, 2, \ldots, n.$$

The elements $b_1, b_2, \ldots, b_n$ are called the *coordinates* of β in the basis $\alpha_1, \alpha_2, \ldots, \alpha_n$.

We shall give some criteria for n elements of $\mathbb{F}_{q^n}$ to be a basis of $\mathbb{F}_{q^n}$ over $\mathbb{F}_q$. First, we give the following definition.

Definition 8.1 *Let $\alpha_1, \alpha_2, \ldots, \alpha_n \in \mathbb{F}_{q^n}$. The determinant*

$$\begin{vmatrix} \mathrm{Tr}(\alpha_1\alpha_1) & \mathrm{Tr}(\alpha_1\alpha_2) & \cdots & \mathrm{Tr}(\alpha_1\alpha_n) \\ \mathrm{Tr}(\alpha_2\alpha_1) & \mathrm{Tr}(\alpha_2\alpha_2) & \cdots & \mathrm{Tr}(\alpha_2\alpha_n) \\ \vdots & \vdots & & \vdots \\ \mathrm{Tr}(\alpha_n\alpha_1) & \mathrm{Tr}(\alpha_n\alpha_2) & \cdots & \mathrm{Tr}(\alpha_n\alpha_n) \end{vmatrix}, \tag{8.1}$$

where Tr *represents* $\mathrm{Tr}_{\mathbb{F}_{q^n}/\mathbb{F}_q}$, *is called the discriminant of the elements* $\alpha_1, \alpha_2, \ldots, \alpha_n$ *and is denoted by* $\Delta_{\mathbb{F}_{q^n}/\mathbb{F}_q}(\alpha_1, \alpha_2, \ldots, \alpha_n)$ *or, simply, by* $\Delta(\alpha_1, \alpha_2, \ldots, \alpha_n)$.

Then we have

Theorem 8.3 *Let $\alpha_1, \alpha_2, \ldots, \alpha_n \in \mathbb{F}_{q^n}$. Then $\{\alpha_1, \alpha_2, \ldots, \alpha_n\}$ is a basis of $\mathbb{F}_{q^n}$ over $\mathbb{F}_q$ if and only if their discriminant $\Delta(\alpha_1, \alpha_2, \ldots, \alpha_n) \ne 0$.*

Proof. Let $\{\alpha_1, \alpha_2, \ldots, \alpha_n\}$ be a basis of $\mathbb{F}_{q^n}$ over $\mathbb{F}_q$. If we can show that the n row vectors of (8.1) are linearly independent over $\mathbb{F}_q$, then $\Delta(\alpha_1, \alpha_2, \ldots, \alpha_n) \ne 0$. Suppose

$$b_1\mathrm{Tr}(\alpha_1\alpha_j) + b_2\mathrm{Tr}(\alpha_2\alpha_j) + \cdots + b_n\mathrm{Tr}(\alpha_n\alpha_j) = 0 \text{ for } j = 1, 2, \ldots, n,$$

where $b_1, b_2, \ldots, b_n \in \mathbb{F}_q$. Let $\beta = b_1\alpha_1 + b_2\alpha_2 + \cdots + b_n\alpha_n$, then $\mathrm{Tr}(\beta\alpha_j) = 0$ for $j = 1, 2, \ldots, n$. For any $\alpha \in \mathbb{F}_{q^n}$, we can express α as $\alpha = a_1\alpha_1 + a_2\alpha_2 + \cdots + a_n\alpha_n$, where $a_1, a_2, \ldots, a_n \in \mathbb{F}_q$. Then $\mathrm{Tr}(\beta\alpha) = 0$ for all $\alpha \in \mathbb{F}_{q^n}$. If $\beta \neq 0$, substituting $\beta^{-1}\alpha$ for α in $\mathrm{Tr}(\beta\alpha) = 0$, we obtain $\mathrm{Tr}(\alpha) = 0$ for all $\alpha \in \mathbb{F}_{q^n}$. By Theorem 7.16(i) Tr : $\mathbb{F}_{q^n} \to \mathbb{F}_q$ is surjective, thus we arrive at a contradiction. Therefore we must have $\beta = 0$ and, hence, $b_1 = b_2 = \cdots = b_n = 0$. This proves that the n row vectors of (8.1) are linearly independent.

Conversely, suppose $\Delta(\alpha_1, \alpha_2, \ldots, \alpha_n) \neq 0$. Assume $c_1\alpha_1 + c_2\alpha_2 + \cdots + c_n\alpha_n = 0$ for some $c_1, c_2, \ldots, c_n \in \mathbb{F}_q$. Then

$$c_1\alpha_1\alpha_j + c_2\alpha_2\alpha_j + \cdots + c_n\alpha_n\alpha_j = 0 \text{ for } j = 1, 2, \ldots, n.$$

Applying the trace map, we obtain

$$c_1\mathrm{Tr}(\alpha_1\alpha_j) + c_2\mathrm{Tr}(\alpha_2\alpha_j) + \cdots + c_n\mathrm{Tr}(\alpha_n\alpha_j) = 0 \text{ for } j = 1, 2, \ldots, n.$$

The determinant of the above system of linear homogeneous equations in $c_1, c_2, \ldots, c_n$ is $\Delta(\alpha_1, \alpha_2, \ldots, \alpha_n)$, which is supposed to be nonzero, therefore we must have $c_1 = c_2 = \cdots = c_n = 0$. Hence $\alpha_1, \alpha_2, \cdots, \alpha_n$ are linearly independent over $\mathbb{F}_q$. $\square$

Corollary 8.4 *Let $\alpha_1, \alpha_2, \ldots, \alpha_n \in \mathbb{F}_{q^n}$. Then $\{\alpha_1, \alpha_2, \ldots, \alpha_n\}$ is a basis of $\mathbb{F}_{q^n}$ over $\mathbb{F}_q$ if and only if*

$$\begin{vmatrix} \alpha_1 & \alpha_2 & \cdots & \alpha_n \\ \alpha_1^q & \alpha_2^q & \cdots & \alpha_n^q \\ \vdots & \vdots & & \vdots \\ \alpha_1^{q^{n-1}} & \alpha_2^{q^{n-1}} & \cdots & \alpha_n^{q^{n-1}} \end{vmatrix} \neq 0. \tag{8.2}$$

Proof. Denote the determinant of (8.2) by D. Computing D^2 column by column, we obtain $D^2 = \Delta(\alpha_1, \alpha_2, \ldots, \alpha_n)$. Thus Corollary 8.4 follows from Theorem 8.3. $\square$

Now let us study how to use bases to perform algebraic operations in finite fields.

Let $\{\alpha_1, \alpha_2, \ldots, \alpha_n\}$ be a basis of $\mathbb{F}_{q^n}$ over $\mathbb{F}_q$. Let $\beta, \gamma \in \mathbb{F}_{q^n}$. We express β and γ in the basis $\{\alpha_1, \alpha_2, \ldots, \alpha_n\}$ as follows:

$$\beta = b_1\alpha_1 + b_2\alpha_2 + \cdots + b_n\alpha_n, \text{ where } b_i \in \mathbb{F}_q \text{ for } i = 1, 2, \ldots, n,$$

$$\gamma = c_1\alpha_1 + c_2\alpha_2 + \cdots + c_n\alpha_n, \text{ where } c_i \in \mathbb{F}_q \text{ for } i = 1, 2, \ldots, n.$$

The addition and substraction are easy:

$$\beta \pm \gamma = (b_1 \pm c_1)\alpha_1 + (b_2 \pm c_2)\alpha_2 + \cdots + (b_n \pm c_n)\alpha_n.$$

That is, to obtain the coordinates of $\beta \pm \gamma$ in the basis $\{\alpha_1, \alpha_2, \ldots, \alpha_n\}$, we need only to perform the addition or substraction of the coordinates: $b_i \pm c_i$, $i = 1, 2, \ldots, n$. For the multiplication, we have

$$\beta\gamma = \sum_{i=1}^{n} b_i\alpha_i \sum_{j=1}^{n} c_j\alpha_j$$

$$= \sum_{i=1}^{n}\sum_{j=1}^{n} b_i c_j \alpha_i \alpha_j.$$

We can express $\alpha_i\alpha_j$ as

$$\alpha_i\alpha_j = \sum_{k=1}^{n} c_{ijk}\alpha_k, \text{ where } c_{ijk} \in \mathbb{F}_q,$$

then

$$\beta\gamma = \sum_{i=1}^{n}\sum_{j=1}^{n}\sum_{k=1}^{n} b_i c_j c_{ijk}\alpha_k$$

$$= \sum_{k=1}^{n}(\sum_{i=1}^{n}\sum_{j=1}^{n} b_i c_j c_{ijk})\alpha_k.$$

Therefore we have to compute n^3 constant $c_{ijk}'s$ first and then evaluate

$$\sum_{i=1}^{n}\sum_{j=1}^{n} b_i c_j c_{ijk}.$$

Of course we hope to have some bases for which the n^3 constants $c_{ijk}'s$ are easier to compute.

Let us choose an element $\alpha \in \mathbb{F}_{q^n}$ which is of degree n over $\mathbb{F}_q$. Then

$$\{1, \alpha, \alpha^2, \ldots, \alpha^{n-1}\}$$

is a basis of $\mathbb{F}_{q^n}$ over $\mathbb{F}_q$. Such a basis is called a *polynomial basis*. Let $\beta, \gamma \in \mathbb{F}_{q^n}$. Express them in terms of the polynomial basis $\{1, \alpha, \alpha^2, \ldots, \alpha^{n-1}\}$:

$$\beta = b_0 + b_1\alpha + b_2\alpha^2 + \cdots + b_{n-1}\alpha^{n-1}, \text{ where } b_i \in \mathbb{F}_q \text{ for } i = 0, 1, 2, \ldots, n-1,$$

$\gamma = c_0 + c_1\alpha + c_2\alpha^2 + \cdots + c_{n-1}\alpha^{n-1}$, where $c_i \in \mathbb{F}_q$ for $i = 0, 1, 2, \ldots, n-1$.

To perform the multiplication $\beta\gamma$, we have to compute

$$\alpha^i \alpha^j = \sum_{k=1}^{n} c_{ijk}\alpha^k.$$

Clearly, some $c_{ijk}'s$ are zeros, so performing the multiplication in a polynomial basis is easier than in a general basis. Usually we can perform the multiplication much easier as follows. We introduce

$$b(x) = b_0 + b_1 x + b_2 x^2 + \cdots + b_{n-1}x^{n-1},$$

$$c(x) = c_0 + c_1 x + c_2 x^2 + \cdots + c_{n-1}x^{n-1}.$$

Then $b(\alpha) = \beta$ and $c(\alpha) = \gamma$. Let $f(x)$ be the minimal polynomial of α over $\mathbb{F}_q$. By Theorem 7.5, $f(x)$ is irreducible over $\mathbb{F}_q$. Since α is of degree n over $\mathbb{F}_q$, $\deg f(x) = n$. Dividing $b(x)c(x)$ by $f(x)$, we obtain

$$b(x)c(x) = q(x)f(x) + r(x),$$

where $q(x), r(x) \in \mathbb{F}_q[x]$ and $\deg r(x) < \deg f(x)$. Substituting $x = \alpha$ into the above equation, we obtain

$$\beta\gamma = b(\alpha)c(\alpha) = r(\alpha).$$

Let $\beta \neq 0$. To compute the inverse of β, notice that $\gcd(b(x), f(x)) = 1$. Performing the Euclidean algorithm to $b(x)$ and $f(x)$, we obtain

$$g(x)b(x) + h(x)f(x) = 1,$$

where $g(x), h(x) \in \mathbb{F}_q[x]$ and $\deg g(x) < \deg f(x)$, $\deg h(x) < \deg b(x)$. Substituting $x = \alpha$ into the above equation, we obtain $g(\alpha)b(\alpha) = 1$. Hence

$$b(\alpha)^{-1} = g(\alpha).$$

Example 8.1 Consider the finite field $\mathbb{F}_{2^3}$. We regard $\mathbb{F}_{2^3} = \mathbb{F}_2[x]/(x^3 + x + 1)$. Denote the residue class of $x \bmod x^3 + x + 1$ by α. Then α is an element of degree 3 over $\mathbb{F}_2$, $\{1, \alpha, \alpha^2\}$ is a polynomial basis of $\mathbb{F}_{2^3}$ over $\mathbb{F}_2$, and $x^3 + x + 1$ is the minimal polynomial of α over $\mathbb{F}_2$. Let $b(\alpha) = 1 + \alpha^2$ and $c(\alpha) = 1 + \alpha + \alpha^2$. Let us compute $b(\alpha)c(\alpha)$. Let $b(x) = 1 + x^2$ and $c(x) = 1 + x + x^2$, then $b(x)c(x) = 1 + x + x^3 + x^4$. Dividing $x^4 + x^3 + x + 1$ by $x^3 + x + 1$, we obtain

$$x^4 + x^3 + x + 1 = (x + 1)(x^3 + x + 1) + x^2 + x.$$

Thus
$$b(\alpha)c(\alpha) = \alpha + \alpha^2.$$
To compute $b(\alpha)^{-1}$, we perform the Euclidean algorithm to $b(x) = x^2 + 1$ and $x^3 + x + 1$. We have
$$x(x^2 + 1) + 1 \cdot (x^3 + x + 1) = 1.$$
Thus
$$b(\alpha)^{-1} = \alpha.$$

Definition 8.2 *Let $\{\alpha_1, \alpha_2, \ldots, \alpha_n\}$ and $\{\beta_1, \beta_2, \ldots, \beta_n\}$ be two bases of $\mathbb{F}_{q^n}$ over $\mathbb{F}_q$. They are said to be equivalent if there exists an element $c \in \mathbb{F}_q^*$ such that $\alpha_i = c\beta_i$ for all $i = 1, 2, \ldots, n$, and they are said to be weakly equivalent if there exists an element $\delta \in \mathbb{F}_{q^n}^*$ such that $\alpha_i = \delta\beta_i$ for all $i = 1, 2, \ldots, n$.*

8.2   Dual Bases

Definition 8.3 *Let $\{\alpha_1, \alpha_2, \ldots, \alpha_n\}$ and $\{\beta_1, \beta_2, \ldots, \beta_n\}$ be two bases of $\mathbb{F}_{q^n}$ over $\mathbb{F}_q$. If*
$$\mathrm{Tr}(\alpha_i\beta_j) = \delta_{i,j} \text{ for all } i, j = 1, 2, \ldots, n, \tag{8.3}$$
then the bases $\{\alpha_1, \alpha_2, \ldots, \alpha_n\}$ and $\{\beta_1, \beta_2, \ldots, \beta_n\}$ are said to be dual to each other, $\{\beta_1, \beta_2, \ldots, \beta_n\}$ is called the dual basis of $\{\alpha_1, \alpha_2, \ldots, \alpha_n\}$, and $\{\alpha_1, \alpha_2, \ldots, \alpha_n\}$ is called the dual basis of $\{\beta_1, \beta_2, \ldots, \beta_n\}$.

Theorem 8.5 *Any basis of $\mathbb{F}_{q^n}$ over $\mathbb{F}_q$ has a dual basis which is unique.*

Proof.   Let $\{\alpha_1, \alpha_2, \ldots, \alpha_n\}$ be a basis of $\mathbb{F}_{q^n}$ over $\mathbb{F}_q$. By Theorem 8.3, $(\mathrm{Tr}(\alpha_i\alpha_j))_{1\le i,j\le n}$ is a nonsingular matrix over $\mathbb{F}_q$. Let $a_{ij} = \mathrm{Tr}(\alpha_i\alpha_j)$, $A = (a_{ij})_{1\le i,j\le n}$, and $B = A^{-1}$. Then B is also a nonsingular matrix over $\mathbb{F}_q$. Let $B = (b_{ij})_{1\le i,j\le n}$ and
$$\beta_j = \sum_{k=1}^{n} b_{kj}\alpha_k \text{ for } j = 1, 2, \ldots, n.$$
By Theorem 8.2 $\{\beta_1, \beta_2, \ldots, \beta_n\}$ is also a basis of $\mathbb{F}_{q^n}$ over $\mathbb{F}_q$, and
$$\mathrm{Tr}(\alpha_i\beta_j) = \sum_{k=1}^{n} b_{kj}\mathrm{Tr}(\alpha_i\alpha_k)$$
$$= \sum_{k=1}^{n} b_{kj}a_{ik}$$
$$= \delta_{ij}, \ i, j = 1, 2, \ldots, n.$$

Therefore $\{\beta_1, \beta_2, \ldots, \beta_n\}$ is a dual basis of $\{\alpha_1, \alpha_2, \ldots, \alpha_n\}$.

We may write (8.3) in matrix form

$$
\begin{pmatrix}
\alpha_1 & \alpha_1^q & \cdots & \alpha_1^{q^{n-1}} \\
\alpha_2 & \alpha_2^q & \cdots & \alpha_2^{q^{n-1}} \\
\vdots & \vdots & & \vdots \\
\alpha_n & \alpha_n^q & \cdots & \alpha_n^{q^{n-1}}
\end{pmatrix}^t
\begin{pmatrix}
\beta_1 & \beta_1^q & \cdots & \beta_1^{q^{n-1}} \\
\beta_2 & \beta_2^q & \cdots & \beta_2^{q^{n-1}} \\
\vdots & \vdots & & \vdots \\
\beta_n & \beta_n^q & \cdots & \beta_n^{q^{n-1}}
\end{pmatrix}
= I^{(n)},
$$

where t denotes the transpose of a matrix, which implies that the dual basis $\{\beta_1, \beta_2, \ldots, \beta_n\}$ of $\{\alpha_1, \alpha_2, \ldots, \alpha_n\}$ is uniquely determined. $\quad\square$

The dual basis of a polynomial basis can be easily computed by the following theorem which is due to Imamura (1983), in which computing the inverse matrix is avoided.

Theorem 8.6 Let $\{1, \alpha, \alpha^2, \ldots, \alpha^{n-1}\}$ be a polynomial basis of $\mathbb{F}_{q^n}$ over $\mathbb{F}_q$ and $\{\beta_0, \beta_1, \beta_2, \ldots, \beta_{n-1}\}$ be its dual basis. Let $f(x)$ be the minimal polynomial of α over $\mathbb{F}_q$ and let $f(x) = (x - \alpha)(\gamma_0 + \gamma_1 x + \cdots + \gamma_{n-1} x^{n-1})$, where $\gamma_{n-1} = 1$, then

$$
\beta_i = \frac{\gamma_i}{f'(\alpha)}, \quad i = 0, 1, 2, \ldots, n-1.
$$

Proof. Let

$$
A = \begin{pmatrix}
1 & 1 & \cdots & 1 \\
\alpha & \alpha^q & \cdots & \alpha^{q^{n-1}} \\
\vdots & \vdots & & \vdots \\
\alpha^{n-1} & \alpha^{(n-1)q} & \cdots & \alpha^{(n-1)q^{n-1}}
\end{pmatrix}
$$

and

$$
B = \begin{pmatrix}
\beta_0 & \beta_1 & \cdots & \beta_{n-1} \\
\beta_0^q & \beta_1^q & \cdots & \beta_{n-1}^q \\
\vdots & \vdots & & \vdots \\
\beta_0^{q^{n-1}} & \beta_1^{q^{n-1}} & \cdots & \beta_{n-1}^{q^{n-1}}
\end{pmatrix}.
$$

By the definition of dual bases, $AB = I^{(n)}$. Thus $BA = I^{(n)}$. Let $\beta(x) = \beta_0 + \beta_1 x + \cdots + \beta_{n-1} x^{n-1}$, then $BA = I^{(n)}$ yields

$$
\beta(\alpha^{q^i}) = \begin{cases} 1, & \text{for } i = 0, \\ 0, & \text{for } 1 \le i \le n-1. \end{cases}
$$

By Lagrange's interpolation formula, the polynomial

$$
\beta_1(x) = \frac{(x - \alpha^q)(x - \alpha^{q^2}) \cdots (x - \alpha^{q^{n-1}})}{(\alpha - \alpha^q)(\alpha - \alpha^{q^2}) \cdots (\alpha - \alpha^{q^{n-1}})} = \frac{f(x)}{(x - \alpha)f'(\alpha)}
$$

is the unique polynomial of degree $\leq n - 1$ such that

$$\beta_1(\alpha^{q^i}) = \begin{cases} 1, & \text{for } i = 0, \\ 0, & \text{for } 1 \leq i \leq n - 1. \end{cases}$$

Therefore $\beta_1(x) = \beta(x)$. Hence

$$\beta_i = \frac{\gamma_i}{f'(\alpha)}, \quad i = 0, 1, 2, \ldots, n - 1.$$

$\square$

Let $\{\alpha_1, \alpha_2, \ldots, \alpha_n\}$ be a basis of $\mathbb{F}_{q^n}$ and $\{\beta_1, \beta_2, \ldots, \beta_n\}$ be its dual basis. We call $\{\alpha_1, \alpha_2, \ldots, \alpha_n\}$ the *primal basis* and $\{\beta_1, \beta_2, \ldots, \beta_n\}$ the *dual basis*. Any $x \in \mathbb{F}_{q^n}$ can be expressed as

$$x = \sum_{i=1}^{n} x_i \alpha_i = \sum_{i=1}^{n} x_i' \beta_i.$$

We call $x_1, x_2, \ldots, x_n$ the *primal coordinates* of x and $x_1', x_2', \ldots, x_n'$ the *dual coordinates* of x.

Theorem 8.7 *Let $\{\alpha_1, \alpha_2, \ldots, \alpha_n\}$ and $\{\beta_1, \beta_2, \ldots, \beta_n\}$ be the primal and dual bases of $\mathbb{F}_{q^n}$ over $\mathbb{F}_q$, respectively. For any $x \in \mathbb{F}_{q^n}$ let $x = \sum_{i=1}^{n} x_i \alpha_i = \sum_{i=1}^{n} x_i' \beta_i$. Then*

$$x_i = \mathrm{Tr}(x\beta_i), \quad i = 1, 2, \ldots, n \tag{8.4}$$

and

$$x_i' = \mathrm{Tr}(x\alpha_i), \quad i = 1, 2, \ldots, n. \tag{8.5}$$

Proof. Clearly,

$$\mathrm{Tr}(x\beta_i) = \mathrm{Tr}((\sum_{j=1}^{n} x_j \alpha_j)\beta_i) = \sum_{j=1}^{n} x_j \delta_{ji} = x_i.$$

and

$$\mathrm{Tr}(x\alpha_i) = \mathrm{Tr}((\sum_{j=1}^{n} x_j' \beta_j)\alpha_i) = \sum_{j=1}^{n} x_j' \delta_{ij} = x_i'.$$

$\square$

Example 8.2 As in Examples 7.1, let $\mathbb{F}_{2^3} = \mathbb{F}_2[x]/(x^3 + x + 1)$ and denote the residue class of $x \bmod x^3 + x + 1$ by α. Then $\alpha^3 + \alpha + 1 = 0$ and $\{1, \alpha, \alpha^2\}$ is a polynomial basis of $\mathbb{F}_{2^3}$ over $\mathbb{F}_2$. Let $\{1, \alpha, \alpha^2\}$ be the primal basis. We are going to compute the corresponding dual basis $\{\beta_0, \beta_1, \beta_2\}$. By Example

7.5 $\text{Tr}(1) = 1, \text{Tr}(\alpha) = \text{Tr}(\alpha^2) = 0$, and $\text{Tr}(a_0 + a_1\alpha + a_2\alpha^2) = a_0$ for $a_0, a_1, a_2 \in \mathbb{F}_2$. Thus

$$A = (\text{Tr}(\alpha^i \alpha^j))_{0 \leq i,j \leq 2}$$

$$= \begin{pmatrix} \text{Tr}(1) & \text{Tr}(\alpha) & \text{Tr}(\alpha^2) \\ \text{Tr}(\alpha) & \text{Tr}(\alpha^2) & \text{Tr}(\alpha^3) \\ \text{Tr}(\alpha^2) & \text{Tr}(\alpha^3) & \text{Tr}(\alpha^4) \end{pmatrix}$$

$$= \begin{pmatrix} 1 & 0 & 0 \\ 0 & 0 & 1 \\ 0 & 1 & 0 \end{pmatrix}$$

and

$$B = A^{-1} = \begin{pmatrix} 1 & 0 & 0 \\ 0 & 0 & 1 \\ 0 & 1 & 0 \end{pmatrix}.$$

Then

$$(\beta_0, \beta_1, \beta_2) = (1, \alpha, \alpha^2)B = (1, \alpha^2, \alpha).$$

That is, the dual basis is $\{1, \alpha^2, \alpha\}$.

The dual basis $\{\beta_0, \beta_1, \beta_2\}$ can also be computed by Theorem 8.6. The minimal polynomial of α is

$$f(x) = x^3 + x + 1 = (x - \alpha)((\alpha^2 + 1) + \alpha x + x^2)$$

and $f'(\alpha) = \alpha^2 + 1$. Thus $\beta_0 = 1, \beta_1 = \alpha^2, \beta_2 = \alpha$.

For any $x \in \mathbb{F}_{2^3}$ express x as $x = x_0 + x_1\alpha + x_2\alpha^2 = x_0'\beta_0 + x_1'\beta_1 + x_2'\beta_2$, then $x_0' = x_0$, $x_1' = x_2$, $x_2' = x_1$.

Example 8.3 As in Example 7.2, let $\mathbb{F}_{2^4} = \mathbb{F}_2[x]/(x^4 + x + 1)$ and denote the residue class of $x \bmod x^4 + x + 1$ by α. Then $\alpha^4 + \alpha + 1 = 0$. Let $\{1, \alpha, \alpha^2, \alpha^3\}$ be the primal basis of $\mathbb{F}_{2^4}$ over $\mathbb{F}_2$. We are going to compute the corresponding dual basis $\{\beta_0, \beta_1, \beta_2, \beta_3\}$. By Example 7.6, $\text{Tr}(1) = \text{Tr}(\alpha) = \text{Tr}(\alpha^2) = 0$, $\text{Tr}(\alpha^3) = 1$, and $\text{Tr}(a_0 + a_1\alpha + a_2\alpha^2 + a_3\alpha^3) = a_3$ for $a_0, a_1, a_2, a_3 \in \mathbb{F}_2$. Thus

$$A = (\text{Tr}(\alpha_i \alpha_j))_{0 \leq i,j \leq 3}$$

$$= \begin{pmatrix} 0 & 0 & 0 & 1 \\ 0 & 0 & 1 & 0 \\ 0 & 1 & 0 & 0 \\ 1 & 0 & 0 & 1 \end{pmatrix}$$

and

$$B = A^{-1} = \begin{pmatrix} 1 & 0 & 0 & 1 \\ 0 & 0 & 1 & 0 \\ 0 & 1 & 0 & 0 \\ 1 & 0 & 0 & 0 \end{pmatrix}.$$

Then
$$(\beta_0, \beta_1, \beta_2, \beta_3) = (1, \alpha, \alpha^2, \alpha^3)B = (1 + \alpha^3, \alpha^2, \alpha, 1).$$

That is, the dual basis is $\{1 + \alpha^3, \alpha^2, \alpha, 1\}$.

Let us compute the dual basis by Theorem 8.6. The minimal polynomial of α is

$$f(x) = x^4 + x + 1 = (x - \alpha)((\alpha^3 + 1) + \alpha^2 x + \alpha x^2 + x^3)$$

and $f'(\alpha) = 1$. Thus $\beta_0 = \alpha^3 + 1$, $\beta_1 = \alpha^2$, $\beta_2 = \alpha$, $\beta_3 = 1$.

For any $x \in \mathbb{F}_{2^4}$ express x as $x = \sum_{i=0}^{3} x_i \alpha^i = \sum_{i=0}^{3} x_i' \beta_i$, then $x_0' = x_3$, $x_1' = x_2$, $x_2' = x_1$, $x_3' = x_0 + x_3$..

Let $\mathbb{F}_{q^n} = \mathbb{F}_q[\alpha]$, where $\alpha \in \mathbb{F}_{q^n}$ and is an element of degree n over $\mathbb{F}_q$. Let $\{1, \alpha, \alpha^2, \ldots, \alpha^{n-1}\}$ be the primal basis of $\mathbb{F}_{q^n}$ over $\mathbb{F}_q$ and the corresponding dual basis be $\{\beta_0, \beta_1, \ldots, \beta_{n-1}\}$. We shall see how the multiplication of elements in $\mathbb{F}_{q^n}$ becomes easy in the dual coordinate system.

Let $x \in \mathbb{F}_{q^n}$ be expressed as

$$x = x_0' \beta_0 + x_1' \beta_1 + x_2' \beta_2 + \cdots + x_{n-1}' \beta_{n-1},$$

where $x_i' \in \mathbb{F}_q$, $i = 0, 1, 2, \ldots, n - 1$. By (8.5),

$$x_i' = \text{Tr}(x\alpha^i), \ i = 0, 1, 2, \ldots, n - 1.$$

We want to compute αx. Express αx as

$$\alpha x = (\alpha x)_0' \beta_0 + (\alpha x)_1' \beta_1 + (\alpha x)_2' \beta_2 + \cdots + (\alpha x)_{n-1}' \beta_{n-1}$$

where $(\alpha x)_i' \in \mathbb{F}_q$, $i = 0, 1, 2, \ldots, n - 1$. Also by (8.5),

$$\begin{aligned}(\alpha x)_i' &= \text{Tr}(\alpha x \cdot \alpha^i) \\ &= \text{Tr}(x\alpha^{i+1}), \ i = 0, 1, 2, \ldots, n - 1.\end{aligned}$$

Let $m(x) = x^n + c_1 x^{n-1} + c_2 x^{n-2} + \cdots + c_n$ be the minimal polynomial of α over $\mathbb{F}_q$. Then

$$\begin{aligned}(\alpha x)_{n-1}' &= \text{Tr}(x\alpha^n) = -\text{Tr}(c_1 x\alpha^{n-1} + c_2 x\alpha^{n-2} + \cdots + c_n x) \\ &= -(c_1 x_{n-1}' + c_2 x_{n-2}' + \cdots + c_n x_0').\end{aligned}$$

Therefore

$$(\alpha x)_i' = \begin{cases} x_{i+1}', & i = 0, 1, 2, \ldots, n - 2, \\ -(c_1 x_{n-1}' + c_2 x_{n-2}' + \cdots + c_n x_0'), & i = n - 1. \end{cases}$$

When $q = 2$, $\mathbb{F}_{q^n} = \mathbb{F}_{2^n} = \mathbb{F}_2[\alpha]$, where α is an element of degree n over $\mathbb{F}_2$. The above formula becomes

$$(\alpha x)'_i = \begin{cases} x'_{i+1}, & i = 0, 1, 2, \ldots, n-2, \\ c_1 x'_{n-1} + c_2 x'_{n-2} + \cdots + c_n x'_0, & i = n-1, \end{cases}$$

which suggests a simple shift-register shown in Figure 8.1 that can multiply x by α. Let the n flip-flops be numbered as the first, the second, $\cdots$, and the n-th from left to right and let the i-th flip-flop be filled with the dual coordinate x'_{n-i}, $i = 1, 2, \ldots, n$. Let the box with the entry XOR stand for an adder in $\mathbb{F}_2$, called an XOR-gate, and the circle with entry $c_i = 1$ or 0 indicate that there is a line or no line connecting the i-th flip-flop and box. Then when the shift register is clocked, the new contents of the n flip-flops are equal to the dual coordinates of αx.

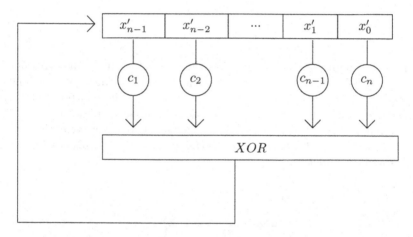

Figure 8.1

More generally, let $x, y \in \mathbb{F}_{q^n}$ and we are going to perform the multiplication of x by y. Express x in dual coordinate system as $x = x'_0 \beta_0 + x'_1 \beta_1 + \cdots + x'_{n-1} \beta_{n-1}$, where $x_i \in \mathbb{F}_q$, $i = 0, 1, \ldots, n-1$ and express y in primal coordinate system as $y = y_0 + y_1 \alpha + \cdots + y_{n-1} \alpha^{n-1}$, where $y_i \in \mathbb{F}_q$, $i = 1, 2, \ldots, n-1$. Then

$$yx = \sum_{i=0}^{n-1} y_i \alpha^i x = \sum_{i=0}^{n-1} y_i \left(\sum_{j=0}^{n-1} (\alpha^i x) \beta_j \right)'_j = \sum_{j=0}^{n-1} \left(\sum_{i=0}^{n-1} y_i (\alpha^i x)'_j \right) \beta_j.$$

Thus

$$(yx)'_j = \sum_{i=0}^{n-1} y_i (\alpha^i x)'_j.$$

When $q = 2$, the device shown in the lower part of Figure 8.2, (i.e., the device shown in Figure 8.1) gives us $x, \alpha x, \alpha^2 x, \ldots, \alpha^{n-1} x$ in dual coordinates successively when the shift register is clocked. Suppose that the upper shift register is filled with the primal coordinates $y_0, y_1, \ldots, y_{n-1}$ of y from right to left. Let the device

represent a multiplier which produces the product ab in $\mathbb{F}_2$ when the inputs are a and b. Then the output of the circuit will be the dual coordinates $(yx)'_0, (yx)'_1, \ldots, (yx)'_{n-1}$ of yx successively as the shift register is clocked. Since the input of coordinates of x and y is done in parallel, while the coordinates of yx are computed bitwise in successive clock cycle, the device shown in Figure 8.2 is called a *bit-serial dual basis multiplier*, which was suggested by Berlekamp (1982).

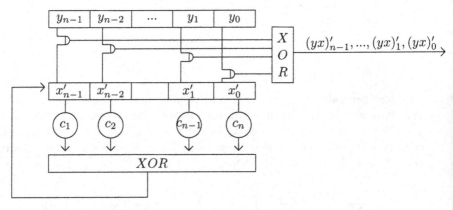

Figure 8.2

8.3 Self-dual Bases

Definition 8.4 *A basis $\{\alpha_1, \alpha_2, \ldots, \alpha_n\}$ of $\mathbb{F}_{q^n}$ over $\mathbb{F}_q$ is called a self-dual basis if*

$$\mathrm{Tr}_{\mathbb{F}_{q^n}/\mathbb{F}_q}(\alpha_i \alpha_j) = \delta_{ij} \text{ for all } i, j = 1, 2, \ldots, n.$$

If $\{\alpha_1, \alpha_2, \ldots, \alpha_n\}$ is a self-dual basis of $\mathbb{F}_{q^n}$ over $\mathbb{F}_q$, then any permutation of $\alpha_1, \alpha_2, \ldots, \alpha_n$ gives also a self-dual basis and we regard them as the same self-dual basis.

We recall that an $n \times n$ matrix S over $\mathbb{F}_q$ is said to be *symmetric* if

$$^tS = S.$$

If we write

$$S = (s_{ij})_{1 \leq i, j \leq n},$$

then the condition $^tS = S$ is equivalent to

$$s_{ij} = s_{ji} \text{ for all } i, j = 1, 2, \ldots, n.$$

Two $n \times n$ symmetric matrices S and T over $\mathbb{F}_q$ are said to be *cogredient* over $\mathbb{F}_q$ if there is an $n \times n$ nonsingular matrix P over $\mathbb{F}_q$ such that

$$^tPSP = T.$$

The transformation

$$S \longmapsto {}^tPSP,$$

from the set of $n \times n$ symmetric matrices over $\mathbb{F}_q$ to itself, where P is an $n \times n$ nonsingular matrix over $\mathbb{F}_q$, is called a *cogredience transformation*.

We begin by studying the condition for the existence of self-dual basis of $\mathbb{F}_{q^n}$ over $\mathbb{F}_q$. Consider first the case when q is even. We need the following lemma.

Lemma 8.8 *Let q be a power of 2. Let S be an $n \times n$ nonsingular symmetric matrix over $\mathbb{F}_q$ and assume that one of the diagonal elements of S is nonzero. Then S is cogredient to the $n \times n$ identity matrix $I^{(n)}$.*

Proof. We apply induction on n. If $n = 1$, clearly our lemma follows from Theorem 6.20(i).

Now assume $n > 1$. Let

$$S = (s_{ij})_{1 \leq i, j \leq n}.$$

Suppose $s_{ii} \neq 0$ for some i. Interchanging the 1st row and the ith row, and the 1st column and the ith column simultaneously, (which is clearly a cogredience transformation), we can assume that $s_{11} \neq 0$. Write

$$S = \begin{pmatrix} s_{11} & u \\ {}^t u & S_1 \end{pmatrix},$$

where u is an $(n-1)$-dimensional row vector and S_1 is an $(n-1) \times (n-1)$ symmetric matrix. Then

$$ {}^t\begin{pmatrix} 1 & s_{11}^{-1} u \\ 0 & I^{(n-1)} \end{pmatrix} \begin{pmatrix} s_{11} & u \\ {}^t u & S_1 \end{pmatrix} \begin{pmatrix} 1 & s_{11}^{-1} u \\ 0 & I^{(n-1)} \end{pmatrix} = \begin{pmatrix} s_{11} & 0 \\ 0 & S_1 + s_{11}^{-1}\, {}^t u u \end{pmatrix}. $$

By Theorem 6.20(i), $\mathbb{F}_q^* = \mathbb{F}_q^{*2}$. Thus, there is an element $s_{11}^{-1/2} \in \mathbb{F}_q^*$ such that $(s_{11}^{-1/2})^2 = s_{11}^{-1}$. Then

$$ {}^t\begin{pmatrix} s_{11}^{-1/2} & 0 \\ 0 & I^{(n-1)} \end{pmatrix} \begin{pmatrix} s_{11} & 0 \\ 0 & S_1 + s_{11}^{-1}\, {}^t u u \end{pmatrix} \begin{pmatrix} s_{11}^{-1/2} & 0 \\ 0 & I^{(n-1)} \end{pmatrix} $$

$$ = \begin{pmatrix} 1 & 0 \\ 0 & S_1 + s_{11}^{-1}\, {}^t u u \end{pmatrix}. $$

Let $T_1 = S_1 + s_{11}^{-1}\, {}^t u u$. Since S is nonsingular, T_1 is also nonsingular. Let

$$T_1 = (t_{ij})_{2 \le i,j \le n}.$$

If one of the diagonal elements of T_1 is nonzero, our lemma follows from induction. If all the diagonal elements of T_1 are zeros, then there is a nonzero non-diagonal element. Suppose $t_{ij} \neq 0$ where $i \neq j$. Interchanging the 1st row and the $(i-1)$-th row of T_1 and 1st column and the $(i-1)$-th column of T_1 simultaneously and then interchanging the 2nd row and the $(j-1)$-th row and the 2nd column and the $(j-1)$-th column of the matrix just obtained simultaneously, we can assume that $t_{23} \neq 0$. Then

$$ {}^t\begin{pmatrix} 1 & t_{23} & 0 \\ 0 & 1 & 0 \\ 1 & t_{23} & 1 \end{pmatrix} \begin{pmatrix} 1 & 0 & 0 \\ 0 & 0 & t_{23} \\ 0 & t_{23} & 0 \end{pmatrix} \begin{pmatrix} 1 & t_{23} & 0 \\ 0 & 1 & 0 \\ 1 & t_{23} & 1 \end{pmatrix} = \begin{pmatrix} 1 & 0 & 0 \\ 0 & t_{23}^2 & t_{23} \\ 0 & t_{23} & 0 \end{pmatrix}. $$

This reduces to the previous case. $\square$

Theorem 8.9 *Let q be a power of 2. Then there exists a self-dual basis of $\mathbb{F}_{q^n}$ over $\mathbb{F}_q$.*

Proof. Let $\{\alpha_1, \alpha_2, \ldots, \alpha_n\}$ be a basis of $\mathbb{F}_{q^n}$ over $\mathbb{F}_q$, and let

$$T = (\text{Tr}(\alpha_i \alpha_j))_{1 \le i,j \le n},$$

where Tr is an abbreviation for $\text{Tr}_{\mathbb{F}_{q^n}/\mathbb{F}_q}$. Then T is an $n \times n$ symmetric matrix over $\mathbb{F}_q$. If all the diagonal elements of T are zeros, i.e., $\text{Tr}(\alpha_1{}^2) = \text{Tr}(\alpha_2{}^2) = \cdots = \text{Tr}(\alpha_n{}^2) = 0$, then $\text{Tr}(\alpha_1) = \text{Tr}(\alpha_2) = \cdots = \text{Tr}(\alpha_n) = 0$, and, hence, $\text{Tr}(\alpha) = 0$ for every $\alpha \in \mathbb{F}_{q^n}$, which contradicts to Theorem 7.16(i). Therefore $\text{Tr}(\alpha_i{}^2) \ne 0$ for some i. By Lemma 8.8 there is an $n \times n$ nonsingular matrix P over $\mathbb{F}_q$ such that ${}^tPTP = I^{(n)}$.

Write

$$P = (p_{ij})_{1 \le i,j \le n}$$

and define

$$\beta_j = \sum_{i=1}^{n} p_{ij}\alpha_i, \quad j = 1, 2, \ldots, n.$$

Then $\{\beta_1, \beta_2, \ldots, \beta_n\}$ is a basis of $\mathbb{F}_{q^n}$ over $\mathbb{F}_q$. Let

$$A = \begin{pmatrix} \alpha_1 & \alpha_2 & \cdots & \alpha_n \\ \alpha_1^q & \alpha_2^q & \cdots & \alpha_n^q \\ \vdots & \vdots & & \vdots \\ \alpha_1^{q^{n-1}} & \alpha_2^{q^{n-1}} & \cdots & \alpha_n^{q^{n-1}} \end{pmatrix} \tag{8.6}$$

and

$$B = \begin{pmatrix} \beta_1 & \beta_2 & \cdots & \beta_n \\ \beta_1^q & \beta_2^q & \cdots & \beta_n^q \\ \vdots & \vdots & & \vdots \\ \beta_1^{q^{n-1}} & \beta_2^{q^{n-1}} & \cdots & \beta_n^{q^{n-1}} \end{pmatrix}.$$

Then ${}^tAA = T$ and $B = AP$. Thus

$${}^tBB = {}^tP\,{}^tAAP = {}^tPTP = I^{(n)},$$

i.e., $\text{Tr}(\beta_i \beta_j) = \delta_{ij}$ for $i, j = 1, 2, \ldots, n$. Therefore $\{\beta_1, \beta_2, \ldots, \beta_n\}$ is a self-dual basis of $\mathbb{F}_{q^n}$ over $\mathbb{F}_q$. $\square$

Now let us come to the case when q is odd. We need the following lemma.

Lemma 8.10 *Let q be a power of an odd prime. Let S be an $n \times n$ nonsingular symmetric matrix over $\mathbb{F}_q$. Then S is cogredient to either the $n \times n$ identity matrix $I^{(n)}$ or the diagonal matrix $\text{diag}[1, 1, \ldots, 1, z]$, where z is a fixed non-square element of $\mathbb{F}_q^*$. Moreover, $I^{(n)}$ and $\text{diag}[1, 1, \ldots, 1, z]$ are not cogredient.*

Proof. First we prove that S is cogredient to a diagonal matrix. We apply induction on n. If $n = 1$ this is trivial. Now assume $n > 1$. Let

$$S = (s_{ij})_{1 \le i,j \le n}.$$

Suppose $s_{ii} \ne 0$ for some i. As in the proof of Lemma 8.8, we can show that S is cogredient to an $n \times n$ symmetric matrix of the form

$$\begin{pmatrix} s_{11} & 0 \\ 0 & T_1 \end{pmatrix}$$

Since S is nonsingular, T_1 is also nonsingular. By induction hypothesis T_1 is cogredient to a diagonal matrix, so is S. Now suppose all the diagonal elements of S are zeros. Then there is a nonzero non-diagonal element. Let $s_{ij} \ne 0$, where $i \ne j$. Interchanging the 1st row and the i-th row of S and the 1st column and the i-th column of S simultaneously and then interchanging the 2nd row and the j-th row and the 2nd column and the j-th column of the matrix just obtained simultaneously, we can assume that $s_{12} \ne 0$. Then

$$^t\begin{pmatrix} 1 & 0 \\ 1 & 1 \end{pmatrix} \begin{pmatrix} 0 & s_{12} \\ s_{12} & 0 \end{pmatrix} \begin{pmatrix} 1 & 0 \\ 1 & 1 \end{pmatrix} = \begin{pmatrix} 2s_{12} & s_{12} \\ s_{12} & 0 \end{pmatrix}.$$

Since $\mathbb{F}_q$ is of characteristic $\ne 2$, $2s_{12} \ne 0$. This reduces to the previous case.

Let $S = \mathrm{diag}[s_1, s_2, \ldots, s_n]$. Since S is nonsingular, $s_1, s_2, \ldots, s_n$ are all nonzero. After permuting the elements $s_1, s_2, \ldots, s_n$, (which is clearly a cogredience transformation,) we can assume that $s_1, s_2, \ldots, s_r \in \mathbb{F}_q^{*2}$ and $s_{r+1}, \ldots, s_n \notin \mathbb{F}_q^{*2}$. Let $s_i = t_i^2$, $1 \le i \le r$, and $s_j = zt_j^2$, $r+1 \le j \le n$, where $t_i, t_j \in \mathbb{F}_q^*$ and z is fixed non-square element of $\mathbb{F}_q$. Then S is cogredient to

$$\mathrm{diag}[t_1^{-1}, t_2^{-1}, \ldots, t_n^{-1}] \, S \, \mathrm{diag}[t_1^{-1}, t_2^{-1}, \ldots, t_n^{-1}] = \mathrm{diag}[\underbrace{1, \ldots, 1}_{r \ 1's}, \underbrace{z, \ldots, z}_{n-r \ z's}].$$

We distinguish the cases when $-1 \notin \mathbb{F}_q^{*2}$ and $-1 \in \mathbb{F}_q^{*2}$. First consider the case $-1 \notin \mathbb{F}_q^{*2}$. As x runs through $\mathbb{F}_q$, $1 + x^2$ runs through $\frac{1}{2}(q+1)$ elements of $\mathbb{F}_q^*$. Then there is an element $x \in \mathbb{F}_q$ such that $1 + x^2$ is a non-square element of $\mathbb{F}_q^*$. By Theorem 6.20(ii) $\mathbb{F}_q^* = \mathbb{F}_q^{*2} \bigcup z\mathbb{F}_q^{*2}$. Thus $1 + x^2 \in z\mathbb{F}_q^{*2}$ and there is an element $y \in \mathbb{F}_q^*$ such that $(1 + x^2)y^2 = z$. Then

$$^t\begin{pmatrix} xy & -y \\ y & xy \end{pmatrix} \begin{pmatrix} 1 & 0 \\ 0 & 1 \end{pmatrix} \begin{pmatrix} xy & -y \\ y & xy \end{pmatrix} = \begin{pmatrix} z & 0 \\ 0 & z \end{pmatrix}.$$

Then consider the case $-1 \in \mathbb{F}_q^{*2}$. Let $-1 = w^2$, where $w \in \mathbb{F}_q^*$. Then

$${}^t\left(\begin{array}{cc} \frac{1}{2}(1+z) & \frac{w}{2}(1-z) \\ \frac{1}{2w}(1-z) & \frac{1}{2}(1+z) \end{array} \right) \left(\begin{array}{cc} 1 & 0 \\ 0 & 1 \end{array} \right) \left(\begin{array}{cc} \frac{1}{2}(1+z) & \frac{w}{2}(1-z) \\ \frac{1}{2w}(1-z) & \frac{1}{2}(1+z) \end{array} \right)$$

$$= \left(\begin{array}{cc} z & 0 \\ 0 & z \end{array} \right).$$

It follows that S is congruent to either $I^{(n)}$ or $\mathrm{diag}[1,1,\ldots,1,z]$. Clearly, the determinants of two cogredient nonsingular symmetric matrix differ by a square element of $\mathbb{F}_q^*$. Therefore $I^{(n)}$ and $\mathrm{diag}[1,1,\ldots,1,z]$ are not cogredient. $\qquad\square$

Theorem 8.11 *Let q be an odd prime power, then there exists a self-dual basis of $\mathbb{F}_{q^n}$ over $\mathbb{F}_q$ if and only if n is also odd.*

Proof. Let $\{\alpha_1, \alpha_2, \ldots, \alpha_n\}$ be a basis of $\mathbb{F}_{q^n}$ over $\mathbb{F}_q$ and let $\{\beta_1, \beta_2, \ldots, \beta_n\}$ be a self-dual basis of $\mathbb{F}_{q^n}$ over $\mathbb{F}_q$. Then

$$\mathrm{Tr}(\beta_i\beta_j) = \delta_{ij} \text{ for } i, j = 1, 2, \ldots, n.$$

We can express $\beta_1, \beta_2, \ldots, \beta_n$ linearly in terms of $\alpha_1, \alpha_2, \ldots, \alpha_n$, say

$$\beta_j = \sum_{i=1}^{n} p_{ij}\alpha_i, \; j = 1, 2, \ldots, n, \text{ where } p_{ij} \in \mathbb{F}_q.$$

Then

$$\delta_{ij} = \mathrm{Tr}(\beta_i\beta_j) = \mathrm{Tr}\left(\sum_{k=1}^{n} p_{ki}\alpha_k \sum_{l=1}^{n} p_{lj}\alpha_l \right)$$

$$= \sum_{k=1}^{n}\sum_{l=1}^{n} p_{ki}\mathrm{Tr}(\alpha_k\alpha_l)p_{lj}.$$

Let $P = (p_{lj})_{1\le l,j\le n}$, then P is nonsingular. Let $T = (\mathrm{Tr}(\alpha_k\alpha_l))_{1\le k,l\le n}$ and write the above equation in matrix language, we have

$$I^{(n)} = {}^tPTP.$$

That is, T is cogredient to $I^{(n)}$ over $\mathbb{F}_q$. The above reasoning can be reversed. Therefore we proved that there exists a self-dual basis of $\mathbb{F}_{q^n}$ over $\mathbb{F}_q$ if and only if T is cogredient to $I^{(n)}$ over $\mathbb{F}_q$, and, by Lemma 8.10, this is equivalent to requiring that $\det T \in \mathbb{F}_q^{*2}$.

Let A be the matrix (8.6), then $T = {}^t A A$ and $\det T = (\det A)^2$. Thus $\det T \in \mathbb{F}_q^{*2}$ if and only if $\det A \in \mathbb{F}_q^*$. We have

$$(\det A)^q = \begin{vmatrix} \alpha_1^q & \alpha_2^q & \cdots & \alpha_n^q \\ \alpha_1^{q^2} & \alpha_2^{q^2} & \cdots & \alpha_n^{q^2} \\ \vdots & \vdots & & \vdots \\ \alpha_1 & \alpha_2 & \cdots & \alpha_n \end{vmatrix} = (-1)^{n-1} \det A.$$

Therefore $\det A \in \mathbb{F}_q^*$ if and only if n is odd. $\qquad\qquad\square$

We can combine Theorems 8.9 and 8.11 into a single theorem.

Theorem 8.12 *There exists a self-dual basis of $\mathbb{F}_{q^n}$ over $\mathbb{F}_q$ if and only if either q is even or both q and n are odd.*

Theorem 8.12 is due to Seroussi and Lempel (1980).

Now let us enumerate the number of self-dual bases of $\mathbb{F}_{q^n}$ over $\mathbb{F}_q$.

Theorem 8.13 *Let $\{\alpha_1, \alpha_2, \ldots, \alpha_n\}$ be a self-dual basis of $\mathbb{F}_{q^n}$ over $\mathbb{F}_q$. Let $C = (c_{ij})_{1 \le i,j \le n}$ be an $n \times n$ nonsingular matrix over $\mathbb{F}_q$, and define*

$$\beta_j = \sum_{i=1}^n c_{ij} \alpha_i, \quad j = 1, 2, \ldots, n.$$

Then $\{\beta_1, \beta_2, \ldots, \beta_n\}$ is also a self-dual basis of $\mathbb{F}_{q^n}$ over $\mathbb{F}_q$ if and only if ${}^t C C = I^{(n)}$.

Proof. $\{\beta_1, \beta_2, \ldots, \beta_n\}$ is self-dual if and only if

$$\delta_{ij} = \mathrm{Tr}(\beta_i \beta_j) = \mathrm{Tr}\left(\sum_{k=1}^n c_{ki} \alpha_k \sum_{l=1}^n c_{lj} \alpha_l \right)$$

$$= \sum_{k=1}^n \sum_{l=1}^n c_{ki} \mathrm{Tr}(\alpha_k \alpha_l) c_{lj}$$

$$= \sum_{k=1}^n \sum_{l=1}^n c_{ki} \delta_{kl} c_{lj} = \sum_{k=1}^n c_{ki} c_{kj}, \text{ for } i, j = 1, 2, \ldots, n,$$

which is equivalent to ${}^t C C = I^{(n)}$. $\qquad\qquad\square$

An $n \times n$ matrix C over $\mathbb{F}_q$ satisfying the condition ${}^t C C = I^{(n)}$ is called an $n \times n$ *orthogonal* or *pseudo-symplectic matrix* with respect to $I^{(n)}$, according as q is odd or even, respectively. When q is odd and $n = 2\nu + 1$ is also odd, the set of $n \times n$ orthogonal matrices over $\mathbb{F}_q$ with respect to $I^{(n)}$ forms a group with respect to the matrix multiplication, which is called the *orthogonal group* of degree $n = 2\nu + 1$ over $\mathbb{F}_q$ with respect to $I^{(n)}$ and is denoted by $O_{2\nu+1}(\mathbb{F}_q)$. When q is even, the set of $n \times n$ pseudo-symplectic matrices over $\mathbb{F}_q$ with respect to $I^{(n)}$ forms also a group with respect to the matrix multiplication, which is called the *pseudo-symplectic group* of degree n over $\mathbb{F}_q$ and is denoted by $Ps_{2\nu+\delta}(\mathbb{F}_q)$, where $n = 2\nu + \delta$ and $\delta = 1$ or 2 according as n is odd or even, respectively.

Denote by $sd(n, q)$ the number of self-dual bases of $\mathbb{F}_{q^n}$ over $\mathbb{F}_q$. Then we have

Corollary 8.14

$$
sd(n, q) = \begin{cases} \frac{1}{n!} |Ps_{2\nu+\delta}(\mathbb{F}_q)|, & \text{if } q \text{ is even, } n = 2\nu + \delta \text{ and } \delta = 1 \text{ or } 2, \\ \frac{1}{n!} |O_{2\nu+1}(\mathbb{F}_q)|, & \text{if } q \text{ is odd and } n = 2\nu + 1 \text{ is also odd,} \\ 0, & \text{if } q \text{ is odd and } n \text{ is even.} \end{cases}
$$

This Corollary is due to Lempel and Seroussi (1991).

The order $|O_{2\nu+1}(\mathbb{F}_q)|$ and $|Ps_{2\nu+\delta}(\mathbb{F}_q)|$, where $\delta = 1$ or 2, were computed by Dickson (1900) and Macwilliams (1969), respectively, and proofs can also be found in Wan (2002). From Theorems 4.7 and 6.21 of Wan (2002) we have

$$
|O_{2\nu+1}(\mathbb{F}_q)| = 2q^{\nu^2} \prod_{i=1}^{\nu} (q^{2i} - 1), \qquad \text{if } q \text{ is odd,}
$$

$$
|Ps_{2\nu+1}(\mathbb{F}_q)| = q^{\nu^2} \prod_{i=1}^{\nu} (q^{2i} - 1), \qquad \text{if } q \text{ is even,}
$$

$$
|P_{2\nu+2}(\mathbb{F}_q)| = q^{(\nu+1)^2} \prod_{i=1}^{\nu} (q^{2i} - 1), \qquad \text{if } q \text{ is even.}
$$

Remark: When q is odd and n is even there are two different orthogonal groups, denoted by $O_{2\nu}(\mathbb{F}_q)$ and $O_{2\nu+2}(\mathbb{F}_q)$, when $n = 2\nu$ and $n = 2\nu + 2$, respectively, depending on whether $(-1)^{n/2} \in \mathbb{F}_q^{*2}$ or not. For completeness we quote the following results from Theorem 6.21 of Wan (2002).

$$
|O_{2\nu}(\mathbb{F}_q)| = 2q^{\nu(\nu-1)}(q^\nu - 1) \prod_{i=1}^{\nu-1} (q^{2i} - 1),
$$

$$
|O_{2\nu+2}(\mathbb{F}_q)| = 2q^{\nu(\nu+1)}(q^{\nu+1} + 1) \prod_{i=1}^{\nu} (q^{2i} - 1).
$$

8.4 Normal Bases

Definition 8.5 *A basis* $N = \{\alpha_0, \alpha_1, \alpha_2, \ldots, \alpha_{n-1}\}$ *of* $\mathbb{F}_{q^n}$ *over* $\mathbb{F}_q$ *is called a normal basis if there is an element* $\alpha \in \mathbb{F}_{q^n}$ *such that*

$$\alpha_0 = \alpha, \quad \alpha_1 = \sigma(\alpha), \quad \alpha_2 = \sigma^2(\alpha), \ldots, \quad \alpha_{n-1} = \sigma^{n-1}(\alpha),$$

up to a rearrangement of $\alpha_0, \alpha_1, \alpha_2, \ldots, \alpha_{n-1}$, *where* σ *is the Frobenius automorphism of* $\mathbb{F}_{q^n}$ *over* $\mathbb{F}_q$. *The element* $\alpha \in \mathbb{F}_{q^n}$ *is called a normal basis generator of* $\mathbb{F}_{q^n}$ *over* $\mathbb{F}_q$ *and is said to generate the normal basis* N.

Clearly, if α is a normal basis generator of $\mathbb{F}_{q^n}$ over $\mathbb{F}_q$, so are $\sigma(\alpha)$, $\sigma^2(\alpha)$, $\ldots$, and $\sigma^{n-1}(\alpha)$, and they generate the same normal basis. Moreover, if α is a normal basis generator of $\mathbb{F}_{q^n}$ over $\mathbb{F}_q$, α is of degree n over $\mathbb{F}_q$ and $\mathrm{Tr}_{\mathbb{F}_{q^n}/\mathbb{F}_q}(\alpha) \neq 0$.

Example 8.4 We may regard $\mathbb{F}_{2^3} = \mathbb{F}_2[x]/(x^3 + x^2 + 1)$. Denote the residue class of $x \bmod x^3 + x^2 + 1$ by α, then $\mathbb{F}_{2^3} = \mathbb{F}_2[\alpha]$ and $\alpha^3 = \alpha^2 + 1$. Let σ be the Frobenius automorphism of $\mathbb{F}_{2^3}$ over $\mathbb{F}_2$. Then

$$\{\alpha, \ \sigma(\alpha) = \alpha^2, \ \sigma^2(\alpha) = \alpha^4 = \alpha^2 + \alpha + 1\}$$

is a normal basis of $\mathbb{F}_{2^3}$ over $\mathbb{F}_2$ and α is a normal basis generator.

But if we regard $\mathbb{F}_{2^3} = \mathbb{F}_2[x]/(x^3 + x + 1)$ and denote the residue class of $x \bmod x^3 + x + 1$ by β, then $\mathbb{F}_{2^3} = \mathbb{F}_2[\beta]$ and $\beta^3 = \beta + 1$. Now,

$$\{\beta, \ \sigma(\beta) = \beta^2, \ \sigma^2(\beta) = \beta^4 = \beta^2 + \beta\}$$

is not a normal basis of $\mathbb{F}_{2^3}$ over $\mathbb{F}_2$. However,

$$\{\beta + 1, \ \sigma(\beta + 1) = \beta^2 + 1, \ \sigma^2(\beta + 1) = \beta^4 + 1 = \beta^2 + \beta + 1\}$$

is a normal basis of $\mathbb{F}_{2^3}$ over $\mathbb{F}_2$.

Theorem 8.15 *For any prime power* q *and positive integer* n, *there exists a normal basis of* $\mathbb{F}_{q^n}$ *over* $\mathbb{F}_q$.

Proof. Let σ be the Frobenius automorphism of $\mathbb{F}_{q^n}$ over $\mathbb{F}_q$. We regard $\mathbb{F}_{q^n}$ as an n-dimensional vector space over $\mathbb{F}_q$, then σ can be regarded as a linear transformation of the vector space structure of $\mathbb{F}_{q^n}$ over $\mathbb{F}_q$. Clearly $\sigma^n = 1$. If we can show that $x^n - 1$ is the minimal polynomial of the linear transformation σ, then by Cayley-Hamilton Theorem, $x^n - 1$ is also the characteristic polynomial of σ. By a well-known result in Linear Algebra, there exists a vector $\alpha \in \mathbb{F}_{q^n}$ such that $\{\alpha, \sigma(\alpha), \ldots, \sigma^{n-1}(\alpha)\}$ is a basis of

$\mathbb{F}_{q^n}$ over $\mathbb{F}_q$ (Exercise 8.12). Then $\{\alpha, \sigma(\alpha), \ldots, \sigma^{n-1}(\alpha)\}$ is a normal basis of $\mathbb{F}_{q^n}$ over $\mathbb{F}_q$.

Suppose there is a polynomial

$$g(x) = g_0 + g_1 x + \cdots + g_m x^m,$$

where $m < n$, $g_0, g_1, \ldots, g_m \in \mathbb{F}_{q^n}$, and $g_m \neq 0$ such that $g(\sigma) = 0$. We may assume that $g(x)$ is such a polynomial of least degree. Of course $\sigma^m \neq 0$, so there is a j where $0 \leq j \leq m - 1$ such that $g_j \neq 0$. From $g(\sigma) = 0$ we deduce $g(\sigma)(\beta) = 0$ for all $\beta \in \mathbb{F}_{q^n}$, i.e.,

$$g_0 \beta + g_1 \sigma(\beta) + \cdots + g_m \sigma^m(\beta) = 0 \quad \text{for all} \quad \beta \in \mathbb{F}_{q^n}. \tag{8.7}$$

Since $\sigma^m \neq \sigma^j$, there is an element $\eta \in \mathbb{F}_{q^n}$ such that $\sigma^m(\eta) \neq \sigma^j(\eta)$. We also have $g(\sigma)(\eta\beta) = 0$, i.e.,

$$g_0 \eta \beta + g_1 \sigma(\eta)\sigma(\beta) + \cdots + g_m \sigma^m(\eta)\sigma^m(\beta) = 0. \tag{8.8}$$

Multiplying (8.7) by $\sigma^m(\eta)$, we have

$$g_0 \sigma^m(\eta)\beta + g_1 \sigma^m(\eta)\sigma(\beta) + \cdots + g_m \sigma^m(\eta)\sigma^m(\beta) = 0. \tag{8.9}$$

Substracting (8.8) from (8.9) we obtain

$$g_0(\sigma^m(\eta) - \eta)\beta + \cdots + g_{m-1}(\sigma^m(\eta) - \sigma^{m-1}(\eta))\sigma^{m-1}(\beta) = 0.$$

Let

$$h(x) = g_0(\sigma^m(\eta) - \eta) + \cdots + g_{m-1}(\sigma^m(\eta) - \sigma^{m-1}(\eta))x^{m-1},$$

then $h(\sigma)(\beta) = 0$ for all $\beta \in \mathbb{F}_{q^n}$, thus $h(\sigma) = 0$. The coefficient of x^j in $h(x)$ is $g_j(\sigma^m(\eta) - \sigma^j(\eta)) \neq 0$, so $h(x) \neq 0$. But $\deg h(x) < m$, we obtain a contradiction. Therefore there exists no polynomial $g(x) \in \mathbb{F}_{q^n}[x]$ of degree $< n$ such that $g(\sigma) = 0$. In particular, there exists no polynomial $g(x) \in \mathbb{F}_q[x]$ of degree $< n$ such that $g(\sigma) = 0$. Hence $x^n - 1$ is the minimal polynomial of σ. $\qquad \square$

Theorem 8.16 *The dual basis of a normal basis of $\mathbb{F}_{q^n}$ over $\mathbb{F}_q$ is also a normal basis.*

Proof. Let $\{\alpha, \sigma(\alpha), \sigma^2(\alpha), \ldots, \sigma^{n-1}(\alpha)\}$ be a normal basis of $\mathbb{F}_{q^n}$ over $\mathbb{F}_q$ and $\{\beta_0, \beta_1, \beta_2, \ldots, \beta_{n-1}\}$ be its dual. By definition

$$\text{Tr}(\sigma^i(\alpha)\beta_j) = \delta_{ij} \quad \text{for all } i, j = 0, 1, 2, \ldots, n - 1. \tag{8.10}$$

Let

$$A = \begin{pmatrix} \alpha & \sigma(\alpha) & \cdots & \sigma^{n-1}(\alpha) \\ \sigma(\alpha) & \sigma^2(\alpha) & \cdots & \alpha \\ \vdots & \vdots & & \vdots \\ \sigma^{n-1}(\alpha) & \alpha & \cdots & \sigma^{n-2}(\alpha) \end{pmatrix}$$

and

$$B = \begin{pmatrix} \beta_0 & \beta_1 & \cdots & \beta_{n-1} \\ \sigma(\beta_0) & \sigma(\beta_1) & \cdots & \sigma(\beta_{n-1}) \\ \vdots & \vdots & & \vdots \\ \sigma^{n-1}(\beta_0) & \sigma^{n-1}(\beta_1) & \cdots & \sigma^{n-1}(\beta_{n-1}) \end{pmatrix}.$$

Then (8.10) can be written in matrix notation as $AB = I^{(n)}$. Thus $BA = I^{(n)}$ and $I^{(n)} = {}^t(AB) = {}^tB{}^tA = {}^tBA$, since A is a symmetric matrix. From $BA = I^{(n)} = {}^tBA$ we deduce $B = {}^tB$. It follows that $\beta_i = \sigma^i(\beta_0)$ for $i = 0$, $1, 2, \ldots, n-1$ and hence $\{\beta_0, \beta_1, \ldots, \beta_{n-1}\} = \{\beta_0, \sigma(\beta_0), \ldots, \sigma^{n-1}(\beta_0)\}$ is a normal basis.   □

The following theorem which is due to Gao (1993) suggests a method of computing the dual basis of a normal basis.

Theorem 8.17 *Let α be a normal basis generator of $\mathbb{F}_{q^n}$ over $\mathbb{F}_q$, $\alpha_i = \alpha^{q^i}$, $i = 0, 1, \ldots, n-1$ and $N = \{\alpha_0, \alpha_1, \ldots, \alpha_{n-1}\}$. Let $t_i = \mathrm{Tr}_{\mathbb{F}_{q^n}/\mathbb{F}_q}(\alpha_0\alpha_i)$, $i = 0, 1, \ldots, n-1$, and $t(x) = \sum_{i=0}^{n-1} t_i x^i$. Furthermore, let $d(x) = \sum_{i=0}^{n-1} d_i x^i \in \mathbb{F}_q[x]$ be the unique polynomial of degree $\leq n-1$ such that $t(x)d(x) \equiv 1 \pmod{x^n - 1}$ and $\beta = \sum_{i=0}^{n-1} d_i\alpha_i$. Then β is a normal basis generator of $\mathbb{F}_{q^n}$ over $\mathbb{F}_q$ which generates the dual basis of N.*

Proof. We have

$$t(x)d(x) = \sum_{0 \leq i,j \leq n-1} t_i d_j x^{i+j} \equiv \sum_{i=0}^{n-1}\sum_{k=0}^{n-1} d_k t_{i-k} x^i \pmod{x^n - 1},$$

where the subscripts are reduced modulo n. From $t(x)d(x) \equiv 1 \pmod{x^n-1}$ we deduce

$$\sum_{k=0}^{n-1} d_k t_{i-k} = \delta_{i0}.$$

Then

$$\mathrm{Tr}_{\mathbb{F}_{q^n}/\mathbb{F}_q}(\alpha_i\beta^{q^j}) = \mathrm{Tr}_{\mathbb{F}_{q^n}/\mathbb{F}_q}(\alpha_i(\textstyle\sum_{k=0}^{n-1} d_k\alpha_{j+k}))$$
$$= \textstyle\sum_{k=0}^{n-1} d_k\mathrm{Tr}_{\mathbb{F}_{q^n}/\mathbb{F}_q}(\alpha_i\alpha_{j+k})$$
$$= \textstyle\sum_{k=0}^{n-1} d_k\mathrm{Tr}_{\mathbb{F}_{q^n}/\mathbb{F}_q}(\alpha_{i-j-k}\alpha_0)$$
$$= \textstyle\sum_{k=0}^{n-1} d_k t_{i-j-k} = \delta_{ij}.$$

Therefore $\{\beta, \beta^q, \ldots, \beta^{q^{n-1}}\}$ is the dual basis of N. □

Now let us compute the number of normal bases of $\mathbb{F}_{q^n}$ over $\mathbb{F}_q$. We introduce the circulant matrices first. Let

$$C = (c_{ij})_{0 \le i,j \le n-1}$$

be an $n \times n$ matrix with all the entries c_{ij} in a finite field $\mathbb{F}_q$. C is called a *circulant matrix* if its rows are generated by successive shifts of its first row. In other words, $c_{i+1,j+1} = c_{ij}$ for all i, $j = 0, 1, 2, \ldots, n-1$, where the subscripts are reduced modulo n. Thus if we let $c_j = c_{0j}$ for $j = 0, 1, 2, \ldots, n-1$, then

$$C = \begin{pmatrix} c_0 & c_1 & c_2 & \cdots & c_{n-1} \\ c_{n-1} & c_0 & c_1 & \cdots & c_{n-2} \\ c_{n-2} & c_{n-1} & c_0 & \cdots & c_{n-3} \\ \vdots & \vdots & \vdots & \ddots & \vdots \\ c_1 & c_2 & c_3 & \cdots & c_0 \end{pmatrix}. \tag{8.11}$$

We denote by $C(c_0, c_1, c_2, \ldots, c_{n-1})$ the above $n \times n$ circulant matrix with $(c_0, c_1, c_2, \ldots, c_{n-1})$ as its first row; for example, let

$$P = \begin{pmatrix} 0 & 1 & 0 & \cdots & 0 \\ 0 & 0 & 1 & \cdots & 0 \\ 0 & 0 & 0 & \cdots & 0 \\ \vdots & \vdots & \vdots & & \vdots \\ 0 & 0 & 0 & \cdots & 1 \\ 1 & 0 & 0 & \cdots & 0 \end{pmatrix},$$

then $P = C(0, 1, 0, \ldots, 0)$.

Theorem 8.18 *Let σ be the Frobenius automorphism of $\mathbb{F}_{q^n}$ over $\mathbb{F}_q$ and $\{\alpha, \sigma(\alpha), \sigma^2(\alpha), \ldots, \sigma^{n-1}(\alpha)\}$ be a normal basis of $\mathbb{F}_{q^n}$ over $\mathbb{F}_q$. Let $C = (c_{ij})_{0 \le i,j \le n-1}$ be a nonsingular matrix over $\mathbb{F}_q$ and define*

$$\beta_j = \sum_{i=0}^{n-1} c_{ij}\sigma^i(\alpha), \quad i = 0, 1, 2, \ldots, n-1. \tag{8.12}$$

Then $\{\beta_0, \beta_1, \beta_2, \ldots, \beta_{n-1}\}$ is a normal basis of $\mathbb{F}_{q^n}$ over $\mathbb{F}_q$ and $\beta_j = \sigma^j(\beta_0)$ for $j = 0, 1, 2, \ldots, n-1$, if and only if C is a circulant matrix.

Proof. Since C is nonsingular, by Theorem 8.2 $\{\beta_0, \beta_1, \beta_2, \ldots, \beta_{n-1}\}$ is a basis of $\mathbb{F}_{q^n}$ over $\mathbb{F}_q$.

Suppose C is a circulant matrix. Let $\beta = \beta_0$, then

$$\beta = \sum_{i=0}^{n-1} c_{i0}\sigma^i(\alpha).$$

Thus

$$\sigma(\beta) = \sum_{i=0}^{n-1} c_{i0}\sigma^{i+1}(\alpha) = \sum_{i=0}^{n-1} c_{i+1,1}\sigma^{i+1}(\alpha) = \sum_{i=0}^{n-1} c_{i1}\sigma^i(\alpha) = \beta_1,$$

and, more generally, for any j, $0 \le j \le n-1$,

$$\sigma^j(\beta) = \sum_{i=0}^{n-1} c_{i0}\sigma^{i+j}(\alpha) = \sum_{i=0}^{n-1} c_{i+j,j}\sigma^{i+j}(\alpha) = \sum_{i=0}^{n-1} c_{ij}\sigma^i(\alpha) = \beta_j.$$

Therefore $\{\beta_0, \beta_1, \beta_2, \ldots, \beta_{n-1}\}$ is a normal basis of $\mathbb{F}_{q^n}$ over $\mathbb{F}_q$.

Conversely, let $\{\beta_0, \beta_1, \beta_2, \ldots, \beta_{n-1}\}$ be a normal basis of $\mathbb{F}_{q^n}$ over $\mathbb{F}_q$ and $\beta_j = \sigma^j(\beta_0)$ for $j = 0, 1, 2, \ldots, n-1$. Then

$$\sum_{i=0}^{n-1} c_{ij}\sigma^i(\alpha) = \sum_{i=0}^{n-1} c_{i0}\sigma^{j+i}(\alpha), \quad j = 0, 1, 2, \ldots, n-1.$$

Comparing the coefficients of $\sigma^i(\alpha)$ of both sides of the j-th equation, we obtain $c_{ij} = c_{i-j,0}$. Comparing the coefficients of $\sigma^{i+1}(\alpha)$ of both sides of the $(j+1)$-th equation, we obtain $c_{i+1,j+1} = c_{i-j,0}$. Therefore $c_{i+1,j+1} = c_{ij}$. Hence C is a circulant matrix. $\qquad\square$

For example, let $\{\alpha, \sigma(\alpha), \sigma^2(\alpha), \ldots, \sigma^{n-1}(\alpha)\}$ be a normal basis of $\mathbb{F}_{q^n}$ over $\mathbb{F}_q$. Take $c_j = 1$ and $c_i = 0$ for $i \ne j$ in (8.11), and define β_j by (8.12) for $j = 0, 1, 2, \ldots, n-1$. Then $\{\beta_0, \beta_1, \beta_2, \ldots, \beta_{n-1}\}$ is a cyclic permutation of $\{\alpha, \sigma(\alpha), \sigma^2(\alpha), \ldots, \sigma^{n-1}(\alpha)\}$, or more precisely, $\beta_0 = \sigma^j(\alpha), \beta_1 = \sigma^{j+1}(\alpha), \ldots, \beta_{n-1} = \sigma^{j+n-1}(\alpha)$, where the indices are reduced modulo n.

Corollary 8.19 *The number of normal bases of $\mathbb{F}_{q^n}$ over $\mathbb{F}_q$ is equal to the number of non-singular circulant $n \times n$ matrices over $\mathbb{F}_q$ divided by n.*

Denote the set of $n \times n$ circulant matrices over $\mathbb{F}_q$ by $R(n, q)$ and the set of $n \times n$ non-singular circulant matrix over $\mathbb{F}_q$ by $G(n, q)$. Corollary 8.19 says that the number of normal bases of $\mathbb{F}_{q^n}$ over $\mathbb{F}_q$ is equal to $(1/n)|G(n, q)|$.

Theorem 8.20 $R(n, q)$ *is a commutative ring with respect to the matrix addition and matrix multiplication, and the map*

$$\phi: \quad \begin{aligned} R(n, q) &\rightarrow \mathbb{F}_q[x]/(x^n - 1) \\ C(c_0, c_1, \ldots, c_{n-1}) &\mapsto c_0 + c_1 x + \cdots + c_{n-1} x^{n-1} \end{aligned}$$

is a ring isomorphism. Moreover, $G(n, q)$ is a group and is isomorphic to the group of units of $\mathbb{F}_q[x]/(x^n - 1)$ under the map ϕ. In particular, $C(c_0, c_1, \ldots, c_{n-1})$ is nonsingular if and only if the polynomial $c_0 + c_1 x + \cdots + c_{n-1} x^{n-1}$ and $x^n - 1$ are coprime in $\mathbb{F}_q[x]$.

Proof. Clearly, ϕ is a bijection and

$$C(c_0, c_1, c_2, \ldots, c_{n-1}) = c_0 I + c_1 P + c_2 P^2 + \cdots + c_{n-1} P^{n-1},$$

where $P = C(0, 1, 0, \ldots, 0)$. We have $P^n = I$ and $x^n \equiv 1 \pmod{x^n - 1}$. Therefore ϕ preserves the addition and multiplication. By Theorem 3.21, $R(n, q)$ is a ring and ϕ is a ring isomorphism. The remaining statements of the theorem follows immediately. $\square$

Let $f(x)$ be any polynomial in $\mathbb{F}_q[x]$. Denote the group of units of $\mathbb{F}_q[x]/(f(x))$ by $(\mathbb{F}_q[x]/(f(x)))^*$ and let $\psi_q(f) = |(\mathbb{F}_q[x]/(f(x)))^*|$. By Theorem 8.20 $\psi_q(x^n - 1)$ is the number of $n \times n$ non-singular circulant matrices over $\mathbb{F}_q$. Then by Corollary 8.19 we have

Corollary 8.21 *The number of normal bases of $\mathbb{F}_{q^n}$ over $\mathbb{F}_q$ is equal to $\psi_q(x^n - 1)/n$.*

As the Euler's function $\phi(n)$, there is an explicit expression for $\psi_q(f)$. At first, parallel to Corollary 2.25 we have

Theorem 8.22 *Let $p(x)$ be an irreducible polynomial over $\mathbb{F}_q$, $\deg p(x) = n \geq 1$, and e be a positive integer. Then*

$$\psi_q(p(x)^e) = q^{n(e-1)}(q^n - 1).$$

Proof. Let $g(x) \in \mathbb{F}_q[x]$ and $\deg g(x) < ne$. Then $\overline{g(x)} \in (\mathbb{F}_q[x]/(p(x)^e))^*$ if and only if $\gcd(g(x), p(x)) = 1$. Thus $\overline{g(x)} \notin (\mathbb{F}_q[x]/(p(x)^e))^*$ if and only if $g(x) = h(x)p(x)$, where $h(x) \in \mathbb{F}_q[x]$, $\deg h(x) < n(e-1)$. The number of polynomials of degree $< ne$ in $\mathbb{F}_q[x]$ is q^{ne} and that of degree $< n(e-1)$ in $\mathbb{F}_q[x]$ is $q^{n(e-1)}$. Therefore $\psi_q(p(x)^e) = q^{ne} - q^{n(e-1)} = q^{n(e-1)}(q^n - 1)$. □

Completely parallel to Corollary 2.27, from the Chinese Remainder Theorem for polynomials over a field (Theorem 4.13) we deduce

Corollary 8.23 *Let* $f_1(x), f_2(x), \ldots, f_r(x)$ *be* r *polynomials of degree* ≥ 1 *in* $\mathbb{F}_q[x]$, *which are pairwise coprime, and let* $f(x) = f_1(x)f_2(x)\cdots f_r(x)$. *Then*

$$\psi_q(f(x)) = \psi_q(f_1(x))\psi_q(f_2(x))\cdots\psi_q(f_r(x)).$$

From Theorem 8.22 and Corollary 8.23 we deduce

Theorem 8.24 *Let* $f(x)$ *be a polynomial of degree* $n \geq 1$ *in* $\mathbb{F}_q[x]$ *and*

$$f(x) = cp_1(x)^{e_1}p_2(x)^{e_2}\cdots p_r(x)^{e_r},$$

where $c \in \mathbb{F}_q^*$, $p_1(x), p_2(x), \ldots, p_r(x)$ *are distinct monic irreducible polynomials over* $\mathbb{F}_q$, $\deg p_i(x) = n_i$, $i = 1, 2, \ldots, r$, *and* $e_1, e_2, \ldots, e_r$ *are positive integers. Then*

$$\psi_q(f(x)) = \prod_{i=1}^{r} q^{n_i(e_i-1)}(q^{n_i} - 1) = q^n \prod_{i=1}^{r}\left(1 - \frac{1}{q^{n_i}}\right). \qquad (8.13)$$

Thus if we factorize $x^n - 1$ into a product of powers of distinct monic irreducible polynomials, we can obtain an explicit expression of $\psi_q(x^n - 1)$ by Theorem 8.24. However, using results of the next chapter, viz., formula (9.20) and Theorem 9.16, we can obtain a closed expression of $\psi_q(x^n - 1)$ without factorizing $x^n - 1$ as follows.

Theorem 8.25 *Let* q *be a power of a prime* p, n *be a positive integer, and* $n = p^e m$ *where* $p \nmid m$. *Then*

$$\psi_q(x^n - 1) = q^n \prod_{d|n}\left(1 - q^{-\mathrm{ord}_d(q)}\right)^{\phi(d)/\mathrm{ord}_d(q)}, \qquad (8.14)$$

where $\mathrm{ord}_d(q)$ *represents the order of* q *in the group* $\mathbb{Z}_d^*$.

Proof. By (9.20),

$$x^n - 1 = \prod_{d|n} \Phi_d(x).$$

By Theorem 9.16 $\Phi_d(x)$ is a product of $\phi(d)/\mathrm{ord}_d(q)$ monic irreducible polynomials of degree $\mathrm{ord}_d(q)$. Then (8.14) follows immediately from (8.13). □

Corollary 8.26 *Let q be a power of a prime p, n be a positive integer and $n = p^e m$ where $p \nmid m$. Then the number of normal bases of $\mathbb{F}_{q^n}$ over $\mathbb{F}_q$ is equal to*

$$\frac{1}{n}\psi_q(x^n - 1) = \frac{q^n}{n} \prod_{d|n}(1 - q^{-\mathrm{ord}_d(q)})^{\phi(d)/\mathrm{ord}_d(q)}.$$

Knowing a normal basis of $\mathbb{F}_{q^n}$ over $\mathbb{F}_q$ we can obtain all normal bases of $\mathbb{F}_{q^n}$ over $\mathbb{F}_q$ by Theorem 8.18. To find a normal basis generator is the same thing as to find a normal basis. Thus it is interesting to have some criterions for normal basis generators. First, we have the following easy consequence of Corollary 8.4.

Theorem 8.27 *Let α be an element of $\mathbb{F}_{q^n}$. Then α is a normal basis generator of $\mathbb{F}_{q^n}$ over $\mathbb{F}_q$ if and only if*

$$\begin{vmatrix} \alpha & \alpha^q & \cdots & \alpha^{q^{n-1}} \\ \alpha^q & \alpha^{q^2} & \cdots & \alpha \\ \vdots & \vdots & & \vdots \\ \alpha^{q^{n-1}} & \alpha & \cdots & \alpha^{q^{n-2}} \end{vmatrix} \neq 0. \tag{8.15}$$

Then we deduce

Theorem 8.28 *Let α be an element of $\mathbb{F}_{q^n}$. Then α is a normal basis generator of $\mathbb{F}_{q^n}$ over $\mathbb{F}_q$ if and only if the polynomials $x^n - 1$ and $\alpha^{q^{n-1}}x^{n-1} + \cdots + \alpha^q x + \alpha$ are coprime in $\mathbb{F}_{q^n}[x]$.*

Proof. Let

$$C = \begin{pmatrix} \alpha & \alpha^q & \cdots & \alpha^{q^{n-1}} \\ \alpha^{q^{n-1}} & \alpha & \cdots & \alpha^{q^{n-2}} \\ \vdots & \vdots & & \vdots \\ \alpha^q & \alpha^{q^2} & \cdots & \alpha \end{pmatrix}.$$

Then C is a circulant matrix and the rows of C is a permutation of the rows of (8.15). By Theorem 8.27 α is a normal basis generator of $\mathbb{F}_{q^n}$ over $\mathbb{F}_q$ if and only if C is nonsingular. Therefore our theorem follows from Theorem 8.20. $\square$

Theorem 8.28 and the following Corollary are due to Perlis (1942).

Corollary 8.29 *Let α be an element of $\mathbb{F}_{q^n}$ and assume that $n = p^e$ is a power of the characteristic p of $\mathbb{F}_{q^n}$. Then an element α of $\mathbb{F}_{q^n}$ is a normal basis generator of $\mathbb{F}_{q^n}$ over $\mathbb{F}_q$ if and only if $\mathrm{Tr}_{\mathbb{F}_{q^n}/\mathbb{F}_q}(\alpha) \neq 0$.*

Proof. Let σ be the Frobenius automorphism of $\mathbb{F}_{q^n}$ over $\mathbb{F}_q$. Then

$$\mathrm{Tr}_{\mathbb{F}_{q^n}/\mathbb{F}_q}(\alpha) = \alpha + \sigma(\alpha) + \cdots + \sigma^{n-1}(\alpha).$$

If α is a normal basis generator of $\mathbb{F}_{q^n}$ over $\mathbb{F}_q$, then $\alpha, \sigma(\alpha), \ldots, \sigma^{n-1}(\alpha)$ are linearly independent, and, hence, $\mathrm{Tr}_{\mathbb{F}_{q^n}/\mathbb{F}_q}(\alpha) \neq 0$. Conversely, suppose $\mathrm{Tr}_{\mathbb{F}_{q^n}/\mathbb{F}_q}(\alpha) \neq 0$, then 1 is not a root of $\alpha^{q^{n-1}}x^{n-1} + \cdots + \alpha^q x + \alpha = 0$. By assumption $n = p^e$, thus $x^n - 1 = (x-1)^{p^e}$ and 1 is the only root of $x^n - 1$. Therefore $x^n - 1$ and $\alpha^{q^{n-1}}x^{n-1} + \cdots + \alpha^q x + \alpha$ are coprime in $\mathbb{F}_{q^n}[x]$. By Theorem 8.28, α is a normal basis generator of $\mathbb{F}_{q^n}$ over $\mathbb{F}_q$. $\square$

Corollary 8.30 *Let* $f(x) = x^2 + a_1 x + a_2$ *be an irreducible quadratic polynomial over* $\mathbb{F}_q$ *and* α *be one of its roots. Then* α *is a normal basis generator of* $\mathbb{F}_{q^2}$ *over* $\mathbb{F}_q$ *if and only if* $a_1 \neq 0$, *which is equivalent to* $Tr_{\mathbb{F}_{q^2}/\mathbb{F}_q}(\alpha) \neq 0$.

Proof. Note that $x^2 - 1 = (x+1)(x-1)$. Since $\alpha^q \neq \alpha$, $x + 1$ and $\alpha^q x + \alpha$ are coprime. Clearly, $x - 1$ and $\alpha^q x + \alpha$ are coprime if and only if $\alpha^q + \alpha \neq 0$. But $\alpha^q + \alpha = -a_1$. $\square$

Let σ be the Frobenius automorphism of $\mathbb{F}_{q^n}$ over $\mathbb{F}_q$, $\alpha \in \mathbb{F}_{q^n}$ and $f(x) = a_0 + a_1 x + \cdots + a_m x^m \in \mathbb{F}_q[x]$. Define $f(\sigma) = a_0 + a_1 \sigma + \cdots + a_m \sigma^m$ and

$$f(\sigma)(\alpha) = a_0 \alpha + a_1 \sigma(\alpha) + \cdots + a_m \sigma^m(\alpha).$$

If $f(\sigma)(\alpha) = 0$, we say that $f(x)$ *annihilates* α. By the proof of Theorem 8.15, the minimal polynomial of σ, which is $x^n - 1$, annihilates every element of $\mathbb{F}_{q^n}$. Thus there is a monic polynomial of least degree which annihilates α; moreover, such a polynomial is unique and divides any polynomial which annihilates α. We have

Theorem 8.31 *Let* q *be a power of a prime* p, $n = n_1 p^e$ *with* $\gcd(p, n_1) = 1$ *and* $e \geq 0$. *Assume*

$$x^{n_1} - 1 = \phi_1(x)\phi_2(x)\cdots\phi_r(x),$$

where $\phi_1(x), \phi_2(x), \ldots, \phi_r(x)$ *are irreducible polynomials in* $\mathbb{F}_q[x]$, *and let*

$$\Psi_i(x) = (x^n - 1)/\phi_i(x), \quad i = 1, 2, \ldots, r.$$

Then an element $\alpha \in \mathbb{F}_{q^n}$ *is a normal basis generator of* $\mathbb{F}_{q^n}$ *over* $\mathbb{F}_q$ *if and only if*

$$\Psi_i(\sigma)\alpha \neq 0, \quad i = 1, 2, \ldots, r.$$

Proof. α is a normal basis generator of $\mathbb{F}_{q^n}$ over $\mathbb{F}_q$ if and only if $\alpha, \sigma(\alpha)$, $\sigma^2(\alpha), \ldots, \sigma^{n-1}(\alpha)$ are linearly independent over $\mathbb{F}_q$, i.e., the monic polynomial of least degree which annihilates α is $x^n - 1$. Clearly, this is true if and only if $\Psi_i(\sigma)\alpha \neq 0$ for $i = 1, 2, \ldots, r$. $\square$

Theorem 8.31 is due to Schwarz (1988) and the following Corollary is due to Pei et al. (1986).

Corollary 8.32 *Let q be a power of a prime p, n be a prime different from p, α be an element of $\mathbb{F}_{q^n}$ not belonging to $\mathbb{F}_q$. Suppose q is a primitive element modulo n. Then α is a normal basis generator of $\mathbb{F}_{q^n}$ over $\mathbb{F}_q$ if and only if $\mathrm{Tr}_{\mathbb{F}_{q^n}/\mathbb{F}_q}(\alpha) \neq 0$.*

Proof. We have

$$x^n - 1 = (x - 1)(x^{n-1} + \cdots + x + 1).$$

Let ξ be a root of $x^{n-1} + \cdots + x + 1$ in an extension of $\mathbb{F}_q$, then ξ^q, ξ^{q^2}, $\ldots$ are its roots in the extension, too. We assert that $\xi, \xi^q, \xi^{q^2}, \ldots, \xi^{q^{n-2}}$ are distinct. Suppose $\xi^{q^i} = \xi^{q^j}$, where $0 \leq i, j \leq n - 2$, then $\xi^{q^j - q^i} = 1$. But $\xi^n = 1$ and n is a prime, so the order of ξ in $\mathbb{F}_q[\xi]^*$ is n. Therefore $n \mid (q^j - q^i)$, i.e., $q^j \equiv q^i \pmod{n}$. Since q is an primitive element modulo n, we must have $i = j$. Thus $\xi, \xi^q, \xi^{q^2}, \ldots, \xi^{q^{n-2}}$ are $n - 1$ distinct roots of $x^{n-1} + \cdots + x + 1$, which implies $x^{n-1} + \cdots + x + 1$ is irreducible over $\mathbb{F}_q$. Let $\Psi_1(x) = (x^n - 1)/(x - 1)$ and $\Psi_2(x) = (x^n - 1)/(x^{n-1} + \cdots + x + 1)$. By Theorem 8.31 α is a normal basis generator of $\mathbb{F}_{q^n}$ over $\mathbb{F}_q$ if and only if $\Psi_1(\sigma)(\alpha) \neq 0$ and $\Psi_2(\sigma)(\alpha) \neq 0$. But $\Psi_2(\sigma)(\alpha) = \alpha^q - \alpha \neq 0$, since $\alpha \notin \mathbb{F}_q$, and

$$\Psi_1(\sigma)(\alpha) = \alpha^{q^{n-1}} + \cdots + \alpha^q + \alpha = \mathrm{Tr}_{\mathbb{F}_{q^n}/\mathbb{F}_q}(\alpha).$$

$\square$

Combining Corollaries 8.29 and 8.32, we have

Theorem 8.33 *Let q be a power of a prime p and n be a positive integer. Assume that either $n = p^e$ or n is a prime different from p having q as a primitive element modulo n. Then any element α of degree n over $\mathbb{F}_q$ with $\mathrm{Tr}_{\mathbb{F}_{q^n}/\mathbb{F}_q}(\alpha) \neq 0$ is a normal basis generator of $\mathbb{F}_{q^n}$ over $\mathbb{F}_q$.*

Chang et al. (1999) proved that the converse of Theorem 8.33 also holds.

Theorem 8.34 *Let q be a power of a prime p and n be a positive integer. Assume that every element α of degree n over $\mathbb{F}_q$ with $\mathrm{Tr}_{\mathbb{F}_{q^n}/\mathbb{F}_q}(\alpha) \neq 0$ is a normal basis generator of $\mathbb{F}_{q^n}$ over $\mathbb{F}_q$. Then $n = p^e$ or n is a prime different from p having q as a primitive element modulo n.*

A simpler proof of Theorem 8.34 due to Hachenberger (2004) will be given in Section 12.4.

We shall see how the multiplication of elements in $\mathbb{F}_{q^n}$ becomes easier in the normal basis coordinate system.

Let $\{\alpha, \alpha^q, \alpha^{q^2}, \ldots, \alpha^{q^{n-1}}\}$ be a normal basis of $\mathbb{F}_{q^n}$ over $\mathbb{F}_q$. Let $\xi, \eta \in \mathbb{F}_{q^n}$ and express them as

$$\xi = x_0\alpha + x_1\alpha^q + \cdots + x_{n-1}\alpha^{q^{n-1}}, \quad \text{where } x_i \in \mathbb{F}_q, \qquad (8.16)$$

$$\eta = y_0\alpha + y_1\alpha^q + \cdots + y_{n-1}\alpha^{q^{n-1}}, \quad \text{where } y_i \in \mathbb{F}_q. \qquad (8.17)$$

Then

$$\xi\eta = \sum_{i,j=0}^{n-1} x_iy_j\alpha^{q^i}\alpha^{q^j} = \sum_{i,j=0}^{n-1} x_iy_j\alpha^{q^i+q^j} = \sum_{i,j=0}^{n-1} x_iy_j(\alpha^{q^{i-j}+1})^{q^j}. \qquad (8.18)$$

Thus if we know how to express the product of an arbitrary basis element α^{q^m} with α in the normal basis $\{\alpha, \alpha^q, \alpha^{q^2}, \ldots, \alpha^{q^{n-1}}\}$ we can evaluate arbitrary products. Let

$$\alpha\alpha^{q^j} = \alpha^{q^j+1} = \sum_{i=0}^{n-1} t_{ij}\alpha^{q^i}, \quad \text{where } t_{ij} \in \mathbb{F}_q. \qquad (8.19)$$

Substituting (8.19) into (8.18), we obtain

$$\xi\eta = \sum_{i,j,k=0}^{n-1} x_iy_jt_{k,i-j}\alpha^{q^{k+j}} = \sum_{i,j,m=0}^{n-1} x_iy_jt_{m-j,i-j}\alpha^{q^m}. \qquad (8.20)$$

In contrast to the procedure of computing products in arbitrary basis coordinate system given in Section 8.1, where n^3 constants $c_{ijk}'s$ $(i,j,k = 1, 2, \ldots, n)$ of $\mathbb{F}_q$ are required, we now only need one matrix $T = (t_{ij})_{0 \le i,j \le n-1}$ over $\mathbb{F}_q$, which was pointed out by Massey and Omura (1986).

Furthermore, we have

Theorem 8.35 *Let α be a normal basis generator of $\mathbb{F}_{q^n}$ over $\mathbb{F}_q$, and the matrix $T = (t_{ij})_{0 \le i,j \le n}$ be determined by formula (8.19). Define*

$$A = (a_{ij})_{0 \le i,j \le n-1} \quad \text{with } a_{ij} = t_{-j,i-j},$$

where the subscripts are reduced modulo n, and let

$$f(\xi, \eta) = (x_0, x_1, \ldots, x_{n-1})A\,{}^t(y_0, y_1, \ldots, y_{n-1}),$$

where ξ and η are as in (8.16) and (8.17). Then A is a symmetric matrix and the product $\xi\eta$ can be computed as follows:

$$\xi\eta = f(\xi,\eta)\alpha + f(\xi^{q^{n-1}},\eta^{q^{n-1}})\alpha^q + f(\xi^{q^{n-2}},\eta^{q^{n-2}})\alpha^{q^2} + \cdots + f(\xi^q,\eta^q)\alpha^{q^{n-1}}.$$

$$(8.21)$$

Proof. Let $\alpha_i = \alpha^{q^i}$, $i = 0, 1, \ldots, n-1$, then (8.19) can be written as

$$\alpha_0\alpha_j = \sum_{i=0}^{n-1} t_{ij}\alpha_i. \quad \text{where } t_{ij} \in \mathbb{F}_q. \tag{8.22}$$

In particular,

$$\alpha_0\alpha_{i-j} = \sum_{k=0}^{n-1} t_{k,i-j}\alpha_k.$$

Raising to the q^j-th power, we obtain

$$\alpha_j\alpha_i = \sum_{k=0}^{n-1} t_{k,i-j}\alpha_{k+j}.$$

Similarly,

$$\alpha_i\alpha_j = \sum_{k=0}^{n-1} t_{k,j-i}\alpha_{k+i}.$$

Thus

$$\sum_{k=0}^{n-1} t_{k,i-j}\alpha_{k+j} = \sum_{k=0}^{n-1} t_{k,j-i}\alpha_{k+i}.$$

Equating the coefficients of α_0, we get

$$t_{-j,i-j} = t_{-i,j-i}, \tag{8.23}$$

i.e., $a_{ij} = a_{ji}$. Therefore A is symmetric.

Let

$$\xi\eta = \sum_{m=0}^{n-1} p_m\alpha^{q^m},$$

then

$$p_m = \sum_{i,j=0}^{n-1} t_{m-j,i-j}x_iy_j = \sum_{i,j=0}^{n-1} t_{-j,i-j}x_{m+i}y_{m+j} = \sum_{i,j=0}^{n-1} a_{ij}x_{m+i}y_{m+j}.$$

In particular,

$$p_0 = \sum_{i,j=0}^{n-1} a_{ij} x_i y_j = f(\xi, \eta).$$

From (8.16) we deduce

$$\xi^q = x_{n-1}\alpha + x_0\alpha^q + x_1\alpha^{q^2} + \cdots + x_{n-2}\alpha^{q^{n-1}}.$$

Thus the coordinates of ξ^q in the normal basis are obtained by a cyclic shift of the coordinates of ξ to the right. In general,

$$\xi^{q^{n-m}} = x_m\alpha + x_{m+1}\alpha^q + x_{m+2}\alpha^{q^2} + \cdots + x_{m-1}\alpha^{q^{m-1}}.$$

Therefore

$$p_m = f(\xi^{q^{n-m}}, \eta^{q^{n-m}}).$$

$\square$

Based on (8.21) Massey and Omura suggested to construct a parallel-input-serial-output multiplier to perform the multiplication of two elements ξ and η in $\mathbb{F}_{2^n}$: The coordinates (with respect to the normal basis generated by α) of ξ and η are loaded into two shift registers. For every $a_{ij} = 1$, the ith and jth flip-flops of the first and second shift register, respectively, are connected to a multiplier. The outputs of all the multipliers are summed up (modulo 2) via XOR gates. Then, with each clock cycle, the output of the device gives one coordinate of the product $\xi\eta$; thus one successfully obtains all the coordinates in clock cycles.

Example 8.5 Let α be a root of the irreducible polynomial $x^3 + x^2 + 1 \in \mathbb{F}_2[x]$, then $\mathbb{F}_{2^3} = \mathbb{F}_2[\alpha]$ and $\alpha^3 = \alpha^2 + 1$. Clearly,

$$\alpha, \ \alpha^2, \ \alpha^{2^2} = \alpha^4 = \alpha^2 + \alpha + 1$$

is a normal basis of $\mathbb{F}_{2^3}$ over $\mathbb{F}_2$. We compute the products of α with its conjugates:

$$\alpha \cdot \alpha = \alpha^2, \ \alpha \cdot \alpha^2 = \alpha + \alpha^{2^2}, \ \alpha \cdot \alpha^{2^2} = \alpha^2 + \alpha^{2^2},$$

and we obtain the matrix

$$T = \begin{pmatrix} 0 & 1 & 0 \\ 1 & 0 & 1 \\ 0 & 1 & 1 \end{pmatrix}$$

and the matrix $A = T$. The corresponding Massey-Omura multiplier is given in Figure 8.3.

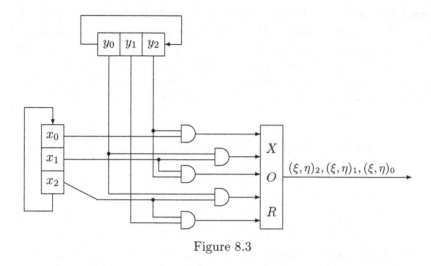

Figure 8.3

8.5 Optimal Normal Bases

Let α be a normal basis generator of $\mathbb{F}_{q^n}$ over $\mathbb{F}_q$, $\alpha_i = \alpha^{q^i}$, $i = 0, 1, \ldots, n-1$, and $N = \{\alpha_0, \alpha_1, \ldots, \alpha_{n-1}\}$ be the normal basis of $\mathbb{F}_{q^n}$ over $\mathbb{F}_q$ generated by α. Let (8.22)

$$\alpha\alpha_j = \sum_{i=0}^{n-1} t_{ij}\alpha_i, \qquad 0 \le i \le n-1,$$

where $t_{ij} \in \mathbb{F}_q$ and $T = (t_{ij})_{0 \le i,j \le n-1}$. As we mentioned in last section that we can evaluate arbitrary products in $\mathbb{F}_{q^n}$, once knowing the matrix T. Thus it is natural to introduce the following definition.

Definition 8.6 *The number of nonzero entries in the matrix T is called the complexity of the normal basis $N = \{\alpha_0, \alpha_1, \ldots, \alpha_{n-1}\}$, which is denoted by C_N, and also by C_α.*

Theorem 8.36 *For any normal basis N of $\mathbb{F}_{q^n}$ over $\mathbb{F}_q$, $C_N \ge 2n - 1$.*

Proof. By (8.22) the jth column of T is the vector of coefficients of the product $\alpha\alpha^{q^j}$ with respect to the normal basis N. Thus the sum of all columns of T equals the vector of coefficients of $\alpha(\alpha + \alpha^q + \cdots + \alpha^{q^{n-1}}) = \alpha\,\mathrm{Tr}_{\mathbb{F}_{q^n}/\mathbb{F}_q}(\alpha)$ and is therefore the n-dimensional column vector ${}^t(\mathrm{Tr}_{\mathbb{F}_{q^n}/\mathbb{F}_q}(\alpha), 0, \ldots, 0)$.

Since the elements of N are linearly independent, $\text{Tr}_{\mathbb{F}_{q^n}/\mathbb{F}_q}(\alpha) \neq 0$. Clearly, $\alpha\alpha, \alpha\alpha^q, \ldots, \alpha\alpha^{q^{n-1}}$ are linearly independent over $\mathbb{F}_q$, thus the columns of T are linearly independent, which implies that the rows of T are linearly independent, too. In particular, T has no zero row. We have seen that the sum of entries in each row but the first is zero; hence each row (except possibly for the first) contains at least two nonzero entries. This proves $C_N \geq 2n - 1$. $\qquad\qquad\square$

Definition 8.7 *A normal basis N of $\mathbb{F}_{q^n}$ over $\mathbb{F}_q$ is called optimal if $C_N = 2n - 1$. A normal basis generator α of $\mathbb{F}_{q^n}$ over $\mathbb{F}_q$ is called an optimal normal basis generator if α generates an optimal normal basis.*

Clearly, equivalent normal bases (cf. Definition 8.2) have the same complexity and if one of them is optimal so is the other.

In this section we shall first present two types of optimal normal bases and then prove that any optimal normal basis is equivalent to one of the two types.

Theorem 8.37 *Let q be a prime power and $n+1$ be a prime. Assume that $\gcd(q, n+1) = 1$ and q is a primitive element in $\mathbb{Z}_{n+1}$. Then $x^n + \cdots + x + 1$ is irreducible over $\mathbb{F}_q$, its n roots are primitive $(n+1)$-th root of unity in $\mathbb{F}_{q^n}$ and form an optimal normal basis of $\mathbb{F}_{q^n}$ over $\mathbb{F}_q$.*

Proof. Since q is a primitive element in $\mathbb{Z}_{n+1}$, $q, q^2, \ldots, q^n \equiv 1 \pmod{n+1}$ are the n distinct elements of $\mathbb{Z}_{n+1}^*$. From $q^n \equiv 1 \pmod{n+1}$ we deduce $(n+1) \mid (q^n - 1)$. By Theorem 6.3, $\mathbb{F}_{q^n}$ is a cyclic group of order $q^n - 1$, and by Theorem 2.29 $\mathbb{F}_{q^n}$ has a unique cyclic subgroup of order $n + 1$, which will be denoted by G. Let α be a generator of G, then $\alpha, \alpha^2, \ldots, \alpha^n$ are n roots of $x^{n+1} - 1$ not equal to 1, and, hence, they are all the roots of $x^n + \cdots + x + 1$. α being a generator of G is a primitive $(n+1)$-th root of unity in $\mathbb{F}_{q^n}$. Since $n + 1$ is a prime, $\alpha, \alpha^2, \ldots, \alpha^n$ are primitive roots of unity in $\mathbb{F}_{q^n}$, too.

Clearly, $\alpha, \alpha^q, \alpha^{q^2}, \ldots, \alpha^{q^{n-1}}$ are conjugates of α. Suppose $\alpha^{q^i} = \alpha^{q^j}$, where $0 \leq i < j \leq n-1$, then $\alpha^{q^{j-i}} = \alpha$ and $\alpha^{q^{j-i}-1} = 1$. Since α is of order $n + 1$, $(n+1) \mid (q^{j-i} - 1)$. Thus $q^{j-i} \equiv 1 \pmod{n+1}$. But q is a primitive element in $\mathbb{Z}_{n+1}$, we get a contradiction. Therefore $\alpha, \alpha^q, \alpha^{q^2}, \ldots, \alpha^{q^{n-1}}$ are n distinct conjugates of α. Since they are all the roots of $x^n + \cdots + x + 1$, $x^n + \cdots + x + 1$ is irreducible over $\mathbb{F}_q$. Therefore

$$N = \{\alpha, \alpha^q, \alpha^{q^2}, \cdots, \alpha^{q^{n-1}}\} = \{\alpha, \alpha^2, \cdots, \alpha^n\}$$

and N is a normal basis of $\mathbb{F}_{q^n}$ over $\mathbb{F}_q$.

Since α is a root of $x^n + \cdots + x + 1$, $\alpha^n + \cdots + \alpha + 1 = 0$. Consequently,

$$\alpha \cdot \alpha^n = \alpha^{n+1} = 1 = -\sum_{i=1}^{n} \alpha^i.$$

But

$$\alpha \cdot \alpha^i = \alpha^{i+1} \quad \text{for } 1 \leq i \leq n - 1.$$

Therefore there are $2n - 1$ nonzero entries in T, i.e., $C_N = 2n - 1$. $\qquad \square$

Any optimal normal basis obtained by the construction of Theorem 8.37 is called a *Type I optimal normal basis.*

Theorem 8.38 *Let $q = 2^v$ and $2n + 1$ be a prime, where $\gcd(v, n) = 1$. Assume that either*

(i) *2 generates the group $\mathbb{F}_{2n+1}^*$, or*

(ii) *$2n + 1 \equiv 3 \, (\text{mod} \, 4)$ and 2 generates the group $\mathbb{F}_{2n+1}^{*2}$.*

Let ζ be a primitive $(2n+1)$-th root of unity in $\mathbb{F}_{2^{2n}}$ and $\alpha = \zeta + \zeta^{-1}$. Then α is an optimal normal basis generator of $\mathbb{F}_{q^n}$ over $\mathbb{F}_q$.

Proof. Assume that (i) holds. By Theorem 8.37, ζ is an optimal normal basis generator of $\mathbb{F}_{2^{2n}}$ over $\mathbb{F}_2$. Since 2 generates the group $\mathbb{F}_{2n+1}^*$, $2^n \equiv -1 \, (\text{mod} \, 2n + 1)$. Thus

$$\alpha^{2^n} = (\zeta + \zeta^{-1})^{2^n} = \zeta^{2^n} + \zeta^{-2^n} = \zeta^{-1} + \zeta = \alpha.$$

Hence $\alpha \in \mathbb{F}_{2^n}$. We also have

$$\alpha^{2^k} = (\zeta + \zeta^{-1})^{2^k} = \zeta^{2^k} + \zeta^{-2^k} \quad \text{for } k = 0, 1, 2, \ldots, n - 1.$$

Since ζ is an optimal normal basis generator of $\mathbb{F}_{2^{2n}}$ over $\mathbb{F}_2$, the $2n$ conjugates of ζ: $\zeta, \zeta^2, \zeta^{2^2}, \ldots, \zeta^{2^{2n-1}}$ are linearly independent over $\mathbb{F}_2$. It follows that the n conjugates of α: $\alpha, \alpha^2, \alpha^{2^2}, \ldots, \alpha^{2^{n-1}}$ are also linearly independent over $\mathbb{F}_2$ and $N = \{\alpha, \alpha^2, \alpha^{2^2}, \ldots, \alpha^{2^{n-1}}\}$ form a basis of $\mathbb{F}_{2^n}$ over $\mathbb{F}_2$. Thus α is a normal basis generator of $\mathbb{F}_{2^n}$ over $\mathbb{F}_2$. Since $\gcd(v, n) = 1$, $\{\alpha, \alpha^2, \alpha^{2^2}, \ldots, \alpha^{2^{n-1}}\}$ is also a basis of $\mathbb{F}_{q^n} = \mathbb{F}_{2^{vn}}$ over $\mathbb{F}_q = \mathbb{F}_{2^v}$ (Exercise 8.3). From $\gcd(v, n) = 1$ we deduce also that $\{v, 2v, \ldots, (n-1)v \, (\text{mod} \, n)\}$ is a permutation of $\{1, 2, \ldots, n-1\}$, which implies $\{\alpha, \alpha^q, \alpha^{q^2}, \ldots, \alpha^{q^{n-1}}\} = \{\alpha, \alpha^{2^v}, \alpha^{2^{2v}}, \ldots, \alpha^{2^{(n-1)v}}\} = \{\alpha, \alpha^2, \alpha^{2^2}, \ldots, \alpha^{2^{n-1}}\} = N$. Therefore α is a normal basis generator of $\mathbb{F}_{q^n}$ over $\mathbb{F}_q$.

Let us compute the complexity of N. We have

$$\alpha\alpha = \alpha^2,$$
$$\alpha\alpha^{2^k} = (\zeta + \zeta^{-1})(\zeta^{2^k} + \zeta^{-2^k})$$
$$= (\zeta^{1+2^k} + \zeta^{-(1+2^k)}) + (\zeta^{1-2^k} + \zeta^{-(1-2^k)}), \quad 1 \le k \le n-1.$$

Since 2 is primitive in $\mathbb{Z}_{2n+1}$ and ζ is a primitive $(2n+1)$-th root of unity in $\mathbb{F}_{2^{2n}}$, $\zeta^{1+2^k}, \zeta^{-(1+2^k)}, \zeta^{1-2^k}, \zeta^{-(1-2^k)}$ are distinct in pairs for any k such that $1 \le k \le n-1$. By the proof of Theorem 8.37, $\{\zeta, \zeta^2, \zeta^{2^2}, \ldots, \zeta^{2^{2n-1}}\} = \{\zeta, \zeta^2, \zeta^3, \ldots, \zeta^{2n}\}$. Hence

$$\zeta^{1+2^k} + \zeta^{-(1+2^k)} = \alpha^{2^s}, \quad \zeta^{1-2^k} + \zeta^{-(1-2^k)} = \alpha^{2^t}$$

for suitable values of s and t with $s \ne t$. Then each product $\alpha\alpha^{2^k}$ ($1 \le k \le n-1$) is a sum of two elements of N. Therefore $C_N = 2n - 1$. Hence α is an optimal normal basis generator of $\mathbb{F}_{q^n}$ over $\mathbb{F}_q$.

Now consider the case when (ii) holds. As the above case ζ is also an optimal normal basis generator of $\mathbb{F}_{2^{2n}}$ over $\mathbb{F}_2$. By assumption $2n + 1 \equiv 3 \pmod 4$, thus $-1 \notin \mathbb{F}_{2n+1}^{*2}$ (Exercise 6.20). Therefore $\mathbb{F}_{2n+1}^* = \mathbb{F}_{2n+1}^{*2} \cup (-1)\mathbb{F}_{2n+1}^{*2}$. Also by assumption $\mathbb{F}_{2n+1}^{*2} = \langle 2 \rangle = \langle 1, 2, 2^2, \ldots, 2^{n-1} \rangle$ and $2^n \equiv 1 \pmod{2n+1}$. Then $(-1)\mathbb{F}_{2n+1}^{*2} = \langle -1, -2, -2^2, \ldots, -2^{n-1} \rangle$. Let

$$m_\zeta(x) = \prod_{i=0}^{n-1}(x - \zeta^{2^i}) \quad \text{and} \quad m_{\zeta^{-1}}(x) = \prod_{i=0}^{n-1}(x - \zeta^{-2^i}).$$

Clearly, both $m_\zeta(x)$ and $m_{\zeta^{-1}}(x)$ are irreducible over $\mathbb{F}_2$ and

$$x^{2n} + \cdots + x + 1 = m_\zeta(x)m_{\zeta^{-1}}(x).$$

Since $2^n \equiv 1 \pmod{2n+1}$, $\zeta^{2^n} = \zeta$ and $\alpha^{2^n} = (\zeta + \zeta^{-1})^{2^n} = \zeta^{2^n} + \zeta^{-2^n} = \zeta + \zeta^{-1} = \alpha$. Hence $\zeta \in \mathbb{F}_{2^n}$ and $\alpha \in \mathbb{F}_{2^n}$.

We claim that $N = \{\alpha, \alpha^2, \alpha^{2^2}, \ldots, \alpha^{2^{n-1}}\}$ is a normal basis of $\mathbb{F}_{2^n}$ over $\mathbb{F}_2$. Suppose that

$$a_0\alpha + a_1\alpha^2 + \cdots + a_{n-1}\alpha^{2^{n-1}} = 0, \quad \text{where } a_i \in \mathbb{F}_2 \ (0 \le i \le n-1).$$

Substituting $\alpha = \zeta + \zeta^{-1}$ into the above equation, we obtain

$$a_0(\zeta + \zeta^{-1}) + a_1(\zeta^2 + \zeta^{-2}) + \cdots + a_{n-1}(\zeta^{2^{n-1}} + \zeta^{-2^{n-1}}) = 0.$$

Define b_j ($1 \le j \le 2n$) by

$$b_j = a_i \quad \text{if } j \equiv 2^i \text{ or } -2^i \pmod{2n+1}.$$

Since $\mathbb{F}^{*2}_{2n+1} = \langle 2 \rangle$ and $-1 \notin \mathbb{F}^{*2}_{2n+1}$, all b_j are uniquely determined. Let

$$f(x) = b_1 + b_2 x + \cdots + b_{2n} x^{2n-1}.$$

It can be verified that $f(\zeta) = 0$ and $f(\zeta^{-1}) = 0$. Hence $m_\zeta(x) \mid f(x)$, $m_{\zeta^{-1}}(x) \mid f(x)$, and $(x^{2n} + \cdots + x + 1) \mid f(x)$. But $\deg f(x) = 2n - 1 < 2n$, therefore $b_1 = b_2 = \cdots = b_{2n} = 0$, thus $a_i = 0$ for $i = 0, 1, \ldots, n - 1$. This proves that N is a normal basis of $\mathbb{F}_{2^n}$ over $\mathbb{F}_2$. As the case when (i) holds we can prove that α is also a normal basis generator of $\mathbb{F}_{q^n}$ over $\mathbb{F}_q$.

The computation of the complexity of N is similar to the case when (i) holds. Since $2n + 1 \equiv 3 \,(\mathrm{mod}\,4)$ and 2 generates the quadratic residues modulo $2n+1$, ζ^{1+2^k}, $\zeta^{-(1+2^k)}$, ζ^{1-2^k}, $\zeta^{-(1-2^k)}$ are distinct in pairs for any k such that $1 \leq k \leq n - 1$.

The theorem is completely proved. $\qquad\square$

An optimal normal basis obtained by the construction of Theorem 8.38 is called a *Type II optimal normal basis*.

Both Theorems 8.37 and 8.38 are due to Mullin, Onyszchuk, Vanstone and Wilson (1988/1989).

Example 8.6 We know that $11 = 2 \cdot 5 + 1$ is a prime $\equiv 3 \,(\mathrm{mod}\,4)$ and 2 generates the group $\mathbb{F}^*_{11}$. Let ζ be a primitive 11-th root of unity over $\mathbb{F}_2$ and $\alpha = \zeta + \zeta^{-1}$. By Theorem 8.38,

$$\alpha_0 = \zeta + \zeta^{-1}, \alpha_1 = \zeta^2 + \zeta^{-2}, \alpha_2 = \zeta^4 + \zeta^{-4}, \alpha_3 = \zeta^8 + \zeta^{-8}, \alpha_4 = \zeta^5 + \zeta^{-5}$$

form a type II optimal normal basis N of $\mathbb{F}_{2^5}$ over $\mathbb{F}_2$. We can compute the values $\alpha\alpha_i$ by only calculating with powers of ζ as follows:

$$\begin{aligned}
\alpha\alpha &= (\zeta + \zeta^{-1})^2 = \zeta^2 + \zeta^{-2} = \alpha_1, \\
\alpha\alpha_1 &= (\zeta + \zeta^{-1})(\zeta^2 + \zeta^{-2}) = \zeta^3 + \zeta^{-3} + \zeta + \zeta^{-1} = \alpha_0 + \alpha_3, \\
\alpha\alpha_2 &= (\zeta + \zeta^{-1})(\zeta^4 + \zeta^{-4}) = \zeta^5 + \zeta^{-5} + \zeta^3 + \zeta^{-3} = \alpha_3 + \alpha_4, \\
\alpha\alpha_3 &= (\zeta + \zeta^{-1})(\zeta^8 + \zeta^{-8}) = \zeta^9 + \zeta^{-9} + \zeta^7 + \zeta^{-7} = \alpha_2 + \alpha_4.
\end{aligned}$$

Therefore $C_N = 9 = 2 \cdot 5 - 1$.

The following theorem gives the minimal polynomials of the normal basis generators constructed in Theorem 8.38.

Theorem 8.39 *Under the hypothesis of Theorem 8.38, the polynomial $f_n(x) \in \mathbb{F}_q[x]$ defined recursively by*

$$f_0(x) = 1, \quad f_1(x) = x + 1 \ and \ f_k(x) = x f_{k-1}(x) + f_{k-2}(x) \ for \ k \geq 2$$

is irreducible of degree n over $\mathbb{F}_q$ and its n roots form an optimal normal basis of $\mathbb{F}_{q^n}$ over $\mathbb{F}_q$.

Proof. It is easy to prove by induction that

$$f_k(y + y^{-1}) = 1 + \sum_{i=1}^{k}(y^i + y^{-i}),$$

where y is an indeterminate over $\mathbb{F}_q$. In particular,

$$f_n(y + y^{-1}) = 1 + \sum_{i=1}^{n}(y^i + y^{-i}).$$

Let $\alpha = \zeta + \zeta^{-1}$, where ζ is a primitive $(2n + 1)$-th root of unity in $\mathbb{F}_{2^{2n}}$, and substituting $y = \zeta$ in the above equation, we obtain

$$f_n(\alpha) = 1 + \sum_{i=1}^{n}(\zeta^i + \zeta^{-i}) = 1 + \sum_{j=1}^{2n}\zeta^j = 0.$$

Since α is of degree n over $\mathbb{F}_q$, $f_n(x)$ is irreducible over $\mathbb{F}_q$. □

For practical applications it would be desirable to have a characterization of the primes $2n + 1$ for which 2 generates the group $\mathbb{F}^*_{2n+1}$ or the group $\mathbb{F}^{*2}_{2n+1}$. This is an unsolved problem. Nevertheless there are several sufficient conditions which are collected in the following theorem.

Theorem 8.40 *Let $2n + 1$ be a prime.*

(a) *If $n = 2s$, where s is an odd prime, then 2 generates the group $\mathbb{F}^*_{2n+1}$.*

(b) *If n is a prime congruent to $1 \bmod 4$, then 2 generates the group $\mathbb{F}^*_{2n+1}$.*

(c) *If n is a prime congruent to $3 \bmod 4$, then 2 generates the group $\mathbb{F}^{*2}_{2n+1}$.*

Proof. We prove (a) only. The proofs of (b) and (c) are similar.

We write $p = 2n + 1$ for simplicity. Then $p = 4s + 1$. Since s is odd, $p \equiv 5 \,(\text{mod } 8)$. Let ξ be a root of $x^4 + 1$ in some extension field of $\mathbb{F}_p$ and put $\beta = \xi + \xi^{-1}$. Since $\xi^4 = -1$, we have $\xi^2 + \xi^{-2} = 0$. Thus $\beta^2 = 2$. Since $p \equiv 5 \,(\text{mod } 8)$, $\beta^p = \xi^p + \xi^{-p} = \xi^4\beta = -\beta$, i.e., $\beta \notin \mathbb{F}_p$. Therefore $2 \notin \mathbb{F}^{*2}_p$.

Let t be the order of 2 in $\mathbb{F}_p$, then $t \,|\, (p - 1)$, i.e., $t \,|\, 4s$. We want to show that t is not a proper divisor of $4s$. We prove by contradiction. Suppose $t = 4$, then $2^4 \equiv 1 \,(\text{mod } p)$, which implies $p = 5$ and $s = 1$, contradicting to s being an odd prime. Therefore $t \neq 4$. Suppose $t \,|\, 2s$, then $1 = 2^{2s} = \beta^{4s} = \beta^{p-1}$, which implies $\beta^p = \beta$, i.e., $\beta \in \mathbb{F}_p$, contradicting to $\beta \notin \mathbb{F}_p$. Therefore $t = 4s$. This proves that $\langle 2 \rangle = \mathbb{F}^*_p$. □

Now let us come to the characterization of the optimal normal basis, which is due to Gao and Lenstra (1992).

Theorem 8.41 *Let* $N = \{\alpha, \alpha^q, \ldots, \alpha^{q^{n-1}}\}$ *be an optimal normal basis of* $\mathbb{F}_{q^n}$ *over* $\mathbb{F}_q$. *Then* N *is equivalent to either a Type I optimal normal basis or a Type II optimal normal basis. In other words, let* $b = \mathrm{Tr}_{\mathbb{F}_{q^n}/\mathbb{F}_q}(\alpha)$, *then either*

(i) $n + 1$ *is a prime,* q *is primitive in* $\mathbb{Z}_{n+1}$, *and* $-\alpha/b$ *is a primitive* $(n + 1)$-*th root of unity, or*

(ii) (a) $q = 2^v$ *for some integer* v *such that* $\gcd(v, n) = 1$,

(b) $2n + 1$ *is a prime, 2 and* -1 *generate the multiplicative group* $\mathbb{Z}_{2n+1}^*$ *and*

(c) $\alpha/b = \zeta + \zeta^{-1}$ *for some primitive* $(2n + 1)$-*th root* ζ *of unity.*

Proof. Write Tr for $\mathrm{Tr}_{\mathbb{F}_{q^n}/\mathbb{F}_q}$. Then $\mathrm{Tr}(\alpha) = b \in \mathbb{F}_q$. Let $\alpha_i = \alpha^{q^i}$, $i = 1, 2, \ldots, n - 1$. Since $\alpha_0, \alpha_1, \ldots, \alpha_{n-1}$ are linearly independent, $b \neq 0$. Replacing α by $-\alpha/b$, we have $\mathrm{Tr}(\alpha) = -1$. Let $\beta \in \mathbb{F}_{q^n}$ be the normal basis generator of the dual basis M of N and let $\beta_i = \beta^{q^i}$, $i = 1, 2, \ldots, n - 1$. Then

$$\mathrm{Tr}(\alpha)\mathrm{Tr}(\beta) = \sum_{i,j=0}^{n-1} \alpha_i \beta_j = \sum_{j=0}^{n-1} \mathrm{Tr}(\alpha_0 \beta_j) = 1.$$

Consequently $\mathrm{Tr}(\beta) = -1$.

Now we give the following remark which is of independent interest. Assume

$$\alpha \beta_j = \sum_{i=0}^{n-1} s_{ij} \beta_i, \quad 0 \leq i \leq n - 1,$$

where $s_{ij} \in \mathbb{F}_q$. Since N and M are dual to each other,

$$s_{ij} = \mathrm{Tr}\left(\alpha_i \left(\sum_{k=0}^{n-1} s_{kj}\beta_k\right)\right) = \mathrm{Tr}(\alpha_i \alpha \beta_j) = \mathrm{Tr}\left(\left(\sum_{k=0}^{n-1} t_{ki}\alpha_k\right)\beta_j\right) = t_{ji}.$$

Therefore

$$\alpha \beta_j = \sum_{i=0}^{n-1} t_{ji}\beta_i, \quad 0 \leq i \leq n - 1. \tag{8.24}$$

Since N is optimal, from the proof of Theorem 8.36 we see that the first row of T contains exactly one nonzero entry (in the mth column, say), which necessarily equals $\mathrm{Tr}(\alpha) = -1$, whereas every other row contains exactly two nonzero entries, the sum of which is zero. Thus, by (8.24), there is an integer m, $0 \leq m \leq n - 1$ and for each i, $1 \leq i \leq n - 1$, there are two integers k_i and l_i, $0 \le k_i, l_i \le n-1$, $k_i \ne l_i$ and an element $a_i \in \mathbb{F}_q$ such that

$$\alpha\beta_0 = -\beta_m, \tag{8.25}$$
$$\alpha\beta_i = a_i\beta_{k_i} - a_i\beta_{l_i}, \quad i = 1, 2, \ldots, n-1. \tag{8.26}$$

If $m = 0$, from (8.25) we deduce $\alpha = -1$, so $n = 1$, then clearly, when q is odd (i) holds and when q is even (ii) holds. Henceforth we assume $m \ne 0$.

We distinguish the cases $n = 2m$ and $n \ne 2m$.

Case 1. $n = 2m$. Raising (8.25) to q^m-th power, we see that

$$\alpha_m\beta_m = -\beta_{2m} = -\beta_0 = \beta_m/\alpha.$$

Therefore

$$\alpha\alpha_m = 1 = -\text{Tr}(\alpha) = -\alpha_0 - \alpha_1 - \cdots - \alpha_{n-1}.$$

Thus $t_{im} = -1$ for all $i = 0, 1, \ldots, n-1$, which together with (8.26) implies that for each $i \ne 0$ there is a unique $i^* \ne m$ such that

$$\alpha\beta_i = \beta_{i^*} - \beta_m, \quad i = 1, 2, \ldots, n-1.$$

Clearly $i^* = j^*$ if and only if $i = j$. Therefore $i \mapsto i^*$ is a bijective map from $\{0, 1, \ldots, n-1\} \setminus \{0\}$ to $\{0, 1, \ldots, n-1\} \setminus \{m\}$. So $t_{ii^*} = 1$ for $i^* \ne m$ and, hence,

$$\alpha\alpha_{i^*} = \alpha_i \quad \text{for } i^* \ne m. \tag{8.27}$$

Multiplying (8.27), where i^* runs from 0 to $n-1$ but $i^* \ne m$, together with $\alpha\alpha_m = 1$ and $\alpha 1 = \alpha_0$, we obtain

$$\left(\prod_{i=0}^{n-1} \alpha\alpha_i \right) \cdot \alpha 1 = \prod_{i=0}^{n-1} \alpha_i.$$

Therefore $\alpha^{n+1} = 1$. But $\alpha \ne 1$, α is a root of $x^n + \cdots + x + 1$. Since $\mathbb{F}_{q^n} = \mathbb{F}_q[\alpha]$ and $[\mathbb{F}_{q^n} : \mathbb{F}_q] = n$, α has degree n over $\mathbb{F}_q$ and the polynomial $x^n + \cdots + x + 1$ is irreducible over $\mathbb{F}_q$. Hence $n + 1$ is a prime number, α is a primitive $(n+1)$-th root of unity, and q is primitive in $\mathbb{Z}_{n+1}$. This shows that we are in Case (i) of Theorem 8.41.

Case 2. $n \ne 2m$. First, from (8.25) $\alpha\beta_0 = -\beta_m$, we see $t_{0m} = -1$ and $t_{0i} = 0$ for $i \ne m$.

Next, we compute $\alpha\beta_{-m}$. From (8.23) we deduce

$$t_{-m,-m} = t_{0m} = -1 \quad \text{and} \quad t_{ii} = t_{0,-i} = 0 \quad \text{for } i \ne -m. \tag{8.28}$$

Therefore $\alpha\beta_{-m}$ involves the term $-\beta_{-m}$. Since every row of T, not the first row, contains only two nonzero entries, the sum of which is zero, there is a unique index $l \neq -m$ such that

$$\alpha\beta_{-m} = \beta_l - \beta_{-m}. \tag{8.29}$$

Then we claim that

$$\alpha\beta_{l+m} = \beta_{2m} - \beta_m. \tag{8.30}$$

and that $l \neq 0$. We compute $\alpha_m\alpha\beta$ in two different ways.

$$\alpha_m\alpha\beta = \alpha_m(\alpha\beta) = -\alpha_m\beta_m = -(\alpha\beta)^{q^m} = \beta_m^{q^m} = \beta_{2m},$$
$$\alpha_m\alpha\beta = \alpha(\alpha_m\beta) = \alpha(\alpha\beta_{-m})^{q^m} = \alpha(\beta_l - \beta_{-m})^{q^m}$$
$$= \alpha(\beta_{l+m} - \beta) = \alpha\beta_{l+m} + \beta_m.$$

From the above two identities we deduce (8.30). Since $m \neq -m$, by (8.28) $t_{mm} = 0$. Then from (8.30) we deduce $l \neq 0$.

Then we compute $\alpha\beta_{-l-m}$. From (8.23) and (8.29) we deduce

$$t_{-l-m,-l} = t_{-m,l} = 1.$$

Thus there is a k, $0 \leq k \leq n - 1$ such that

$$\alpha\beta_{-l-m} = \beta_{-l} - \beta_k.$$

Clearly $k \neq -l$. Since $l \neq 0$, we have $-l - m \neq -m$, which together with (8.28) implies $k \neq -l - m$.

Now we are going to prove that $\mathbb{F}_q$ is of characteristic 2 and

$$\alpha\beta_m = \beta_{m+l} - \beta_0. \tag{8.31}$$

We compute $\alpha\alpha_l\beta_{-m}$ in two different ways.

$$\alpha(\alpha_l\beta_{-m}) = \alpha(\alpha\beta_{-l-m})^{q^l} = \alpha(\beta_{-l} - \beta_k)^{q^l} = \alpha(\beta_0 - \beta_{k+l})$$
$$= -\beta_m - \alpha\beta_{k+l},$$
$$\alpha_l(\alpha\beta_{-m}) = \alpha_l(\beta_l - \beta_{-m}) = (\alpha(\beta_0 - \beta_{-l-m}))^{q^l} = (-\beta_m - \beta_{-l} + \beta_k)^{q^l}$$
$$= -\beta_{m+l} - \beta_0 + \beta_{k+l}.$$

Equating the right hand sides of the above two identities, then transforming, we obtain

$$\alpha\beta_{k+l} = -\beta_m + \beta_{m+l} + \beta_0 - \beta_{k+l}.$$

$\alpha\beta_{k+l}$ has exactly two terms and $t_{k+l,k+l} = 0$ since $k \neq -l - m$, the term $-\beta_{k+l}$ in the right hand side of the above equation must be canceled with one of the three preceding terms. Since $k+l \neq 0$, it cannot be canceled with β_0. If $k+l = m+l$ then $k = m$ and, hence, $\alpha\beta_{l+m} = -\beta_m + \beta_0$. Comparing with (8.30) we obtain $2m \equiv 0 \pmod n$, which is a contradiction. Therefore $-\beta_m - \beta_{k+l} = 0$ which implies $k + l = m$ and $2\beta_m = 0$. Hence $\mathbb{F}_q$ is of characteristic 2 and (8.31) holds.

Next we are going to prove that α has degree n over $\mathbb{F}_2$. Raising (8.29) to the q^m-th power, we obtain $\alpha_m\beta = \beta_{l+m} - \beta_0$. Comparing with (8.31), we have $\alpha\beta_m = \alpha_m\beta$. By (8.25) $\alpha\beta = \beta_m$, thus $\alpha^2\beta = \alpha_m\beta$. Consequently, $\alpha^2 = \alpha_m = \alpha^{q^m}$. By induction on k, $\alpha^{2^k} = \alpha^{q^{mk}}$ for every nonnegative integer k. Let k be the least positive integer such that $\alpha^{q^{mk}} = \alpha$, then $k \leq n$ and k is the least positive integer such that $\alpha^{2^k} = \alpha$, which implies $\alpha \in \mathbb{F}_{2^k}$ and α has degree $\leq k$ over $\mathbb{F}_2$. But α has degree n over $\mathbb{F}_q$. Thus

$$k \leq n = [\mathbb{F}_{q^n} : \mathbb{F}_q] = [\mathbb{F}_q(\alpha) : \mathbb{F}_q] \leq [\mathbb{F}_2(\alpha) : \mathbb{F}_2] \leq k.$$

Then we have equality signs throughout. In particular, $k = n$ and $[\mathbb{F}_2(\alpha) : \mathbb{F}_2] = n$. Since α has the same degree n over $\mathbb{F}_2$ and over $\mathbb{F}_q$, by Theorem 6.19 we must have $q = 2^v$ with $\gcd(v, n) = 1$ and the n conjugates of α over $\mathbb{F}_q$ are the same as those over $\mathbb{F}_2$, namely, $\alpha, \alpha^2, \ldots, \alpha^{2^{n-1}}$.

Then we prove that $N = \{\alpha, \alpha^q, \alpha^{q^2}, \ldots, \alpha^{q^{n-1}}\}$ is a self-dual normal basis of $\mathbb{F}_{q^n}$ over $\mathbb{F}_q$. We have already seen that $n = k$, thus n is the least positive integer such that $\alpha^{q^{mn}} = \alpha$. It follows that $\gcd(m, n) = 1$. Let m_1 be a positive integer such that $mm_1 \equiv 1 \pmod n$. From $\alpha_m\beta = \alpha\beta_m$ we deduce

$$\frac{\alpha}{\beta} = \frac{\alpha_m}{\beta_m} = \left(\frac{\alpha}{\beta}\right)^{q^m} = \left(\frac{\alpha}{\beta}\right)^{q^{mm_1}} = \left(\frac{\alpha}{\beta}\right)^q$$

which implies $\alpha/\beta \in \mathbb{F}_q$. There is a $c \in \mathbb{F}_q$ such that $\alpha = c\beta$. Taking traces of both sides of $\alpha = c\beta$ we obtain $-1 = \text{Tr}(\alpha) = c\text{Tr}(\beta) = -c$. Thus $c = 1$ and $\alpha = \beta$. Therefore $N = \{\alpha, \alpha^q, \alpha^{q^2}, \ldots, \alpha^{q^{n-1}}\}$ is a self-dual optimal normal basis, and, hence, T is symmetric (Exercise 8.6(2)(iii)).

Let ζ be a root of $x^2 + \alpha x + 1$ in a suitable extension field of $\mathbb{F}_{q^n}$, for example $\mathbb{F}_{q^{2n}}$, then $\alpha = \zeta + \zeta^{-1}$. We know that $|\mathbb{F}_{q^{2n}}^*| = q^{2n} - 1$, so $\text{ord}(\zeta)$ is odd and we may let $\text{ord}(\zeta) = 2t + 1$, where t is a nonnegative integer. For each integer i, let

$$\gamma_i = \zeta^i + \zeta^{-i}.$$

For example $\gamma_0 = 0, \gamma_1 = \alpha, \gamma_2 = \alpha^2$, and more generally $\gamma_{2^i} = \alpha^{2^i}$ for every non-negative integer i. We assert that there are exactly t distinct nonzero γ_i's, namely, $\gamma_1, \gamma_2, \ldots, \gamma_t$. When $i = \pm j$, clearly $\gamma_i = \gamma_j$. Conversely, assume that $\gamma_i = \gamma_j$, i.e., $\zeta^i + \zeta^{-i} = \zeta^j + \zeta^{-j}$. Since ζ^i and ζ^{-i} are the two roots of $x^2 + \gamma_i x + 1$, and ζ^j and ζ^{-j} are the two roots of $x^2 + \gamma_j x + 1$, we must have $\zeta^i = \zeta^j, \zeta^{-i} = \zeta^{-j}$, or $\zeta^i = \zeta^{-j}$, $\zeta^{-i} = \zeta^j$, which implies $i \equiv \pm j \,(\mathrm{mod}\ 2t + 1)$. This proves our assertion. We know that the n conjugates of α occur among the nonzero γ_i's, so $n \leq t$. Conversely, we show by induction that each nonzero γ_i is conjugate to α, which implies $n = t$. We have already seen that $\gamma_1 = \alpha, \gamma_2 = \alpha^2$. Assume that each nonzero $\gamma_k\,(1 \leq k \leq i - 1)$ is conjugate to α. We distinguish the cases $\gamma_{i-2} = 0$ and $\gamma_{i-2} \neq 0$. If $\gamma_{i-2} = 0$, then $i - 2 \equiv 0\,(\mathrm{mod}\ 2t + 1)$ and $i \equiv 2\,(\mathrm{mod}\ 2t + 1)$, which implies $\gamma_i = \gamma_2$. Since $\gamma_2 = \alpha^2$ is conjugate to α, γ_i is also conjugate to α. Then consider the case $\gamma_{i-2} \neq 0$. We distinguish further the cases $\gamma_{i-1} = 0$ or $\gamma_{i-1} \neq 0$. If $\gamma_{i-1} = 0$, then $i - 1 \equiv 0\,(\mathrm{mod}\ 2t + 1)$ and $i \equiv 1\,(\mathrm{mod}\ 2t + 1)$, which implies $\gamma_i = \gamma_1$. Since $\gamma_1 = \alpha$, we have $\gamma_i = \alpha$. If $\gamma_{i-1} \neq 0$, we have

$$\alpha\gamma_{i-2} = (\zeta + \zeta^{-1})(\zeta^{i-2} + \zeta^{-(i-2)}) = \gamma_{i-1} + \gamma_{i-3}. \tag{8.32}$$

By induction hypothesis both γ_{i-2} and γ_{i-1} are conjugates of α, and γ_{i-3} is either conjugate to α or equal to 0 (when $i \equiv 3\,(\mathrm{mod}\ 2t + 1)$). (8.32) tells us when $\alpha\gamma_{i-2}$ is expressed as a linear combination of the normal basis $\alpha, \alpha^2, \alpha^{2^2}, \ldots, \alpha^{2^{n-1}}$, γ_{i-1} appears with coefficient 1. Since T is symmetric, when $\alpha\gamma_{i-1}$ is expressed as a linear combination of $\alpha, \alpha^2, \alpha^{2^2}, \ldots, \alpha^{2^{n-1}}$, γ_{i-2} appears with coefficient 1, i.e., $\alpha\gamma_{i-1}$ is equal to the sum of γ_{i-2} and another conjugate of α. But we have

$$\alpha\gamma_{i-1} = (\zeta + \zeta^{-1})(\zeta^{i-1} + \zeta^{-(i-1)}) = \gamma_i + \gamma_{i-2},$$

thus the other conjugate must be γ_i. This completes the induction proof.

For each integer i not divisible by $2n + 1$, γ_i is conjugate to α. Suppose $\gamma_i = \alpha^{2^j}$. But $\alpha^{2^j} = \zeta^{2^j} + \zeta^{-2^j} = \gamma_{2^j}$, so $\gamma_i = \gamma_{2^j}$, which implies $i \equiv \pm 2^j\,(\mathrm{mod}\ 2n + 1)$. In particular, every integer i not divisible by $2n + 1$ is coprime with $2n + 1$, which implies $2n + 1$ is a prime and either 2 or -2 is a primitive root of $\mathbb{Z}_{2n+1}$, i.e., $\mathbb{Z}_{2n+1}^*$ is generated by 2 and -1. Therefore all assertions of (ii) have been proved. $\square$

From the proof of Theorem 8.42 we have

Corollary 8.42 *An optimal normal basis of Type II is self-dual.*

For optimal normal basis of Type I we have the following result of Wan and Zhou (2007).

Theorem 8.43 *Let $N = \{\alpha, \alpha^q, \ldots, \alpha^{q^{n-1}}\}$ be a type I optimal normal basis of $\mathbb{F}_{q^n}$ over $\mathbb{F}_q$, where $n \geqslant 2$. Then the complexity of the dual basis M*

of N is either $3n - 3$ or $3n - 2$, according as the characteristic of $\mathbb{F}_q$ is 2 or not 2, respectively.

Proof. Since N is a type I optimal normal basis, by Theorem 8.37, $n + 1$ is a prime, q is a primitive element in $\mathbb{Z}_{n+1}$, $x^n + \cdots + x + 1$ is irreducible over $\mathbb{F}_q$, its roots $\alpha, \alpha^q, \ldots, \alpha^{q^{n-1}}$ are all the $(n + 1)$-th primitive roots of unity in $\mathbb{F}_q$ and they form N. Therefore $\text{Tr}(\alpha) = -1$. As in the proof of Theorem 8.41 we deduce $\text{Tr}(\beta) = -1$, where β is a normal basis generator of the dual basis M of N such that $\text{Tr}(\alpha\beta) = 1$. Let $\alpha_i = \alpha^{q^i}$ and $\beta_i = \beta^{q^i}$ for $i = 1, 2, \ldots, n - 1$. Then there is an integer m, $0 \leq m \leq n - 1$ such that (8.25) holds. We assumed that $n \geq 2$, thus from the proof of Theorem 8.41 we must have $n = 2m$, $m \geqslant 1$. Moreover, by the proof of Theorem 8.41, $\alpha\alpha_m = 1$, for each $i \neq 0$ there is a unique $i^* \neq m$ such that

$$\alpha\beta_i = \beta_{i^*} - \beta_m, \quad i = 1, 2, \ldots, n - 1, \tag{8.33}$$

and $\tau : i \mapsto i^*$ is a bijective map from $\{0, 1, \ldots, n-1\} \setminus \{0\}$ to $\{0, 1, \ldots, n-1\} \setminus \{m\}$ such that

$$\alpha\alpha_{i^*} = \alpha_i \text{ for } i^* \neq m.$$

Since $\tau(i) = i^*$, we have

$$\alpha\alpha_{\tau(i)} = \alpha_i \text{ for } i \neq 0.$$

Therefore

$$\alpha\alpha_i = \alpha_{\tau^{-1}(i)} \text{ for } i \neq m.$$

Let $t(x) = t_{n-1}x^{n-1} + \cdots + t_1 x + t_0$, where $t_i = \text{Tr}(\alpha\alpha_i)$. Then $t_i = \text{Tr}(\alpha_{\tau^{-1}(i)}) = \text{Tr}(\alpha) = -1$ for $i \neq m$, and $t_m = \text{Tr}(1) = n$. So

$$t(x) = -(x^{n-1} + \cdots + x + 1) + rx^m, \tag{8.34}$$

where $r = n + 1 \in \mathbb{F}_q$. Let $d(x) = d_{n-1}x^{n-1} + \cdots + d_1 x + d_0$ be the unique polynomial of degree $\leqslant n - 1$ satisfying $d(x)t(x) \equiv 1 \pmod{x^n - 1}$. By Theorem 8.17, $\beta = d_0\alpha_0 + \cdots + d_{n-1}\alpha_{n-1}$. Because $\text{Tr}(\alpha) = \text{Tr}(\beta) = -1$, $d_0 + d_1 + \cdots + d_{n-1} = 1$. Thus

$$t(x)d(x) = [-(x^{n-1} + \cdots + x + 1) + rx^m][d_{n-1}x^{n-1} + \cdots + d_1 x + d_0]$$
$$= [-(x^{n-1} + \cdots + x + 1) + rx^m][d_{n-1}(x^{n-1} - 1) + \cdots + d_1(x - 1) + 1].$$

Because $x^n - 1 = (x - 1)(x^{n-1} + \cdots + x + 1)$ and $(x - 1)|(x^k - 1)$ for $k \in \mathbb{N}, k \geqslant 1$,

$$
\begin{aligned}
t(x)d(x) &\equiv -(x^{n-1} + \cdots + x + 1) + rx^m(d_{n-1}x^{n-1} + \cdots + d_1x + d_0) \\
&= -(x^{n-1} + \cdots + x + 1) + r[d_{n-1}x^{m-1+n} + d_{n-2}x^{m-2+n} \\
&\quad + \cdots + d_mx^n + \cdots + d_1x^{m+1} + d_0x^m] \\
&\equiv -(x^{n-1} + \cdots + x + 1) + r(d_{n-1}x^{m-1} + d_{n-2}x^{m-2} + \cdots \\
&\quad + d_mx^0 + d_{m-1}x^{n-1} + \cdots + d_1x^{m+1} + d_0x^m) \\
&= (-1 + rd_{m-1})x^{n-1} + (-1 + rd_{m-2})x^{n-2} + \cdots + \\
&\quad + (-1 + rd_0)x^m + (-1 + rd_{n-1})x^{m-1} + (-1 + rd_{n-2})x^{m-2} \\
&\quad + \cdots + (-1 + rd_m).
\end{aligned}
$$

Thus $t(x)d(x) \equiv 1 \pmod{x^n - 1}$ implies $d_i = r^{-1}$ for $i \neq m$, and $d_m = 2r^{-1}$. Therefore

$$
\begin{aligned}
\beta &= d_0\alpha_0 + \cdots + d_{n-1}\alpha_{n-1} \\
&= r^{-1}(\alpha_0 + \alpha_1 + \cdots + \alpha_{n-1}) + r^{-1}\alpha_m \\
&= r^{-1}\mathrm{Tr}(\alpha) + r^{-1}\alpha_m \\
&= -r^{-1} + r^{-1}\alpha_m.
\end{aligned}
$$

So

$$
\beta_i = \beta^{q^i} = [-r^{-1} + r^{-1}\alpha_m]^{q^i} = -r^{-1} + r^{-1}\alpha_{m+i}.
$$

When $i \neq m$,

$$
\begin{aligned}
\beta\beta_i &= (-r^{-1} + r^{-1}\alpha_m)(-r^{-1} + r^{-1}\alpha_{m+i}) \\
&= r^{-2} - r^{-2}\alpha_m - r^{-2}\alpha_{m+i} + r^{-2}(\alpha\alpha_i)^{q^m} \\
&= r^{-2} - r^{-2}\alpha_m - r^{-2}\alpha_{m+i} + r^{-2}\alpha_{\tau^{-1}(i)+m} \\
&= -r^{-1}(-r^{-1} + r^{-1}\alpha_m) - r^{-1}(-r^{-1} + r^{-1}\alpha_{m+i}) \\
&\quad + r^{-1}(-r^{-1} + r^{-1}\alpha_{\tau^{-1}(i)+m}) \\
&= -r^{-1}\beta - r^{-1}\beta_i + r^{-1}\beta_{\tau^{-1}(i)}.
\end{aligned}
$$

Since $\alpha\alpha_i = \alpha_{\tau^{-1}(i)}$ for $i \neq m$, we have $i \neq \tau^{-1}(i)$. But τ is not defined on 0, so $\tau^{-1}(i) \neq 0$. In particular,

$$
\begin{aligned}
\beta\beta &= -r^{-1}\beta - r^{-1}\beta + r^{-1}\beta_{\tau^{-1}(0)} \\
&= \begin{cases} r^{-1}\beta_{\tau^{-1}(0)}, & \text{when } \mathbb{F}_q \text{ is of characteristic 2,} \\ -2r^{-1}\beta + r^{-1}\beta_{\tau^{-1}(0)}, & \text{when } \mathbb{F}_q \text{ is of characteristic not 2.} \end{cases}
\end{aligned}
$$

When $i = m$,

$$\beta\beta_m = (-r^{-1} + r^{-1}\alpha_m)(-r^{-1} + r^{-1}\alpha_{m+m})$$
$$= r^{-2} - r^{-2}\alpha_m - r^{-2}\alpha_{m+m} + r^{-2}(\alpha\alpha_m)^{q^m}$$
$$= r^{-2} - r^{-2}\alpha_m - r^{-2}\alpha_{m+m} + r^{-2}$$
$$= -r^{-1}\beta - r^{-1}\beta_m.$$

Now we conclude that when the characteristic of $\mathbb{F}_q$ is 2, the complexity of M is $1 + 2 + 3(n-2) = 3n - 3$, and that when the characteristic of $\mathbb{F}_q$ is not 2, the complexity of M is $2 + 2 + 3(n-2) = 3n - 2$. $\square$

A partial converse of Theorem 8.43 was obtained by Wan and Zhou (2007), which will not be included here.

8.6   Exercises

8.1 If a different ordering of a basis is regarded as a different basis, prove that the number of different bases of $\mathbb{F}_{q^n}$ over $\mathbb{F}_q$ is

$$q^{n(n-1)/2} \prod_{i=1}^{n}(q^i - 1),$$

which is equal to the order of the group $\mathrm{GL}_n(\mathbb{F}_q)$ consisting of all $n \times n$ nonsingular matrices over $\mathbb{F}_q$. However, if any different ordering of a basis is regarded as the same basis, the number of different bases of $\mathbb{F}_{q^n}$ over $\mathbb{F}_q$ is

$$(1/n!)q^{n(n-1)/2} \prod_{i=1}^{n}(q^i - 1).$$

8.2 Prove that the number of polynomial bases of $\mathbb{F}_{q^n}$ over $\mathbb{F}_q$ is equal to

$$\sum_{d|n} \mu(d)q^d,$$

where d runs through all positive divisors of n.

8.3 Let m and n both be positive integers and assume $\gcd(m, n) = 1$. If $\{\alpha_1, \alpha_2, \ldots, \alpha_n\}$ is a basis of $\mathbb{F}_{q^n}$ over $\mathbb{F}_q$, it is also a basis of $\mathbb{F}_{q^{mn}}$ over $\mathbb{F}_{q^m}$.

8.4 Let m and n both be positive integers and assume $\gcd(m,n) = 1$. Let $\{\alpha_1, \alpha_2, \ldots, \alpha_m\}$ be a basis of $\mathbb{F}_{q^m}$ over $\mathbb{F}_q$ and $\{\beta_1, \beta_2 \ldots, \beta_n\}$ be a basis of $\mathbb{F}_{q^n}$ over $\mathbb{F}_q$. Prove that $\{\alpha_i\beta_j : i = 1, 2, \ldots, m; j = 1, 2, \ldots, n\}$ is a basis of $\mathbb{F}_{q^{mn}}$ over $\mathbb{F}_q$.

8.5 (1) If we view $\mathbb{F}_{q^n}$ as a vector space over $\mathbb{F}_q$ and for each $\alpha \in \mathbb{F}_{q^n}$ define $L_\alpha(\xi) = \alpha\xi$ for all $\xi \in \mathbb{F}_{q^n}$, prove that L_α is a linear transformation of the vector space $\mathbb{F}_{q^n}$. (2) Let $\{\alpha_1, \alpha_2, \ldots, \alpha_n\}$ be a basis of $\mathbb{F}_{q^n}$ over $\mathbb{F}_q$. Assume $\alpha\alpha_j = \sum_{i=1}^n a_{ij}\alpha_i$, where $a_{ij} \in \mathbb{F}_q$ and write $T_\alpha = (a_{ij})_{1 \le i,j \le n}$. Let $F = \{T_\alpha : \alpha \in \mathbb{F}_{q^n}\}$. Prove that F is a field with q^n elements with respect to matrix addition and multiplication and the map $\alpha \mapsto T_\alpha$ is an isomorphism of fields from $\mathbb{F}_{q^n}$ to F. (3) Let C be an $n \times n$ nonsingular matrix such that $CT_\alpha = T_\alpha C$ for all $\alpha \in \mathbb{F}_{q^n}$. Prove that $C = \gamma I^{(n)}$, for some $\gamma \in \mathbb{F}_{q^n}$.

8.6 (1) Let $\{\alpha_1, \alpha_2, \ldots, \alpha_n\}$ and $\{\beta_1, \beta_2 \ldots, \beta_n\}$ be two bases of $\mathbb{F}_{q^n}$ over $\mathbb{F}_q$ and $\beta_j = \sum_{i=1}^n c_{ij}\alpha_i$, $j = 1, \ldots, n$, where $c_{ij} \in \mathbb{F}_q$. For $\alpha \in \mathbb{F}_{q^n}$ let $\alpha\alpha_j = \sum_{i=1}^n t_{ij}\alpha_i$, $\alpha\beta_j = \sum_{i=1}^n s_{ij}\beta_i$, where $t_{ij}, s_{ij} \in \mathbb{F}_q$ and let $C = (c_{ij})$, $T_\alpha = (t_{ij})$, $S_\alpha = (s_{ij})$. Prove that $S_\alpha = C^{-1}T_\alpha C$. (2) Assume that $\{\alpha_1, \alpha_2, \ldots, \alpha_n\}$ and $\{\beta_1, \beta_2 \ldots, \beta_n\}$ are dual to each other. Prove that

  (i) $S_\alpha = {}^t T_\alpha$ for all $\alpha \in \mathbb{F}_{q^n}$.

  (ii) $\{\alpha_1, \alpha_2, \ldots, \alpha_n\}$ and $\{\beta_1, \beta_2 \ldots, \beta_n\}$ are weakly equivalent, if and only if ${}^t T_\alpha = T_\alpha$ for all $\alpha \in \mathbb{F}_{q^n}$.

  (iii) $\{\alpha_1, \alpha_2, \ldots, \alpha_n\}$ is self-dual, (i.e., $\{\alpha_1, \alpha_2, \ldots, \alpha_n\} = \{\beta_1, \beta_2 \ldots, \beta_n\}$) if and only if ${}^t T_\alpha = T_\alpha$ for all $\alpha \in \mathbb{F}_{q^n}$ and $\alpha_1 = \beta_1$.

8.7 Let $\{\alpha, \alpha^q, \alpha^{q^2}, \ldots, \alpha^{q^{n-1}}\}$ be a self-dual normal basis of $\mathbb{F}_{q^n}$ over $\mathbb{F}_q$ and let $\alpha\alpha^{q^j} = \sum_{i=0}^{n-1} t_{ij}\alpha^{q^i}$, where $t_{ij} \in \mathbb{F}_q$. Let $T = (t_{ij})$, $A = (a_{ij})$, where $a_{ij} = t_{-j,n-j}$. Prove that $A = T$.

8.8 (Imamura) Prove that there does not exist a self-dual polynomial basis of $\mathbb{F}_{q^n}$ over $\mathbb{F}_q$ when $n \ge 2$.

8.9 (Gollman) Let $\{1, \alpha, \alpha^2, \ldots, \alpha^{n-1}\}$ be a polynomial basis of $\mathbb{F}_{q^n}$ over $\mathbb{F}_q$, where $n \ge 2$. Prove that the dual basis of $\{1, \alpha, \alpha^2, \ldots, \alpha^{n-1}\}$ is also a polynomial basis if and only if $n \equiv 1 \pmod{p}$ where p is the characteristic of $\mathbb{F}_q$ and the minimal polynomial of α is of the form $x^n + a$.

8.10 Let n be a positive integer and d be a positive divisor of n. Assume α is a normal basis generator of $\mathbb{F}_{q^n}$ over $\mathbb{F}_q$. Prove that $\mathrm{Tr}_{\mathbb{F}_{q^n}/\mathbb{F}_{q^d}}(\alpha)$ is a normal basis generator of $\mathbb{F}_{q^d}$ over $\mathbb{F}_q$. Conversely, for any normal basis generator β of $\mathbb{F}_{q^d}$ over $\mathbb{F}_q$, prove that there is a normal basis generator α of $\mathbb{F}_{q^n}$ over $\mathbb{F}_q$ such that $\beta = \mathrm{Tr}_{\mathbb{F}_{q^n}/\mathbb{F}_{q^d}}(\alpha)$.

8.11 Let m and n both be positive integers and assume $\gcd(m,n) = 1$. Let $\alpha \in \mathbb{F}_{q^m}$ and $\beta \in \mathbb{F}_{q^n}$, then $\alpha\beta \in \mathbb{F}_{q^{mn}}$ is a normal basis generator of $\mathbb{F}_{q^{mn}}$ over $\mathbb{F}_q$ if and only if α and β are normal basis generator of $\mathbb{F}_{q^m}$ and $\mathbb{F}_{q^n}$, respectively, over $\mathbb{F}_q$. Moreover, $C_{\alpha\beta} = C_\alpha C_\beta$.

8.12 Let V be an n-dimensional vector space over $\mathbb{F}_q$ and σ be a linear transformation of V. Assume that the minimal polynomial of σ coincides with the characteristic polynomial of σ. Prove that there is a vector $v \in V$ such that v, $\sigma(v)$, $\sigma^2(v)$, $\ldots, \sigma^{n-1}(v)$ form a basis of the vector space V.

8.13 (Koenig-Rados) Prove that the number of distinct non-zero roots of $f(x) = \sum_{i=0}^{q-2} \alpha_i x^i \in \mathbb{F}_q[x]$ is equal to $q - 1 + r$, where r is the rank of the circulant matrix $C(a_0, a_1, \ldots, a_{q-2})$.

8.14 If the number of normal bases of $\mathbb{F}_{q^n}$ over $\mathbb{F}_q$ is odd, there is a self-dual normal basis. In particular, if n is odd there is a self-dual normal basis of $\mathbb{F}_{2^n}$ over $\mathbb{F}_2$.

8.15 Write down the detailed proof of Corollary 8.22 and Theorem 8.23.

8.16 (Gao) Let $\alpha \in \mathbb{F}_{q^n}$, $\alpha_i = \alpha^{q^i}$, and $t_i = \mathrm{Tr}_{\mathbb{F}_{q^n}/\mathbb{F}_q}(\alpha_0 \alpha_i)$, $i = 0, 1, \ldots,$ $n - 1$. Then α is a normal basis generator of $\mathbb{F}_{q^n}$ over $\mathbb{F}_q$ if and only if the polynomial $x^n - 1$ and $t_0 + t_1 x + \cdots + t_{n-1} x^{n-1}$ in $\mathbb{F}_q[x]$ are coprime.

8.17 Let α be a normal basis generator of $\mathbb{F}_{q^n}$ over $\mathbb{F}_q$, and $a, b \in \mathbb{F}_q^*$. Then $a = b\alpha$ is also a normal basis generator of $\mathbb{F}_{q^n}$ over $\mathbb{F}_q$ if and only if $na + b\mathrm{Tr}(\alpha) \neq 0$.

Chapter 9

Factoring Polynomials over Finite Fields

9.1 Factoring Polynomials over Finite Fields

In the following we will present an algorithm of Berlekamp by which we can always factorize a polynomial over a finite field into a product of powers of irreducible polynomials. The essence of the algorithm is the following theorem.

Theorem 9.1 *Let $f(x)$ be a monic polynomial of $\mathbb{F}_q[x]$ and $g(x)$ be a polynomial of $\mathbb{F}_q[x]$ which satisfies the congruence*

$$g(x)^q \equiv g(x) \pmod{f(x)}. \tag{9.1}$$

Then

$$f(x) = \prod_{s \in \mathbb{F}_q} \gcd(f(x), g(x) - s) \tag{9.2}$$

and the $\gcd(f(x), g(x) - s)'s$, $s \in \mathbb{F}_q$, are pairwise coprime.

As a preparation we prove the following lemmas first.

Lemma 9.2 *Let $g(x)$ be a polynomial of $\mathbb{F}_q[x]$ and s_1, s_2 be two distinct elements of $\mathbb{F}_q$, then*

$$\gcd(g(x) - s_1, \ g(x) - s_2) = 1.$$

Proof. Clearly,

$$-(s_1 - s_2)^{-1}(g(x) - s_1) + (s_1 - s_2)^{-1}(g(x) - s_2) = 1.$$

Therefore $\gcd(g(x) - s_1, g(x) - s_2) = 1$.   □

Lemma 9.3 *Let $f(x)$ be a monic polynomial of $\mathbb{F}_q[x]$ and $g(x)$ be a polynomial of $\mathbb{F}_q[x]$. Then $g(x)$ satisfies the congruence (9.1)*

$$g(x)^q \equiv g(x) \,(\mathrm{mod}\, f(x)).$$

if and only if

$$f(x) \mid \prod_{s \in \mathbb{F}_q} (g(x) - s), \tag{9.3}$$

and if and only if

$$f(x) \mid \prod_{s \in \mathbb{F}_q} \gcd(f(x), g(x) - s). \tag{9.4}$$

Proof.   By Corollary 4.5 and Theorem 6.1

$$X^q - X = \prod_{s \in \mathbb{F}_q} (X - s), \tag{9.5}$$

where X denotes an indeterminate over $\mathbb{F}_q[x]/(f(x))$. Substituting $g(x)$ for X in (9.5), we have

$$g(x)^q - g(x) = \prod_{s \in \mathbb{F}_q} (g(x) - s).$$

Thus $g(x)$ satisfies the congruence (9.1) if and only if

$$\prod_{s \in \mathbb{F}_q} (g(x) - s) \equiv 0 \,(\mathrm{mod}\, f(x)),$$

i.e., if and only if (9.3) holds, and, hence, if and only if (9.4) holds.   □

Proof of Theorem 9.1. It is clear that for any $s \in \mathbb{F}_q$

$$\gcd(f(x), g(x) - s) \mid f(x).$$

For $s_1, s_2 \in \mathbb{F}_q$ with $s_1 \neq s_2$, by Lemma 9.2 we have

$$\gcd(g(x) - s_1, g(x) - s_2) = 1.$$

So

$$\gcd(\gcd(f(x), g(x) - s_1), \gcd(f(x), g(x) - s_2)) = 1.$$

Hence

$$\prod_{s \in \mathbb{F}_q} \gcd(f(x), g(x) - s) \mid f(x). \tag{9.6}$$

On the other hand, by Lemma 9.3 we have (9.4).

From (9.4) and (9.6) we deduce (9.2). $\square$

Remark: By Theorem 6.1 all elements of $\mathbb{F}_q$ satisfy (9.1). Thus if $g(x) \equiv s_0 \pmod{f(x)}$ for some $s_0 \in \mathbb{F}_q$, then the factorization (9.2) degenerates to

$$f(x) = \gcd(f(x), 0) \prod_{s \in \mathbb{F}_q^*} \gcd(f(x), s)$$
$$= f(x) \prod_{s \in \mathbb{F}_q^*} 1.$$

But if $g(x)$ satisfies (9.1) and $g(x) \not\equiv s \pmod{f(x)}$ for any $s \in \mathbb{F}_q$, then $f(x)$ can not appear as a factor on the RHS of (9.2) and the factorization (9.2) is called a *proper factorization* of $f(x)$ into pairwise coprime factors.

Theorem 9.4 *Let $f(x)$ be a monic polynomial of degree ≥ 1 over $\mathbb{F}_q$, then the number of solutions of $x^q - x$ in the residue class ring $\mathbb{F}_q[x]/(f(x))$ is a power q^r of q if and only if $f(x)$ factorizes into a product of powers of r distinct monic irreducible polynomials of $\mathbb{F}_q[x]$.*

Proof. Let

$$f(x) = p_1(x)^{e_1} p_2(x)^{e_2} \cdots p_r(x)^{e_r},$$

where $p_1(x), p_2(x), \ldots, p_r(x)$ are r distinct monic irreducible polynomials of $\mathbb{F}_q[x]$ and $e_1, e_2, \ldots, e_r$ are r positive integers. By Lemma 9.3 the polynomial $g(x) \in \mathbb{F}_q[x]$ is a solution of the congruence (9.1) if and only if (9.3)

$$f(x) \mid \prod_{s \in \mathbb{F}_q} (g(x) - s)$$

holds. Since the factors in the product $\prod_{s \in \mathbb{F}_q} (g(x) - s)$ are pairwise coprime, for each $i \in \{1, 2, \ldots, r\}$ there exists a unique $s_i \in \mathbb{F}_q$ such that

$$g(x) \equiv s_i \pmod{p_i(x)^{e_i}}. \tag{9.7}$$

This establishes a mapping from the set S of solutions of the congruence (9.1) to the set r-tuples $\mathbb{F}_q^r = \{(s_1, s_2, \ldots, s_r) : s_i \in \mathbb{F}_q, \, i = 1, 2, \ldots, r\}$ over $\mathbb{F}_q$

$$
\begin{aligned}
S &\to \mathbb{F}_q^r \\
g(x) &\mapsto (s_1, s_2, \ldots, s_r)
\end{aligned}
\tag{9.8}
$$

where s_i $(i = 1, 2, \ldots, r)$ satisfies (9.7).

Conversely, since $p_1(x)^{e_1}, p_2(x)^{e_2}, \ldots, p_r(x)^{e_r}$ are pairwise coprime, by the Chinese Remainder Theorem, for any r-tuple $(s_1, s_2, \ldots, s_r) \in \mathbb{F}_q^r$ there exists a polynomial $g(x) \in \mathbb{F}_q[x]$ such that (9.7) hold for $i = 1, 2, \ldots, r$. Clearly, (9.7) implies (9.3), which is, as mentioned above, a necessary and sufficient condition for $g(x) \in \mathbb{F}_q[x]$ to be a solution of the congruence (9.1). Again by the Chinese Remainder Theorem two solutions $g(x)$ and $g_1(x)$ of the congruence (9.1) correspond to the same r-tuple in $\mathbb{F}_q^r$ if and only if $g(x) \equiv g_1(x) \pmod{f(x)}$. Then the conclusion of the theorem follows immediately. $\qquad\qquad\square$

Remark: Let $f(x)$ be a monic polynomial of degree n over $\mathbb{F}_q$. Then $\mathbb{F}_q[x]/(f(x))$ is a vector space over $\mathbb{F}_q$ with $\{1, x, x^2, \ldots, x^{n-1}\}$ as a basis and $\dim_{\mathbb{F}_q} \mathbb{F}_q[x]/(f(x)) = n$. Define

$$V = \{\overline{g(x)} \in \mathbb{F}_q[x]/(f(x)) : g(x)^q \equiv g(x) \pmod{f(x)}\}.$$

where $\overline{g(x)}$ denotes the residue class containing $g(x)$ mod $f(x)$. It is easy to verify that V is a subspace of $\mathbb{F}_q[x]/(f(x))$ which consists of all roots of $X^q - X$ in $\mathbb{F}_q[x]/(f(x))$. Then $\dim_{\mathbb{F}_q} V = r$ if and only if the number of roots of $X^q - X$ in $\mathbb{F}_q[x]/(f(x))$ is q^r.

Example 9.1 Consider the factorization of the polynomial

$$f(x) = x^3 + x + 1$$

over $\mathbb{F}_2$. Let $g(x) = g_0 + g_1 x + g_2 x^2$. Then

$$\begin{aligned} g(x)^2 &= g_0 + g_1 x^2 + g_2 x^4 \\ &\equiv g_0 + g_2 x + (g_1 + g_2)x^2 \pmod{f(x)}. \end{aligned}$$

Thus the condition $g(x)^2 \equiv g(x) \pmod{f(x)}$ is equivalent to

$$\begin{aligned} g_0 &= g_0, \\ g_2 &= g_1, \\ g_1 + g_2 &= g_2. \end{aligned}$$

We must have $g_1 = g_2 = 0$. Therefore $\overline{g(x)} = 0$ and $\overline{g(x)} = 1$ are all the solution of $X^2 - X$ in $\mathbb{F}_2[x]/(f(x))$. By Theorem 9.4 $x^3 + x + 1$ is power of a monic irreducible polynomial over $\mathbb{F}_2$. But we already knew that $x^3 + x + 1$ is irreducible.

Example 9.2 Consider the factorization of the polynomial

$$f(x) = x^9 + x + 1$$

over $\mathbb{F}_2$. Let

$$g(x) = g_0 + g_1 x + g_2 x^2 + g_3 x^3 + g_4 x^4 + g_5 x^5 + g_6 x^6 + g_7 x^7 + g_8 x^8.$$

In matrix notation the condition $g(x)^2 \equiv g(x) \pmod{f(x)}$ is

$$(x^0, x^2, x^4, x^6, x^8, x^{10}, x^{12}, x^{14}, x^{16}) \,^t(g_0, g_1, g_2, g_3, g_4, g_5, g_6, g_7, g_8)$$
$$\equiv (x^0, x^1, x^2, x^3, x^4, x^5, x^6, x^7, x^8) \,^t(g_0, g_1, g_2, g_3, g_4, g_5, g_6, g_7, g_8)$$
$$(\mathrm{mod}\, f(x)). \tag{9.9}$$

First let us compute

$$
\begin{aligned}
x^0 &\equiv x^0 && (\mathrm{mod}\, f(x)), \\
x^2 &\equiv x^2 && (\mathrm{mod}\, f(x)), \\
x^4 &\equiv x^4 && (\mathrm{mod}\, f(x)), \\
x^6 &\equiv x^6 && (\mathrm{mod}\, f(x)), \\
x^8 &\equiv x^8 && (\mathrm{mod}\, f(x)), \\
x^{10} &\equiv x^2 + x && (\mathrm{mod}\, f(x)), \\
x^{12} &\equiv x^4 + x^3 && (\mathrm{mod}\, f(x)), \\
x^{14} &\equiv x^6 + x^5 && (\mathrm{mod}\, f(x)), \\
x^{16} &\equiv x^8 + x^7 && (\mathrm{mod}\, f(x)).
\end{aligned}
$$

Then we have

$$(x^0, x^2, x^4, x^6, x^8, x^{10}, x^{12}, x^{14}, x^{16}) = (x^0, x^1, x^2, x^3, x^4, x^5, x^6, x^7, x^8)A,$$

where

$$A = \begin{pmatrix}
1 & 0 & 0 & 0 & 0 & 0 & 0 & 0 & 0 \\
0 & 0 & 0 & 0 & 0 & 1 & 0 & 0 & 0 \\
0 & 1 & 0 & 0 & 0 & 1 & 0 & 0 & 0 \\
0 & 0 & 0 & 0 & 0 & 0 & 1 & 0 & 0 \\
0 & 0 & 1 & 0 & 0 & 0 & 1 & 0 & 0 \\
0 & 0 & 0 & 0 & 0 & 0 & 0 & 1 & 0 \\
0 & 0 & 0 & 1 & 0 & 0 & 0 & 1 & 0 \\
0 & 0 & 0 & 0 & 0 & 0 & 0 & 0 & 1 \\
0 & 0 & 0 & 0 & 1 & 0 & 0 & 0 & 1
\end{pmatrix}.$$

Let

$$B = A - I = \begin{pmatrix}
0 & 0 & 0 & 0 & 0 & 0 & 0 & 0 & 0 \\
0 & 1 & 0 & 0 & 0 & 1 & 0 & 0 & 0 \\
0 & 1 & 1 & 0 & 0 & 1 & 0 & 0 & 0 \\
0 & 0 & 0 & 1 & 0 & 0 & 1 & 0 & 0 \\
0 & 0 & 1 & 0 & 1 & 0 & 1 & 0 & 0 \\
0 & 0 & 0 & 0 & 0 & 1 & 0 & 1 & 0 \\
0 & 0 & 0 & 1 & 0 & 0 & 1 & 1 & 0 \\
0 & 0 & 0 & 0 & 0 & 0 & 0 & 1 & 1 \\
0 & 0 & 0 & 0 & 1 & 0 & 0 & 0 & 0
\end{pmatrix}.$$

Then condition (9.9) becomes

$$B\,^t(g_0, g_1, g_2, g_3, g_4, g_5, g_6, g_7, g_8) = 0. \tag{9.10}$$

Solving (9.10), we get two roots of $X^2 - X$ in $\mathbb{F}_2[x]/(f(x))$. They are $\overline{g(x)} = 0$ and $\overline{g(x)} = 1$. By Theorem 9.4 $f(x) = x^9 + x + 1$ is a power of an irreducible polynomial over $\mathbb{F}_2$.

Clearly, $f'(x) = x^8 + 1$ and $f(x) + xf'(x) = 1$. Thus $\gcd(f(x), f'(x)) = 1$. Therefore $f(x) = x^9 + x + 1$ is an irreducible polynomial over $\mathbb{F}_2$.

Example 9.3 Consider the factorization of the polynomial

$$f(x) = x^5 + x + 1$$

over $\mathbb{F}_2$. Let

$$g(x) = g_0 + g_1 x + g_2 x^2 + g_3 x^3 + g_4 x^4.$$

In matrix notation the condition $g(x)^2 \equiv g(x) \pmod{f(x)}$ is

$$(x^0, x^2, x^4, x^6, x^8)\ {}^t(g_0, g_1, g_2, g_3, g_4)$$
$$\equiv (x^0, x^1, x^2, x^3, x^4)\ {}^t(g_0, g_1, g_2, g_3, g_4) \pmod{f(x)}.$$

First we compute

$$\begin{aligned}
x^0 &\equiv x^0 &\pmod{f(x)}, \\
x^2 &\equiv x^2 &\pmod{f(x)}, \\
x^4 &\equiv x^4 &\pmod{f(x)}, \\
x^6 &\equiv x^2 + x &\pmod{f(x)}, \\
x^8 &\equiv x^4 + x^3 &\pmod{f(x)}.
\end{aligned}$$

Then we have

$$(x^0, x^2, x^4, x^6, x^8) \equiv (x^0, x^1, x^2, x^3, x^4)A \pmod{f(x)}, \qquad (9.11)$$

where

$$A = \begin{pmatrix} 1 & 0 & 0 & 0 & 0 \\ 0 & 0 & 0 & 1 & 0 \\ 0 & 1 & 0 & 1 & 0 \\ 0 & 0 & 0 & 0 & 1 \\ 0 & 0 & 1 & 0 & 1 \end{pmatrix}.$$

Let

$$B = A - I = \begin{pmatrix} 0 & 0 & 0 & 0 & 0 \\ 0 & 1 & 0 & 1 & 0 \\ 0 & 1 & 1 & 1 & 0 \\ 0 & 0 & 0 & 1 & 1 \\ 0 & 0 & 1 & 0 & 0 \end{pmatrix}.$$

Then condition (9.11) becomes

$$B\ {}^t(g_0, g_1, g_2, g_3, g_4) = 0. \qquad (9.12)$$

Solving (9.12), we get two linearly independent roots of $X^2 - X$ in the residue class ring $\mathbb{F}_2[x]/(f(x))$. Let them be $\overline{g_1(x)}$ and $\overline{g_2(x)}$, where

$$g_1(x) = 1 \text{ and } g_2(x) = x^4 + x^3 + x.$$

Therefore $f(x)$ has two coprime factors. It is easy to check $\gcd(x^5 + x + 1, x^4 + x^3 + x) = x^3 + x^2 + 1$ and $\gcd(x^5 + x + 1, x^4 + x^3 + x + 1) = x^2 + x + 1$. Hence

$$x^5 + x + 1 = (x^3 + x^2 + 1)(x^2 + x + 1).$$

We know that both $x^3 + x^2 + 1$ and $x^2 + x + 1$ are irreducible over $\mathbb{F}_2$, so this is the complete factorization of $x^5 + x + 1$.

Now we prove that a set of linearly independent roots of $X^q - X$ in $\mathbb{F}_q[x]/(f(x))$ can help us to achieve the factorization of the monic polynomial $f(x)$ into a product of powers of distinct irreducible factors. Now let

$$\overline{g_1(x)} = 1, \overline{g_2(x)}, \ldots, \overline{g_r(x)}$$

be such a set. Clearly, deg $g_i(x) \geq 1$ for $i = 2, \ldots, r$. By Theorem 9.1 we have the proper factorization

$$f(x) = \prod_{s \in \mathbb{F}_q} \gcd(f(x), g_2(x) - s).$$

Deleting those factors in this factorization which are equal to 1, we can write $f(x)$ as

$$f(x) = f_1(x)f_2(x) \cdots f_m(x), \tag{9.13}$$

where $m > 1$ and each $f_i(x)$ $(1 \leq i \leq m)$ is a monic polynomial of the form $\gcd(f(x), g_2(x) - s)$ for some $s \in \mathbb{F}_q$ and is not equal to 1. Then $f_1(x), f_2(x), \ldots, f_m(x)$ are pairwise coprime. If $m = r$, by Theorem 9.4 we are finished. For the case $m < r$, we need the following lemma.

Lemma 9.5 *Assume that $m < r$, then there exist integers $i\,(3 \leq i \leq r)$ and $j\,(1 \leq j \leq m)$ such that $g_i(x) \not\equiv s \pmod{f_j(x)}$ for all $s \in \mathbb{F}_q$.*

Proof. Suppose on the contrary that for all $i = 3, \ldots, r$ and $j = 1, 2, \ldots, m$, there are elements $s_{ij} \in \mathbb{F}_q$ such that

$$g_i(x) \equiv s_{ij} \pmod{f_j(x)}. \tag{9.14}$$

Since $m < r$, some of the $f_j(x)'s$, say $f_1(x)$, can be further factorized into products of coprime factors. Then there is a root $\overline{h_1(x)}$ of $X^q - X$ in

$\mathbb{F}_q[x]/(f_1(x))$, where $\deg h_1(x) \geq 1$. By Chinese Remainder Theorem, there is an element $g(x) \in \mathbb{F}_q[x]/(f(x))$ such that

$$g(x) \equiv h_1(x) \pmod{f_1(x)},$$
$$g(x) \equiv 0 \qquad \pmod{f_j(x)}, \; j = 2, 3, \ldots, m.$$

It follows that

$$g(x)^q \equiv h_1(x)^q \equiv h_1(x) \pmod{f_1(x)},$$
$$g(x)^q \equiv 0 \qquad \pmod{f_j(x)}, \; j = 2, 3, \ldots, m.$$

By the uniqueness part of the Chinese Remainder Theorem, we must have $g(x)^q \equiv g(x) \pmod{f(x)}$. That is, $\overline{g(x)} \in V$. Then $g(x)$ can be expressed as

$$g(x) \equiv a_1 + a_2 g_2(x) + \cdots + a_r g_r(x) \pmod{f(x)},$$

where $a_1, a_2, \ldots, a_r \in \mathbb{F}_q$. It follows that

$$g(x) \equiv a_1 + a_2 g_2(x) + \cdots + a_r g_r(x) \pmod{f_1(x)}.$$

By (9.14), $g_i(x) \equiv s_{i1} \pmod{f_1(x)}$ for $i = 3, \ldots, r$. We know that $f_1(x) = \gcd(f(x), g_2(x) - s)$ for some $s \in \mathbb{F}_q$, so $g_2(x) \equiv s \pmod{f_1(x)}$. Write s_{21} for s, then $g_2(x) \equiv s_{21} \pmod{f_1(x)}$. Thus

$$g(x) \equiv a_1 + a_2 s_{21} + \cdots + a_r s_{r1} \pmod{f_1(x)}.$$

Clearly, $a_1 + a_2 s_{21} + \cdots + a_r s_{r1} \in \mathbb{F}_q$. But $g(x) \equiv h_1(x) \pmod{f_1(x)}$ and $\deg h_1(x) \geq 1$. This is a contradiction. □

Therefore where $m < r$ there exist integers $i \, (3 \leq i \leq r)$ and $j \, (1 \leq j \leq m)$ such that $g_i(x) \not\equiv s \pmod{f_j(x)}$ for all $s \in \mathbb{F}_q$. After rearranging $g_3(x), \ldots, g_r(x)$, we can assume that $i = 3$. Then $g_3(x) \not\equiv s$ for all $s \in \mathbb{F}_q$. From $g_3(x)^q \equiv g_3(x) \pmod{f(x)}$ we deduce $g_3(x)^q \equiv g_3(x) \pmod{f_i(x)}$ for all $i = 1, 2, \ldots, m$. By Theorem 9.1 we have the factorizations

$$f_i(x) = \prod_{s \in \mathbb{F}_q} \gcd(f_i(x), g_3(x) - s), \; i = 1, 2, \ldots, m. \qquad (9.15)$$

Substituting (9.15) for each $i = 1, 2, \ldots, m$ into (9.13) and deleting those factors which are equal to 1, we obtain

$$f(x) = f_{21}(x) f_{22}(x) \ldots f_{2m'}(x),$$

where each $f_{2i}(x) \, (1 \leq i \leq m')$ is a monic polynomial of the form

$$\gcd(\gcd(f(x), g_2(x) - s), g_3(x) - s') \text{ for some } s, s' \in \mathbb{F}_q$$

and is $\neq 1$. Since the factorization of $f_j(x)$ in (9.15) is a proper factorization, $m' > m$. If $m' = r$, by Theorem 9.4, we are finished. For the case $m' < r$, we have the following generalization of Lemma 9.5.

Lemma 9.6 *Assume that $m' < r$, then there exist integers i $(4 \leq i \leq r)$ and j $(1 \leq j \leq m')$ such that $g_i(x) \not\equiv s$ (mod $f_{2j}(x)$) for all $s \in \mathbb{F}_q$.*

The proof is similar to Lemma 9.5 and is left as an exercise.

Then proceeding as above we obtain a finer factorization of $f(x)$. Finally, $f(x)$ is factored into a product of powers of distinct monic irreducible polynomials over $\mathbb{F}_q$.

Factorizing a polynomial $f(x)$ in $\mathbb{F}_q[x]$ and of degree ≥ 1 into a product of factors and knowing that each factor is a power of an irreducible polynomial, it is still necessary to find these irreducible polynomials and calculate the number of times that they appear in the factors. In order to solve this problem, we proceed as follows.

Theorem 9.7 *Let $f(x) \in \mathbb{F}_q[x]$, where q is a power of a prime number p, and assume that $f'(x) = 0$. Then $f(x) = h(x)^p$ where $h(x) \in \mathbb{F}_q[x]$.*

Proof. Let

$$f(x) = a_0 + a_1 x + a_2 x^2 + \cdots + a_n x^n,$$

where $a_0, a_1, a_2, \ldots, a_n \in \mathbb{F}_q$ and $a_n \neq 0$. Then

$$f'(x) = a_1 + 2a_2 x + \cdots + na_n x^{n-1}.$$

From $f'(x) = 0$ we deduce

$$a_1 = 2a_2 = \cdots = na_n = 0.$$

It follows that $a_i = 0$ if $p \nmid i$. Since $a_n \neq 0$, $p \mid n$. Let $n = pn_1$. For those i with $p \mid i$, by Theorem 6.19 let $b_i \in \mathbb{F}_q$ be such that $a_i = b_i^p$. Then

$$
\begin{aligned}
f(x) &= a_0 + a_p x^p + a_{2p} x^{2p} + \cdots + a_{n_1 p} x^{n_1 p} \\
&= b_0^p + b_p^p x^p + b_{2p}^p x^{2p} + \cdots + b_{n_1 p}^p x^{n_1 p} \\
&= (b_0 + b_p x + b_{2p} x^2 + \cdots + b_{n_1 p} x^{n_1})^p.
\end{aligned}
$$

Let $h(x) = b_0 + b_p x + b_{2p} x^2 + \cdots + b_{n_1 p} x^{n_1}$, then $h(x) \in \mathbb{F}_q[x]$ and $f(x) = h(x)^p$. $\square$

Corollary 9.8 *Let $f(x) \in \mathbb{F}_q[x]$, where q is a power of a prime number p. Then there is a power p^m such that $f(x) = h(x)^{p^m}$ and $h'(x) \neq 0$.*

Theorem 9.9 *Let $f(x)$ be a power of a monic irreducible polynomial over $\mathbb{F}_q$ and assume that $f'(x) \neq 0$. If $\gcd(f(x), f'(x)) = 1$, then $f(x)$ is irreducible. If $\gcd(f(x), f'(x)) \neq 1$, then $f(x)/\gcd(f(x), f'(x))$ is irreducible. Denote*

$$p(x) = f(x)/\gcd(f(x), f'(x)),$$

then

$$f(x) = p(x)^e,$$

where $e = \deg f(x)/\deg p(x)$.

Proof. Let $f(x) = p(x)^e$, where $p(x)$ is a monic irreducible polynomial in $\mathbb{F}_q[x]$. Then $f'(x) = ep(x)^{e-1}p'(x)$. Since $f'(x) \neq 0$, $\gcd(f(x), f'(x)) = p(x)^{e-1}$. If $\gcd(f(x), f'(x)) = 1$, then $e = 1$, hence $f(x) = p(x)$ is irreducible. If $\gcd(f(x), f'(x)) \neq 1$, then $p(x) = f(x)/\gcd(f(x), f'(x))$. Hence $f(x)/\gcd(f(x), f'(x))$ is irreducible. □

Using Corollary 9.8 and Theorem 9.9 if we know that the monic polynomial $f(x)$ of degree ≥ 1 over $\mathbb{F}_q$ is a power of an irreducible polynomial, then we can find the irreducible polynomial and the power.

Example 9.4 Consider the factorization of the polynomial

$$f(x) = x^8 + x^4 + x^3 + x^2 + x + 1$$

over $\mathbb{F}_2$. Let

$$g(x) = g_0 + g_1 x + g_2 x^2 + g_3 x^3 + g_4 x^4 + g_5 x^5 + g_6 x^6 + g_7 x^7.$$

Then the condition $g(x)^2 \equiv g(x) \pmod{f(x)}$ is

$$(x^0, x^2, x^4, x^6, x^8, x^{10}, x^{12}, x^{14})\,{}^t(g_0, g_1, g_2, g_3, g_4, g_5, g_6, g_7)$$
$$\equiv (x^0, x^1, x^2, x^3, x^4, x^5, x^6, x^7)\,{}^t(g_0, g_1, g_2, g_3, g_4, g_5, g_6, g_7) \pmod{f(x)}.$$

By computation, we have

$$(x^0, x^2, x^4, x^6, x^8, x^{10}, x^{12}, x^{14}) \equiv (x^0, x^1, x^2, x^3, x^4, x^5, x^6, x^7)A \pmod{f(x)},$$

where

$$A = \begin{pmatrix} 1 & 0 & 0 & 0 & 1 & 0 & 1 & 1 \\ 0 & 0 & 0 & 0 & 1 & 0 & 1 & 0 \\ 0 & 1 & 0 & 0 & 1 & 1 & 1 & 1 \\ 0 & 0 & 0 & 0 & 1 & 1 & 1 & 1 \\ 0 & 0 & 1 & 0 & 1 & 1 & 0 & 1 \\ 0 & 0 & 0 & 0 & 0 & 1 & 1 & 0 \\ 0 & 0 & 0 & 1 & 0 & 1 & 1 & 0 \\ 0 & 0 & 0 & 0 & 0 & 0 & 1 & 1 \end{pmatrix}.$$

Let

$$B = A - I = \begin{pmatrix} 0 & 0 & 0 & 0 & 1 & 0 & 1 & 1 \\ 0 & 1 & 0 & 0 & 1 & 0 & 1 & 0 \\ 0 & 1 & 1 & 0 & 1 & 1 & 1 & 1 \\ 0 & 0 & 0 & 1 & 1 & 1 & 1 & 1 \\ 0 & 0 & 1 & 0 & 0 & 1 & 0 & 1 \\ 0 & 0 & 0 & 0 & 0 & 0 & 1 & 0 \\ 0 & 0 & 0 & 1 & 0 & 1 & 0 & 0 \\ 0 & 0 & 0 & 0 & 0 & 0 & 1 & 0 \end{pmatrix}.$$

Then the condition $g(x)^2 \equiv g(x) \pmod{f(x)}$ is equivalent to

$$B^t(g_0, g_1, g_2, g_3, g_4, g_5, g_6, g_7) = 0.$$

Solving, we get three independent roots $\overline{g_1(x)}$, $\overline{g_2(x)}$, and $\overline{g_3(x)}$ of $x^2 - x$ in $\mathbb{F}_2[x]/(f(x))$, where

$$g_1(x) = 1, \; g_2(x) = x^5 + x^3 + x^2, \; g_3(x) = x^7 + x^4 + x^2 + x.$$

Now let us use the Euclidean algorithm to calculate $\gcd(f(x), g_2(x))$ and $\gcd(f(x), g_2(x) + 1)$. We obtain

$$\gcd(f(x), g_2(x)) = x^3 + x + 1,$$

$$\gcd(f(x), g_2(x) + 1) = x^5 + x^3 + x^2 + 1.$$

Hence

$$f(x) = (x^5 + x^3 + x^2 + 1)(x^3 + x + 1).$$

Calculating the greatest common divisor of $x^5 + x^3 + x^2 + 1$ and $g_3(x)$ and also that of $x^5 + x^3 + x^2 + 1$ and $g_3(x) + 1$, we have

$$\gcd(x^5 + x^3 + x^2 + 1, g_3(x)) = x^3 + x^2 + x + 1,$$

$$\gcd(x^5 + x^3 + x^2 + 1, g_3(x) + 1) = x^2 + x + 1.$$

Hence

$$f(x) = (x^3 + x^2 + x + 1)(x^2 + x + 1)(x^3 + x + 1).$$

Clearly, both $x^2 + x + 1$ and $x^3 + x + 1$ are irreducible over $\mathbb{F}_2$. Denote

$$f_1(x) = x^3 + x^2 + x + 1.$$

Then

$$f_1'(x) = x^2 + 1 = (x + 1)^2.$$

So

$$(f_1(x), f_1'(x)) = (x + 1)^2,$$

and

$$p_1(x) = f_1(x)/(f_1(x), f_1'(x)) = x + 1.$$

Thus

$$f_1(x) = (x + 1)^3.$$

Hence

$$f(x) = (x + 1)^3(x^2 + x + 1)(x^3 + x + 1).$$

$\square$

9.2   Factorization of $x^n - 1$

Let $f(x) = x^n - 1$, where n is a positive integer. We want to factorize $f(x)$ in $\mathbb{F}_q[x]$, where q is a power of a prime number p. Assume that $n = n_1 p^e$, where $\gcd(n_1, p) = 1$. Then $x^n - 1 = (x^{n_1} - 1)^{p^e}$. Thus the factorization of $x^n - 1$ reduces to the factorization of $x^{n_1} - 1$ with $\gcd(n_1, p) = 1$.

Now we assume that $f(x) = x^n - 1$, where $\gcd(n, p) = 1$.

Theorem 9.10 *Let $\mathbb{F}_q$ be a finite field with q elements and n be a positive integer coprime with q, then $x^n - 1$ has no multiple factors. Moreover, $x^n - 1$ has n distinct roots in $\mathbb{F}_{q^m}$, where m is the order of q in $\mathbb{Z}_n^*$ and they form a cyclic group of order n.*

Proof. First, since $\gcd(n, p) = 1$, $f'(x) = n x^{n-1} \neq 0$. Hence $\gcd(f(x), f'(x)) = 1$. From this we deduce that $f(x)$ has no multiple factors.

Next, since $\gcd(n, q) = 1$, $q \in \mathbb{Z}_n^*$. Assume that the order of q in $\mathbb{Z}_n^*$ is m, then $q^m \equiv 1 \pmod{n}$. Hence $n \mid (q^m - 1)$. Let ξ be a primitive element of $\mathbb{F}_{q^m}$. By Theorem 2.21,

$$\alpha = \xi^{(q^m - 1)/n}$$

is an element of order n of $\mathbb{F}_{q^m}^*$. Then the n elements

$$\alpha^0 = 1, \alpha, \alpha^2, \ldots, \alpha^{n-1}$$

are distinct, they are the n roots of the polynomial $x^n - 1$ in $\mathbb{F}_{q^m}$, and they form a cyclic group of order n.   □

By Theorem 9.10 the n distinct roots of $x^n - 1$ form a cyclic group of order n. There are $\phi(n)$ elements of order n in this group and each of them is called a primitive n-th root of unity.

Let α be a primitive n-th root of unity in $\mathbb{F}_{q^m}$, where m is the order of q in $\mathbb{Z}_n^*$. Let σ be the Frobenius automorphism of $\mathbb{F}_{q^m}$ over $\mathbb{F}_q$. Let i be a nonnegative integer such that $0 \leq i \leq n - 1$. Assume that l is the smallest positive integer such that $\sigma^l(\alpha^i) = \alpha^i$, i.e., $\alpha^{iq^l} = \alpha^i$. Then

$$\alpha^i, \alpha^{iq}, \alpha^{iq^2}, \ldots, \alpha^{iq^{l-1}} \tag{9.16}$$

are distinct and all of them are roots of $x^n - 1$. Denote

$$h_1(x) = (x - \alpha^i)(x - \alpha^{iq})(x - \alpha^{iq^2}) \cdots (x - \alpha^{iq^{l-1}}). \tag{9.17}$$

By Theorem 7.5 $h_1(x)$ is the minimal polynomial of α^i over $\mathbb{F}_q$ and $h_1(x)$ is irreducible over $\mathbb{F}_q$. Clearly $h_1(x)|(x^n-1)$. Thus $h_1(x)$ is a monic irreducible factor of $x^n - 1$.

Now let $f_1(x) \in \mathbb{F}_q[x]$ be a monic irreducible factor of $x^n - 1$, then its roots must be some powers of α. Assume that α^i is a root of $f_1(x)$, $0 \le i \le n - 1$, then $f_1(\alpha^i) = 0$. Thus

$$f_1(\alpha^{iq^j}) = [f_1(\alpha^i)]^{q^j} = \sigma^j(f_1(\alpha^i)) = 0.$$

If l is the smallest positive integer such that $\alpha^{iq^l} = \alpha^i$, then the l powers of α in (9.16) are distinct. Let $h_1(x)$ be the polynomial (9.17). Then $h_1(x)$ is a monic irreducible factor of $x^n - 1$ in $\mathbb{F}_q[x]$. But it is clear that $h_1(x)|f_1(x)$. Since $f_1(x)$ is irreducible, $h_1(x) = f_1(x)$. Thus the l powers of α in (9.16) are all the roots of $f_1(x)$.

We give the following definition.

Definition 9.1 *Let* $a_0, a_1, a_2, \ldots, a_{l-1}$ *be* l *distinct numbers chosen from* 0 *to* $n - 1$. *Assume that they have the properties:*

$$a_i q \equiv a_{i+1} \quad (\text{mod } n), \, i = 0, 1, 2, \ldots, l - 2,$$
$$a_{l-1} q \equiv a_0 \quad (\text{mod } n).$$

Then we say that they form a q*-cycle* $\text{mod } n$, *denoted by*

$$(a_0, a_1, a_2, \ldots, a_{l-1}),$$

and call l *the length of the* q*-cycle* $\text{mod } n$.

Thus, we can restate the previous conclusion as follows.

Theorem 9.11 *Let* $\mathbb{F}_q$ *be a finite field with* q *elements and* n *be a positive integer coprime with* q. *Assume that* α *is a primitive* n-*th root of unity* (*if the order of* q *in* $\mathbb{Z}_n^*$ *is* m, *then there must exist primitive* n-*th roots of unity in* $\mathbb{F}_{q^m}$). *If* $(a_0, a_1, a_2, \ldots, a_{l-1})$ *is a* q*-cycle* $\text{mod } n$, *then*

$$f_1(x) = (x - \alpha^{a_0})(x - \alpha^{a_1})(x - \alpha^{a_2}) \cdots (x - \alpha^{a_{l-1}})$$

is a monic irreducible factor of $x^n - 1$ *in* $\mathbb{F}_q[x]$. *Conversely, if* $f_1(x)$ *is a monic irreducible factor of* $x^n - 1$ *in* $\mathbb{F}_q[x]$, *then all the roots of* $f_1(x)$ *are powers of* α *whose exponents form a* q*-cycle* $\text{mod } n$.

Corollary 9.12 *Let* $\mathbb{F}_q$ *be a finite field with* q *elements and* n *be a positive integer coprime with* q, *then the number of irreducible factors of* $x^n - 1$ *in* $\mathbb{F}_q[x]$ *is equal to the number of* q*-cycles* $\text{mod } n$ *formed by the* n *numbers* $0, 1, 2, \ldots, n - 1$. $\square$

It seems that when $\gcd(n, p) = 1$, the complete factorization of $x^n - 1$ over $\mathbb{F}_q$, is completely solved by Theorem 9.11 and Corollary 9.12, but to express the polynomial $f_1(x)$ in Theorem 9.11 as a polynomial with coefficients in $\mathbb{F}_q$ needs a lot of computation. So we have to go back to the method of factorization described in the previous section. According to that method, in order to factorize $x^n - 1$ into a product of irreducible factors, at first it is necessary to determine a basis of V where

$$V = \{\overline{g(x)} : \deg g(x) < n \text{ and } g(x)^q \equiv g(x) \pmod{x^n - 1}\}.$$

The method presented in last section to determine a basis of V is by matrix calculation. Now we give a simpler method which is only valid for $x^n - 1$.

Theorem 9.13 *Let $\mathbb{F}_q$ be a finite field with q elements and n be a positive integer coprime with q. Assume that the n numbers $0, 1, 2, \ldots, n - 1$ are partitioned into r q-cycles $\mathrm{mod}\, n$. For each q-cycle $(a_0, a_1, a_2, \ldots, a_{l-1})$ $\mathrm{mod}\, n$, let*

$$g(x) = x^{a_0} + x^{a_1} + x^{a_2} + \cdots + x^{a_{l-1}},$$

then $\overline{g(x)} \in V$. We obtain altogether r elements of V and they form a basis of V.

Proof. Clearly, $g(x)^q \equiv g(x) \pmod{x^n - 1}$, therefore $\overline{g(x)} \in V$. It is clear that each power $x^i (0 \le i \le n - 1)$ of x appears in one and only one of these r elements. Hence they are linearly independent over $\mathbb{F}_q$. By Theorem 9.4, $\dim_{\mathbb{F}_q} V$ is equal to the number of irreducible factors of $x^n - 1$ and by Theorem 9.13 the latter is also equal to the number of q-cycles $\mathrm{mod}\, n$. Therefore these r polynomials form a basis of V. $\square$

Example 9.5 Factorize the polynomial

$$x^{15} - 1$$

over $\mathbb{F}_2$ into a product of irreducible polynomials.

First, partition the 15 numbers $0, 1, 2, \ldots, 14$ into 2-cycles mod 15:

$$(0), (1, 2, 4, 8), (3, 6, 12, 9), (5, 10), (7, 14, 13, 11).$$

Corresponding to each of these 2-cycles there is an irreducible factor $x^{15} - 1$ over $\mathbb{F}_2$. The order of 2 in $\mathbb{Z}_{15}^*$ is 4. Then $15 \mid (2^4 - 1)$. Let ξ be a primitive element of $\mathbb{F}_{2^4}$. Then the irreducible factors of $x^{15} - 1$ over $\mathbb{F}_2$ are

$$\begin{aligned}
f_0(x) &= (x - 1) \\
f_1(x) &= (x - \xi)(x - \xi^2)(x - \xi^4)(x - \xi^8) \\
f_3(x) &= (x - \xi^3)(x - \xi^6)(x - \xi^{12})(x - \xi^9) \\
f_5(x) &= (x - \xi^5)(x - \xi^{10}) \\
f_7(x) &= (x - \xi^7)(x - \xi^{14})(x - \xi^{13})(x - \xi^{11}).
\end{aligned}$$

Then we have the complete factorization of $x^{15} - 1$ over $\mathbb{F}_2$:

$$x^{15} - 1 = f_0(x)f_1(x)f_3(x)f_5(x)f_7(x).$$

To express $f_1(x), f_3(x), f_5(x), f_7(x)$ as polynomial with coefficient in $\mathbb{F}_2$ it needs a lot of computation. We will use Berlekamp's algorithm to factorize $x^{15} - 1$ instead.

Let

$$g_1(x) = 1,$$
$$g_2(x) = x + x^2 + x^4 + x^8,$$
$$g_3(x) = x^3 + x^6 + x^9 + x^{12},$$
$$g_4(x) = x^5 + x^{10},$$
$$g_5(x) = x^7 + x^{11} + x^{13} + x^{14},$$

then $\overline{g_1(x)}, \overline{g_2(x)}, \overline{g_3(x)}, \overline{g_4(x)}, \overline{g_5(x)}$ form a basis of V. Using Euclidean algorithm, we can compute

$$\gcd(x^{15} - 1, g_2(x)) = x^7 + x^3 + x + 1,$$
$$\gcd(x^{15} - 1, g_2(x) + 1) = x^9 + x^4 + x^2 + x + 1.$$

Therefore

$$x^{15} - 1 = (x^7 + x^3 + x + 1)(x^9 + x^4 + x^2 + x + 1).$$

Using Euclidean algorithm, we can also compute

$$\gcd(x^7 + x^3 + x + 1, g_3(x)) = x^3 + 1,$$
$$\gcd(x^7 + x^3 + x + 1, g_3(x) + 1) = x^4 + x + 1.$$

Clearly,

$$x^3 + 1 = (x + 1)(x^2 + x + 1).$$

Therefore

$$x^{15} - 1 = (x + 1)(x^2 + x + 1)(x^4 + x + 1)(x^9 + x^4 + x^2 + x + 1).$$

We can also compute

$$\gcd(x^9 + x^4 + x^2 + x + 1, g_4(x)) = x^4 + x^3 + x^2 + x + 1,$$
$$\gcd(x^9 + x^4 + x^2 + x + 1, g_4(x) + 1) = x^4 + x + 1.$$

Hence

$$x^{15} - 1 = (x + 1)(x^2 + x + 1)(x^4 + x + 1)(x^4 + x^3 + 1)(x^4 + x^3 + x^2 + x + 1).$$

This is the complete factorization of $x^{15} - 1$ into a product of monic irreducible polynomials over $\mathbb{F}_2$.

9.3   Cyclotomic Polynomials

Let $\mathbb{C}$ be the complex field, n a positive integer, and ξ a primitive complex n-th root of unity. We have the complete factorization of $x^n - 1$ into linear factors over $\mathbb{C}$:

$$x^n - 1 = \prod_{i=0}^{n-1} (x - \xi^i). \tag{9.18}$$

By Theorem 2.21 $\operatorname{ord}(\xi^i) = n/\gcd(n, i)$, thus the order of each ξ^i is a divisor of n. Let $d > 0$ be a divisor of n. Define

$$\Phi_d(x) = \prod_{0 \le i \le n-1, \operatorname{ord}(\xi^i)=d} (x - \xi^i). \tag{9.19}$$

Clearly, $\Phi_d(x)$ is independent of n and $\deg \Phi_n(x) = \phi(n)$. We call $\Phi_d(x)$ the d-th cyclotomic polynomial. From (9.18) and (9.19) we deduce

$$x^n - 1 = \prod_{d \mid n} \Phi_d(x). \tag{9.20}$$

By the multiplicative version of Möbius inversion formula (Exercise 6.11), from (9.20) we deduce

$$\Phi_n(x) = \prod_{d \mid n} (x^d - 1)^{\mu(n/d)}, \tag{9.21}$$

which implies the following theorem.

Theorem 9.14 *For any positive integer n, $\Phi_n(x)$ is a polynomial with integer coefficients.*

Proof. By (9.21) $\Phi_n(x)$ is expressed as the quotient of two monic polynomial with integer coefficients; the numerator in this quotient is the product of all the terms $x^d - 1$ for which $\mu(n/d) = +1$ and the denominator is the product of all the terms for which $\mu(n/d) = -1$. The process of dividing one polynomial with integer coefficients by a monic polynomial with integer coefficients can only result integer coefficients in the quotient. □

It follows from Theorem 9.14 that (9.20) is a factorization of $x^n - 1$ in $\mathbb{Z}[x]$. Clearly, when $d \ne d'$, $\Phi_d(x)$ and $\Phi_{d'}(x)$ have no common divisor in $\mathbb{C}[x]$ and also in $\mathbb{Z}[x]$ (Exercise 9.7).

Example 9.6 By (9.21) we have

$$\Phi_{15}(x) = \frac{(x^{15}-1)(x-1)}{(x^5-1)(x^3-1)}$$
$$= \frac{x^{10}+x^5+1}{x^2+x+1}$$
$$= x^8 - x^7 + x^5 - x^4 + x^3 - x + 1.$$

We also have

$$\Phi_{20}[x] = \frac{(x^{20}-1)(x^2-1)}{(x^{10}-1)(x^4-1)}$$
$$= \frac{x^{10}+1}{x^2+1}$$
$$= x^8 - x^6 + x^4 - x^2 + 1.$$

The following is a table of the first few cyclotomic polynomials

n	$\Phi_n(x)$
1	$x - 1$
2	$x + 1$
3	$x^2 + x + 1$
4	$x^2 + 1$
5	$x^4 + x^3 + x^2 + x + 1$
6	$x^2 - x + 1$
7	$x^6 + x^5 + x^4 + x^3 + x^2 + x + 1$
8	$x^4 + 1$
9	$x^6 + x^3 + 1$
10	$x^4 - x^3 + x^2 - x + 1$

By Theorem 9.14 we define $\Phi_n(x)$ also to be the n-th cyclotomic polynomial over any finite field $\mathbb{F}_q$; of course the integer coefficients of $\Phi_n(x)$ should be interpreted as elements in the prime field $\mathbb{F}_p$ of $\mathbb{F}_q$. Then (9.20) is also a factorization of $x^n - 1$ in any finite field $\mathbb{F}_q$.

It is a classical result that for any positive integer n, $\Phi_n(x)$ is irreducible over the rational field $\mathbb{Q}$. However, over finite fields the situation is quite different. For example, the polynomial $\Phi_4(x) = x^2 + 1$ is irreducible over the rational field $\mathbb{Q}$. But we have

$$x^2 + 1 = (x + 1)^2 \text{ over } \mathbb{F}_2,$$
$$x^2 + 1 \text{ is irreducible over } \mathbb{F}_3,$$
$$x^2 + 1 = (x + 1)^2 \text{ over } \mathbb{F}_4,$$
$$x^2 + 1 = (x - 2)(x - 3) \text{ over } \mathbb{F}_5,$$
$$x^2 + 1 \text{ is irreducible over } \mathbb{F}_7.$$

Now let us study the factorization of $\Phi_n(x)$ over a given finite field $\mathbb{F}_q$. First we have

Theorem 9.15 *Let p be a prime number and assume that $\gcd(p,n) = 1$. Then*
$$\Phi_{np^k}(x) = \Phi_n(x)^{p^k - p^{k-1}}$$
in any field of characteristic p.

Proof. By (9.21),
$$
\begin{aligned}
\Phi_{np^k}(x) &= \prod_{d \mid np^k} (x^d - 1)^{\mu(np^k/d)} = \prod_{d \mid np^k} (x^{p^k n/d} - 1)^{\mu(d)} \\
&= \prod_{d \mid n} (x^{p^k n/d} - 1)^{\mu(d)} \prod_{dp \mid np} (x^{p^k n/dp} - 1)^{\mu(dp)} \\
&= (\prod_{d \mid n} (x^{n/d} - 1)^{\mu(d)})^{p^k} (\prod_{d \mid n} (x^{n/d} - 1)^{-\mu(d)})^{p^{k-1}} \\
&= \Phi_n(x)^{p^k} \Phi_n(x)^{-p^{k-1}} \\
&= \Phi_n(x)^{p^k - p^{k-1}}.
\end{aligned}
$$
$\square$

Example 9.7 Consider $\Phi_{36}(x)$. We have $36 = 9 \cdot 2^2 = 4 \cdot 3^2$. By Theorem 9.15, over a field of characteristic 2

$$\Phi_{36}(x) = \Phi_{9 \cdot 2^2}(x) = \Phi_9(x)^{2^2 - 2} = \Phi_9(x)^2 = (x^6 + x^3 + 1)^2$$

and over a field of characteristic 3

$$\Phi_{36}(x) = \Phi_{4 \cdot 3^2}(x) = \Phi_4(x)^{3^2 - 3} = \Phi_4(x)^6 = (x^2 + 1)^6 = (x^6 + 1)^2.$$

By Theorem 9.15, when factoring $\Phi_n(x)$ over a field of characteristic p, we can assume that $p \nmid n$.

Theorem 9.16 *Let $q = p^e$ where p is a prime number and assume that $p \nmid n$. Let m be the least positive integer such that $q^m \equiv 1 \pmod{n}$. Then over the finite field $\mathbb{F}_q$ $\Phi_n(x)$ factors into the product of $\phi(n)/m$ distinct monic irreducible polynomials of degree m.*

Proof. Let ξ be a primitive element of the finite field $\mathbb{F}_{q^m}$. Then the order of ξ is $q^m - 1$. Let $\alpha = \xi^{(q^m - 1)/n}$, then α is of order n and all α^i, where $1 \le i \le n - 1$ and $\gcd(i, n) = 1$, are of order n. They are $\phi(n)$ in number. Therefore over $\mathbb{F}_{q^m}$

$$\Phi_n(x) = \prod_{1 \le i \le n-1, \gcd(i,n)=1} (x - \alpha^i).$$

That is, $\Phi_n(x)$ splits completely into linear factors over $\mathbb{F}_{q^n}$. For each i with $1 \leq i \leq n - 1$ and $\gcd(i, n) = 1$, the q-cycle of i is $(i, iq, iq^2, \ldots, iq^{m-1})$. Corresponding to this q-cycle there is a monic irreducible polynomial over $\mathbb{F}_q$

$$f_i(x) = (x - \alpha^i)(x - \alpha^{iq})(x - \alpha^{iq^2}) \cdots (x - \alpha^{iq^{m-1}}), \qquad (9.22)$$

which is a factor of $\Phi_n(x)$, altogether we get $\phi(n)/m$ monic irreducible polynomials of degree m over $\mathbb{F}_q$, the product of which is $\Phi_n(x)$. $\qquad \square$

Corollary 9.17 *Let $q = p^e$ where p is a prime number and assume that $p \nmid n$. Then $\Phi_n(x)$ is irreducible over $\mathbb{F}_q$ if and only if $\phi(n)$ is the least positive integer such that $q^{\phi(n)} \equiv 1 \pmod{n}$.*

In most cases, $\Phi_n(x)$ is reducible. The irreducible factors of $\Phi_n(x)$ are given by (9.22). But to write (9.22) as a polynomial with coefficients in $\mathbb{F}_q$ there needs a lot of work. It seems to be easier to factorize $\Phi_n(x)$ using Berlekamp's algorithm given in Section 9.1.

Example 9.8 Factorize $\Phi_{20}(x)$ over $\mathbb{F}_3$.

By Example 9.6,

$$\Phi_{20}(x) = x^8 - x^6 + x^4 - x^2 + 1.$$

It is easy to check that 4 is the least positive integer such that $3^4 \equiv 1 \pmod{20}$. By Theorem 9.16 $\Phi_{20}(x)$ is a product of two monic irreducible polynomials of degree 4 over $\mathbb{F}_3$.

Partition the 20 numbers 0, 1, 2, $\ldots$, 19 into 3-cycles mod 20:

$$(0), (1, 3, 9, 7), (2, 6, 18, 14), (4, 12, 16, 8), (5, 15), (10), (11, 13, 19, 17).$$

The residue classes of the polynomials

$$\begin{aligned}
g_1(x) &= 1, \\
g_2(x) &= x + x^3 + x^7 + x^9, \\
g_3(x) &= x^2 + x^6 + x^{14} + x^{18}, \\
g_4(x) &= x^4 + x^8 + x^{12} + x^{16}, \\
g_5(x) &= x^5 + x^{15}, \\
g_6(x) &= x^{10}, \\
g_7(x) &= x^{11} + x^{13} + x^{17} + x^{19}
\end{aligned}$$

mod $\Phi_{20}(x)$ contain a basis of

$$V = \{\overline{g(x)} : \deg g(x) < n \text{ and } g(x)^3 \equiv g(x) \,(\mathrm{mod}\, \Phi_{20}(x))\}.$$

Using the Euclidean algorithm we can compute

$$\gcd\left(\Phi_{20}(x), g_2(x)\right) = 1,$$
$$\gcd\left(\Phi_{20}(x), g_2(x) + 1\right) = x^4 + x^3 + 2x + 1,$$
$$\gcd\left(\Phi_{20}(x), g_2(x) + 2\right) = x^4 + 2x^3 + x + 1.$$

Therefore

$$\Phi_{20}(x) = (x^4 + x^3 + 2x + 1)(x^4 + 2x^3 + x + 1)$$

is the complete factorization of $\Phi_{20}(x)$ over $\mathbb{F}_3$.

9.4   The Period of a Polynomial

Let $f(x)$ be a polynomial over $\mathbb{F}_q$ of degree ≥ 1 and with a nonzero constant term. Denote the group of units of the residue classes ring $\mathbb{F}_q[x]/(f(x))$ by $(\mathbb{F}_q[x]/(f(x)))^*$. For any $a(x) \in \mathbb{F}_q[x]$, denote by $\overline{a(x)}$ the residue class of $a(x)$ modulo $f(x)$

We assumed that $f(x)$ has a nonzero constant term, which implies $\bar{x} \in (\mathbb{F}_q[x]/(f(x)))^*$. Clearly, $(\mathbb{F}_q[x]/(f(x)))^*$ is a finite group and $\bar{x}$ is of finite order. Therefore it is reasonable to give the following definition.

Definition 9.2 *Let $f(x)$ be a polynomial over $\mathbb{F}_q$ of degree ≥ 1 and with a nonzero constant term. The period of $f(x)$ is defined to be the least positive integer l such that $f(x) \mid (x^l - 1)$. We use $p(f(x))$ or simply $p(f)$ to denote the period of $f(x)$.*

Clearly, for a monic irreducible polynomial $\neq x$, its period defined in Section 7.3 coincides with this definition.

Theorem 9.18 *Let $f(x)$ be a polynomial over $\mathbb{F}_q$ of degree ≥ 1 and with a nonzero constant term. Then $p(f) = \text{ord}(x)$, where $\text{ord}(x)$ denotes the order of the residue class of x modulo $f(x)$ in the group $(\mathbb{F}_q[x]/(f(x)))^*$.*

Corollary 9.19 *Let $f(x)$ be a polynomial over $\mathbb{F}_q$ of degree ≥ 1 and with a nonzero constant term. If $f(x) \mid (x^l - 1)$, then $p(f) \mid l$.*

We also have

Lemma 9.20 *Let $f(x)$ be an irreducible polynomial over $\mathbb{F}_q$ of degree n and with a nonzero constant term. Then $p(f) \mid (q^n - 1)$ and $\gcd(p(f), p) = 1$.*

Proof. By Lemma 6.9, $f(x)|(x^{q^n} - x)$. Since $f(x)$ has a nonzero constant term, $x \nmid f(x)$. Therefore $f(x)|(x^{q^n-1}-1)$. By Corollary 9.19, $p(f)|(q^n-1)$. Then from $\gcd(q^n - 1, p) = 1$ we deduce $\gcd(p(f), p) = 1$. $\quad\square$

In the following we will show how to determine the period $p(f)$ of a polynomial $f(x)$ over $\mathbb{F}_q$ of degree ≥ 1 and with a nonzero constant term. We need some lemmas.

Lemma 9.21 *Let $f(x)$ be a polynomial over $\mathbb{F}_q$ of degree ≥ 1 and with a nonzero constant term and assume that $f(x) = f_1(x)f_2(x)$, where $\deg f_1(x)$, $\deg f_2(x) \geq 1$, and $\gcd(f_1(x), f_2(x)) = 1$. Then $p(f) = \mathrm{lcm}[p(f_1), p(f_2)]$.*

Proof. We have $f_1(x)|(x^{p(f_1)} - 1)$ and $f_2(x)|(x^{p(f_2)} - 1)$. Now let $l = \mathrm{lcm}[p(f_1), p(f_2)]$, then $(x^{p(f_1)} - 1)|(x^l - 1)$ and $(x^{p(f_2)} - 1)|(x^l - 1)$. Hence $f_1(x)|(x^l - 1)$ and $f_2(x)|(x^l - 1)$. But $\gcd(f_1(x), f_2(x)) = 1$ and $f(x) = f_1(x)f_2(x)$, so $f(x)|(x^l - 1)$. Therefore by Corollary 9.19 we have $p(f)|l$.

On the other hand, from $f(x)|(x^{p(f)} - 1)$ we deduce $f_1(x)|(x^{p(f)} - 1)$ and $f_2(x)|(x^{p(f)} - 1)$. Again by Corollary 9.19, we have $p(f_1)|p(f)$ and $p(f_2)|p(f)$. Hence the least common multiple l of $p(f_1)$ and $p(f_2)$ is also a factor of $p(f)$, that is, $l|p(f)$.

Then from $p(f)|l$ and $l|p(f)$, we deduce $p(f) = l = \mathrm{lcm}[p(f_1), p(f_2)]$. $\quad\square$

Lemma 9.22 *Assume that $\gcd(j, p) = 1$, then the polynomial $x^j - 1$ in $\mathbb{F}_q[x]$ has no multiple factors.*

Proof. Let $f(x) = x^j - 1$, then $f'(x) = jx^{j-1}$. Since $\gcd(j, p) = 1$ and p is the characteristic of $\mathbb{F}_q$, $f'(x) \neq 0$. The only irreducible factor of $f'(x)$ is x and $x \nmid (x^j - 1)$. Hence $\gcd(f(x), f'(x)) = 1$. Then by Theorem 4.15 $f(x)$ has no multiple factors. $\quad\square$

Lemma 9.23 *Let $f(x)$ be an irreducible polynomial over $\mathbb{F}_q$, $f(0) \neq 0$, and let e be an integer ≥ 0. Then*

$$p(f^e) = p(f)\min\{p^i : p^i \geq e\},$$

where $\min\{p^i : p^i \geq e\}$ denotes the smallest power of p which is $\geq e$.

Proof. Let

$$\min\{p^i : p^i \geq e\} = p^m,$$

that is,

$$p^{m-1} < e \quad \text{and} \quad p^m \geq e.$$

From $f(x) \mid (x^{p(f)} - 1)$ we deduce $f(x)^e \mid (x^{p(f)} - 1)^e$. But

$$(x^{p(f)} - 1)^e \mid (x^{p(f)} - 1)^{p^m},$$

so $f(x)^e \mid (x^{p(f)} - 1)^{p^m}$. Since the characteristic of $\mathbb{F}_q$ is p, $(x^{p(f)} - 1)^{p^m} = x^{p(f)p^m} - 1$. Hence

$$p(f^e) \mid p(f)p^m. \tag{9.23}$$

On the other hand, let $p(f^e) = p^i j$, where $\gcd(j, p) = 1$. Then

$$f(x)^e \mid (x^{p^i j} - 1).$$

But $x^{p^i j} - 1 = (x^j - 1)^{p^i}$. By Lemma 9.22, $x^j - 1$ has no multiple factors. Since $f(x)$ is an irreducible polynomial, by the unique factorization theorem (Theorem 4.14), we deduce $f(x) \mid (x^j - 1)$ and $e \leq p^i$. Then by Corollary 9.20, $p(f) \mid j$. From $e \leq p^i$ we deduce $p^i \geq p^m$. Hence

$$p(f^e) = p^i j \geq p^m p(f). \tag{9.24}$$

From (9.23) and (9.24) we deduce

$$p(f^e) = p(f)p^m = p(f)\min\{p^i : p^i \geq e\}.$$

$\square$

From the above lemmas we can deduce immediately

Theorem 9.24 *Let $f(x)$ be a polynomial over $\mathbb{F}_q$ with a nonzero constant term and assume that*

$$f(x) = f_1(x)^{e_1} f_2(x)^{e_2} \cdots f_r(x)^{e_r},$$

where $f_1(x), f_2(x), \ldots, f_r(x)$ are r distinct irreducible polynomial of $\mathbb{F}_q[x]$ and $e_1, e_2, \ldots, e_r$ are r positive integers, then

$$p(f) = \text{lcm}[p(f_1), p(f_2), \ldots, p(f_r)]\min\{p^t : p^t \geq e_1, e_2, \ldots, e_r\}.$$

Proof. By Lemma 9.21, applying induction on r, we have

$$\begin{aligned} p(f) &= \text{lcm}[p(f_1^{e_1}), p(f_2^{e_2}, \ldots, f_r^{e_r})] \\ &= \text{lcm}[p(f_1^{e_1}), \text{lcm}[p(f_2^{e_2}), \ldots, p(f_r^{e_r})]] \\ &= \text{lcm}[p(f_1^{e_1}), p(f_2^{e_2}), \ldots, p(f_r^{e_r})]. \end{aligned}$$

Next by Lemma 9.23, we have

$$p(f_i^{e_i}) = p(f_i)\min\{p^t : p^t \geq e_i\}, \ i = 1, 2, \ldots, r.$$

Further, by Lemma 9.21, $(p(f_i), p) = 1$, so

$$p(f) = [p(f_1), p(f_2), \ldots, p(f_r)] \min\{p^t : p^t \geq e_1, e_2, \ldots, e_r\}.$$

$\square$

By Theorem 9.24 to determine the period of a polynomial $f(x)$ over $\mathbb{F}_q$ of degree ≥ 1 and with a nonzero constant term, we may proceed in two steps. The first step is to factorize $f(x)$ into a product of irreducible polynomials in $\mathbb{F}_q[x]$, or to express $f(x)$ as a product of powers of a finite number of distinct irreducible polynomial over $\mathbb{F}_q$. Concerning this topic, Berlekamp's algorithm given in Section 9.1 can be applied. The second step is to determine the periods of the irreducible factors of $f(x)$. We have

Theorem 9.25 *Let $f(x)$ be an irreducible polynomial of degree n over $\mathbb{F}_q$ with a nonzero constant term and let*

$$q^n - 1 = p_1^{e_1} p_2^{e_2} \cdots p_r^{e_r},$$

where $p_1, p_2, \ldots, p_r$ are r distinct primes and $e_1, e_2, \ldots, e_r$ are r positive integers. We make the convention that

$$x^{(q^n-1)/p_i^{e_i+1}} \not\equiv 1 \,(\mathrm{mod}\ f(x)). \tag{9.25}$$

Then for each $i = 1, 2, \ldots, r$, there is a nonnegative integer $f_i \leq e_i$ such that

$$x^{(q^n-1)/p_i^{f_i}} \equiv 1 \,(\mathrm{mod}\ f(x)), \quad x^{(q^n-1)/p_i^{f_i+1}} \not\equiv 1 \,(\mathrm{mod}\ f(x)). \tag{9.26}$$

Then

$$p(f) = p_1^{e_1-f_1} p_2^{e_2-f_2} \cdots p_r^{e_r-f_r}. \tag{9.27}$$

Proof. By Lemma 9.20, $p(f) \,|\, (q^n - 1)$, i.e.,

$$x^{q^n-1} \equiv 1 \,(\mathrm{mod}\ f(x)). \tag{9.28}$$

It follows from (9.25) and (9.28) that there is a nonnegative integer $f_i \geq e_i$ such that (9.26) holds. By Corollary 9.19

$$p(f) \,|\, ((q^n - 1)/p_i^{f_i}), \quad p(f) \nmid ((q^n - 1)/p_i^{f_i+1}).$$

It follows that for each $i = 1, 2, \ldots, r$,

$$p_i^{e_i-f_i} \,|\, p(f), \quad p_i^{e_i-f_i+1} \nmid p(f).$$

Since $p_1, p_2, \ldots, p_r$ are distinct primes, (9.27) follows immediately. $\square$

Therefore in order to determine the period of an irreducible polynomial $f(x)$ of degree n over $\mathbb{F}_q$ with a nonzero constant term, one can proceed according to the following procedure.

1) Factorize $q^n - 1$ into a product of prime numbers. Assume that

$$q^n - 1 = p_1^{e_1} p_2^{e_2} \cdots p_r^{e_r},$$

where $p_1, p_2, \ldots, p_r$ are r distinct prime numbers and $e_1, e_2, \ldots, e_r$ are r positive integers.

2) For each $i = 1, 2, \ldots, r$, calculate

$$x^{(q^n-1)/p_i} \ (\mathrm{mod}\ f(x)),\ \ x^{(q^n-1)/p_i^2} \ (\mathrm{mod}\ f(x)),\ \ x^{(q^n-1)/p_i^3} \ (\mathrm{mod}\ f(x)), \ldots$$

until a nonnegative integer $f_i \leq e_i$ is obtained such that

$$x^{(q^n-1)/p_i^{f_i}} \equiv 1 \,(\mathrm{mod}\ f(x)),\ \ x^{(q^n-1)/p_i^{f_i+1}} \not\equiv 1 \,(\mathrm{mod}\ f(x)).$$

Then
$$p(f) = p_1^{e_1-f_1} p_2^{e_2-f_2} \cdots p_r^{e_r-f_r},$$

we recall that $g(x) \ (\mathrm{mod}\ f(x))$ means the polynomial of degree $< \deg f(x)$ which is congruent to $g(x)$ modulo $f(x)$.

In order to facilitate the computation of

$$x^{(q^n-1)/p_i^s} \ (\mathrm{mod}\ f(x)),\ s = 1, 2, \ldots;\ i = 1, 2, \ldots, r,$$

we proceed as follows:

2.1) Calculate

$$x^0 \ (\mathrm{mod}\ f(x)),\ \ x^q \ (\mathrm{mod}\ f(x)),\ \ x^{2q} \ (\mathrm{mod}\ f(x)), \ldots, x^{(n-1)q} \ (\mathrm{mod}\ f(x)).$$

Assume that

$$x^{jq} \equiv \sum_{i=0}^{n-1} a_{ij} x^i \ (\mathrm{mod}\ f(x)),\ j = 0, 1, \ldots, n-1.$$

Let
$$A = \begin{pmatrix} a_{00} & a_{01} & \cdots & a_{0,n-1} \\ a_{10} & a_{11} & \cdots & a_{1,n-1} \\ \vdots & \vdots & & \vdots \\ a_{n-1,0} & a_{n-1,1} & \cdots & a_{n-1,n-1} \end{pmatrix},$$

which is an $n \times n$ matrix over $\mathbb{F}_q$.

2.2) Calculate

$$x \pmod{f(x)}, \ x^q \pmod{f(x)}, \ x^{q^2} \pmod{f(x)}, \ \ldots, \ x^{q^{n-1}} \pmod{f(x)}.$$

Note that when $q^i \leq (n-1)q, x^{q^i} \pmod{f(x)}$ has been already calculated in Step 2.1). Further, if $x^{q^i} \pmod{f(x)}$ has been calculated, assume that

$$x^{q^i} \equiv \sum_{j=0}^{n-1} b_j x^j \pmod{f(x)},$$

then

$$x^{q^{i+1}} \equiv \sum_{i=1}^{n-1} \left(\sum_{j=0}^{n-1} a_{ij} b_j \right) x^i \pmod{f(x)}.$$

That is, if we let

$$x^{q^{i+1}} \equiv \sum_{j=0}^{n-1} b_j' x^j \pmod{f(x)},$$

then

$$\begin{pmatrix} b_0' \\ b_1' \\ \vdots \\ b_{n-1}' \end{pmatrix} = A \begin{pmatrix} b_0 \\ b_1 \\ \vdots \\ b_{n-1} \end{pmatrix}.$$

2.3) Express $(q^n - 1)/p_i^s$ as a q-ary number. Assume that

$$(q^n - 1)/p_i^{s_i} = a_0 + a_1 q + \cdots + a_{n-1} q^{n-1}, \ 0 \leq a_i \leq q - 1.$$

2.4) Calculate

$$x^{(q^n-1)/p_i^{s_i}} \equiv x^{a_0} (x^q)^{a_1} (x^{q^2})^{a_2} \cdots (x^{q^{n-1}})^{a_{n-1}} \pmod{f(x)}.$$

We emphasize that in the above algorithm, except the first step where $q^n - 1$ is to be factorized into a product of primes, all other steps are not complicated and they can all be programmed on digital computers for calculation. The factorization of $q^n - 1$ into a product of primes, however, is a rather difficult task when q and n are relatively large. Even for $q = 2$, the prime factorization of $2^n - 1$, when n is relatively large, is quite difficult.

Example 9.9 Determine the period of the irreducible polynomial

$$f(x) = x^9 + x + 1$$

over $\mathbb{F}_2$.

The irreducibility of the polynomial $x^9 + x + 1$ over $\mathbb{F}_2$ has been proved in Example 9.2. First, we have

$$2^9 - 1 = 511 = 7 \cdot 73.$$

Next, we calculate the matrix A.

$$A = \begin{pmatrix} 1 & 0 & 0 & 0 & 0 & 0 & 0 & 0 & 0 \\ 0 & 0 & 0 & 0 & 0 & 1 & 0 & 0 & 0 \\ 0 & 1 & 0 & 0 & 0 & 1 & 0 & 0 & 0 \\ 0 & 0 & 0 & 0 & 0 & 0 & 1 & 0 & 0 \\ 0 & 0 & 1 & 0 & 0 & 0 & 1 & 0 & 0 \\ 0 & 0 & 0 & 0 & 0 & 0 & 0 & 1 & 0 \\ 0 & 0 & 0 & 1 & 0 & 0 & 0 & 1 & 0 \\ 0 & 0 & 0 & 0 & 0 & 0 & 0 & 0 & 1 \\ 0 & 0 & 0 & 0 & 1 & 0 & 0 & 0 & 1 \end{pmatrix}.$$

By means of the matrix A, we can write down

$$
\begin{aligned}
x^1 &\equiv x, \\
x^2 &\equiv x^2, \\
x^{2^2} &\equiv x^4, \\
x^{2^3} &\equiv x^8, \\
x^{2^4} &\equiv x^7 + x^8,
\end{aligned}
$$

and calculate

$$
\begin{aligned}
x^{2^5} &\equiv x^5 + x^6 + x^7 + x^8, \\
x^{2^6} &\equiv x + x^2 + x^3 + x^4 + x^5 + x^6 + x^7 + x^8,
\end{aligned}
$$

and so on, where all congruences are modulo $f(x)$. Then we calculate

$$
\begin{aligned}
x^{511/7} &\equiv x^{73} \\
&\equiv x x^{2^3} x^{2^6} \\
&\equiv x^9 x^{2^6} \\
&\equiv (x + 1)(x + x^2 + x^3 + x^4 + x^5 + x^6 + x^7 + x^8) \\
&\equiv 1 \pmod{f(x)}
\end{aligned}
$$

and

$$
\begin{aligned}
x^{511/73} &\equiv x^7 \\
&\not\equiv 1 \pmod{f(x)}.
\end{aligned}
$$

Thus it follows that $p(f) \mid 73$ and $p(f) \nmid 7$. Hence $p(f) = 73$.

Example 9.10 Determine the period of the polynomial over $\mathbb{F}_2$

$$f(x) = (x+1)^3 (x^2 + x + 1)(x^3 + x + 1).$$

It is easy to prove that

$$f_1(x) = x + 1, \quad f_2(x) = x^2 + x + 1, \quad f_3(x) = x^3 + x + 1$$

are all irreducible polynomials over $\mathbb{F}_2$. Clearly, $p(f_1) = 1$. Since $2^2 - 1 = 3$ and $2^3 - 1 = 7$ are both primes, $x^2 + x + 1$ and $x^3 + x + 1$ are both primitive polynomials. Hence $p(f_2) = 3$, $p(f_3) = 7$. Then by Theorem 9.24, we have

$$p(f) = [p(f_1), p(f_2), p(f_3)]2^2 = 3 \cdot 7 \cdot 2^2 = 84.$$

9.5 Exercises

9.1 Factor the following polynomials into products of irreducible factors and compute their periods over $\mathbb{F}_2$.

   (i) $x^8 + x^6 + x^4 + x^3 + 1$.

   (ii) $x^{12} + x^8 + x^7 + x^6 + x^2 + x + 1$.

9.2 Factor the polynomial $x^{20} - 1$ over $\mathbb{F}_2$.

9.3 Determine the 2-cycles mod 21 and find the factorization of $x^{21} - 1$ over $\mathbb{F}_2$.

9.4 Factor $x^{16} - x$ into irreducible factors:

   (i) over $\mathbb{F}_{2^4}$.

   (ii) over $\mathbb{F}_2$.

   (iii) over $\mathbb{F}_{2^2}$.

9.5 Find the complete factorization of $x^{24} - 1$:

   (i) over $\mathbb{F}_2$.

   (ii) over $\mathbb{F}_3$.

9.6 Prove that if n is odd, $\Phi_{2n}(x) = \Phi_n(-x)$.

9.7 Let d and d' be two distinct positive integers. Prove that if we regard $\Phi_d(x)$ and $\Phi_{d'}(x)$ as polynomials over $\mathbb{Z}$, then they have no common factors of degree > 0 over $\mathbb{Z}$.

9.8 Let p be a prime number, d and d' be two distinct positive integers, each of which is coprime with p. Prove that if we regard $\Phi_d(x)$ and $\Phi_{d'}(x)$ as polynomials over $\mathbb{F}_q$, then they are coprime in $\mathbb{F}_q[x]$.

9.9 Let $f(x) = a_0 + a_1 x + \cdots + a_m x^m$, where $a_m \neq 0$. Recall $\widetilde{f}(x) = x^m f(1/x) = a_0 x^m + a_1 x^{m-1} + \cdots + a_m$ and call $\widetilde{f}(x)$ the polynomial reciprocal to $f(x)$. If $\widetilde{f}(x) = f(x)$, then $f(x)$ is called a *self-reciprocal* polynomial. Prove that for any $n \geq 2$, $\Phi_n(x)$ is self-reciprocal and that $\widetilde{\Phi_1}(x) = -\Phi_1(x)$.

9.10 Let p be a prime number and $\gcd(p, n) = 1$. Prove that for $k \geq 1$,

$$\Phi_{np^k}(x) = \Phi_{np}(x^{p^{k-1}}) = \Phi_n(x^{p^k})\Phi_n(x^{p^{k-1}})^{-1}.$$

9.11 Calculate the following cyclotomic polynomials:

  (i) $\Phi_{24}(x)$.

  (ii) $\Phi_{40}(x)$.

  (iii) $\Phi_{45}(x)$.

9.12 Factorize the cyclotomic polynomial $\Phi_{11}(x)$:

  (i) over $\mathbb{F}_2$.

  (ii) over $\mathbb{F}_3$.

9.13 Let F be a field of characteristic 2. Calculate

  (i) $\Phi_{10}(x)$.

  (ii) $\Phi_{16}(x)$.

9.14 Let q be a prime power and $q \equiv 2$ or $5 \pmod 9$. Prove that the cyclotomic polynomial $\Phi_q(x)$ is irreducible over $\mathbb{F}_q$.

Chapter 10

Irreducible Polynomials over Finite Fields

10.1 On the Determination of Irreducible Polynomials

In Section 5.3 we see that to construct concretely the finite field $\mathbb{F}_{p^n}$ with p^n elements, where p is a prime and n is a positive integer, we need an irreducible polynomial of degree n over $\mathbb{F}_p$. In Section 7.3 we see that to obtain a primitive element of $\mathbb{F}_{p^n}$, it is helpful to determine a primitive polynomial of degree n over $\mathbb{F}_p$. Both the problem of determining irreducible polynomials and primitive polynomials over $\mathbb{F}_p$ (or $\mathbb{F}_{p^n}$) are also important in various applications.

Since a primitive polynomial of degree n over $\mathbb{F}_q$ is an irreducible polynomial of degree n and of period $q^n - 1$, in order to determine a primitive polynomial of degree n over $\mathbb{F}_q$, we can first determine irreducible polynomials of degree n over $\mathbb{F}_q$, then calculate their periods and see if they are equal to $q^n - 1$. As to the later problem one can use the method presented in Section 9.4 but the difficulty lies in the prime factorization of $q^n - 1$ when n and q are moderately large.

How do we determine irreducible polynomials over $\mathbb{F}_q$? We know that the status of irreducible polynomials in polynomials corresponds to that of primes in positive integers. In Section 1.2 we discussed the sieve method for determining primes. Naturally we can also use the sieve method to determine irreducible polynomials. For example, arrange polynomials of degree ≥ 1 but $\leq n$ in a table by degrees, with the lower degree ones at the beginning and the higher degrees ones at the back and those of equal degrees arranged by some rules. The polynomial at the very beginning is irreducible, we circle it and cross out its multiples in the table. Then among the left uncircled and uncrossed polynomials the first one is irreducible, it is circled and its multiples are crossed out from the table, and so on. But by this method, if n or q is large the amount of calculation becomes tremendously large. Therefore we usually do not use this method.

In order to determine irreducible polynomials over finite fields, people have found some criterions to judge whether a polynomial is irreducible and found some method to construct irreducible polynomials of given degrees.

The method presented in Section 9.1 can be used to judge whether a polynomial over $\mathbb{F}_q$ is irreducible, but it is rather complicated. We have also the following necessary and sufficient condition for a polynomial to be irreducible.

Theorem 10.1 Let $f(x) \in \mathbb{F}_q[x]$ be a polynomial of degree n. Then $f(x)$ is irreducible over $\mathbb{F}_q$ if and only if

(i) $f(x)|(x^{q^n} - x)$

and

(ii) $\gcd(f(x), x^{q^i} - x) = 1$ for $i = 1, 2, \ldots, [n/2]$.

Proof. If $f(x)$ is irreducible over $\mathbb{F}_q$, by Lemma 6.10 both (i) and (ii) hold. If $f(x)$ is reducible over $\mathbb{F}_q$, $f(x)$ has an irreducible factor $g(x)$ over $\mathbb{F}_q$. Suppose $\deg g(x) = m$ and we can assume $m \leq [n/2]$, then by Lemma 6.9 $g(x)|(x^{q^m} - x)$, thus $\gcd(f(x), x^{q^m} - x) \neq 1$. □

We list two necessary and sufficient conditions for a polynomial over $\mathbb{F}_q$ to be irreducible in the following theorem.

Theorem 10.2 Let $f(x)$ be a polynomial of degree $n > 1$ over $\mathbb{F}_q$. Then $f(x)$ is irreducible if and only if one of the following conditions is satisfied.

(i) The reciprocal polynomial $\widetilde{f}(x)$ is irreducible.

(ii) $(cx+d)^{\deg f(x)} f(\frac{ax+b}{cx+d})$ is irreducible for $a, b, c, d \in \mathbb{F}_q$ with $ad - bc \neq 0$.

Several necessary conditions for a polynomial over $\mathbb{F}_q$ to be irreducible are listed in the following two theorems.

Theorem 10.3 Let $f(x)$ be polynomial over $\mathbb{F}_q$. If $f(x)$ is irreducible over $\mathbb{F}_q$, then

(i) *The constant term of $f(x)$ is nonzero.*

(ii) *The sum of the coefficients of $f(x)$ is not equal to 0.*

(iii) $\gcd(f(x), f'(x)) = 1$.

When $q = 2$, we have

Theorem 10.4 *Let $f(x) \in \mathbb{F}_2[x]$. If $f(x)$ is irreducible over $\mathbb{F}_2$ then*

(i) *The number of terms of $f(x)$ with coefficients equal to 1 is odd.*

(ii) *There is a term x^m of $f(x)$ with coefficient equal to 1 such that $2 \nmid m$.*

The proofs of Theorems 10.2–10.4 are easy and omitted.

There are irreducibility criterions for polynomials of special forms, for example, for binomials. We begin with binomials in the next section.

10.2 Irreducibility Criterion of Binomials

A *binomial* is a polynomial with two nonzero terms, one of which is the constant term. Linear binomials are clearly irreducible. Irreducible nonlinear binomials over a finite field has already been characterized by Serret (1866) and will be reproduced as Theorem 10.7 below. The characterization rests on Lemma 10.5.

Let m be an integer ≥ 1. We recall that $\mathbb{Z}_m^*$ denotes the multiplicative group of all residue classes $\bmod\, m$ which are coprime with m and $\mathrm{ord}_m(a)$ denotes the order of the residue class $\bar{a}$ in the group $\mathbb{Z}_m^*$.

Lemma 10.5 *Let $q \geq 2$ and $m \geq 2$ be coprime integers and let $\mathrm{ord}_m(q) = l$. Let $t \geq 2$ be an integer such that the following two conditions are satisfied:*

(i) *Every prime divisor of t divides m, but not $(q^l - 1)/m$.*

(ii) *If $4 \,|\, t$ then $4 \,|\, (q^l - 1)$.*

Then $\mathrm{ord}_{mt}(q) = lt$.

Proof. We apply induction on the number of prime divisors of t, each counted with its multiplicity. First, let t be a prime number. Since $\mathrm{ord}_m(q)$

$= l$, $q^l \equiv 1 \,(\mathrm{mod}\, m)$. Let $d = (q^l - 1)/m$, then d is a positive integer and $q^l = 1 + dm$. Thus

$$q^{lt} = (1 + dm)^t$$
$$= 1 + \binom{t}{1} dm + \binom{t}{2} (dm)^2 + \cdots + \binom{t}{t-1} (dm)^{t-1} + (dm)^t.$$

Clearly,

$$mt \,\Big|\, \binom{t}{i} (dm)^i, \ i = 1, 2, \ldots, t-1 \ \text{ and } \ mt \,|\, (dm)^t.$$

Therefore $q^{lt} \equiv 1 \,(\mathrm{mod}\, mt)$. Assume $\mathrm{ord}_{mt}(q) = k$, then $q^k \equiv 1 \,(\mathrm{mod}\, mt)$ and $k \,|\, lt$. Thus $q^k \equiv 1 \,(\mathrm{mod}\, m)$ and $l \,|\, k$. Since t is a prime number, from $l \,|\, k$ and $k \,|\, lt$ we deduce $k = l$ or lt. If $k = l$, then $q^l \equiv 1 \,(\mathrm{mod}\, mt)$ and $mt \,|\, (q^l - 1)$. But $q^l - 1 = dm$, so $mt \,|\, dm$, then $t \,|\, d$, which contradicts Condition (i). Therefore we must have $k = lt$.

Now suppose that t has at least two prime divisors and we write $t = rt_0$, where r is a prime divisor of t. By what we have proved in the proceeding paragraph, $\mathrm{ord}_{mr}(q) = lr$. If $4 \,|\, t_0$, then $4 \,|\, t$ and by Condition (ii) $4 \,|\, (q^l - 1)$. Clearly, $(q^l - 1) \,|\, (q^{lr} - 1)$. Therefore if $4 \,|\, t_0$, $4 \,|\, (q^{lr} - 1)$. Moreover, if we can prove that the prime divisors of t_0 divide mr, but not $(q^{lr} - 1)/mr$, then the induction hypothesis on t_0 yields $\mathrm{ord}_{mrt_0}(q) = lrt_0$, i.e., $\mathrm{ord}_{mt}(q) = lt$.

Let r_0 be a prime factor of t_0, then r_0 is a prime divisor of t. By Condition (i) $r_0 \,|\, m$ and, hence, $r_0 \,|\, mr$. It remains to prove $r_0 \nmid (q^{lr} - 1)/mr$. We have

$$q^{lr} - 1 = c(q^l - 1),$$

where $c = q^{l(r-1)} + q^{l(r-2)} + \cdots + q^l + 1$. Let $d_0 = (q^{lr} - 1)/mr$ and $d = (q^l - 1)/m$. Then $d_0 = c(q^l - 1)/mr = cd/r$. Since r is a prime divisor of t, by Condition (i) $r \,|\, m$. Thus from $q^l \equiv 1 \,(\mathrm{mod}\, m)$ we deduce $q^l \equiv 1 \,(\mathrm{mod}\, r)$ and

$$c = q^{l(r-1)} + q^{l(r-2)} + \cdots + q^l + 1 \equiv r \ (\mathrm{mod}\ r)$$
$$\equiv 0 \ (\mathrm{mod}\ r).$$

Therefore $c/r \in \mathbb{Z}$. Since r_0 is a prime divisor of t, by Condition (i) $r_0 \nmid d$. If we can prove $r_0 \nmid (c/r)$, our goal will be achieved.

Since $r_0 \,|\, m$ and $q^l - 1 = md$, $q^l \equiv 1 \,(\mathrm{mod}\, r_0)$ and $c \equiv r \,(\mathrm{mod}\, r_0)$. If $r_0 \neq r$, then $c/r \equiv 1 \,(\mathrm{mod}\, r_0)$, so $r_0 \nmid (c/r)$, and we are finished. Now let $r_0 = r$. From $q^l \equiv 1 \,(\mathrm{mod}\, r)$ we deduce $q^l \equiv 1 + br \,(\mathrm{mod}\, r^2)$, where $b \in \mathbb{Z}$. Then $q^{lj} \equiv (1 + br)^j \equiv 1 + jbr \,(\mathrm{mod}\, r^2)$ for all $j \geq 1$ and

$$c = q^{l(r-1)} + q^{l(r-2)} + \cdots + q^l + 1 \equiv r + br\Big(\sum_{j=1}^{r-1} j\Big) \equiv r + br\frac{r(r-1)}{2} \ (\mathrm{mod}\ r^2).$$

It follows that

$$\frac{c}{r} \equiv 1 + b\frac{r(r-1)}{2} \pmod{r}.$$

If r is odd, then $c/r \equiv 1 \pmod{r}$, which implies $r_0 = r \nmid (c/r)$. If r is even, then $r_0 = r = 2$. But $rr_0 \mid rt_0$ and $rt_0 = t$, so $4 \mid t$. By Condition (ii) $4 \mid (q^l - 1)$. But in this case $c = q^l + 1$, so $c \equiv 2 \pmod{4}$, and thus $c/r \equiv 1 \pmod{2}$, which implies also $r_0 \nmid (c/r)$. $\square$

Corollary 10.6 *Let q be a prime power and $m \geq 2$ be an integer such that $m \mid (q - 1)$. Let $t \geq 2$ be an integer such that the following two conditions are satisfied:*

(i) *Every prime divisor of t divides m, but not $(q - 1)/m$.*

(ii) *If $4 \mid t$ then $4 \mid (q - 1)$.*

Then $\mathrm{ord}_{mt}(q) = t$.

Proof. Clearly, $\gcd(q, q - 1) = 1$. Since $m \mid (q - 1)$, we have $\gcd(q, m) = 1$ and $\mathrm{ord}_m(q) = 1$. Therefore our Corollary follows immediately from Lemma 10.5. $\square$

We have the following irreducibility criterion of binomials over a finite field $\mathbb{F}_q$, which is due to Serret (1866).

Theorem 10.7 *Let t be an integer ≥ 2. Let $a \in \mathbb{F}_q^*$ and $\mathrm{ord}(a) = m > 1$ in $\mathbb{F}_q^*$. Then the binomial $x^t - a \in \mathbb{F}_q[x]$ is irreducible over $\mathbb{F}_q$ if and only if the following two conditions are satisfied:*

(i) *Every prime divisor of t divides m, but not $(q - 1)/m$.*

(ii) *If $4 \mid t$ then $4 \mid (q - 1)$.*

Proof. We observe that $a^{q-1} = 1$ and $\mathrm{ord}(a) = m$ imply $m \mid (q - 1)$. Thus $(q - 1)/m$ is an integer.

First suppose both Conditions (i) and (ii) are satisfied. We can choose a root θ of $x^t - a$ in some extension field of $\mathbb{F}_q$. We assert that θ has order tm. Clearly, $\theta^{tm} = a^m = 1$. If the order of θ is $< tm$, then there is a prime divisor r of tm such that $\theta^{tm/r} = 1$. Since r is a prime, $r \mid tm$ implies $r \mid t$ or $r \mid m$. If $r \mid t$, by Condition (i) we have also $r \mid m$. Thus $a^{m/r} = (\theta^t)^{m/r} = \theta^{tm/r} = 1$, which contradicts the assumption that a has order m. Therefore θ has order tm. Let $m(x)$ be the minimal polynomial of θ over $\mathbb{F}_q$ and let $\deg m(x) = d$. Then by Theorem 7.5, $m(x) \mid (x^t - a)$ and

$$m(x) = (x - \theta)(x - \theta^q)(x - \theta^{q^2}) \cdots (x - \theta^{q^{d-1}}),$$

where $\theta, \theta^q, \theta^{q^2}, \ldots, \theta^{q^{d-1}}$ are distinct and d is the smallest positive integer such that $\theta^{q^d} = \theta$, i.e., $\theta^{q^d-1} = 1$. d is also the smallest positive integer such that $q^d \equiv 1 \pmod{tm}$. Therefore $d = \mathrm{ord}_{tm}(q)$. Since both Conditions (i) and (ii) are satisfied, by Corollary 10.6 $\mathrm{ord}_{tm}(q) = t$. So $d = t$ and $\deg m(x) = t$. Since $m(x)$ is monic irreducible over $\mathbb{F}_q$ and $m(x) \mid (x^t - a)$, $x^t - a = m(x)$. Hence $x^t - a$ is irreducible over $\mathbb{F}_q$.

Suppose Condition (i) is not satisfied. Then there exists a prime divisor r of t which either divides $(q-1)/m$ or does not divide m. We can assume that $t = rt_1$ for some positive integer t_1. If r divides $(q-1)/m$, then we have $(q-1)/m = rs$ for some positive integer s. By Theorem 6.20 the subgroup $\mathbb{F}_q^{*r}$ of $\mathbb{F}_q^*$ has order $(q-1)/r = ms$ and then by Theorem 2.28 $\mathbb{F}_q^{*r}$ has a subgroup of order m, which must be $\langle a \rangle$. $a \in \mathbb{F}_q^{*r}$ implies $a = b^r$ where $b \in \mathbb{F}_q^*$. Then $x^t - a = x^{rt_1} - b^r$ has a factor $x^{t_1} - b$. Now suppose r does not divide $(q-1)/m$ and m. So r does not divide $q - 1$, Since r is a prime, there is an integer r_1 such that $r_1 r \equiv 1 \pmod{q-1}$. We can assume that $r_1 > 0$. Then $x^t - a = x^{rt_1} - a^{r_1 r}$ has a factor $x^{t_1} - a^{r_1}$. Hence in both cases $x^t - a$ is reducible.

Finally suppose Condition (i) is satisfied but Condition (ii) is not satisfied. Then $t = 4t_2$ for some positive t_2 and $4 \nmid (q-1)$. Thus 2 is a prime divisor of t. By Condition (i) $2 \mid m$, so m must be even. Since $m \mid (q-1)$, q must be odd. But $4 \nmid (q-1)$, so $4 \nmid m$ and $m/2$ is odd. Then $\mathrm{ord}(a) = m$ implies $a^{m/2} = -1$, and so $x^t - a = x^t + a^{(m/2)+1} = x^t + a^d$, where $d = (m/2) + 1$ is even. Then $a^{d/2}/2 \in \mathbb{F}_q^*$, thus $(a^{d/2}/2)^{q-1} = 1$ and $(a^{d/2}/2)^{q+1} = (a^{d/2}/2)^2$. Since q is odd and $4 \nmid (q-1)$, $4 \mid (q+1)$. Therefore we have

$$a^d = 4(a^{d/2}/2)^2 = 4(a^{d/2}/2)^{q+1} = 4c^4 \text{ with } c = (a^{d/2}/2)^{(q+1)/4},$$

which leads to the factorization

$$x^t - a = x^t + a^d = x^{4t_2} + 4c^4 = (x^{2t_2} + 2cx^{t_2} + 2c^2)(x^{2t_2} - 2cx^{t_2} + 2c^2).$$

$\square$

Corollary 10.8 *Let r be a prime divisor of $q - 1$ and k be an integer ≥ 0. Let $a \in \mathbb{F}_q^*$ and $\mathrm{ord}(a) = m > 1$ in $\mathbb{F}_q^*$. Assume that r does not divides $(q-1)/m$, (i.e., $a \notin \mathbb{F}_q^{*r}$, by Exercise 6.21). When $r = 2$ and $k \geq 2$, assume further that $4 \mid (q-1)$. Then $x^{r^k} - a$ is irreducible over $\mathbb{F}_q$.*

Proof. Clearly, $\text{ord}(a) = m$ in $\mathbb{F}_q^*$ implies $m \mid (q-1)$. Let $t = r^k$, then t has only one prime divisor r. From $r \mid (q-1)$ and $r \nmid (q-1)/m$ we deduce $r \mid m$. Therefore Condition (i) of Theorem 10.7 is satisfied. If $4 \mid t$ then $r = 2$ and $k \geq 2$, by assumption $4 \mid (q-1)$. Therefore Condition (ii) is also satisfied. By Theorem 10.7 $x^{r^k} - a$ is irreducible over $\mathbb{F}_q$. $\square$

Example 10.1 Using Corollary 10.8 we deduce immediately that for any integer $k \geq 0$,

(a) $x^{2^k} + 2$ and $x^{2^k} - 2$ are irreducible over $\mathbb{F}_5$.

(b) $x^{3^k} + 3$, $x^{3^k} - 3$, $x^{3^k} + 2$ and $x^{3^k} - 2$ are irreducible over $\mathbb{F}_7$.

(c) $x^{3^k} + \alpha$ is irreducible over $\mathbb{F}_4$, where $\mathbb{F}_4 = \mathbb{F}_2(\alpha)$ and α is a root of $x^2 + x + 1$.

From (c) we deduce that

(d) $x^{2 \cdot 3^k} + x^{3^k} + 1 = (x^{3^k} + \alpha)(x^{3^k} + \alpha^2)$ is irreducible over $\mathbb{F}_2$.

10.3　Some Irreducible Trinomials

A *trinomial* is a polynomial of the form $ax^n + bx^k + c$, where $n > k > 0$ and a, b, c are all nonzero.

First, we have the following irreducibility criterion which is due to Serret (1866) and Pillet (1870).

Theorem 10.9 *Let $q = p^m$, where p is a prime. Then the trinomial*

$$x^p - x - b, \ b \in \mathbb{F}_q,$$

is irreducible over $\mathbb{F}_q$ if and only if $Tr_{\mathbb{F}_q/\mathbb{F}_p}(b) \neq 0$.

Proof. Let θ be a root of $x^p - x - b$. Then

$$\theta^p = \theta + b.$$

By induction we can prove that

$$\theta^{p^i} = \theta + b + b^p + \cdots + b^{p^{i-1}}, \ i = 1, 2, \ldots.$$

In particular,

$$\theta^q = \theta^{p^m} = \theta + b + b^p + \cdots + b^{p^{m-1}} = \theta + \mathrm{Tr}_{\mathbb{F}_q/\mathbb{F}_p}(b). \qquad (10.1)$$

So $\mathrm{Tr}_{\mathbb{F}_q/\mathbb{F}_p}(b) = 0$ if and only if $\theta^q = \theta$, i.e., every root of $x^p - x - b$ is in $\mathbb{F}_q$. Consequently, $\mathrm{Tr}_{\mathbb{F}_q/\mathbb{F}_p}(b) = 0$ if and only if $x^p - x - b$ is a product of linear factors in $\mathbb{F}_q[x]$.

Now suppose $\mathrm{Tr}_{\mathbb{F}_q/\mathbb{F}_p}(b) \neq 0$. Let $\tau = \mathrm{Tr}_{\mathbb{F}_q/\mathbb{F}_p}(b)$, then $\tau \in \mathbb{F}_p^*$ and by (10.1)

$$\theta^{q^i} = \theta + i\tau,\ i = 1, 2, \ldots, p-1;\ \theta^{q^p} = \theta.$$

Thus θ has p distinct conjugates over $\mathbb{F}_q$ and the minimal polynomial of θ over $\mathbb{F}_q$ has degree p. Then it must be equal to $x^p - x - b$ itself. Therefore $x^p - x - b$ is irreducible over $\mathbb{F}_q$.   □

Example 10.2 For any $b \in \mathbb{F}_p^*$ the trinomial $x^p - x - b$ is irreducible over $\mathbb{F}_p$.

Let us introduce linearized polynomials over $\mathbb{F}_q$.

Definition 10.1 *Let q be a power of a prime number p. A polynomial of the form*

$$l_v x^{p^v} + l_{v-1} x^{p^{v-1}} + \cdots + l_1 x^p + l_0 x,$$

where $l_i \in \mathbb{F}_q$, $i = 0, 1, 2, \ldots, v$, is called a linearized polynomial over $\mathbb{F}_q$.

If $l(x)$ is a linearized polynomial over $\mathbb{F}_q$ and $b \in \mathbb{F}_q$, $l(x) - b$ is called an affine polynomial over $\mathbb{F}_q$.

Lemma 10.10 *Let $l(x)$ be a linearized polynomial over $\mathbb{F}_q$ and $\mathbb{F}_q$ be of characteristic p, then*

$$l(x + y) = l(x) + l(y), \forall x, y \in \mathbb{F}_q$$

and

$$l(cx) = cl(x), \forall x \in \mathbb{F}_q,\ c \in \mathbb{F}_p.$$

Conversely, if $l(x) \in \mathbb{F}_q[x]$ be such that both of the above two conditions hold, then $l(x)$ is a linearized polynomial over $\mathbb{F}_q$.

The proof of this lemma is easy and left as an exercise.

Lemma 10.11 *Suppose that the linearized polynomial $l(x)$ over $\mathbb{F}_q$ has no nonzero root in $\mathbb{F}_q$. Then for any $b \in \mathbb{F}_q$, the affine polynomial $l(x) - b$ has a linear divisor $x - a_0$, $a_0 \in \mathbb{F}_q$.*

Proof. $l(x)$ defines a map from the additive group of $\mathbb{F}_q$ to itself

$$l : \mathbb{F}_q \rightarrow \mathbb{F}_q$$
$$x \mapsto l(x).$$

For $\alpha, \beta \in \mathbb{F}_q$, if $l(\alpha) = l(\beta)$, then $l(\alpha - \beta) = l(\alpha) - l(\beta) = 0$. By hypothesis $l(x)$ has no nonzero root in $\mathbb{F}_q$. So, $\alpha = \beta$. Therefore l is injective. Since $\mathbb{F}_q$ is finite, l is also surjective. Thus for any $b \in \mathbb{F}_q$ there is an element $a_0 \in \mathbb{F}_q$ such that $l(a_0) - b = 0$. Hence $(x - a_0) \mid (l(x) - b)$. □

From Theorem 10.9 and Lemma 10.11 we deduce more generally

Theorem 10.12 *Let q be a power of a prime p. For $a, b \in \mathbb{F}_q^*$, the trinomial $x^p - ax - b$ is irreducible over $\mathbb{F}_q$ if and only if $a = a_0^{p-1}$ for some $a_0 \in \mathbb{F}_q^*$ and $\mathrm{Tr}_{\mathbb{F}_q/\mathbb{F}_p}(b/a_0^p) \neq 0$.*

Proof. Assume $a = a_0^{p-1}$ for some $a_0 \in \mathbb{F}_q^*$ and $\mathrm{Tr}_{\mathbb{F}_q/\mathbb{F}_p}(b/a_0^p) \neq 0$. Then

$$x^p - ax - b = a_0^p((\frac{x}{a_0})^p - (\frac{x}{a_0}) - (\frac{b}{a_0^p})).$$

Let $y = x/a_0$. By Theorem 10.9 $y^p - y - b/a_0^p$ is irreducible over $\mathbb{F}_q$ and so is $x^p - ax - b$.

Now assume $a \neq a_0^{p-1}$ for any $a_0 \in \mathbb{F}_q^*$. Then the linearized polynomial $x^p - ax$ has no nonzero root in $\mathbb{F}_q$. By Lemma 10.11, the affine polynomial $x^p - ax - b$ has a linear factor in $\mathbb{F}_q[x]$ and, hence, is reducible.

Finally assume $a = a_0^{p-1}$ for some $a_0 \in \mathbb{F}_q^*$, but $\mathrm{Tr}_{\mathbb{F}_q/\mathbb{F}_p}(b/a_0^p) = 0$. By Theorem 10.9, $y^p - y - b/a_0^p$ is reducible over $\mathbb{F}_q$ and so is $x^p - ax - b$. □

When p is a prime and $p \equiv 3 \,(\mathrm{mod}\,4)$, Blake et al.(1993) suggested the following construction of irreducible trinomials over $\mathbb{F}_p$.

Theorem 10.13 *Let p be a prime and $p \equiv 3 \,(\mathrm{mod}\,4)$, then $2^2|(p+1)$. Let r be an integer ≥ 2 such that $2^r\|(p+1)$, then $2^{r+1}\|(p^2 - 1)$. Define elements $a_1, a_2, \ldots, a_r$ of $\mathbb{F}_p$ recursively as follows:*

$$a_1 = 0,$$
$$a_i = ((a_{i-1} + 1)/2)^{(p+1)/4} \quad \text{for } i = 2, 3, \ldots, r - 1,$$
$$a_r = ((a_{r-1} - 1)/2)^{(p+1)/4}.$$

Then for any integer $k \geq 1$ the trinomial

$$x^{2^k} - 2a_r x^{2^{k-1}} - 1 \tag{10.2}$$

is irreducible over $\mathbb{F}_p$, *hence, over* $\mathbb{F}_{p^m}$ *for any odd integer* m, *and divides*
$x^{2^{r+k-1}} + 1$.

Proof. The first two statements are clear. Before we go to prove the main statement let us show by induction on l the following proposition (P_l).

(P_l) The trinomial $x^2 - 2a_l x + 1$ is irreducible over $\mathbb{F}_p$ and divides $x^{2^l} + 1$ for $l = 1, 2, \ldots, r - 1$.

First we give a remark. Since $2^{r+1} | (p^2 - 1)$, $\mathbb{F}_{p^2}$ contains all 2^{k+1}th roots of unity for $0 \leq k \leq r$. But $2^2 \nmid (p - 1)$, so for $1 \leq k \leq r$ every primitive 2^{k+1}th root of unity is not contained in $\mathbb{F}_p$ and, hence, is of degree 2 over $\mathbb{F}_p$.

When $l = 1$, since $p \equiv 3 \pmod{4}$, -1 is a quadratic nonresidue mod p. Hence $x^2 + 1$ is irreducible over $\mathbb{F}_p$ and (P_1) is trivial in this case.

Now let l be such that $1 \leq l < r$, and assume that (P_l) holds. From $(x^2 - 2a_l x + 1) | (x^{2^l} + 1)$, we deduce $(x^4 - 2a_l x^2 + 1) | (x^{2^{l+1}} + 1)$. Let β be any root of $(x^4 - 2a_l x^2 + 1)$ in some extension field of $\mathbb{F}_p$. Then $\mathrm{ord}(\beta) = 2^{l+2}$. As $l + 1 \leq r$, by the above remark $\beta \in \mathbb{F}_{p^2}$ and β is of degree 2 over $\mathbb{F}_p$, which implies that $(x^4 - 2a_l x^2 + 1)$ is a product of two irreducible factors of degree 2 over $\mathbb{F}_p$. We can assume that the minimum polynomial of β is of the form $x^2 - 2sx + t$, where $s, t \in \mathbb{F}_p$. Thus

$$\beta^2 + t = 2s\beta \tag{10.3}$$

and

$$\beta^4 = 2a_l \beta^2 - 1. \tag{10.4}$$

Squaring (10.3) gives

$$\beta^4 = (4s^2 - 2t)\beta^2 - t^2. \tag{10.5}$$

From (10.4) and (10.5) we deduce

$$(4s^2 - 2t)\beta^2 - t^2 = 2a_l \beta^2 - 1. \tag{10.6}$$

Since $\mathrm{ord}(\beta) = 2^{l+2}$, we have $\mathrm{ord}(\beta^2) = 2^{l+1}$. But $2^{l+1} \nmid (p-1)$, so $\beta^2 \notin \mathbb{F}_p$. Thus $\{1, \beta^2\}$ is a basis of $\mathbb{F}_{p^2}$ over $\mathbb{F}_p$, and it follows from (10.6) that

$$t^2 = 1 \tag{10.7}$$

and

$$4s^2 - 2t = 2a_l. \tag{10.8}$$

So,

$$t = \pm 1. \tag{10.9}$$

We distinguish the cases $l < r - 1$ and $l = r - 1$. For the inductive proof we need only consider the case $l < r - 1$, but for the proof of the main statement of the theorem we will also consider the case $l = r - 1$. We prove that $t = 1$ if $l < r - 1$ and $t = -1$ if $l = r - 1$.

Case 1. $l < r - 1$. Then $l + 1 \leq r - 1$ and $l + 2 \leq r$. Suppose $t = -1$ in (10.9). Then the minimum polynomial of β is $x^2 - 2sx - 1$, which is irreducible over $\mathbb{F}_p$ and its roots are primitive 2^{l+2}-th roots of unity. Hence the roots of $x^4 - 2sx^2 - 1$ are primitive 2^{l+3}-th roots of unity. As $l + 2 \leq r$, by the above remark every primitive 2^{l+3}-th root of unity is contained in $\mathbb{F}_{p^2}$ and is of degree 2 over $\mathbb{F}_p$. Thus $x^4 - 2sx^2 - 1$ is a product of two irreducible factors of degree 2 over $\mathbb{F}_p$. Assume $x^2 - 2\bar{s}x + \bar{t}$ is one of them. Then by a similar argument leading to (10.7) and (10.8) we have

$$\bar{t}^2 = -1$$

and

$$4\bar{s}^2 - 2\bar{t} = 2s.$$

Since $p \equiv 3 \pmod 4$, $-1 \notin \mathbb{F}_p^{*2}$, and we obtain a contradiction.

Therefore $t = 1$ in (10.9). By (10.8) $2s^2 = a_l + 1$, thus $(a_l + 1)/2 \in \mathbb{F}_p^{*2}$ and, hence, $((a_l + 1)/2)^{(p-1)/2} = 1$. Let $a_{l+1} = ((a_l + 1)/2)^{(p+1)/4}$. Then

$$a_{l+1}^2 = ((a_l + 1)/2)^{(p+1)/2} = ((a_l + 1)/2)^{(p-1)/2}((a_l + 1)/2) = ((a_l + 1)/2).$$

Thus

$$x^4 - 2a_l x^2 + 1 = (x^2 - 2a_{l+1}x + 1)(x^2 + 2a_{l+1}x + 1)$$

and $(x^2 - 2a_{l+1}x + 1)$ is irreducible over $\mathbb{F}_p$. Clearly, we also have

$$(x^2 - 2a_{l+1}x + 1)|(x^{2^{l+1}} + 1).$$

The induction proof is now complete and (P_l) holds for $l = 1, 2, \ldots, r-1$.

Now let us go to prove the main statement of the theorem. We consider the case $l = r - 1$.

Case 2. $l = r - 1$. In this case $l + 2 = r + 1$, $l + 3 = r + 2 > r + 1$. Suppose $t = 1$ in (10.9). By (10.8) $2s^2 = a_{r-1} + 1$. Then

$$x^4 - 2a_{r-1}x^2 + 1 = (x^2 - 2sx + 1)(x^2 + 2sx + 1),$$

both $x^2 - 2sx + 1$ and $x^2 + 2sx + 1$ are irreducible over $\mathbb{F}_p$ and have roots being primitive 2^{r+1}th roots of unity. It follows that the roots of $x^4 - 2sx^2 + 1$ and $x^4 + 2sx^2 + 1$ are primitive 2^{r+2}th roots of unity. From $2^r \| (p + 1)$ we deduce $2^{r+1} \| (p^2 - 1)$ and $2^{r+2} \| (p^4 - 1)$. It follows that a primitive 2^{r+2}th root of unity is of degree 4 over $\mathbb{F}_p$. So both the polynomials $x^4 - 2sx^2 + 1$ and $x^4 + 2sx^2 + 1$ are irreducible over $\mathbb{F}_p$.

It is easy to see that if $(s + 1)/2 = \bar{s}^2$ for some $\bar{s} \in \mathbb{F}_p$, then $(x^2 - 2\bar{s}x + 1) | (x^4 - 2sx^2 + 1)$. But $x^4 - 2sx^2 + 1$ is irreducible over $\mathbb{F}_p$, so $(s+1)/2 \notin \mathbb{F}_p^2$. Similarly, if $(-s-1)/2 = \bar{s}^2$, then $(x^2 - 2\bar{s}x + 1) | (x^4 + 2sx^2 + 1)$. But $x^4 + 2sx^2 + 1$ is irreducible over $\mathbb{F}_p$, so $(-s - 1)/2 \notin \mathbb{F}_p^2$. Since $p \equiv 3 \pmod 4$, $-1 \notin \mathbb{F}_p^2$. Then one of $(s + 1)/2$ and $(-s - 1)/2 \in \mathbb{F}_p^2$ and the other one $\notin \mathbb{F}_p^2$. This is a contradiction.

Therefore $t = -1$ in (10.9). By (10.8) $2s^2 = a_{r-1} - 1$, thus $(a_{r-1} - 1)/2 \in \mathbb{F}_p^{*2}$ and, hence, $((a_{r-1} - 1)/2)^{(p-1)/2} = 1$. Let $a_r = ((a_{r-1} - 1)/2)^{(p+1)/4}$, then $a_r^2 = (a_{r-1} - 1)/2$ and

$$x^4 - 2a_{r-1}x^2 + 1 = (x^2 - 2a_r x - 1)(x^2 + 2a_r x - 1).$$

Hence $x^2 - 2a_r x - 1$ is irreducible over $\mathbb{F}_p$ and divides $x^{2^r} + 1$. Therefore the main statement of the theorem for the case $k = 1$ is proved.

Now assume that the main statement is true for $k \geq 1$, let us prove that it is also true for $k + 1$. From

$$(x^{2^k} - 2a_r x^{2^{k-1}} - 1) | (x^{2^{r+k-1}} + 1),$$

we deduce immediately

$$(x^{2^{k+1}} - 2a_r x^{2^k} - 1) | (x^{2^{r+k}} + 1).$$

We have also to prove that $(x^{2^{k+1}} - 2a_r x^{2^k} - 1)$ is irreducible over $\mathbb{F}_p$. Let α_1, α_2 be the two roots of $x^2 - 2a_r x - 1$. We know that $\alpha_1, \alpha_2 \in \mathbb{F}_{p^2}$ and have order 2^{r+1}. By Corollary 10.8, both $x^{2^k} - \alpha_1$ and $x^{2^k} - \alpha_2$ are irreducible over $\mathbb{F}_{p^2}$ for any integer $k \geq 1$. Hence

$$(x^{2^k} - \alpha_1)(x^{2^k} - \alpha_2) = x^{2^{k+1}} - 2a_r x^{2^k} - 1$$

is irreducible over $\mathbb{F}_p$. This completes the proof. $\square$

Example 10.3 For any integer $k \geq 1$, the following polynomials are irreducible over the respective fields:

(a) $x^{2^k} + 2x^{2^{k-1}} - 1$ over $\mathbb{F}_3$.

(b) $x^{2^k} + 3x^{2^{k-1}} - 1$ over $\mathbb{F}_7$.

10.4 Compositions of Polynomials

Let $f(x), g(x) \in \mathbb{F}_q[x]$ and let $P(x) = \sum_{i=0}^{n} c_i x^i \in \mathbb{F}_q[x]$ be of degree n. Then the following *composition*

$$g(x)^n P(f(x)/g(x)) = \sum_{i=0}^{n} c_i f(x)^i g(x)^{n-i}$$

is again a polynomial. Obviously, for $g(x)^n P(f(x)/g(x))$ to be irreducible over $\mathbb{F}_q$, $P(x)$ must be irreducible and $f(x)$ and $g(x)$ be coprime. Moreover, we have the following general result due to Cohen (1969).

Theorem 10.14 *Let $f(x)$ and $g(x)$ be two nonzero polynomials in $\mathbb{F}_q[x]$ and $P(x)$ be an irreducible polynomial in $\mathbb{F}_q[x]$ of degree $n > 0$. Then $g(x)^n P(f(x)/g(x))$ is irreducible over $\mathbb{F}_q$ if and only if $f(x) - \lambda g(x)$ is irreducible over $\mathbb{F}_{q^n}$ for some root $\lambda \in \mathbb{F}_{q^n}$ of $P(x)$. Moreover, if $P(x)$ is not of the form cx, where $c \in \mathbb{F}_q^*$, then $g(x)^n P(f(x)/g(x))$ is of degree hn, where $h = \max\{\deg f(x), \deg g(x)\}$.*

Proof. Consider first the case $P(x) = cx$, where $c \in \mathbb{F}_q^*$. Then $n = 1$, $P(x)$ has a unique root 0, and $g(x)P(f(x)/g(x)) = cf(x)$. Clearly, $g(x)P(f(x)/g(x)) = cf(x)$ is irreducible if and only if $f(x) - 0g(x) = f(x)$ is irreducible.

Then assume that $P(x)$ is not of the form cx, where $c \neq 0$. Clearly, $g(x)^n P(f(x)/g(x))$ is of degree hn.

If $n = 1$ we can assume that $P(x) = x + c$. Then $P(x)$ is irreducible over $\mathbb{F}_q$, $-c$ is the unique root of $P(x)$, and $g(x)P(f(x)/g(x)) = f(x) + cg(x) = f(x) - (-c)g(x)$. Hence our theorem trivially holds.

Now assume $n > 1$. Let $d(x) = \gcd(f(x), g(x))$. If $d(x) \neq 1$, then $d(x)|g(x)^n P(f(x)/g(x))$ and $d(x)|(f(x) - \lambda g(x))$ over $\mathbb{F}_{q^n}$, thus our theorem is true. Now assume that $d(x) = 1$. Let γ be a root of $g(x)^n P(f(x)/g(x))$ in some extension field of $\mathbb{F}_q$, then $f(\gamma) \neq 0$ and $g(\gamma) \neq 0$, for otherwise, $d(x) \neq 1$. Let $\lambda = f(\gamma)/g(\gamma)$, then λ is a root of $P(x)$ and $\lambda \in \mathbb{F}_{q^n}$. Obviously, γ is a root of $f(x) - \lambda g(x)$, which is a polynomial of degree h over $\mathbb{F}_{q^n}$. Clearly, $\lambda \in \mathbb{F}_q[\gamma]$, so $\mathbb{F}_q[\lambda] \subseteq \mathbb{F}_q[\gamma]$. Of course $[\mathbb{F}_q[\lambda] : \mathbb{F}_q] = n$. Therefore

$$g(x)^n P(f(x)/g(x)) \text{ is irreducible over } \mathbb{F}_q$$

$$\Leftrightarrow \quad [\mathbb{F}_q[\gamma] : \mathbb{F}_q] = hn$$

$$\Leftrightarrow \quad [\mathbb{F}_q[\gamma] : \mathbb{F}_q[\lambda]] = h$$

$$\Leftrightarrow \quad f(x) - \lambda g(x) \text{ is irreducible over } \mathbb{F}_{q^n}.$$

This completes the proof.   □

We consider some special cases:

(a) $f(x) = ax + b$ and $g(x) = cx + d$ are linear polynomials, and assume that $P(x)$ is irreducible of degree $n > 1$ over $\mathbb{F}_q$. By Theorem 10.14 $(cx + d)^n P((ax + b)/(cx + d))$ is irreducible if and only is $ax + b$ and $cx + d$ are coprime, and if and only if $ad - bc \neq 0$. This result has already been stated in Theorem 10.2(ii).

(b) $f(x) = x^t$ and $g(x) = 1$. We have

Theorem 10.15 *Let t be a positive integer, and $P(x) \in \mathbb{F}_q[x]$ be irreducible over $\mathbb{F}_q$ of degree $n > 0$, not of the form cx, and of period m (which equals the order of any root of $P(x)$). Then $P(x^t)$ is irreducible over $\mathbb{F}_q$ if and only if*

(i) *Each prime divisor of t divides m, but not $(q^n - 1)/m$, and*

(ii) *If $4|t$ then $4|(q^n - 1)$.*

Proof. This is an immediate consequence of Theorems 10.14 and 10.7. □

(c) $f(x) = x^2 + 1$ and $g(x) = x$. Then $f(x)/g(x) = x + x^{-1}$. We distinguish the cases: q even and q odd.

Theorem 10.16 *Let $q = 2^m$, and let $P(x) = \sum_{i=0}^{n} c_i x^i \in \mathbb{F}_q[x]$ be irreducible over $\mathbb{F}_q$ of degree $n > 0$ and with $c_0 \neq 0$. Then $x^n P(x + x^{-1})$ is a self-reciprocal polynomial of degree $2n$ over $\mathbb{F}_q$, and it is irreducible over $\mathbb{F}_q$ if and only if $\mathrm{Tr}_{\mathbb{F}_q/\mathbb{F}_2}(c_1/c_0) \neq 0$.*

Proof. Let $Q(x) = x^n P(x + x^{-1})$. Clearly, $Q(x)$ is of degree $2n$ and

$$x^{2n} Q(x^{-1}) = x^{2n} x^{-n} P(x^{-1} + x) = Q(x).$$

Thus $Q(x)$ is self-reciprocal. Moreover, by Theorem 10.14, $Q(x)$ is irreducible over $\mathbb{F}_q$ if and only if $x^2 + 1 - \alpha x$ is irreducible over $\mathbb{F}_{q^n}$ for some root $\alpha \in \mathbb{F}_{q^n}$ of $P(x)$. By Theorem 10.12, the last condition is equivalent to $\mathrm{Tr}_{\mathbb{F}_{q^n}/\mathbb{F}_2}(\alpha^{-2}) \neq 0$. Since

$$\mathrm{Tr}_{\mathbb{F}_{q^n}/\mathbb{F}_2}(\alpha^{-2}) = (\mathrm{Tr}_{\mathbb{F}_{q^n}/\mathbb{F}_2}(\alpha^{-1}))^2 = (\mathrm{Tr}_{\mathbb{F}_q/\mathbb{F}_2}(\mathrm{Tr}_{\mathbb{F}_{q^n}/\mathbb{F}_q}(\alpha^{-1})))^2$$
$$= (\mathrm{Tr}_{\mathbb{F}_q/\mathbb{F}_2}(-c_1/c_0))^2 = (\mathrm{Tr}_{\mathbb{F}_q/\mathbb{F}_2}(c_1/c_0))^2,$$

it is also equivalent to $(\mathrm{Tr}_{\mathbb{F}_q/\mathbb{F}_2}(c_1/c_0)) \neq 0$.   □

Theorem 10.17 *Let q be an odd prime power, and $P(x)$ be an irreducible polynomial of degree $n > 0$ over $\mathbb{F}_q$. Then $x^n P(x + x^{-1})$ is a self-reciprocal polynomial of degree m and it is irreducible over $\mathbb{F}_q$ if and only if $P(2)P(-2) \notin \mathbb{F}_q^{*2}$.*

Proof. As in the above theorem, $x^n P(x + x^{-1})$ is self-reciprocal of degree $2n$ and by Theorem 10.14 $x^n P(x + x^{-1})$ is irreducible over $\mathbb{F}_q$ if and only if $x^2 - \alpha x + 1$ is irreducible over $\mathbb{F}_{q^n}$ for some root $\alpha \in \mathbb{F}_{q^n}$ of $P(x)$. This is equivalent to the condition $\alpha^2 - 4 \notin \mathbb{F}_{q^n}^{*2}$, and, hence, is equivalent to

$$
\begin{aligned}
-1 &= (\alpha^2 - 4)^{(q^n - 1)/2} \\
&= \{[(2 - \alpha)(-2 - \alpha)]^{(q^n - 1)/(q-1)}\}^{(q-1)/2} \\
&= \{[\prod_{i=0}^{n-1}(2 - \alpha)(-2 - \alpha)]^{q^i}\}^{(q-1)/2} \\
&= \{\prod_{i=0}^{n-1}(2 - \alpha^{q^i})(-2 - \alpha^{q^i})\}^{(q-1)/2} \\
&= \{P(2)P(-2)\}^{(q-1)/2},
\end{aligned}
$$

that is, $P(2)P(-2) \notin \mathbb{F}_q^{*2}$. $\qquad\square$

Both Theorems 10.16 and 10.17 are due to Meyn (1990).

Corollary 10.18 *Let q be an odd prime power and $P(x)$ be an irreducible polynomial of degree n over $\mathbb{F}_q$. Then $2^n x^n P((x + x^{-1})/2)$ is irreducible over $\mathbb{F}_q$ if and only if $P(1)P(-1) \notin \mathbb{F}_q^{*2}$.*

Proof. Let $P_0(x) = 2^n P(x/2)$ and apply Theorem 10.17 to $P_0(x)$. $\qquad\square$

(d) $f(x) = x^p - x - b$ and $g(x) = 1$. We have the following result which is due to Varshamov (1973, 1973a)

Theorem 10.19 *Let q be a power of a prime p, $P(x) = x^n + \sum_{i=0}^{n-1} c_i x^i$ be an irreducible polynomial over $\mathbb{F}_q$, and $b \in \mathbb{F}_q$. Then $P(x^p - x - b)$ is irreducible over $\mathbb{F}_q$ if and only if $\mathrm{Tr}_{\mathbb{F}_q/\mathbb{F}_p}(nb - c_{n-1}) \neq 0$.*

Proof. By Theorem 10.14, $P(x^p - x - b)$ is irreducible over $\mathbb{F}_q$ if and only if $x^p - x - b - \alpha$ is irreducible over $\mathbb{F}_{q^n}$ for some root $\alpha \in \mathbb{F}_{q^n}$ of $P(x)$. By Theorem 10.9 this is equivalent to the condition

$$
\begin{aligned}
\mathrm{Tr}_{\mathbb{F}_{q^n}/\mathbb{F}_p}(b + \alpha) &= \mathrm{Tr}_{\mathbb{F}_q/\mathbb{F}_p}(\mathrm{Tr}_{\mathbb{F}_{q^n}/\mathbb{F}_q}(b + \alpha)) \\
&= \mathrm{Tr}_{\mathbb{F}_q/\mathbb{F}_p}(nb - c_{n-1}) \neq 0.
\end{aligned}
$$

$\square$

(e) $f(x) = l(x)$ is a linearized polynomial and $g(x) = 1$. The condition of the irreducibility of this type of polynomials was established by Agou (1977, 1978, 1980). Following Cohen (1984) we consider the simple case $l(x) = x^p - ax$, where $a \in \mathbb{F}_q^*$, first.

Theorem 10.20 *Let* $\mathbb{F}_q$ *be of characteristic* p, $P(x) = x^n + \sum_{i=0}^{n-1} c_i x^i$ *be an irreducible polynomial over* $\mathbb{F}_q$, *and* α *be a root of* $P(x)$ *in* $\mathbb{F}_{q^n}$. *Then for any* $a \in \mathbb{F}_q^*$, $P(x^p - ax)$ *is irreducible over* $\mathbb{F}_q$ *if and only if*

$$a^{n_1(q-1)/(p-1)} = 1 \ and \ \mathrm{Tr}_{\mathbb{F}_{q^n}/\mathbb{F}_p}(\alpha/a_0^p) \neq 0,$$

where $n_1 = \gcd(n, p-1)$ *and* $a_0 \in \mathbb{F}_{q^n}^*$ *such that* $a_0^{p-1} = a$. *In particular, if* $a_0 \in \mathbb{F}_q^*$, *then* $P(x^p - a_0^{p-1}x)$ *is irreducible over* $\mathbb{F}_q$ *if and only if* $\mathrm{Tr}_{\mathbb{F}_q/\mathbb{F}_p}(c_{n-1}/a_0^p) \neq 0$.

Proof. By Theorem 10.14, $P(x^p - ax)$ is irreducible over $\mathbb{F}_q$ if and only if $x^p - ax - \alpha$ is irreducible over $\mathbb{F}_{q^n}$. By Theorem 10.12 this is equivalent to $a = a_0^{p-1}$ for some $a_0 \in \mathbb{F}_{q^n}^*$ and $\mathrm{Tr}_{\mathbb{F}_{q^n}/\mathbb{F}_p}(\alpha/a_0^p) \neq 0$. Clearly, $a = a_0^{p-1}$ for some $a_0 \in \mathbb{F}_{q^n}^*$ if and only if

$$a^{(q^n - 1)/(p-1)} = 1. \tag{10.10}$$

Since $a \in \mathbb{F}_q^*$, $a^{q-1} = 1$. Thus (10.10) holds if and only if $a^h = 1$, where

$$h = \gcd\left(\frac{q^n - 1}{p - 1}, q - 1\right) = \frac{q - 1}{p - 1}\gcd\left(\frac{q^n - 1}{q - 1}, p - 1\right).$$

But $(q^n - 1)/(q - 1) = q^{n-1} + q^{n-2} + \ldots + 1 \equiv n \pmod{p - 1}$. Hence $h = n_1(q - 1)/(p - 1)$.

Moreover, if $a_0 \in \mathbb{F}_q^*$, then $a^{n_1(q-1)/(p-1)} = a_0^{n_1(q-1)} = 1$ holds automatically and

$$\mathrm{Tr}_{\mathbb{F}_{q^n}/\mathbb{F}_p}(\alpha/a_0^p) = \mathrm{Tr}_{\mathbb{F}_q/\mathbb{F}_p}\mathrm{Tr}_{\mathbb{F}_{q^n}/\mathbb{F}_q}(\alpha/a_0^p)$$
$$= \mathrm{Tr}_{\mathbb{F}_q/\mathbb{F}_p}(\mathrm{Tr}_{\mathbb{F}_{q^n}/\mathbb{F}_q}(\alpha)/a_0^p)$$
$$= -\mathrm{Tr}_{\mathbb{F}_q/\mathbb{F}_p}(c_{n-1}/a_0^p).$$

Therefore the last assertion also holds. $\square$

Now we turn to the general case, i.e., $l(x)$ is any linearized polynomial. We need some preparation.

Lemma 10.21 *Let $l(x)$ be a linearized polynomial over $\mathbb{F}_q$. Then there exists another linearized polynomial $g(x)$ over $\mathbb{F}_q$ and an element r in $\mathbb{F}_q$ such that*

$$l(x) = g(x^p - x) + rx.$$

Proof. Let $l(x) = l_v x^{p^v} + l_{v-1} x^{p^{v-1}} + \cdots + l_0 x$. We apply induction on v. The case $v = 0$ is trivial. Suppose $v \geq 1$ and put

$$\bar{l}(x) = l(x) - l_v(x^p - x)^{p^{v-1}} = (l_{v-1} + l_v)x^{p^{v-1}} + \cdots,$$

which is another linearized polynomial but of degree $\leq p^{v-1}$. By induction hypothesis, there is a linearized polynomial $\bar{g}(x)$ and an element $r \in \mathbb{F}_q$ such that $\bar{l}(x) = \bar{g}(x^p - x) + rx$. Then $l(x) = \bar{g}(x^p - x) + rx + l_v(x^p - x)^{p^{v-1}}$. Let $g(x) = l_v x^{p^{v-1}} + \bar{g}(x)$, then $g(x)$ is a linearized polynomial and $l(x) = g(x^p - x) + rx$. $\qquad\square$

Lemma 10.22 *Suppose that the linearized polynomial $l(x)$ over $\mathbb{F}_q$ has a nonzero root a_0 in $\mathbb{F}_q$. Then there exists a linearized polynomial $g(x)$ over $\mathbb{F}_q$ such that $l(x) = g(x^p - a_0^{p-1}x)$.*

Proof. $l(a_0 x)$ is a linearized polynomial over $\mathbb{F}_q$ with 1 as a root. By Lemma 10.21, there exists another linearized polynomial $g_1(x)$ over $\mathbb{F}_q$ and $r \in \mathbb{F}_q$ such that $l(a_0 x) = g_1(x^p - x) + rx$. Substituting $x = 1$ into this equation, we obtain $0 = g_1(0) + r = r$. Thus $l(a_0 x) = g_1(x^p - x)$. Let $g(x) = g_1(x/a_0^p)$, then $l(x) = g(x^p - a_0^{p-1}x)$. $\qquad\square$

Lemma 10.23 *Let $l(x)$ be a monic linearized polynomial over $\mathbb{F}_q$ of degree p^v with $v \geq 2$, and let $b \in \mathbb{F}_q^*$. Then $l(x) - b$ is irreducible over $\mathbb{F}_q$ if and only if (i) $p=v=2$, and (ii) $l(x)$ has the form*

$$l(x) = x(x + a_0)(x^2 + a_0 x + b_0), \tag{10.11}$$

where $a_0, b_0 \in \mathbb{F}_q^*$, such that the quadratic polynomials $x^2 + a_0 x + b_0$ and $x^2 + b_0 x + b$ are both irreducible over $\mathbb{F}_q$.

Proof. Let us prove the "only if" part first. Assume that for $b \in \mathbb{F}_q^*$, $l(x) - b$ is irreducible over $\mathbb{F}_q$. Then by Lemma 10.11 $l(x)$ has a nonzero root $a_0 \in \mathbb{F}_q^*$. By Lemma 10.22 there exists a linearized polynomial $g(x)$ over $\mathbb{F}_q$ such that $l(x) = g(x^p - a_0^{p-1}x)$. Put $\bar{g}(x) = g(x) - b$, then $\bar{g}(x^p - a_0^{p-1}x) = l(x) - b$. Since $l(x) - b$ is irreducible over $\mathbb{F}_q$, $\bar{g}(x)$ is also irreducible over $\mathbb{F}_q$. By assumption, $l(x)$ is monic, so both $g(x)$ and $\bar{g}(x)$ are monic.

Also by assumption $\deg l(x) = p^v$, so $\deg \bar{g}(x) = \deg g(x) = p^{v-1}$. For simplicity let $n = \deg \bar{g}(x) = p^{v-1}$ and write $\bar{g}(x) = x^n + \sum_{i=0}^{n-1} c_i x^i$, where $c_i \in \mathbb{F}_q$, $i = 1, 2, \ldots, n-1$. Since both $\bar{g}(x)$ and $\bar{g}(x^p - a_0^{p-1} x) = l(x) - b$ are irreducible over $\mathbb{F}_q$, applying the last statement of Theorem 10.20 to $\bar{g}(x)$, we obtain

$$\mathrm{Tr}_{\mathbb{F}_q/\mathbb{F}_p}(c_{n-1}/a_0^p) \neq 0, \tag{10.12}$$

which implies $c_{n-1} \neq 0$. But $\bar{g}(x)$ is an affine polynomial of degree p^{v-1}, so $n = p^{v-1}$, $n - 1 = p^{v-1} - 1 = p^{v-2}$. Thus $p = v = 2$, $n = 2$, and $g(x) = x^2 + b_0 x$, where $b_0 \in \mathbb{F}_q$, whence $\bar{g}(x) = x^2 + b_0 x + b$ and $l(x) = g(x^2 - a_0 x) = x(x - a_0)(x^2 + a_0 x + b_0)$. Comparing the coefficients of the two expressions of $\bar{g}(x)$, we obtain $b_0 = c_{n-1}$ and (10.12) becomes $\mathrm{Tr}_{\mathbb{F}_q/\mathbb{F}_p}(b_0/a_0^2) \neq 0$ which, by Theorem 10.12, is equivalent to $x^2 + a_0 x + b_0$ being irreducible over $\mathbb{F}_q$. We know already that $x^2 + b_0 x + b = \bar{g}(x)$ is irreducible over $\mathbb{F}_q$. The "only if" part is completely proved.

Now let us prove the "if" part. We have (10.11)

$$l(x) = x(x + a_0)(x^2 + a_0 x + b_0),$$

where $a_0, b_0 \in \mathbb{F}_q^*$ such that both $x^2 + a_0 x + b_0$ and $x^2 + b_0 x + b$ are irreducible over $\mathbb{F}_q$. Then

$$l(x) - b = (x^2 + a_0 x)^2 + b_0(x^2 + a_0 x) + b.$$

Let $\bar{g}(x) = x^2 + b_0 x + b$, then $l(x) - b = \bar{g}(x^2 + a_0 x)$. Since $x^2 + a_0 x + b_0$ is irreducible over $\mathbb{F}_q$, by Theorem 10.12 $\mathrm{Tr}_{\mathbb{F}_q/\mathbb{F}_p}(b_0/a_0^2) \neq 0$. Then by Theorem 10.20 $\bar{g}(x^2 + a_0 x)$ is irreducible over $\mathbb{F}_q$. Therefore $l(x) - b$ is irreducible over $\mathbb{F}_q$. $\qquad\square$

The following Theorem is due to Agou (1980).

Theorem 10.24 *Let $P(x) = x^n + \sum_{i=0}^{n-1} c_i x^i$ be a monic irreducible polynomial of degree n over $\mathbb{F}_q$, and $l(x)$ be a monic linearized polynomial of degree p^v with $v \geq 2$ over $\mathbb{F}_q$ of degree p^v with $v \geq 2$. Then $P(l(x))$ is irreducible over $\mathbb{F}_q$ if and only if (i) $p = v = 2$, (ii) n is odd, and (iii) $l(x)$ has the form (10.11) where $a_0, b_0 \in \mathbb{F}_q^*$ and both $x^2 + a_0 x + b_0$ and $x^2 + b_0 x + c_{n-1}$ are irreducible over $\mathbb{F}_q$.*

Proof. By Theorem 10.14, $P(l(x))$ is irreducible over $\mathbb{F}_q$ if and only if $l(x) - \alpha$ is irreducible over $\mathbb{F}_{q^n}$ for some $\alpha \in \mathbb{F}_{q^n}$ such that $P(\alpha) = 0$. Applying Lemma 10.23 to $l(x) - \alpha$, we conclude that $P(l(x))$ is irreducible over $\mathbb{F}_q$ if and only if $p = v = 2$, and $l(x)$ has the form (10.11) where $a_0, b_0 \in \mathbb{F}_{q^n}^*$ and both $x^2 + a_0 x + b_0$ and $x^2 + b_0 x + \alpha$ are irreducible over $\mathbb{F}_{q^n}$.

Assume now that $p = v = 2$. Then $\deg l(x) = 4$ and $\deg (l(x)/x) = 3$. If $l(x)/x$ is irreducible over $\mathbb{F}_q$ and has a root γ in $\mathbb{F}_{q^n}$, then $\mathbb{F}_q[\gamma] = \mathbb{F}_{q^3} \subseteq \mathbb{F}_{q^n}$, which implies $3 \mid n$. By Theorem 7.6 $l(x)/x$ is a product of 3 linear factors over $\mathbb{F}_{q^n}$. Thus if $l(x)/x$ is irreducible over $\mathbb{F}_q$ or a product of three linear factors over $\mathbb{F}_q$, then it remains so over $\mathbb{F}_{q^n}$. Therefore if $l(x)/x$ has an irreducible quadratic factor over $\mathbb{F}_{q^n}$ it must have an irreducible quadratic factor over $\mathbb{F}_q$. Now assume further that $l(x)$ is of form (10.11) where $a_0, b_0 \in \mathbb{F}_{q^n}^*$ and both $x^2 + a_0 x + b_0$ and $x^2 + b_0 x + \alpha$ are irreducible over $\mathbb{F}_{q^n}$. Then $a_0, b_0 \in \mathbb{F}_q^*$, $x^2 + a_0 x + b_0$ is irreducible over $\mathbb{F}_q$, and n is odd.

Finally, by Theorem 7.20 or Theorem 10.12 $x^2 + b_0 x + \alpha$ is irreducible over $\mathbb{F}_{q^n}$ if and only if $\mathrm{Tr}_{\mathbb{F}_{q^n}/\mathbb{F}_p}(\alpha/b_0^2) \neq 0$. But

$$\mathrm{Tr}_{\mathbb{F}_{q^n}/\mathbb{F}_p}(\alpha/b_0^2) = \mathrm{Tr}_{\mathbb{F}_q/\mathbb{F}_p}\mathrm{Tr}_{\mathbb{F}_{q^n}/\mathbb{F}_q}(\alpha/b_0^2)$$
$$= \mathrm{Tr}_{\mathbb{F}_q/\mathbb{F}_p}(\mathrm{Tr}_{\mathbb{F}_{q^n}/\mathbb{F}_q}(\alpha)/b_0^2)$$
$$= -\mathrm{Tr}_{\mathbb{F}_q/\mathbb{F}_p}(c_{n-1}/b_0^2).$$

By Theorem 7.20 or Theorem 10.12 again, $\mathrm{Tr}_{\mathbb{F}_q/\mathbb{F}_p}(c_{n-1}/b_0^2) \neq 0$ if and only if $x^2 + b_0 x + c_{n-1}$ is irreducible over $\mathbb{F}_q$. This completes the proof of Theorem 10.24.   □

10.5   Recursive Constructions

Based on the irreducibility criteria developed in the previous section, let us study how to construct recursively irreducible polynomials of arbitrarily large degrees.

First we introduce the following recursive construction of Varshamov (1984).

Theorem 10.25 Let p be a prime and let $f(x) = x^n + \sum_{i=0}^{n-1} c_i x^i$ be irreducible over $\mathbb{F}_p$. Suppose that there exists an element $a \in \mathbb{F}_p^*$, such that $(na + c_{n-1})f'(a) \neq 0$. Let $g(x) = x^p - x + a$ and define $f_k(x)$ for $k = 0, 1, 2, \ldots$ recursively by

$$f_0(x) = f(g(x)),$$
$$f_k(x) = \widetilde{f}_{k-1}(g(x)) \text{ for } k \geq 1,$$

where $\widetilde{f}_{k-1}(x)$ is the reciprocal polynomial of $f_{k-1}(x)$. Then for all $k \geq 0$, $f_k(x)$ is irreducible over $\mathbb{F}_p$ of degree np^{k+1}.

Proof. For any $k \geq 0$, let deg $f_k(x) = n_k$ and

$$f_k(x) = \sum_{i=0}^{n_k} f_{ki} x^i.$$

Denote by P_k the aggregate of assertions: $f_{k1} = f_k'(a) \neq 0$, both $f_k(x)$ and $f_k'(x)$ are constant on $\mathbb{F}_p$, $f_k(x)$ is irreducible over $\mathbb{F}_p$, and $n_k = np^{k+1}$. We prove P_k by induction on k.

When $k = 0$, we have

$$f_0'(x) = f'(g(x))g'(x)$$

Then

$$\begin{aligned}
f_{01} &= (f_0'(x)) \mid_{x=0} \\
&= (f'(g(x))g'(x)) \mid_{x=0} \\
&= -f'(a) \ (\because g(0) = a, \ g'(0) = -1)
\end{aligned}$$

and

$$\begin{aligned}
f_0'(a) &= (f'(g(x))g'(x)) \mid_{x=a} \\
&= -f'(a). \ (\because g(a) = a, \ g'(a) = -1).
\end{aligned}$$

Thus $f_{01} = f_0'(a) = -f'(a) \neq 0$, by assumption. Clearly, $g(x)$ is constant on $\mathbb{F}_p$ and $g'(x) = -1$, hence both $f_0(x) = f(g(x))$ and $f_0'(x)$ are constant on $\mathbb{F}_p$. Obviously, deg $f_0(x) = np$, thus $n_0 = np$. From Theorem 10.19, $f_0(x) = f(g(x))$ is irreducible over $\mathbb{F}_p$ if and only if $\mathrm{Tr}_{\mathbb{F}_p/\mathbb{F}_p}(na + c_{n-1}) = na + c_{n-1} \neq 0$. By assumption $na + c_{n-1} \neq 0$, hence $f_0(x)$ is irreducible over $\mathbb{F}_p$. P_0 is completely proved.

Now assume that P_k is true for $k \geq 0$. We prove that P_{k+1} is also true. By induction hypothesis $f_k(x)$ is irreducible over $\mathbb{F}_p$ and $n_k = np^{k+1}$, thus $f_{k+1}(x) = \tilde{f}_k(g(x))$ is of degree np^{k+2} and $n_{k+1} = np^{k+2}$. Since $f_k(x)$ is irreducible, $f_{k0} \neq 0$. Thus $f_{k0}^{-1}\tilde{f}_k(x)$ is monic and its coefficient of x^{n_k-1} is $f_{k0}^{-1} f_{k1} \neq 0$. Then

$$\mathrm{Tr}_{\mathbb{F}_p/\mathbb{F}_p}(n_k a + f_{k0}^{-1} f_{k1}) = f_{k0}^{-1} f_{k1} \neq 0,$$

for $n_k = np^{k+1} = 0$ in $\mathbb{F}_p$. It follows from Theorem 10.19 that $f_{k+1}(x) = \tilde{f}_k(g(x))$ is irreducible over $\mathbb{F}_p$. By definition

$$f_{k+1}(x) = \tilde{f}_k(g(x)) = \sum_{i=0}^{n_k} f_{ki} g(x)^{n_k-i}.$$

Thus

$$f'_{k+1}(x) = \sum_{i=0}^{n_k-1} f_{ki}(n_k - i)g(x)^{n_k-i-1}g'(x)$$
$$= \sum_{i=0}^{n_k-1} f_{ki}ig(x)^{n_k-i-1},$$

for $g'(x) = -1$. Since $g(x)$ is constant on $\mathbb{F}_p$, so are $f_{k+1}(x)$ and $f'_{k+1}(x)$. Moreover,

$$f_{k+1,1} = (f'_{k+1}(x))\mid_{x=0} = (\widetilde{f_k}{}'(g(x))g'(x))\mid_{x=0}$$

$$= -\widetilde{f_k}{}'(a) = -f'_k(a^{-1})a^{n_k-1} = -f'_k(a),$$

which is non-zero by the induction hypothesis. Similarly,

$$f'_{k+1}(a) = (f'_{k+1}(x))\mid_{x=a} = (\widetilde{f_k}{}'(g(x))g'(x))\mid_{x=a} = -\widetilde{f_k}{}'(a) \neq 0,$$

as above. This completes the proof of P_{k+1}.

By mathematical induction P_k holds for all $k \geq 0$. In particular, for all $k \geq 0$, $f_k(x)$ is irreducible over $\mathbb{F}_p$ of degree np^{k+1}. □

Example 10.4 Let p be any prime and $f(x) = x$. Then $n = 1$, $c_{n-1} = c_0 = 0$, and $f'(x) = 1$. Let $a = -1$. Then $(na + c_{n-1})f'(a) \neq 0$. Let $f_0(x) = f(x^p - x - 1)$, $f_k(x) = \widetilde{f}_{k-1}(x^p - x - 1)$ for $k \geq 1$. Then by Theorem 10.25 $f_k(x)$ is irreducible over $\mathbb{F}_p$ of degree p^{k+1} for every $k \geq 0$.

For q being a power of 2, based on Theorem 10.16 we have the following recursive construction.

Theorem 10.26 *Let* $q = 2^m$ *and let* $f(x) = \sum_{i=0}^{n} c_i x^i \in \mathbb{F}_q[x]$ *be irreducible over* $\mathbb{F}_q$ *of degree* n *with* $c_0 c_n \neq 0$. *For all* $k \geq 0$, *define polynomials recursively:*

$$f_0(x) = f(x),$$
$$f_k(x) = x^{n2^{k-1}} f_{k-1}(x + x^{-1}), k \geq 1.$$

Then $f_k(x)$ *is a self-reciprocal polynomial of degree* $n2^k$ *over* $\mathbb{F}_q$ *for each* $k > 0$, *and*

(*i*) $f_1(x)$ *is irreducible if* $\mathrm{Tr}_{\mathbb{F}_q/\mathbb{F}_2}(c_1/c_0) \neq 0$,

(*ii*) $f_k(x)$ *is irreducible for every* $k > 1$, *if both* $\mathrm{Tr}_{\mathbb{F}_q/\mathbb{F}_2}(c_1/c_0) \neq 0$ *and* $\mathrm{Tr}_{\mathbb{F}_q/\mathbb{F}_2}(c_{n-1}/c_n) \neq 0$.

Proof. By Theorem 10.16 it is easy to see by induction on k that $f_k(x)$ is of degree $n2^k$ for every $k \geq 0$ and $f_k(x)$ is a self-reciprocal polynomial for every $k > 0$.

Clearly, (i) follows immediately from Theorem 10.16. It remains to prove (ii). We apply induction on k to prove that $f_k(x)$ is irreducible for every $k \geq 1$. Since $\mathrm{Tr}_{\mathbb{F}_q/\mathbb{F}_2}(c_1/c_0) \neq 0$ by assumption, $f_1(x)$ is irreducible. Let $k \geq 1$ and assume that $f_k(x)$ is irreducible. Let $n_k = n2^k$ and $f_k(x) = \sum_{i=0}^{n_k} c_{ki} x^i$, $k \geq 0$. We have

$$
\begin{aligned}
f_k(x) &= x^{n_{k-1}} f_{k-1}((1 + x^2)/x) \\
&= x^{n_{k-1}} \sum_{i=0}^{n_{k-1}} c_{k-1,i}((1 + x^2)/x)^i \\
&= \sum_{i=0}^{n_{k-1}} c_{k-1,i}(1 + x^2)^i x^{n_{k-1}-i} \\
&= \sum_{i=0}^{n_k} c_{ki} x^i.
\end{aligned}
$$

Thus

$$c_{k0} = c_{k-1,n_{k-1}} \quad \text{and} \quad c_{k1} = c_{k-1,n_{k-1}-1}. \tag{10.13}$$

By Theorem 10.16 $f_{k+1}(x)$ is irreducible over $\mathbb{F}_q$ if and only if

$$\mathrm{Tr}_{\mathbb{F}_q/\mathbb{F}_2}(c_{k1}/c_{k0}) \neq 0. \tag{10.14}$$

Since $f_j(x)$ is self-reciprocal for $j \geq 1$, (10.13) implies

$$
\begin{aligned}
c_{k0} &= c_{k-1,n_{k-1}} = c_{k-1,0} = \cdots = c_{10} = c_{0,n_0} = c_n \\
c_{k1} &= c_{k-1,n_{k-1}-1} = c_{k-1,1} = \cdots = c_{11} = c_{0,n_0-1} = c_{n-1}.
\end{aligned}
$$

Since $\mathrm{Tr}_{\mathbb{F}_q/\mathbb{F}_2}(c_{n-1}/c_n) \neq 0$ by assumption, (10.14) is true for $k \geq 1$, and so $f_{k+1}(x)$ is irreducible over $\mathbb{F}_q$ for $k \geq 1$. $\qquad\square$

Example 10.5 Let $q = 2$ and $f(x) = x + 1 \in \mathbb{F}_2$. Define $f_0(x) = f(x)$ and $f_k(x) = x^{2^{k-1}} f_{k-1}(x + x^{-1})$ for $k \geq 1$. Then by Theorem 10.26 $f_k(x)$ is irreducible over $\mathbb{F}_2$ of degree 2^k.

For q odd, based on Corollary 10.18 we have the following construction which is due to Cohen (1992).

Theorem 10.27 *Let q be odd and let $f(x)$ be an irreducible polynomial of degree $n \geq 1$ over $\mathbb{F}_q$, where n is even if $q \equiv 3 \pmod 4$. Suppose that $f(1)f(-1) \notin \mathbb{F}_q^{*2}$. Define*

$$
\begin{aligned}
f_0(x) &= f(x), \\
f_k(x) &= (2x)^{n_{k-1}} f_{k-1}((x + x^{-1})/2) \text{ for } k \geq 1,
\end{aligned}
$$

where $n_k = n2^k$ denotes the degree of $f_k(x)$. Then $f_k(x)$ is an irreducible polynomial over $\mathbb{F}_q$ of degree $n2^k$ for every $k \geq 1$.

Proof. It is easy to see by induction on k that $f_k(x)$ is of degree $n_k = n2^k$ for every $k \geq 0$. For $k \geq 1$ we have

$$
\begin{aligned}
f_k(1)f_k(-1) &= 2^{n_{k-1}}f_{k-1}(1)(-2)^{n_{k-1}}f_{k-1}(-1) \\
&= (-1)^{n_{k-1}}2^{2n_{k-1}}f_{k-1}(1)f_{k-1}(-1) \\
&= \cdots \\
&= (-1)^n d_k^2 f_0(1)f_0(-1),
\end{aligned}
$$

where $d_k \in \mathbb{F}_q^*$. By assumption either $-1 \in \mathbb{F}_q^{*2}$ (when $q \equiv 1 \pmod 4$) or n is even, so $(-1)^n \in \mathbb{F}_q^{*2}$. Thus

$$
f_k(1)f_k(-1) \in f_0(1)f_0(-1)\mathbb{F}_q^{*2}.
$$

But $f_0(1)f_0(-1) = f(1)f(-1) \notin \mathbb{F}_q^{*2}$, therefore $f_k(1)f_k(-1) \notin \mathbb{F}_q^{*2}$ for every $k \geq 0$. Applying induction on k, we can prove, by Corollary 10.18, that $f_k(x)$ is irreducible over $\mathbb{F}_q$ for every $k \geq 1$. $\qquad\square$

Example 10.6 Applying Theorem 10.27 to the following special cases gives several infinite families of irreducible polynomials over the respective fields:

(a) $q = 3, f(x) = x^2 \pm x - 1$.

(b) $q = 5, f(x) = x \pm 2$ or $x^2 \pm x + 2$.

(c) $q \equiv 1 \pmod 4, f(x) = x - c$ or $x^2 + 2cx + 1$, where c is such that $c^2 - 1$ is a non-square in $\mathbb{F}_q$.

(d) $q \equiv 3 \pmod 4, f(x) = x^2 + 2cx - 1$, where c is such that $c^2 + 1$ is a non-square in $\mathbb{F}_q$. $\qquad\square$

10.6 Composed Product and Sum of Polynomials

Now let us study how to construct irreducible polynomials of degree mn from irreducible polynomials of degree m and n with $\gcd(m, n) = 1$. We have the following theorem which is due to Brawley and Carlitz (1987).

Theorem 10.28 *Let $f(x)$ and $g(x)$ be irreducible polynomials of degrees m and n, respectively, over $\mathbb{F}_q$. Assume that $\gcd(m, n) = 1$. Let $\alpha_1, \alpha_2 \dots, \alpha_m$ be the m distinct roots of $f(x)$ in $\mathbb{F}_{q^m}$ and $\beta_1, \beta_2, \dots \beta_n$ be the n distinct roots of $g(x)$ in $\mathbb{F}_{q^n}$. We may regard both $\mathbb{F}_{q^m}$ and $\mathbb{F}_{q^n}$ as subfields of $\mathbb{F}_{q^{mn}}$. Let*

$$
(f \odot g)(x) = \prod_{i=1}^{m}\prod_{j=1}^{n}(x - \alpha_i\beta_j)
$$

and

$$(f \oplus g)(x) = \prod_{i=1}^{m} \prod_{j=1}^{n} (x - (\alpha_i + \beta_j)).$$

Then both $(f \odot g)$ and $(f \oplus g)$ are irreducible polynomials of degree mn over $\mathbb{F}_q$.

Proof.   Let σ be the Frobenius automorphism of $\mathbb{F}_{q^{mn}}$ over $\mathbb{F}_q$. The restriction of σ to $\mathbb{F}_{q^m}$ is the Frobenius automorphism of $\mathbb{F}_{q^m}$ over $\mathbb{F}_q$, which will be denoted simply also by σ. Similarly, the restriction of σ to $\mathbb{F}_{q^n}$ is the Frobenius automorphism of $\mathbb{F}_{q^n}$ over $\mathbb{F}_q$ and will also be denoted by σ. Since $f(x)$ is irreducible of degree m over $\mathbb{F}_q$, σ permutes the roots $\alpha_1, \alpha_2, \ldots, \alpha_m$ of $f(x)$ cyclically, i.e., we can assume that $\sigma(\alpha_1) = \alpha_2, \sigma(\alpha_2) = \alpha_3, \ldots, \sigma(\alpha_m) = \alpha_1$. Similarly, σ also permutes the roots $\beta_1, \beta_2, \ldots, \beta_n$ of $g(x)$ cyclically. Since $\gcd(m, n) = 1$, σ permutes the roots $\alpha_i \beta_j (i = 1, 2, \ldots, m; j = 1, 2, \ldots, n)$ of $f \odot g$ cyclically and permutes the roots $\alpha_i + \beta_j (i = 1, 2, \ldots, m; j = 1, 2, \ldots, n)$ of $f \oplus g$ cyclically (Exercise 10.18). By Theorem 7.5 $f \odot g$ is the minimum polynomial of $\alpha_1 \beta_1$ over $\mathbb{F}_q$ and, hence, is of degree mn and is irreducible over $\mathbb{F}_q$. Similarly, $f \oplus g$ is the minimum polynomial of $\alpha_1 + \beta_1$ over $\mathbb{F}_q$, is of degree mn, and is irreducible over $\mathbb{F}_q$.   □

$f \odot g$ and $f \oplus g$ are called the *composed product* and *composed sum* of f and g, respectively.

In Theorem 10.28 the computation of $f \odot g$ and $f \oplus g$ requires finding out the roots of f and g and computing their products, which lie in an extension field of $\mathbb{F}_q$. However, if we use Kronecker product of matrices the computation can be achieved in $\mathbb{F}_q$.

Let A be an $m \times m$ matrix of $\mathbb{F}_q$ and B be an $n \times n$ matrix over $\mathbb{F}_q$. Let

$$A = (a_{ij})_{1 \le i, j \le m}$$

and

$$B = (b_{kl})_{1 \le k, l \le n}.$$

Define

$$A \otimes B = (a_{ij}B)_{1 \le i, j \le n} = \begin{pmatrix} a_{11}B & a_{12}B & \ldots & a_{1m}B \\ a_{21}B & a_{22}B & \ldots & a_{2m}B \\ \vdots & \vdots & & \vdots \\ a_{m1}B & a_{m2}B & \ldots & a_{mm}B \end{pmatrix},$$

where

$$a_{ij}B = \begin{pmatrix} a_{ij}b_{11} & a_{ij}b_{12} & \cdots & a_{ij}b_{1n} \\ a_{ij}b_{21} & a_{ij}b_{22} & \cdots & a_{ij}b_{2n} \\ \vdots & \vdots & & \vdots \\ a_{ij}b_{n1} & a_{ij}b_{n2} & \cdots & a_{ij}b_{nn} \end{pmatrix}, \ i,j = 1, 2, \ldots, m.$$

Clearly $A \otimes B$ is an $mn \times mn$ matrix over $\mathbb{F}_q$, called the *Kronecker product* of A and B. It is known from linear algebra that if A, B, C, D are matrices of appropriate sizes, then

$$(A \otimes B)(C \otimes D) = (AC) \otimes (BD)$$

and

$$(A \otimes B)^{-1} = A^{-1} \otimes B^{-1}$$

provided that A^{-1} and B^{-1} exist (Exercise 10.19). It is also known from linear algebra that if the characteristic roots of A in some extension field of $\mathbb{F}_q$ are $\alpha_1, \alpha_2, \ldots, \alpha_m$ and that of B are $\beta_1, \beta_2, \ldots, \beta_n$, then the characteristic roots of $A \otimes B$ are $\alpha_i\beta_j (i = 1, 2, \ldots, m; j = 1, 2, \ldots, n)$ and that of $A \otimes I^{(n)} + I^{(m)} \otimes B$ are $\alpha_i + \beta_j (i = 1, 2, \ldots, m, j = 1, 2, \ldots, n)$ (Exercise 10.20). Moreover, it is known from linear algebra (Exercise 10.21) that the characteristic polynomial of the $m \times m$ matrix

$$A = \begin{pmatrix} 0 & & & & -a_m \\ 1 & 0 & & & \vdots \\ & 1 & \ddots & & \vdots \\ & & \ddots & 0 & -a_2 \\ & & & 1 & -a_1 \end{pmatrix}, \tag{10.15}$$

where the blanks in the matrix represent omitted zeros, is

$$x^m + a_1 x^{m-1} + a_2 x^{m-2} + \cdots + a_m$$

and the characteristic polynomial of the $n \times n$ matrix

$$B = \begin{pmatrix} 0 & & & & -b_n \\ 1 & 0 & & & \vdots \\ & 1 & \ddots & & \vdots \\ & & \ddots & 0 & -b_2 \\ & & & 1 & -b_1 \end{pmatrix} \tag{10.16}$$

is
$$x^n + b_1 x^{n-1} + b_2 x^{n-2} + \ldots + b_n.$$

Therefore we have

Theorem 10.29 *Let $f(x) = x^m + a_1 x^{m-1} + a_2 x^{m-2} + \cdots + a_m$ and $g(x) = x^n + b_1 x^{n-1} + b_2 x^{n-2} + \cdots + b_n$ be two irreducible polynomials of degrees m and n, respectively, over $\mathbb{F}_q$. Assume that $\gcd(m,n) = 1$. Let A be the $m \times m$ matrix (10.15) and B be the $n \times n$ matrix (10.16). Then*

$$f \odot g = \det(x I^{(mn)} - A \otimes B) \quad and \quad f \oplus g = \det(x I^{(mn)} - (A \otimes I^{(n)} + I^{(m)} \otimes B))$$

are both irreducible polynomials over $\mathbb{F}_q$ of degree mn.

The method of computing $f \odot g$ and $f \oplus g$ in the above theorem involves computing the determinant of a matrix of size $mn \times mn$. However it can be reduced to computing the determinant of a matrix of size $m \times m$ or $n \times n$ (cf. Blake et al.(1991)).

Theorem 10.30 *Let $f(x) = \sum_{i=0}^{m} a_i x^{m-i}$ and $g(x) = \sum_{j=0}^{n} b_j x^{n-j}$ be two monic irreducible polynomials of degrees m and n, respectively, over $\mathbb{F}_q$. Assume that $\gcd(m,n) = 1$. Let A be the $m \times m$ matrix (10.15) and B be the $n \times n$ matrix (10.16). Then*

$$f \odot g = \det\left(\sum_{j=0}^{n} b_j x^{n-j} A^j \right) = \det\left(\sum_{i=0}^{m} a_i x^{m-i} B^i \right)$$

and

$$f \oplus g = \det\left(\sum_{j=0}^{n} b_j (xI - A)^{n-j} \right) = \det\left(\sum_{i=0}^{m} a_i (xI - B)^{m-i} \right)$$

are both irreducible polynomials over $\mathbb{F}_q$ of degree mn.

Proof. Since $f(x)$ is irreducible, $f(x)$ has m distinct roots, say $\alpha_1, \alpha_2, \ldots, \alpha_m$. Then there is an invertible matrix P such that $P^{-1}AP$ is the diagonal matrix

$$\begin{pmatrix} \alpha_1 & & & \\ & \alpha_2 & & \\ & & \ddots & \\ & & & \alpha_m \end{pmatrix},$$

which will be denoted by D. Then $A = PDP^{-1}$ and thus

$$
\begin{aligned}
f \oplus g &= \det(xI - (A \otimes I + I \otimes B)) \\
&= \det(xI - [(PDP^{-1}) \otimes I + (PIP^{-1}) \otimes B]) \\
&= \det((P \otimes I)[xI - ((D \otimes I) + (I \otimes B))](P^{-1} \otimes I)) \\
&= \det(xI - [D \otimes I + I \otimes B]) \\
&= \det \begin{pmatrix} (xI - \alpha_1 I - B) & & \\ & \ddots & \\ & & (xI - \alpha_m I - B) \end{pmatrix} \\
&= \textstyle\prod_{i=1}^{m} \det(xI - B - \alpha_i I) \\
&= \det \prod_{i=1}^{m} ((xI - B) - \alpha_i I) \\
&= \det \textstyle\sum_{i=0}^{m} a_i (xI - B)^{m-i}.
\end{aligned}
$$

Similarly,

$$
f \oplus g = \det \left(\sum_{j=0}^{n} b_j (xI - A)^{n-j} \right).
$$

The formula for $f \odot g$ can be proved in the same way. $\qquad\square$

10.7 Irreducible Polynomials of Any Degree

Now let n be a positive integer and assume that n has the following complete factorization

$$
n = r_1^{e_1} r_2^{e_2} \dots r_l^{e_l},
$$

where $r_1, r_2, \dots r_l$ are distinct primes and $e_1, e_2, \dots, e_l$ are positive integers. If for each i an irreducible polynomial of degree $r_i^{e_i}$ over $\mathbb{F}_q$ can be constructed, then by Theorems 10.28, 10.29 or 10.30 an irreducible polynomial of degree n over $\mathbb{F}_q$ can be constructed.

Now let r be a prime and e be a positive integer. Let us study how to construct an irreducible polynomial of degree r^e over $\mathbb{F}_p$. We distinguish the following cases.

(a) $r = p$ and p is odd. An irreducible polynomial of degree p^e for every $e \geq 1$ over $\mathbb{F}_p$ has been constructed in Example 10.4.

(b) $r = p = 2$. An irreducible polynomial of degree 2^e over $\mathbb{F}_2$ has been constructed in Example 10.5.

(c) $r = 2$ and $p \equiv 1 \,(\mathrm{mod}\,4)$. Let $a \in \mathbb{F}_p^* \setminus \mathbb{F}_p^{*2}$. For any integer $k \geq 0$, by Corollary 10.8 the polynomial $x^{2^k} - a$ is an irreducible polynomial over $\mathbb{F}_p$.

(d) $r = 2$ and $p \equiv 3 \,(\mathrm{mod}\,4)$. An irreducible polynomial over $\mathbb{F}_p$ of degree 2^k for any integer $k \geq 1$ can be constructed by Theorem 10.13.

(e) r is odd and $r \neq p$. The following theorem due to Shoup (1990) can be used to construct irreducible polynomials over $\mathbb{F}_p$ of degree r^e for any $e \geq 0$.

Theorem 10.31 *Let p be a prime, $r \neq p$ an odd prime, and let $n = \mathrm{ord}_r(p)$ be the order of p in $\mathbb{Z}_r^*$. Assume that $f(x)$ is an irreducible polynomial in $\mathbb{F}_p[x]$ of degree n, α is a root of $f(x)$ in $\mathbb{F}_{p^n}$, $a \in \mathbb{F}_p[\alpha] = \mathbb{F}_{p^n}$, but $a \notin \mathbb{F}_{p^n}^{*r}$. Then for any positive integer e, $x^{r^e} - a$ is an irreducible polynomial in $\mathbb{F}_{p^n}[x]$. Moreover, let β be a root of $x^{r^e} - a$, then $\beta \in \mathbb{F}_{p^{nr^e}}$ and*

$$\gamma = \mathrm{Tr}_{\mathbb{F}_{p^{nr^e}}/\mathbb{F}_{p^{r^e}}}(\beta) = \sum_{i=0}^{n-1} \beta^{p^{ir^e}} \tag{10.17}$$

has degree r^e over $\mathbb{F}_p$. Thus the minimal polynomial of γ over $\mathbb{F}_p$ is an irreducible polynomial over $\mathbb{F}_p$ of degree r^e.

Proof. Let m be the order of a in $\mathbb{F}_{p^n}^*$. Since $a \notin \mathbb{F}_{p^n}^{*r}$, r does not divide $(p^n - 1)/m$. By Corollary 10.8, $g(x) = x^{r^e} - a$ is irreducible over $\mathbb{F}_{p^n}$ for any e. So $\mathbb{F}_{p^n}(\beta) = \mathbb{F}_{p^{nr^e}}$ and γ is in $\mathbb{F}_{p^{r^e}}$.

We assert that $\gcd(n, r) = 1$. Suppose $r \mid n$, then let $n = rn_0$, where n_0 is a positive integer $< n$. We have $p^r \equiv p \,(\mathrm{mod}\,r)$, thus $p^{n_0} \equiv p^{rn_0} \equiv p^n \equiv 1 \,(\mathrm{mod}\,r)$, which contradicts that n is the order of p in $\mathbb{Z}_r^*$. This proves our assertion. Consequently, $\gcd(n, r^t) = 1$ for any $t > 0$.

Suppose to the contrary that γ has degree r^t over $\mathbb{F}_p$, where $t < e$. Since $\gcd(n, r^t) = 1$, γ has degree r^t over $\mathbb{F}_{p^n}$. Now $[\mathbb{F}_{p^n}(\beta) : \mathbb{F}_{p^n}(\beta^r)] = r$, and so in particular, γ lies in $\mathbb{F}_{p^n}(\beta^r)$. For each i, $0 \leq i \leq n - 1$, by division algorithm let $p^{ir^e} = x_i r + y_i$, where $0 < y_i < r$. From the fact that p, and hence also p^{r^e}, has order $n \,(\mathrm{mod}\,r)$, it follows that the y_i's are distinct. Then (10.17) yields the equation

$$(\beta^r)^{x_0}\beta^{y_0} + \ldots + (\beta^r)^{x_{n-1}}\beta^{y_{n-1}} - \gamma = 0.$$

Thus, β is a root of a non-zero polynomial over $\mathbb{F}_{p^n}(\beta^r)$ of degree less than r. But this contradicts the fact that β has degree r over $\mathbb{F}_{p^n}(\beta^r)$, and so the theorem is proved. $\square$

After the foregoing preparation, we are now ready to give the proof of Theorem 8.34.

10.8 Exercises

10.1 Prove Theorem 10.2.

10.2 Prove Theorem 10.3.

10.3 Prove Theorem 10.4.

10.4 Let $f(x) \in \mathbb{F}_q[x]$ be an irreducible polynomial of degree n and let $k > 0$, $d = \gcd(n, k)$. Prove that $f(x)$ factors into d irreducible polynomials of the same degree n/d over $\mathbb{F}_{q^k}$ and that $f(x)$ is irreducible over $\mathbb{F}_{q^k}$ if and only if $\gcd(n, k) = 1$.

10.5 Prove Example 10.1.

10.6 Prove Lemma 10.10.

10.7 Let p be a prime and assume that $p \equiv 3 \pmod 4$. Let r be a positive integer such that $2^r \| (p + 1)$. Define subsets $H_1, H_2, \cdots, H_r$ of $\mathbb{F}_p$ recursively as follows:

$$H_1 = \{0\},$$
$$H_l = \{\pm((u + 1)/2)^{(p+1)/4} : u \in H_{l-1}\}, l = 2, 3, \cdots, r - 1,$$
$$H_r = \{\pm((u - 1)/2)^{(p+1)/4} : u \in H_{r-1}\}.$$

Then both

$$x^{2^l} + 1 = \prod_{u \in H_l} (x^2 - 2ux + 1), \ 1 \le l \le r - 1,$$

and

$$x^{2^{r+k}} + 1 = \prod_{u \in H_r} (x^{2^{k+1}} - 2ux^{2^k} - 1) \text{ for any integer } k \ge 0,$$

are complete factorizations over $\mathbb{F}_p$.

10.8 Prove Example 10.3.

10.9 Let p be a prime. Then the trinomial $x^p - x - a \in \mathbb{F}_p[x]$ is a primitive polynomial over $\mathbb{F}_p$ if and only if a is a primitive element of $\mathbb{F}_p$ and $p(x^p - x - 1) = (p^p - 1)/(p - 1)$.

10.10 Let $q = 2^m$ and let $P(x) = \sum_{i=0}^n c_i x^i \in \mathbb{F}_q[x]$ be irreducible over $\mathbb{F}_q$ of degree n. Prove that $x^n \widetilde{P}(x + x^{-1})$ is a self-reciprocal polynomial of degree $2n$ over $\mathbb{F}_q$, and that it is irreducible over $\mathbb{F}_q$ if and only if

$$\mathrm{Tr}_{\mathbb{F}_q/\mathbb{F}_2}(c_{n-1}/c_n) \ne 0.$$

10.11 Let q be an odd prime power and $P(x)$ be an irreducible polynomial of degree $n \geq 1$ over $\mathbb{F}_q$. Then $x^n P(x + x^{-1})$ is a self-reciprocal polynomial of degree $2n$ and it is irreducible over $\mathbb{F}_q$ if and only if $\widetilde{P}(2)\widetilde{P}(-2) \notin \mathbb{F}_q^{*2}$.

10.12 Let r be an odd prime and q a prime power. Suppose that q is a primitive root modulo r and $r^2 \nmid (q^{r-1} - 1)$. Then the polynomial

$$x^{(r-1)r^k} + x^{(r-2)r^k} + \dots + x^{r^k} + 1$$

is irreducible over $\mathbb{F}_q$ for each $k \geq 0$.

10.13 Let p be an odd prime and $f(x) = x^p + x^{p-1} + \dots + x - 1 \in \mathbb{F}_p[x]$. Define $f_{-1}(x) = f(x)$, $f_0(x) = f(x^p - x - 1)$ and $f_k(x) = \widetilde{f}_{k-1}(x^p - x - 1)$ for $k \geq 1$. Prove that for every $k \geq -1$, $f_k(x)$ is irreducible over $\mathbb{F}_p$ of degree p^{k+2}.

10.14 Let $q = 2^m$. Define polynomials $c_k(x), d_k(x) \in \mathbb{F}_q[x]$ recursively by

$$c_0(x) = x, \quad d_0(x) = 1,$$

$$c_{k+1}(x) = c_k^2(x) + d_k^2(x), \quad d_{k+1}(x) = c_k(x)d_k(x) \quad k \geq 0.$$

Prove that

$$\frac{c_{k+1}(x)}{d_{k+1}(x)} = \frac{1 + (c_k(x)/d_k(x))^2}{c_k(x)/d_k(x)} = \frac{c_k(1 + x^{-1})}{d_k(1 + x^{-1})}$$

and

$$\frac{d_{k+1}(x)}{d_k(x + x^{-1})} = x^{2^k}.$$

Then deduce that the $f_k(x)$ defined in Theorem 12.26 can be expressed as

$$f_k(x) = (d_k(x))^k f(c_k(x)/d_k(x)) \text{ for all } k \geq 0.$$

10.15 Let $q = 2^m$ and $f(x) = \sum_{i=0}^n c_i x^i \in \mathbb{F}_q[x]$ be irreducible over $\mathbb{F}_q$ of degree n with $c_0 c_n \neq 0$. For all $k \geq 0$, define polynomials

$$f_0(x) = f(x),$$
$$f_k(x) = (1 + x^2)^{n2^{k-1}} f_{k-1}(x/(1 + x^2)) \text{ for } k \geq 1.$$

Then $f_k(x)$ is a self-reciprocal polynomial of degree $n2^k$ over $\mathbb{F}_q$ for each $k > 0$ and

(i) $f_1(x)$ is irreducible, if $\text{Tr}_{\mathbb{F}_q/\mathbb{F}_2}(c_{n-1}/c_n) \neq 0$.

(ii) $f_k(x)$ is irreducible for every $k \geq 1$, if both $\mathrm{Tr}_{\mathbb{F}_q/\mathbb{F}_2}(c_{n-1}/c_n) \neq 0$ and $\mathrm{Tr}_{\mathbb{F}_q/\mathbb{F}_2}(c_1/c_0) \neq 0$.

10.16 Let $q = 2^m$ and define polynomials $a_k(x), b_k(x) \in \mathbb{F}_q[x]$ recursively by

$$a_0(x) = x, \quad b_0(x) = 1,$$

$$a_{k+1}(x) = a_k(x)b_k(x), \quad b_{k+1} = a_k(x)^2 + b_k(x)^2, \text{ for } k \geq 0.$$

Prove that

$$\frac{a_{k+1}(x)}{b_{k+1}(x)} = \frac{a_k(x)/b_k(x)}{1 + (a_k(x)/b_k(x))^2} = \frac{a_k(x/(1+x^2))}{b_k(x/(1+x^2))}$$

and

$$\frac{b_{k+1}(x)}{b_k(x + x^{-1})} = (1 + x^2)^{2^k}.$$

Then deduce that the $f_k(x)$ defined in the above exercise can be expressed as

$$f_k(x) = (b_k(x))^n f(a_k(x)/b_k(x)) \text{ for all } k \geq 0.$$

10.17 Let q be an odd prime power. Define polynomials $c_k(x), d_k(x) \in \mathbb{F}_q[x]$ by

$$c_0(x) = x, \quad d_0(x) = 1,$$

$$c_{k+1}(x) = c_k(x)^2 + d_k(x)^2, \quad d_{k+1}(x) = 2c_k(x)d_k(x), \quad k \geq 0.$$

Prove that

$$\frac{c_{k+1}(x)}{d_{k+1}(x)} = \frac{1 + (c_k(x)/d_k(x))^2}{2c_k(x)/d_k(x)} = \frac{c_k((x + x^{-1})/2)}{d_k((x + x^{-1})/2)}$$

and

$$\frac{d_{k+1}(x)}{d_k((x + x^{-1})/2)} = (2x)^{2^k}.$$

Let $f(x) \in \mathbb{F}[x]$ be of degree $n \geq 1$. Define

$$f_0(x) = f(x),$$
$$f_k(x) = (2x)^{k2^{k-1}} f((x + x^{-1})/2), \quad k \geq 1.$$

Prove that

$$f_k(x) = d_k(x)^k f(c_k(x)/d_k(x)), \quad k \geq 0.$$

10.18 Let m and n be positive integers such that $\gcd(m,n) = 1$, $f(x)$ and $g(x)$ be irreducible polynomials of degrees m and n, respectively, over $\mathbb{F}_q$, $\alpha_1, \ldots, \alpha_m$ are roots of $f(x)$ in $\mathbb{F}_{q^m}$ and $\beta_1, \ldots, \beta_n$ be roots of $g(x)$ in $\mathbb{F}_{q^n}$, and let σ be the Frobenious automorphism of $\mathbb{F}_{q^{mn}}$ over $\mathbb{F}_q$. Assume σ permutes both $\alpha_1, \ldots, \alpha_m$ and $\beta_1, \ldots, \beta_n$ cyclically. Prove that σ permutes both $\alpha_i \beta_j$ $(1 \leq i \leq m, 1 \leq j \leq n)$ and $\alpha_i + \beta_j$ cyclically.

10.19 Let A, B, C, D be matrices of approximate sizes so that A and C can be multiplied and B and D can also be multiplied, then

$$(A \otimes B)(C \otimes D) = (AC) \otimes (BD).$$

Moreover, if both A and B are nonsingular matrices, then

$$(A \otimes B)^{-1} = A^{-1} \otimes B^{-1}.$$

10.20 Let A be an $m \times m$ matrix and B be an $n \times n$ matrix, both over $\mathbb{F}_q$. Let $\alpha_1, \alpha_2, \ldots, \alpha_m$ be the characteristic roots of A in some extension field of $\mathbb{F}_q$ and $\beta_1, \beta_2, \ldots, \beta_n$ be those of B. Prove that $\alpha_i \beta_j$ $(i = 1, 2, \ldots, m; j = 1, 2, \ldots, n)$ are the characteristic roots of $A \otimes B$ and $\alpha_i + \beta_j$ $(i = 1, 2, \ldots, m; j = 1, 2, \ldots, n)$ are the characteristic roots of $A \otimes I^{(n)} + I^{(m)} \otimes B$.

10.21 Let

$$A = \begin{pmatrix} 0 & & & & & -a_m \\ 1 & 0 & & & & \vdots \\ & 1 & \ddots & & & \vdots \\ & & \ddots & 0 & -a_2 \\ & & & 1 & -a_1 \end{pmatrix}$$

be an $n \times n$ matrix. Prove that the characteristic polynomial $|xI^{(m)} - A|$ of A is $x^m + a_1 x^{m-1} + a_2 x^{m-2} + \ldots + a_m$.

Chapter 11

Quadratic Forms over Finite Fields

11.1 Quadratic Forms over Finite Fields of Characteristic not 2

Let $\mathbb{F}_q$ be a finite field with q elements where q is a prime power. The homogeneous polynomial of degree 2 in n indeterminates $x_1, x_2, \ldots, x_n$ over $\mathbb{F}_q$

$$Q(x_1, x_2, \ldots, x_n) = \sum_{1 \leqslant i \leqslant k \leqslant n} b_{ik} x_i x_k, \quad b_{ik} \in \mathbb{F}_q, \tag{11.1}$$

is called a *quadratic form* in $x_1, x_2, \ldots, x_n$ over $\mathbb{F}_q$. It is said to be *definite* if $Q(a_1, a_2, \ldots, a_n) = 0$ for $a_1, a_2, \ldots, a_n \in \mathbb{F}_q$ implies $a_1 = a_2 = \ldots = a_n = 0$; otherwise, it is said to be *indefinite*.

In this section we shall study quadratic forms over finite fields $\mathbb{F}_q$ of odd characteristic, so we assume q is odd. If we set

$$a_{ik} = a_{ki} = \frac{1}{2} b_{ik}, \quad 1 \leqslant i < k \leqslant n,$$

and

$$a_{ii} = b_{ii}, \quad 1 \leqslant i \leqslant n,$$

then (11.1) can be expressed as

$$Q(x_1, x_2, \ldots, x_n) = \sum_{1 \leqslant i, k \leqslant n} a_{ik} x_i x_k, \tag{11.2}$$

269

where

$$a_{ik} = a_{ki}, \quad 1 \leqslant i, k \leqslant n.$$

Let us introduce the $n \times n$ matrix

$$S = \begin{pmatrix} a_{11} & a_{12} & \cdots & a_{1n} \\ a_{21} & a_{22} & \cdots & a_{2n} \\ \vdots & \vdots & \ddots & \vdots \\ a_{n1} & a_{n2} & \cdots & a_{nn} \end{pmatrix}$$

and call S the *coefficient matrix* of (11.2). Clearly,

$$^tS = S.$$

That is, S is a symmetric matrix. We can write (11.2) in matrix notation as

$$Q(x_1, x_2, \ldots, x_n) = (x_1, x_2, \ldots, x_n) \, S \, {}^t(x_1, x_2, \ldots, x_n). \tag{11.3}$$

We define the *rank* of the quadratic form (11.3) to be the rank of the coefficient matrix S. If the rank of the quadratic form (11.3) is n, it is said to be *non-singular*.

We introduce the n-dimensional column vector space over $\mathbb{F}_q$:

$$\mathbb{F}_q^n = \left\{ {}^t(x_1, x_2, \ldots, x_n) : \ x_i \in \mathbb{F}_q, \ i = 1, 2, \ldots, n \right\}.$$

Consider a nonsingular linear transformation of $\mathbb{F}_q^n$

$$\begin{pmatrix} x_1 \\ x_2 \\ \vdots \\ x_n \end{pmatrix} \longmapsto T \begin{pmatrix} x_1 \\ x_2 \\ \vdots \\ x_n \end{pmatrix}, \tag{11.4}$$

where T is an $n \times n$ nonsingular matrix. Under (11.4) the quadratic form (11.3) is carried into

$$(x_1, x_2, \ldots, x_n) \, {}^tTST \, {}^t(x_1, x_2, \ldots, x_n). \tag{11.5}$$

Thus the coefficient matrix S of (11.3) is subjected to the cogredience transformation

$$S \longmapsto {}^tTST. \tag{11.6}$$

We say that (11.3) and (11.5) are *cogredient* quadratic forms. Conversely, suppose that the coefficient matrix S of the quadratic form (11.3) is subjected to the cogredience transformation (11.6), then the quadratic form

(11.3) is carried to the quadratic form (11.5) under the nonsingular linear transformation (11.4). Therefore studying how a quadratic form is transformed under nonsingular linear transformations is equivalent to studying how a symmetric matrix is transformed under cogredience transformations. We prefer the later problem. To study the cogredience of symmetric matrices over $\mathbb{F}_q$, we need the following lemmas.

Lemma 11.1 *Cogredience transformations preserve the rank of the symmetric matrix.*

This lemma is well-known in linear algebra.

Lemma 11.2 *Let S be an $n \times n$ symmetric matrix of rank r $(1 \leqslant r \leqslant n)$ over $\mathbb{F}_q$, where q is odd. Suppose that S is cogredient to*

$$\begin{pmatrix} S_1 & 0 \\ 0 & 0 \end{pmatrix} \begin{matrix} r \\ n-r \end{matrix}$$
$$\begin{matrix} r & n-r \end{matrix}$$

Then $\det S_1 \neq 0$ and the coset $(\det S_1)\mathbb{F}_q^{*2}$ is uniquely determined by S.

Proof. Suppose that S is cogredient to both

$$\begin{pmatrix} S_1 & 0 \\ 0 & 0 \end{pmatrix} \begin{matrix} r \\ n-r \end{matrix} \quad \text{and} \quad \begin{pmatrix} S_2 & 0 \\ 0 & 0 \end{pmatrix} \begin{matrix} r \\ n-r \end{matrix}.$$
$$\begin{matrix} r & n-r \end{matrix} \qquad\qquad\qquad \begin{matrix} r & n-r \end{matrix}$$

By Lemma 11.1 $\det S_1 \neq 0$ and $\det S_2 \neq 0$. There is an $n \times n$ nonsingular matrix

$$\begin{pmatrix} A & B \\ C & D \end{pmatrix} \begin{matrix} r \\ n-r \end{matrix}$$
$$\begin{matrix} r & n-r \end{matrix}$$

such that

$$^t\!\begin{pmatrix} A & B \\ C & D \end{pmatrix} \begin{pmatrix} S_1 & 0 \\ 0 & 0 \end{pmatrix} \begin{pmatrix} A & B \\ C & D \end{pmatrix} = \begin{pmatrix} S_2 & 0 \\ 0 & 0 \end{pmatrix},$$

from which follows $^t\!AS_1A = S_2$. Taking determinants gives $\det S_1(\det A)^2 = \det S_2$. Therefore $\det A \neq 0$. Hence $\det S_1\,\mathbb{F}_q^{*2} = \det S_2\,\mathbb{F}_q^{*2}$. $\quad\square$

Lemma 11.3 *Any $n \times n$ symmetric matrix S of rank r $(0 \leqslant r \leqslant n)$ over $\mathbb{F}_q$, where q is odd, is cogredient to a diagonal form*

$$\begin{pmatrix} a_1 & & & & & & \\ & a_2 & & & & & \\ & & \ddots & & & & \\ & & & a_r & & & \\ & & & & 0 & & \\ & & & & & \ddots & \\ & & & & & & 0 \end{pmatrix}, \tag{11.7}$$

where $a_1, a_2, \ldots, a_r$ are nonzero elements of $\mathbb{F}_q$.

Proof. If $S = 0$ our lemma holds automatically. Now suppose that $S \neq 0$. We may assume that there is a nonzero diagonal element in S. In fact, if all the diagonal elements of S are zero, there must be a nonzero non-diagonal element in S; suppose that the element a_{ik} in the (i, k) position of S is nonzero, where $i \neq k$. Let E_{ik} be the $n \times n$ matrix which has a 1 in the (i, k) position and 0's in all the other positions. Then S is cogredient to

$$^t(I + E_{ki})S(I + E_{ki}). \tag{11.8}$$

It is easy to compute that the element in the (i, i) position of (11.8) is nonzero. Let

$$P_{1i} = E_{1i} + E_{i1} + \sum_{k \neq 1, i} E_{kk}.$$

Then $^tP_{1i}\,^t(I+E_{ki})S(I+E_{ki})P_{1i}$ has a nonzero element in the $(1, 1)$ position. Therefore we can always assume that the element a_{11} in the $(1, 1)$ position of S is nonzero. Write S as

$$S = \begin{pmatrix} a_{11} & b \\ ^tb & S_1 \end{pmatrix},$$

where b is an $(n-1)$-dimensional row vector and S_1 is an $(n-1) \times (n-1)$ symmetric matrix. Then

$$^t\begin{pmatrix} 1 & -a_{11}^{-1}b \\ & I \end{pmatrix} \begin{pmatrix} a_{11} & b \\ ^tb & S_1 \end{pmatrix} \begin{pmatrix} 1 & -a_{11}^{-1}b \\ & I \end{pmatrix} = \begin{pmatrix} a_{11} & 0 \\ 0 & S_1 - a_{11}^{-1}\,^tb\,b \end{pmatrix},$$

where $S_1 - a_{11}^{-1}\,^tb\,b$ is an $(n-1) \times (n-1)$ symmetric matrix of rank $r-1$. Hence our lemma follows from mathematical induction.   □

Lemma 11.4 *Let z be a fixed non-square element of $\mathbb{F}_q^*$. Then the following two 2×2 diagonal matrices*

$$\begin{pmatrix} 1 & 0 \\ 0 & 1 \end{pmatrix} \quad and \quad \begin{pmatrix} z & 0 \\ 0 & z \end{pmatrix}$$

are cogredient.

This lemma has already been proved in the proof of Lemma 8.10.

Theorem 11.5 *Let q be a power of an odd prime number. Then any $n \times n$ symmetric matrix of rank r over $\mathbb{F}_q$ is cogredient to*

$$\begin{pmatrix} I^{(r)} & \\ & 0^{(n-r)} \end{pmatrix} \quad or \quad \begin{pmatrix} I^{(r-1)} & & \\ & z & \\ & & 0^{(n-r)} \end{pmatrix}, \tag{11.9}$$

where z is a fixed non-square element of $\mathbb{F}_q^*$. Moreover, the two matrices (11.9) are not cogredient.

Proof. Let S be an $n \times n$ symmetric matrix of rank r over $\mathbb{F}_q$. By Lemma 11.3 we can assume that S is the diagonal matrix (11.7). After rearranging $a_1, a_2, \ldots, a_r$ we can assume that $a_1, a_2, \ldots, a_t \in \mathbb{F}_q^{*2}$ and $a_{t+1}, a_{t+2}, \ldots, a_r \notin \mathbb{F}_q^{*2}$. By Theorem 6.20(ii) $\mathbb{F}_q^{*2}$ is a subgroup of $\mathbb{F}_q^*$ of index 2. Let z be a fixed non-square element of $\mathbb{F}_q^*$, then $a_{t+1}, a_{t+2}, \ldots, a_r \in z\mathbb{F}_q^{*2}$. Thus $z^{-1}a_{t+1}, z^{-1}a_{t+2}, \ldots, z^{-1}a_r \in \mathbb{F}_q^{*2}$. For any element $s \in \mathbb{F}_q^{*2}$, there is at least one element in $\mathbb{F}_q^*$ whose square is equal to s, denote such an element by $s^{1/2}$. Let

$$
T = \begin{pmatrix}
a_1^{-\frac{1}{2}} & & & & & & \\
& \ddots & & & & & \\
& & a_t^{-\frac{1}{2}} & & & & \\
& & & (z^{-1}a_{t+1})^{-\frac{1}{2}} & & & \\
& & & & \ddots & & \\
& & & & & (z^{-1}a_r)^{-\frac{1}{2}} & \\
& & & & & & I^{(n-r)}
\end{pmatrix}.
$$

Then

$$
{}^tT \begin{pmatrix}
a_1 & & & & \\
& a_2 & & & \\
& & \ddots & & \\
& & & a_r & \\
& & & & 0^{(n-r)}
\end{pmatrix} T = \begin{pmatrix}
I^{(t)} & & \\
& zI^{(r-t)} & \\
& & 0^{(n-r)}
\end{pmatrix}.
$$

Thus from Lemma 11.4 follows the first statement of the theorem. The second statement of the theorem follows from Lemma 11.2. □

Lemma 8.10 is a special case of Theorem 11.5. The matrices in (11.9) are called the *normal forms* of $n \times n$ symmetric matrices under cogredience transformations.

Now we translate the above theorem on the cogredience of symmetric matrices into the language of quadratic forms as follows.

Theorem 11.6 *Let q be a power of an odd prime number. Any quadratic form of rank r over $\mathbb{F}_q$ can be carried by a nonsingular linear transformation into one of the following normal forms*

$$
\sum_{i=1}^{r} x_i^2 \qquad \text{or} \qquad \sum_{i=1}^{r-1} x_i^2 + zx_r^2.
$$

Moreover, these two forms are not cogredient.

There is another normal form of symmetric matrices relative to cogredience, which is also frequently used.

Theorem 11.7 *Let q be a power of an odd prime number, and let S be a $n \times n$ symmetric matrix of rank r over $\mathbb{F}_q$. If r is odd, we write $r = 2\nu + 1$, then S is cogredient to either*

$$
\begin{pmatrix} 0 & I^{(\nu)} & \\ I^{(\nu)} & 0 & \\ & & 1 \\ & & & 0^{(n-r)} \end{pmatrix}
\qquad or \qquad
\begin{pmatrix} 0 & I^{(\nu)} & \\ I^{(\nu)} & 0 & \\ & & z \\ & & & 0^{(n-r)} \end{pmatrix},
$$

where z is a fixed non-square element of $\mathbb{F}_q^*$, and these two matrices are not cogredient. If r is even, we write $r = 2\nu$, then S is cogredient to either

$$
\begin{pmatrix} 0 & I^{(\nu)} & \\ I^{(\nu)} & 0 & \\ & & 0^{(n-r)} \end{pmatrix}
\qquad or \qquad
\begin{pmatrix} 0 & I^{(\nu-1)} & & \\ I^{(\nu-1)} & 0 & & \\ & & 1 & \\ & & & -z \\ & & & & 0^{(n-r)} \end{pmatrix},
$$

and these two matrices are not cogredient.

Proof. It follows from Theorem 11.5 and Lemma 11.2. $\qquad\square$

The four matrices listed in Theorem 11.7 are also called *normal forms* of $n \times n$ symmetric matrices under cogredience transformations. If an $n \times n$ symmetric matrix is cogredient to any one of the first three normal forms, then ν is called its *index*; and if it is cogredient to the last one, then $\nu - 1$ is called its *index*. Clearly, the quadratic form

$$
(x_1, x_2) \begin{pmatrix} 1 & 0 \\ 0 & -z \end{pmatrix} \begin{pmatrix} x_1 \\ x_2 \end{pmatrix}
$$

is definite. Moreover we have

Corollary 11.8 *The index of an $n \times n$ symmetric matrix is invariant under cogredience transformations.*

The *index* of a quadratic form is defined to be the index of its coefficient matrix. Theorem 11.7 can also be put into the language of quadratic forms.

Theorem 11.9 *Let q be a power of an odd prime number. Any quadratic* *form of index ν over $\mathbb{F}_q$ is cogredient to one of the following normal forms*

$$\sum_{i=1}^{\nu} 2x_i x_{\nu+i},$$

$$\sum_{i=1}^{\nu} 2x_i x_{\nu+i} + x_{2\nu+1}^2,$$

$$\sum_{i=1}^{\nu} 2x_i x_{\nu+i} + z x_{2\nu+1}^2,$$

$$\sum_{i=1}^{\nu} 2x_i x_{\nu+i} + x_{2\nu+1}^2 - z x_{2\nu+2}^2.$$

For Theorems 11.6 and 11.9, cf. Dickson (1900).

As an application of the normal forms, let us compute the number of solutions of the equation $Q(x_1, x_2, \ldots, x_n) = a$, where $Q(x_1, x_2, \ldots, x_n)$ is a quadratic form in n indeterminants $x_1, x_2, \ldots, x_n$ over $\mathbb{F}_q$ and $a \in \mathbb{F}_q$. Here by a solution of $Q(x_1, x_2, \ldots, x_n) = a$, we mean an n-dimensional vector $^t(c_1, c_2, \ldots, c_n) \in \mathbb{F}_q^n$ such that $Q(c_1, c_2, \ldots, c_n) = a$. At the beginning of this section $Q(x_1, x_2, \ldots, x_n)$ has been written in the form (11.3)

$$Q(x_1, x_2, \ldots, x_n) = (x_1, x_2, \ldots, x_n) \, S \, {}^t(x_1, x_2, \ldots, x_n).$$

Let T be an $n \times n$ nonsingular matrix over $\mathbb{F}_q$. Under the nonsingular linear transformation (11.4) the quadratic form (11.3) is carried into (11.5)

$$(x_1, x_2, \ldots, x_n) \, {}^t T S T \, {}^t(x_1, x_2, \ldots, x_n).$$

Moreover, the set of solutions of (11.3) is carried into the set of solutions of (11.5) bijectively by

$$^t(c_1, c_2, \ldots, c_n) \longmapsto T \, {}^t(c_1, c_2, \ldots, c_n).$$

Thus it is enough to determine the number of solutions of $Q(x_1, x_2, \ldots, x_n) = a$ for Q to be one of the normal forms listed in Theorem 11.9. Moreover, it is sufficient to consider the case when $Q(x_1, x_2, \ldots, x_n)$ is nonsingular. In fact, if N is the number of solutions of

$$2 \sum_{i=1}^{\nu} x_i x_{\nu+i} = a \tag{11.10}$$

in $\mathbb{F}_q^{2\nu}$, then $q^{n-2\nu} N$ is the number of solutions of (11.10) in $\mathbb{F}_q^n$. Similar for the other cases. We have

Theorem 11.10 *Let q be a power of an odd prime. Then (i) the number of solutions of* (11.10)

$$2\sum_{i=1}^{\nu} x_i x_{\nu+i} = a$$

in $\mathbb{F}_q^{2\nu}$ is

$$q^{2\nu-1} + q^\nu - q^{\nu-1} \text{ or } q^{2\nu-1} - q^{\nu-1}$$

according to $a = 0$ or $a \neq 0$, respectively, (ii) the number of solutions of

$$2\sum_{i=1}^{\nu} x_i x_{\nu+i} + x_{2\nu+1}^2 = a \qquad (11.11)$$

in $\mathbb{F}_q^{2\nu+1}$ is

$$q^{2\nu}, \ q^{2\nu} + q^\nu, \text{ or } q^{2\nu} - q^\nu$$

according to $a = 0$, $a \in \mathbb{F}_q^{*2}$, or $a \neq 0$ and $a \notin \mathbb{F}_q^{*2}$, respectively, (iii) the number of solutions of

$$2\sum_{i=1}^{\nu} x_i x_{\nu+i} + z x_{2\nu+1}^2 = a \qquad (11.12)$$

in $\mathbb{F}_q^{2\nu+1}$ is

$$q^{2\nu}, \ q^{2\nu} - q^\nu, \text{ or } q^{2\nu} + q^\nu$$

according to $a = 0$, $a \in \mathbb{F}_q^{*2}$, or $a \neq 0$ and $a \notin \mathbb{F}_q^{*2}$, respectively, and (iv) the number of solutions of equation

$$2\sum_{i=1}^{\nu} x_i x_{\nu+i} + x_{2\nu+1}^2 - z x_{2\nu+2}^2 = a \qquad (11.13)$$

(11.13) *in $\mathbb{F}_q^{2\nu+2}$ is*

$$q^{2\nu+1} - q^{\nu+1} + q^\nu \text{ or } q^{2\nu+1} + q^\nu$$

according to $a = 0$ or $a \neq 0$, respectively.

Proof. We begin with equation (11.10). For $x_1 = x_2 = \cdots = x_\nu = 0$, if $^t(x_{\nu+1}, x_{\nu+2}, \ldots, x_{2\nu})$ takes any value in $\mathbb{F}_q^\nu$, $^t(0, \ldots, 0, x_{\nu+1}, x_{\nu+2}, \ldots, x_{2\nu})$ is a solution of (11.10) when $a = 0$, but is not a solution when $a \neq 0$. So there are q^ν or 0 solutions according to $a = 0$ or $a \neq 0$, respectively, with $x_1 = x_2 = \cdots = x_\nu = 0$. Moreover, there are $q^\nu - 1$ nonzero vectors $^t(x_1, x_2, \ldots, x_\nu)$ and for each of them $^t(x_{\nu+1}, x_{\nu+2}, \ldots, x_{2\nu})$ can take $q^{\nu-1}$ values so that (11.10) holds. Altogether there are $q^\nu + (q^\nu - 1)q^{\nu-1} =$

$q^{2\nu-1} + q^\nu - q^{\nu-1}$ or $(q^\nu - 1)q^{\nu-1} = q^{2\nu-1} - q^{\nu-1}$ solutions of (11.10) in $\mathbb{F}_q^{2\nu}$ according to $a = 0$ or $a \neq 0$, respectively.

Next consider the equation (11.11). We distinguish the cases: (i) $a = 0$, (ii) $a \in \mathbb{F}_q^{*2}$, and (iii) $a \in \mathbb{F}_q^* \setminus \mathbb{F}_q^{*2}$.

(i) $a = 0$. By the discussion of equation (11.10) above, we know that there are $q^{2\nu-1} + q^\nu - q^{\nu-1}$ solutions of (11.11) in $\mathbb{F}_q^{2\nu+1}$, which are of the form $(x_1, x_2, \ldots, x_{2\nu}, 0)$ and there are $(q - 1)(q^{2\nu-1} - q^{\nu-1})$ solutions of the form $(x_1, x_2, \ldots, x_{2\nu}, x_{2\nu+1})$ with $x_{2\nu+1} \neq 0$. Altogether there are $q^{2\nu-1} + q^\nu - q^{\nu-1} + (q - 1)(q^{2\nu-1} - q^{\nu-1}) = q^{2\nu}$ solutions of (11.11).

(ii) $a \in \mathbb{F}_q^{*2}$. Also by the above discussion there are $2(q^{2\nu-1} + q^\nu - q^{\nu-1})$ solutions of (11.11) in $\mathbb{F}_q^{2\nu+1}$ which are of the form $(x_1, x_2, \ldots, x_{2\nu}, \pm a^{1/2})$ and there are $(q-2)(q^{2\nu-1} - q^{\nu-1})$ solutions of (11.11) which are of the form $(x_1, x_2, \ldots, x_{2\nu}, x_{2\nu+1})$, where $x_{2\nu+1}^2 \neq a$. Altogether there are $2(q^{2\nu-1} + q^\nu - q^{\nu-1}) + (q - 2)(q^{2\nu-1} - q^{\nu-1}) = q^{2\nu} + q^\nu$ solutions.

(iii) $a \neq 0$ and $a \notin \mathbb{F}_q^{*2}$. Again by the above discussion there are $q(q^{2\nu-1} - q^{\nu-1}) = q^{2\nu} - q^\nu$ solutions.

Similarly, the number of solutions of the equation (11.12) is $q^{2\nu}$, $q^{2\nu} - q^\nu$, or $q^{2\nu} + q^\nu$ according to the cases: $a = 0$, $a \in \mathbb{F}_q^{*2}$, or $a \neq 0$ and $a \notin \mathbb{F}_q^{*2}$.

There remains the equation (11.13) to be considered. We need the following lemma.

Lemma 11.11 *Let q be an odd prime power and $z \in \mathbb{F}_q^* \setminus \mathbb{F}_q^{*2}$. Then for any $a \in \mathbb{F}_q^*$ the number of solutions of the equation*

$$x^2 - zy^2 = a \tag{11.14}$$

is equal to $q + 1$.

Proof. Since $z \in \mathbb{F}_q^* \setminus \mathbb{F}_q^{*2}$, $x^2 - z$ is irreducible over $\mathbb{F}_q$. Let α be a root of $x^2 - z$ in $\mathbb{F}_{q^2}$, then $x^2 - zy^2 = (x - \alpha y)(x - \alpha^q y)$. For any solution $(x, y) \in \mathbb{F}_q^2$ of (11.14),

$$a = x^2 - zy^2 = (x - \alpha y)(x - \alpha^q y) = (x - \alpha y)^{q+1},$$

i.e., $x - \alpha y$ is a solution of $X^{q+1} = a$ in $\mathbb{F}_{q^2}$. Conversely, if $x - \alpha y$ is a solution of $X^{q+1} = a$ in $\mathbb{F}_{q^2}$, (x, y) is a solution of (11.14) in $\mathbb{F}_q^2$. For any $\xi \in \mathbb{F}_{q^2}$, $\xi^{q^2-1} = 1$, then $(\xi^{q+1})^{q-1} = 1$, which implies $(\xi^{q+1})^q = \xi^{q+1}$, hence $\xi^{q+1} \in \mathbb{F}_q^*$. Consider the map

$$\psi: \quad \mathbb{F}_{q^2} \quad \longrightarrow \quad \mathbb{F}_q^*$$
$$\xi \quad \longmapsto \quad \xi^{q+1}.$$

Clearly, it is a group homomorphism, $|\text{Ker } \psi| \leq q + 1$, and $|\text{Im } \psi| \leq q - 1$. But $|\text{Ker } \psi||\text{Im } \psi| = |\mathbb{F}_{q^2}^*| = q^2 - 1$. Hence ψ is necessarily surjective and $|\text{Ker } \psi| = q + 1$. The elements in the same coset of $\mathbb{F}_{q^2}^*$ with respect to Ker ψ are mapped into the same element of $\mathbb{F}_q^*$ and elements in different cosets are mapped into different elements of $\mathbb{F}_q^*$. Since each coset contains the same number of elements of $|\text{Ker } \psi|$, viz., $q + 1$, the equation $X^{q+1} = a$ has $q + 1$ solutions. Consequently, the number of solutions of $x^2 - zy^2 = a$ is equal to $q + 1$.

Let us return to the proof of Theorem 11.10. We distinguish the cases: (i) $a = 0$ and (ii) $a \neq 0$.

(i) $a = 0$. By the discussion of equation (11.10), there are $q^{2\nu-1} + q^\nu - q^{\nu-1}$ choices of ${}^t(x_1, x_2, \ldots, x_{2\nu}) \in \mathbb{F}_q^{2\nu}$, which satisfy (11.10), and for each such choice we must have $x_{2\nu+1} = x_{2\nu+2} = 0$ so that ${}^t(x_1, x_2, \ldots, x_{2\nu}, 0, 0)$ is a solution of (11.13). But there are $q^{2\nu} - (q^{2\nu-1} + q^\nu - q^{\nu-1})$ choices of ${}^t(x_1, x_2, \ldots, x_{2\nu}) \in \mathbb{F}_q^{2\nu}$ such that (11.10) is not satisfied and for each such choice, by Lemma 11.11, there are $q + 1$ choices of $(x_{2\nu+1}, x_{2\nu+2})$ so that (11.13) is satisfied. Altogether we get

$$q^{2\nu-1} + q^\nu - q^{\nu-1} + (q+1)(q^{2\nu} - q^{2\nu-1} - q^\nu + q^{\nu-1}) = q^{2\nu-1} - q^{\nu+1} + q^\nu$$

solutions of (11.13).

(ii) $a \neq 0$. Also by the above discussion of equation (11.10) there are $q^{2\nu-1} - q^{\nu-1}$ choices of ${}^t(x_1, x_2, \ldots, x_{2\nu})$ such that (11.10) is satisfied and for each such choice we must have $x_{2\nu+1} = x_{2\nu+2} = 0$ so that ${}^t(x_1, x_2, \ldots, x_{2\nu}, 0, 0)$ is a solution of (11.13). But there are $q^{2\nu} - (q^{2\nu-1} - q^{\nu-1})$ choices of ${}^t(x_1, x_2, \ldots, x_{2\nu}) \in \mathbb{F}_q^{2\nu}$ such that (11.10) is not satisfied and for each such choice, by Lemma 11.11 there are $q + 1$ choices of $(x_{2\nu+1}, x_{2\nu+2})$ so that (11.13) is satisfied. Altogether we get $q^{2\nu-1} - q^{\nu-1} + (q+1)(q^{2\nu} - q^{2\nu-1} + q^{\nu-1}) = q^{2\nu+1} + q^\nu$ solutions of (11.13).

The Theorem is completely proved.   □

11.2   Alternate Forms over Finite Fields

Let $\mathbb{F}_q$ be a finite field with q elements, where q may be odd or even. The bilinear form in $2n$ indeterminates $x_1, x_2, \ldots, x_n; y_1, y_2, \ldots, y_n$

$$K(x_1, x_2, \ldots, x_n; y_1, y_2, \ldots, y_n) = \sum_{i,j=1}^n k_{ij} x_i y_j, \qquad k_{ij} \in \mathbb{F}_q, \quad (11.15)$$

is called an *alternate form* over $\mathbb{F}_q$, if $K(x_1, x_2, \ldots, x_n; x_1, x_2, \ldots, x_n) = 0$ for all ${}^t(x_1, x_2, \ldots, x_n) \in \mathbb{F}_q^n$. Clearly, $K(x_1, x_2, \ldots, x_n; y_1, y_2, \ldots, y_n)$ is alternate if and only if

$$k_{ij} = -k_{ji}, \qquad 1 \leqslant i, j \leqslant n,$$

and

$$k_{ii} = 0, \qquad 1 \leqslant i \leqslant n.$$

Now let (11.15) be an alternate form. Introduce the matrix

$$K = \begin{pmatrix} 0 & k_{12} & \cdots & k_{1n} \\ -k_{12} & 0 & \cdots & k_{2n} \\ \vdots & \vdots & \ddots & \vdots \\ -k_{1n} & -k_{2n} & \cdots & 0 \end{pmatrix}.$$

K is called an *alternate matrix* and (11.15) can be expressed as

$$K(x_1, x_2, \ldots, x_n; \ y_1, y_2, \ldots, y_n) = (x_1, x_2, \ldots, x_n) K \ {}^t(y_1, y_2, \ldots, y_n), \tag{11.16}$$

thus K is called the *coefficient matrix* of (11.16).

Let T be an $n \times n$ nonsingular matrix. Under the nonsingular linear transformation (11.4)

$$\begin{pmatrix} x_1 \\ x_2 \\ \vdots \\ x_n \end{pmatrix} \longmapsto T \begin{pmatrix} x_1 \\ x_2 \\ \vdots \\ x_n \end{pmatrix},$$

the alternate form (11.16) is carried into

$$(x_1, x_2, \ldots, x_n) \ {}^tTKT \ {}^t(y_1, y_2, \ldots, y_n). \tag{11.17}$$

Then the coefficient matrix K of (11.16) is subjected to the transformation

$$K \longmapsto {}^tTKT. \tag{11.18}$$

We call (11.16) and (11.17) *cogredient* alternate forms and a transformation of the $n \times n$ alternate matrices of the form (11.18) is called a *cogredience transformation*. Now let us study the cogredience of alternate matrices. First, parallel to Lemma 11.1 we have

Lemma 11.12 *Cogredience transformations preserve the rank of the alternate matrix.*

Then we have

Theorem 11.13 *Let K be an $n \times n$ alternate matrix over a finite field of any characteristic. Then the rank of K is necessarily even. Further, if K is of rank 2ν ($\leqslant n$), then K is cogredient to*

$$\left(\begin{array}{c} \begin{matrix} 0 & 1 \\ -1 & 0 \end{matrix} \\ \quad \begin{matrix} 0 & 1 \\ -1 & 0 \end{matrix} \\ \qquad \ddots \\ \qquad\qquad \begin{matrix} 0 & 1 \\ -1 & 0 \end{matrix} \\ \qquad\qquad\qquad 0 \\ \qquad\qquad\qquad\quad \ddots \\ \qquad\qquad\qquad\qquad 0 \end{array} \right)$$

$$\underbrace{\qquad\qquad}_{2\nu} \quad \underbrace{\qquad\qquad}_{n-2\nu}$$

and is also cogredient to

$$\left(\begin{array}{cc} 0 & I^{(\nu)} \\ -I^{(\nu)} & 0 \\ & & 0^{(n-2\nu)} \end{array} \right). \tag{11.19}$$

Moreover, two alternate matrices of the form (11.19) with different ν are not cogredient.

Proof. Clearly, the matrix (11.19) is of rank 2ν. Therefore the second statement follows from Lemma 11.12. Now we apply induction on n to prove the first statement. For $n = 1$, $K = 0$ and the first statement holds automatically. For $n = 2$, K has the form

$$K = \left(\begin{array}{cc} 0 & k_{12} \\ -k_{12} & 0 \end{array} \right).$$

If $k_{12} = 0$, the first statement is true; if $k_{12} \neq 0$, K is cogredient to

$${}^t\!\left(\begin{array}{cc} 1 & 0 \\ 0 & k_{12}^{-1} \end{array} \right) K \left(\begin{array}{cc} 1 & 0 \\ 0 & k_{12}^{-1} \end{array} \right) = \left(\begin{array}{cc} 0 & 1 \\ -1 & 0 \end{array} \right).$$

Now suppose that $n > 2$ and that the first statement holds for $m < n$. Let

$$K = \left(\begin{array}{ccccc} 0 & k_{12} & k_{13} & \cdots & k_{1n} \\ -k_{12} & 0 & k_{23} & \cdots & k_{2n} \\ -k_{13} & -k_{23} & 0 & \cdots & k_{3n} \\ \vdots & \vdots & \vdots & \ddots & \vdots \\ -k_{1n} & -k_{2n} & -k_{3n} & \cdots & 0 \end{array} \right).$$

If $K = 0$, then we are finished. Suppose that $K \neq 0$, then there is a $k_{ij} \neq 0$, $i < j$. Interchanging the first row and the i-th row of K and the first column and the i-th column of K simultaneously, we obtain an alternate matrix K_1 which is cogredient to K and whose element at $(1, j)$ position is non-zero. Then interchanging the second row and the j-th row of K_1 and the second column and the j-th column of K_1 simultaneously, we obtain an alternate matrix K_2 which is cogredient to K_1 and hence also to K and whose element at $(1, 2)$ position is nonzero. Thus we can assume that $k_{12} \neq 0$. Then K is cogredient to

$$
{}^t\!\left(\begin{matrix} 1 & & \\ & k_{12}^{-1} & \\ & & I^{(n-2)} \end{matrix} \right) K \left(\begin{matrix} 1 & & \\ & k_{12}^{-1} & \\ & & I^{(n-2)} \end{matrix} \right),
$$

whose element at $(1, 2)$ position is 1. Write K in block form

$$
K = \left(\begin{matrix} \left(\begin{matrix} 0 & 1 \\ -1 & 0 \end{matrix} \right) & B \\ -{}^t\!B & K_0 \end{matrix} \right),
$$

where B is a $2 \times (n-2)$ matrix and K_0 is an $(n-2) \times (n-2)$ alternate matrix. Then K is cogredient to

$$
{}^t\!\left(\begin{matrix} I & \left(\begin{matrix} 0 & 1 \\ -1 & 0 \end{matrix} \right) B \\ & I \end{matrix} \right) \left(\begin{matrix} \left(\begin{matrix} 0 & 1 \\ -1 & 0 \end{matrix} \right) & B \\ -{}^t\!B & K_0 \end{matrix} \right) \left(\begin{matrix} I & \left(\begin{matrix} 0 & 1 \\ -1 & 0 \end{matrix} \right) B \\ & I \end{matrix} \right)
$$

$$
= \left(\begin{matrix} \left(\begin{matrix} 0 & 1 \\ -1 & 0 \end{matrix} \right) & \\ & K_1 \end{matrix} \right),
$$

where K_1 is an $(n-2) \times (n-2)$ alternate matrix. Therefore our theorem follows from the induction hypothesis. □

Matrices of the form (11.19) are called *normal forms* of alternate matrices and ν in the matrix (11.19) is called the *index* of any $n \times n$ alternate matrix cogredient to (11.19).

Corollary 11.14 *The index of an alternate matrix is invariant under cogredience transformations.*

The *index* of an alternate form is defined to be the index of its coefficient matrix. We can put Theorem 11.13 into the language of alternate forms.

Theorem 11.15 *Any alternate form of index ν over a finite field of any characteristic is cogredient to the alternate form*

$$\sum_{i=0}^{\nu}(x_i y_{\nu+i} - y_i x_{\nu+i}).$$

Moreover, any two alternate forms with different indices are not cogredient.

Alternate forms in Theorem 11.15 are called *normal forms* of alternate forms.

11.3   Quadratic Forms over Finite Fields of Characteristic 2

Now we are going to study the quadratic forms over a finite field $\mathbb{F}_q$ of even characteristic, so from now on we assume that $\mathbb{F}_q$ is of characteristic 2.

Introduce the coefficient matrix of the quadratic form (11.1)

$$G = \begin{pmatrix} b_{11} & b_{12} & \cdots & b_{1n} \\ 0 & b_{22} & \cdots & b_{2n} \\ \vdots & \vdots & \ddots & \vdots \\ 0 & 0 & \cdots & b_{nn} \end{pmatrix},$$

which is an upper triangle matrix. Then (11.1) can be written as

$$Q(x_1, x_2, \ldots, x_n) = (x_1, x_2, \ldots, x_n)\, G\, {}^t(x_1, x_2, \ldots, x_n). \qquad (11.20)$$

Under the nonsingular linear transformation (11.4), (11.20) is carried into

$$(x_1, x_2, \ldots, x_n)\, {}^tTGT\, {}^t(x_1, x_2, \ldots, x_n). \qquad (11.21)$$

(11.20) and (11.21) are called cogredient quadratic forms. Clearly, the coefficient matrix G is subjected to the transformation

$$G \longmapsto {}^tTGT.$$

However ${}^tTGT$ is in general no longer upper triangular. So the first problem we encounter is to find a suitable matrix representation of the quadratic form (11.1). Let $K = (k_{ij})_{1 \leqslant i,j \leqslant n}$ be an $n \times n$ alternate matrix. Then clearly

$$(x_1, \ldots, x_n)(G + K)\, {}^t(x_1, \ldots, x_n) = (x_1, \ldots, x_n)\, G\, {}^t(x_1, \ldots, x_n).$$

Therefore it is natural not to distinguish square matrices differing by an alternate matrix in studying quadratic forms over $\mathbb{F}_q$ of characteristic 2. This leads to the following definition.

Denote the set of all $n \times n$ alternate matrices over $\mathbb{F}_q$ by K_n. Two $n \times n$ matrices A and B over $\mathbb{F}_q$ are said to be *congruent* mod K_n, if $A + B \in K_n$, which is usually denoted by

$$A \equiv B \pmod{K_n},$$

or, simply, $A \equiv B$.

Two $n \times n$ matrices A and B over $\mathbb{F}_q$ are said to be *cogredient*, if there is an $n \times n$ nonsingular matrix T over $\mathbb{F}_q$ such that ${}^t T A T \equiv B$. Thus the set of $n \times n$ matrices over $\mathbb{F}_q$ is partitioned into mutually disjoint classes of cogredient matrices. It can be easily proved that matrices cogredient to an alternate matrix are also alternate and matrices cogredient to a definite matrix are also definite, while a matrix A is said to be *definite* if $xA\,{}^t x = 0$, where ${}^t x \in \mathbb{F}_q^n$, implies $x = 0$, i.e., its associated quadratic form $xA\,{}^t x$ is definite.

We shall study the normal form of $n \times n$ matrices under cogredience. We start with a general result.

Lemma 11.16 *Any $n \times n$ matrix G over the finite field $\mathbb{F}_q$ of characteristic 2 is cogredient to a matrix of the form*

$$\begin{pmatrix} A & I^{(p)} & & \\ & B & & \\ & & C & \\ & & & 0 \end{pmatrix}, \tag{11.22}$$

where A and B are $p \times p$ diagonal matrices, and C is a $d \times d$ definite diagonal matrix. Moreover, p and d are uniquely determined by G.

Proof. Let G be an $n \times n$ matrix over $\mathbb{F}_q$. Then $G + {}^t G \in K_n$. By Theorem 11.13 we can assume that rank $(G + {}^t G) = 2p$ $(0 \leqslant p \leqslant [\frac{n}{2}])$ and there is a nonsingular $n \times n$ matrix Q such that

$$ {}^t Q (G + {}^t G) Q = \begin{pmatrix} 0 & I^{(p)} & \\ I^{(p)} & 0 & \\ & & 0 \end{pmatrix}. \tag{11.23}$$

If we write ^tQGQ in block form

$$^tQGQ = \begin{pmatrix} A_1 & A_2 & A_3 \\ B_1 & B_2 & B_3 \\ C_1 & C_2 & C_3 \end{pmatrix} \begin{matrix} p \\ p \\ n-2p \end{matrix},$$
$$\begin{matrix} p & p & n-2p \end{matrix}$$

then from (11.23) we deduce $^tA_1 = A_1$, $^tB_2 = B_2$, $^tC_3 = C_3$, $^tB_1 + A_2 = I$, $^tC_1 + A_3 = 0$, and $^tC_2 + B_3 = 0$. Thus we have

$$^tQGQ \equiv \begin{pmatrix} A & I & \\ & B & \\ & & C_4 \end{pmatrix} \begin{matrix} p \\ p \\ n-2p \end{matrix},$$
$$\begin{matrix} p & p & n-2p \end{matrix}$$

where A, B and C_4 are diagonal matrices. Let the diagonal elements of C_4 be $\gamma_{2p+1}, \gamma_{2p+2}, \ldots, \gamma_n$ in succession. If there is a nonzero column vector $x = {}^t(x_{2p+1}, x_{2p+2}, \ldots, x_n)$ such that $xC_4\,{}^tx = \gamma_{2p+1}x_{2p+1}^2 + \gamma_{2p+2}x_{2p+2}^2 + \cdots + \gamma_n x_n^2 = 0$, we can assume without loss of generality that $x_n \neq 0$. Let

$$Q_1 = \begin{pmatrix} 1 & & & & x_{2p+1} \\ & 1 & & & x_{2p+2} \\ & & \ddots & & \vdots \\ & & & 1 & x_{n-1} \\ & & & & x_n \end{pmatrix},$$

then

$$^tQ_1C_4Q_1 = \operatorname{diag}[\gamma_{2p+1}, \gamma_{2p+2}, \ldots, \gamma_{n-1}, 0].$$

Proceeding in this way, we finally reduce G to the form (11.22) under cogredience, where C is a $d \times d$ definite diagonal matrix.

Now, let us prove the uniqueness of p and d in (11.22). Let

$$G = \begin{pmatrix} A & I^{(p)} & \\ & B & \\ & & C \\ & & & 0 \end{pmatrix} \quad \text{and} \quad G_1 = \begin{pmatrix} A_1 & I^{(p_1)} & \\ & B_1 & \\ & & C_1 \\ & & & 0 \end{pmatrix}$$

be two $n \times n$ matrices of the form (11.22) and assume that they are cogredient, i.e., there is a nonsingular matrix Q such that $^tQGQ \equiv G_1$. Then $^tQ(G + {}^tG)Q = G_1 + {}^tG_1$, from which we deduce $p = \frac{1}{2}\operatorname{rank}(G + {}^tG) = \frac{1}{2}\operatorname{rank}(G_1 + {}^tG_1) = p_1$. Let $r = n - 2p - d$ and $r_1 = n - 2p - d_1$. Write Q

in block form

$$Q = \begin{pmatrix} Q_{11} & Q_{12} \\ Q_{21} & Q_{22} \end{pmatrix} \begin{matrix} 2p \\ n-2p \end{matrix} \cdot$$
$$\begin{matrix} 2p & n-2p \end{matrix}$$

From ${}^tQ(G+{}^tG)Q = G_1 + {}^tG_1$, we deduce $Q_{12} = 0$, hence Q_{22} is nonsingular and

$$ {}^tQ_{22} \begin{pmatrix} C & \\ & 0 \end{pmatrix} Q_{22} \equiv \begin{pmatrix} C_1 & \\ & 0 \end{pmatrix}. $$
$$ \begin{matrix} d & r \end{matrix} \qquad\qquad \begin{matrix} d_1 & r_1 \end{matrix} $$

Since Q_{22} is nonsingular and both C and C_1 are definite, $d = d_1$. Therefore both p and d are uniquely determined by G. $\qquad\square$

By Lemma 11.16 both p and d are uniquely determined by G. Hence the number $2p+d$ is also uniquely determined by G and is called the "*rank*" of G. If the "rank" of an $n \times n$ matrix G is equal to n, then both G and the quadratic form $(x_1, x_2, \ldots, x_n)G\,{}^t(x_1, x_2, \ldots, x_n)$ are said to be *non-singular*. Clearly, an $n \times n$ matrix G is non-singular if and only if it is cogredient to a matrix of the form

$$ \begin{pmatrix} A & I & \\ & B & \\ & & C \end{pmatrix}, $$

where A and B are $p \times p$ diagonal matrices, C is a $d \times d$ definite diagonal matrix. Then $n = 2p + d$. The number $d = n - 2p$ is called the *defect* of G. If $d = 0$, G is said to be *without defect*.

We introduce a subset of $\mathbb{F}_q$ as follows

$$ N = \{x^2 + x : x \in \mathbb{F}_q\}. $$

Then we have

Lemma 11.17 N is an additive subgroup of $\mathbb{F}_q$ of index 2 and $N^2 = N$.

Proof. Let α and β be any two elements of N. There exist elements x and $y \in \mathbb{F}_q$ such that $\alpha = x^2 + x$ and $\beta = y^2 + y$. Then $\alpha - \beta = \alpha + \beta = x^2 + x + y^2 + y = (x + y)^2 + (x + y) \in N$. This proves that N is a subgroup of the additive group of $\mathbb{F}_q$. For $x, y \in \mathbb{F}_q$ if $x^2 + x = y^2 + y$, then $(x + y)^2 = x + y$, thus $x + y = 0$ or $x + y = 1$. Therefore only x and $x + 1$ give rise to the same element $x^2 + x \in N$. Hence $|N| = q/2$ and $\mathbb{F}_q : N = 2$. Furthermore, from $\mathbb{F}_q^2 = \mathbb{F}_q$ we deduce $N^2 = N$. $\qquad\square$

From now on we choose a fixed element $\alpha \in \mathbb{F}_q$ but $\alpha \notin N$.

Lemma 11.18 *The 2×2 matrix over $\mathbb{F}_q$*

$$\begin{pmatrix} a & 1 \\ & b \end{pmatrix}$$

is cogredient to either

$$\begin{pmatrix} 0 & 1 \\ & 0 \end{pmatrix} \qquad or \qquad \begin{pmatrix} \alpha & 1 \\ & \alpha \end{pmatrix}$$

according to $ab \in N$ or $ab \notin N$, respectively. Moreover, the first one of the above two matrices is not definite and the second one is definite.

Proof. If $a = b = 0$, there is nothing to prove. If $a = 0$ and $b \neq 0$, then

$$\,^t\!\begin{pmatrix} 1 & b \\ 0 & 1 \end{pmatrix} \begin{pmatrix} 0 & 1 \\ & b \end{pmatrix} \begin{pmatrix} 1 & b \\ 0 & 1 \end{pmatrix} = \begin{pmatrix} 0 & 1 \\ & 0 \end{pmatrix}.$$

The case $a \neq 0$ and $b = 0$ we can treat in a similar way.

Now suppose that $ab \neq 0$. Consider first the case that $ab \in N$. There is an $x \in \mathbb{F}_q^*$ such that $ab = (ax)^2 + ax$. Thus

$$\,^t\!\begin{pmatrix} x & 0 \\ 1 & x^{-1} \end{pmatrix} \begin{pmatrix} a & 1 \\ & b \end{pmatrix} \begin{pmatrix} x & 0 \\ 1 & x^{-1} \end{pmatrix} \equiv \begin{pmatrix} 0 & 1 \\ & x^{-2}b \end{pmatrix},$$

which reduces to one of the cases we considered before. Then consider the case $ab \notin N$. Since $\mathbb{F}_q^2 = \mathbb{F}_q$, there is a $\delta \in \mathbb{F}_q^*$ such that $\delta^2 a = b\delta^{-2}$. Let $\lambda = \delta^2 a$, then

$$\,^t\!\begin{pmatrix} \delta & \\ & \delta^{-1} \end{pmatrix} \begin{pmatrix} a & 1 \\ & b \end{pmatrix} \begin{pmatrix} \delta & \\ & \delta^{-1} \end{pmatrix} \equiv \begin{pmatrix} \lambda & 1 \\ & \lambda \end{pmatrix}$$

and $\lambda^2 = ab \notin N$. By Lemma 11.17, from $\alpha \notin N$, we deduce $\alpha^2 \notin N$. Again by the lemma, from $\lambda^2 \notin N$ and $\alpha^2 \notin N$ we deduce $\alpha^2 + \lambda^2 \in N$, i.e., there is an element $t \in \mathbb{F}_q$ such that $\alpha^2 + \lambda^2 = t^2 + t$. Consequently,

$$\,^t\!\begin{pmatrix} (\alpha\lambda^{-1})^{\frac{1}{2}} & (\alpha\lambda)^{-\frac{1}{2}}t \\ 0 & (\alpha^{-1}\lambda)^{\frac{1}{2}} \end{pmatrix} \begin{pmatrix} \lambda & 1 \\ & \lambda \end{pmatrix} \begin{pmatrix} (\alpha\lambda^{-1})^{\frac{1}{2}} & (\alpha\lambda)^{-\frac{1}{2}}t \\ 0 & (\alpha^{-1}\lambda)^{\frac{1}{2}} \end{pmatrix}$$

$$\equiv \begin{pmatrix} \alpha & 1 \\ & \alpha \end{pmatrix}.$$

Clearly $\begin{pmatrix} 0 & 1 \\ & 0 \end{pmatrix}$ is not definite. It remains to show that $\begin{pmatrix} \alpha & 1 \\ & \alpha \end{pmatrix}$ is definite. Suppose that there is a non-zero vector (x, y) such that $\alpha x^2 + xy + \alpha y^2 = 0$. Without loss of generality we can assume that $x \neq 0$, then $\alpha^2 = (\alpha x^{-1}y)^2 + \alpha x^{-1}y \in N$ and hence $\alpha \in N$, contrary to the choice of $\alpha \notin N$. $\square$

Lemma 11.19 *The* 4×4 *matrix*

$$\begin{pmatrix} \alpha & 1 & & \\ & \alpha & 1 & \\ & & \alpha & 1 \\ & & & \alpha \end{pmatrix}$$

is cogredient to

$$\begin{pmatrix} 0 & 1 & & \\ & 0 & 1 & \\ & & 0 & \\ & & & 0 \end{pmatrix}$$

Proof. We compute

$${}^{t}\begin{pmatrix} 1 & 0 & \alpha & 0 \\ 1 & 0 & \alpha & 1 \\ 0 & 1 & 1 & \alpha \\ 0 & 1 & 0 & \alpha \end{pmatrix} \begin{pmatrix} \alpha & & 1 & \\ & \alpha & & 1 \\ & & \alpha & \\ & & & \alpha \end{pmatrix} \begin{pmatrix} 1 & 0 & \alpha & 0 \\ 1 & 0 & \alpha & 1 \\ 0 & 1 & 1 & \alpha \\ 0 & 1 & 0 & \alpha \end{pmatrix}$$

$$\equiv \begin{pmatrix} 0 & 1 & & \\ & 0 & 1 & \\ & & 0 & \\ & & & 0 \end{pmatrix}.$$

$\square$

Lemma 11.20 *The* 3×3 *matrix*

$$\begin{pmatrix} \alpha & 1 & \\ & \alpha & \\ & & 1 \end{pmatrix}$$

is cogredient to

$$\begin{pmatrix} 0 & 1 & \\ & 0 & \\ & & 1 \end{pmatrix}.$$

Proof. We compute

$${}^{t}\begin{pmatrix} 1 & 0 & 0 \\ 0 & 1 & 0 \\ \alpha^{1/2} & \alpha^{1/2} & 1 \end{pmatrix} \begin{pmatrix} \alpha & 1 & \\ 0 & \alpha & \\ & & 1 \end{pmatrix} \begin{pmatrix} 1 & 0 & 0 \\ 0 & 1 & 0 \\ \alpha^{1/2} & \alpha^{1/2} & 1 \end{pmatrix}$$

$$\equiv \begin{pmatrix} 0 & 1 & \\ 0 & 0 & \\ & & 1 \end{pmatrix},$$

where $\alpha^{1/2}$ is the unique square root of α in $\mathbb{F}_q$.

$\square$

Lemma 11.21 *The number of solutions of the equation*

$$\alpha x^2 + xy + \alpha y^2 = \beta,$$

where $\alpha,\ \beta \in \mathbb{F}_q^*$ *and* $\alpha \notin N$ *is* $q + 1$.

Proof. Consider the set of all 2-dimensional non-zero column vectors over $\mathbb{F}_q$, which is denoted by $\mathbb{F}_q^{2*}$. It is known that $|\mathbb{F}_q^{2*}| = q^2 - 1$. Two non-zero vector $^t(x, y)$ and $^t(x_1, y_1)$ are said to be equivalent, if there is a $\lambda \in \mathbb{F}_q^*$ such that $^t(x, y) = \lambda\,^t(x_1, y_1)$. This is an equivalence relation and the set $\mathbb{F}_q^{2*}$ is partitioned into $q + 1$ inequivalent classes, each consisting of $q - 1$ non-zero vectors. By Lemma 11.18,

$$\begin{pmatrix} \alpha & 1 \\ 0 & \alpha \end{pmatrix}$$

is definite. Thus for any non-zero vector $^t(x, y) \in \mathbb{F}_q^{2*}$, $\alpha x^2 + xy + \alpha y^2 \neq 0$. Since $\mathbb{F}_q^2 = \mathbb{F}_q$, as $^t(x, y)$ runs through an equivalence class, $\alpha x^2 + xy + \alpha y^2$ runs through $\mathbb{F}_q^*$. Therefore, for any fixed $\beta \in \mathbb{F}_q^*$ the equation $\alpha x^2 + xy + \alpha y^2 = \beta$ has one and only one solution in each equivalence class. Consequently, the equation $\alpha x^2 + xy + \alpha y^2 = \beta$ has $q + 1$ solutions. $\square$

Lemma 11.22 *The number of solutions of*

$$x_1 x_{\nu+1} + x_2 x_{\nu+2} + \cdots + x_\nu x_{2\nu} = 0 \tag{11.24}$$

in $\mathbb{F}_q^{2\nu}$ *is* $q^{2\nu-1} + q^\nu - q^{\nu-1}$ *and the number of solutions of*

$$x_1 x_\nu + x_2 x_{\nu+1} + \cdots + x_{\nu-1} x_{2\nu-2} + \alpha x_{2\nu-1}^2 + x_{2\nu-1} x_{2\nu} + \alpha x_{2\nu}^2 = 0 \tag{11.25}$$

in $\mathbb{F}_q^{2\nu}$ *is* $q^{2\nu-1} - q^\nu + q^{\nu-1}$.

Proof. Consider first the equation (11.24). For $x_1 = x_2 = \cdots = x_\nu = 0$, $x_{\nu+1}, x_{\nu+2}, \ldots, x_{2\nu}$ can take independent values in $\mathbb{F}_q$. So there are q^ν solutions with $x_1 = x_2 = \cdots = x_\nu = 0$. There are $q^\nu - 1$ nonzero vectors $(x_1, x_2, \cdots, x_\nu)$, and for each one of these nonzero vectors $(x_{\nu+1}, x_{\nu+2}, \ldots, x_{2\nu})$ can take $q^{\nu-1}$ values so that (11.24) holds. Altogether there are $q^\nu + (q^\nu - 1)q^{\nu-1} = q^{2\nu-1} + q^\nu - q^{\nu-1}$ solutions of (11.24) in $\mathbb{F}_q^{2\nu}$.

Then consider the equation (11.25). By the preceding case, there are $q^{2\nu-3} + q^{\nu-1} - q^{\nu-2}$ solutions in $\mathbb{F}_q^{2\nu}$ with $x_{2\nu-1} = x_{2\nu} = 0$. It follows that there are $q^{2\nu-2} - (q^{2\nu-3} + q^{\nu-1} - q^{\nu-2})$ vectors $^t(x_1, x_2, \ldots, x_{2\nu-2})$ such that $x_1 x_\nu + x_2 x_{\nu+1} + \cdots + x_{\nu-1} x_{2\nu-2} \neq 0$. For each such vector, by Lemma 11.21 there are $q + 1$ 2-dimensional vectors $(x_{2\nu-1}, x_{2\nu})$ such that (11.25) holds. Altogether the number of solutions of (11.25) in $\mathbb{F}_q^{2\nu}$ is

$$q^{2\nu-3} + q^{\nu-1} - q^{\nu-2} + \left(q^{2\nu-2} - \left(q^{2\nu-3} + q^{\nu-1} - q^{\nu-2}\right)\right)(q+1)$$

$$= q^{2\nu-1} - q^\nu + q^{\nu-1}.$$

$\square$

Theorem 11.23 *Let q be a power of 2. Any $n \times n$ matrix G over $\mathbb{F}_q$ is cogredient to a matrix of one and only one of the following forms*

$$\begin{pmatrix} 0 & I^{(\nu)} & \\ & 0 & \\ & & 0 \end{pmatrix}, \begin{pmatrix} 0 & I^{(\nu)} & & \\ & 0 & & \\ & & 1 & \\ & & & 0 \end{pmatrix}, \begin{pmatrix} 0 & I^{(\nu)} & & \\ & 0 & & \\ & & \alpha & 1 \\ & & & \alpha \\ & & & & 0 \end{pmatrix}, \quad (11.26)$$

where α is a fixed element of $\mathbb{F}_q$ not belonging to N.

Proof. By Lemma 11.16 G is cogredient to a matrix of the form (11.22)

$$\begin{pmatrix} A & I^{(p)} & \\ & B & \\ & & C & \\ & & & 0 \end{pmatrix},$$

where A and B are $p \times p$ diagonal matrices, and C is a $d \times d$ definite diagonal matrix; moreover, both p and d are uniquely determined by G. We prove further that $d \leqslant 1$. In fact, suppose that $d \geqslant 2$ and let the diagonal elements of C be $\gamma_1, \gamma_2, \ldots, \gamma_d$ in succession. Then we have

$$\left(\gamma_1^{-\frac{1}{2}}, \gamma_2^{-\frac{1}{2}}, 0, \ldots, 0\right) C \,{}^t\!\left(\gamma_1^{-\frac{1}{2}}, \gamma_2^{-\frac{1}{2}}, 0, \ldots, 0\right) = 0,$$

which contradicts the definiteness of C. Furthermore, if $d = 1$, then clearly, $C = (\gamma)$ is cogredient to $\left(\gamma^{-\frac{1}{2}}\right)(\gamma)\left(\gamma^{-\frac{1}{2}}\right) = (1)$. We have proved that G is cogredient to a matrix of the form (11.22), where A and B are $p \times p$ diagonal matrices, and C disappears or C is the 1×1 matrix (1). Then it follows from Lemmas 11.18, 11.19 and 11.20 that G is cogredient to one of the matrices of the form (11.26).

It remains to show that the three forms of matrices in (11.26) are not cogredient to each other. Since the "rank" of the second matrix is odd and the "rank" of the other two are even, the second one can not be cogredient to the other two. Suppose that a matrix of the first form is cogredient to a matrix of the third form, then they must have the same "rank" and we can assume that they are

$$\begin{pmatrix} 0 & I^{(\nu)} & \\ & 0 & \\ & & 0 \end{pmatrix} \text{ and } \begin{pmatrix} 0 & I^{(\nu-1)} & & \\ & 0 & & \\ & & \alpha & 1 \\ & & & \alpha \\ & & & & 0 \end{pmatrix}.$$

Then there is an $n \times n$ matrix

$$Q = \begin{pmatrix} Q_{11} & Q_{12} \\ Q_{21} & Q_{22} \end{pmatrix} \begin{matrix} 2\nu \\ n-2\nu \end{matrix}$$
$$\quad\quad 2\nu \quad n-2\nu$$

such that

$$^t\begin{pmatrix} Q_{11} & Q_{12} \\ Q_{21} & Q_{22} \end{pmatrix} \begin{pmatrix} 0 & I^{(\nu)} \\ 0 & \\ & & 0 \end{pmatrix} \begin{pmatrix} Q_{11} & Q_{12} \\ Q_{21} & Q_{22} \end{pmatrix}$$

$$\equiv \begin{pmatrix} 0 & I^{(\nu-1)} \\ 0 & \\ & & \alpha & 1 \\ & & & \alpha \\ & & & & 0 \end{pmatrix}.$$

It follows that

$$^tQ_{11} \begin{pmatrix} 0 & I^{(\nu)} \\ 0 \end{pmatrix} Q_{11} \equiv \begin{pmatrix} 0 & I^{(\nu-1)} \\ 0 & \\ & & \alpha & 1 \\ & & & \alpha \end{pmatrix},$$

and hence

$$^tQ_{11} \begin{pmatrix} 0 & I^{(\nu)} \\ I^{(\nu)} & 0 \end{pmatrix} Q_{11} = \begin{pmatrix} 0 & I^{(\nu-1)} \\ I^{(\nu-1)} & 0 \\ & & 0 & 1 \\ & & 1 & 0 \end{pmatrix}.$$

Therefore Q_{11} is nonsingular. Then the number of solutions of the equation

$$(x_1, x_2, \ldots, x_{2\nu}) \begin{pmatrix} 0 & I^{(\nu)} \\ 0 \end{pmatrix} {}^t(x_1, x_2, \ldots, x_{2\nu}) = 0$$

or of the equation (11.24)

$$x_1 x_{\nu+1} + x_2 x_{\nu+2} + \cdots + x_\nu x_{2\nu} = 0$$

should be equal to the number of solutions of the equation

$$(x_1, x_2, \ldots, x_{2\nu}) \begin{pmatrix} 0 & I^{(\nu-1)} \\ 0 & \\ & & \alpha & 1 \\ & & & \alpha \end{pmatrix} {}^t(x_1, x_2, \ldots, x_{2\nu}) = 0$$

or of the equation (11.25)

$$x_1 x_\nu + x_2 x_{\nu+1} + \cdots + x_{\nu-1} x_{2\nu-2} + \alpha x_{2\nu-1}^2 + x_{2\nu-1} x_{2\nu} + \alpha x_{2\nu}^2 = 0.$$

By Lemma 11.22 the number of solutions of (11.24) is $q^{2\nu-1} + q^\nu - q^{\nu-1}$ and the number of solutions of (11.25) is $q^{2\nu-1} - q^\nu + q^{\nu-1}$; they are not equal and we obtain a contradiction. □

If an $n \times n$ matrix G over $\mathbb{F}_q$ is cogredient to any of the matrices in (11.26), then ν is called *index* of G. Moreover, if a quadratic form over $\mathbb{F}_q$ is cogredient to a quadratic form whose coefficient matrix is one of the matrices in (11.26), then ν is called the *index* of the quadratic form.

Corollary 11.24 *The index of an $n \times n$ matrix and also the index of a quadratic form are invariant under cogredient transformations.*

Let us now translate the foregoing Theorem 11.23 on the cogredience of square matrices over a finite field of characteristic 2 into a theorem on the cogredience of quadratic forms.

Theorem 11.25 *Any quadratic form $Q(x_1, x_2, \ldots, x_n)$ in n indeterminates $x_1, x_2, \ldots, x_n$ over $\mathbb{F}_q$, where q is a power of 2, is cogredient to a quadratic form of one and only one of the following normal forms*

$$\sum_{i=1}^{\nu} x_i x_{\nu+i},$$

$$\sum_{i=1}^{\nu} x_i x_{\nu+i} + x_{2\nu+1}^2,$$

$$\sum_{i=1}^{\nu} x_i x_{\nu+i} + \alpha x_{2\nu+1}^2 + x_{2\nu+1} x_{2\nu+2} + \alpha x_{2\nu+2}^2,$$

where α is a fixed element of $\mathbb{F}_q$ not belonging to N and ν is the index of the quadratic form; and these forms are not cogredient to each other.

Theorem 11.25 is due to Dickson (1900), and our treatment follows Wan (2002).

Let us compute the number of solutions of the equation $Q(x_1, x_2, \ldots, x_n) = a$, where $Q(x_1, x_2, \ldots, x_n)$ is a quadratic form in n indeterminates $x_1, x_2, \ldots, x_n$ over $\mathbb{F}_q$ and $a \in \mathbb{F}_q$. As in the case when q is odd, it is enough to consider the normal forms given in Theorem 11.25 and to restrict Q to be nonsingular. We have

Theorem 11.26 *Let q be a power of 2 and $a \in \mathbb{F}_q$. Then (i) the number of solutions of*

$$\sum_{i=1}^{\nu} x_i x_{\nu+i} = a \tag{11.27}$$

in $\mathbb{F}_q^{2\nu}$ is

$$q^{2\nu-1} + q^{\nu} - q^{\nu-1} \text{ or } q^{2\nu-1} - q^{\nu-1}$$

according to $a = 0$ or $a \neq 0$, respectively, (ii) the number of solutions of

$$\sum_{i=1}^{\nu} x_i x_{\nu+i} + x_{2\nu+1}^2 = a \tag{11.28}$$

in $\mathbb{F}_q^{2\nu+1}$ is

$$q^{2\nu},$$

and (iii) the number of solutions of

$$\sum_{i=1}^{\nu} x_i x_{\nu+i} + \alpha x_{2\nu+1}^2 + x_{2\nu+1} x_{2\nu+2} + \alpha x_{2\nu+2}^2 = a, \tag{11.29}$$

where $\alpha \in \mathbb{F}_q$ and $\alpha \notin N$, in $\mathbb{F}_q^{2\nu+1}$ is

$$q^{2\nu+1} - q^{\nu+1} + q^{\nu} \text{ or } q^{2\nu+1} + q^{\nu}$$

according to $a = 0$ or $a \neq 0$, respectively.

Proof. Equation (11.27) can be treated in the same way as equation (11.10) in Theorem 11.11, so the details will be omitted.

Now consider the equation (11.28). For any choice of $^t(x_1, x_2, \ldots, x_{2\nu})$ in $\mathbb{F}_q^{2\nu}$, $x_{2\nu+1}$ will be uniquely determined. Therefore there are $q^{2\nu}$ solutions of (11.28) in $\mathbb{F}_q^{2\nu+1}$.

There remains the equation (11.29) to be considered. When $a = 0$, there are $q^{2\nu-1} + q^{\nu} - q^{\nu-1}$ solutions $^t(x_1, x_2, \ldots, x_{2\nu})$ of (11.27) in $\mathbb{F}_q^{2\nu}$, and for each such solution $^t(x_1, x_2, \ldots, x_{2\nu}, x_{2\nu+1}, x_{2\nu+2})$ is a solution of (11.29) if and only if $x_{2\nu+1} = x_{2\nu+2} = 0$, since $\alpha x_{2\nu+1}^2 + x_{2\nu+1} x_{2\nu+2} + \alpha x_{2\nu+2}^2$ is definite, but for each of the remaining vectors $^t(x_1, x_2, \ldots, x_{2\nu})$, which are $q^{2\nu} - (q^{2\nu-1} + q^{\nu} - q^{\nu-1})$ in numbers, by Lemma 11.21 there are $q + 1$ pairs $(x_{2\nu+1}, x_{2\nu+2})$ such that $^t(x_1, x_2, \ldots, x_{2\nu}, x_{2\nu+1}, x_{2\nu+2})$ is a solution of (11.29). Altogether we get

$$q^{2\nu-1} + q^{\nu} - q^{\nu-1} + (q+1)(q^{2\nu} - q^{2\nu-1} - q^{\nu} + q^{\nu-1}) = q^{2\nu+1} - q^{\nu+1} + q^{\nu}$$

solutions of (11.29) in $\mathbb{F}_q^{2\nu+2}$.

When $a \neq 0$, there are $q^{2\nu-1} - q^{\nu-1}$ solutions ${}^t(x_1, x_2, \ldots, x_{2\nu})$ of (11.27) in $\mathbb{F}_q^{2\nu}$, and for each such solution ${}^t(x_1, x_2, \ldots, x_{2\nu}, x_{2\nu+1}, x_{2\nu+2})$ is a solution of (11.29) if and only if $x_{2\nu+1} = x_{2\nu+2} = 0$, but for each of the remaining vectors ${}^t(x_1, x_2, \ldots, x_{2\nu})$, which are $q^{2\nu} - (q^{2\nu-1} - q^{\nu-1})$ in number, by Lemma 11.21 there are $q + 1$ pairs $(x_{2\nu+1}, x_{2\nu+2})$ such that ${}^t(x_1, x_2, \ldots, x_{2\nu}, x_{2\nu+1}, x_{2\nu+2})$ is a solution of (11.29). Altogether we get

$$q^{2\nu-1} - q^{\nu-1} + (q+1)(q^{2\nu} - q^{2\nu-1} + q^{\nu-1}) = q^{2\nu+1} + q^\nu$$

solutions of (11.29). □

11.4 Exercises

11.1 Let $Q(x_1, x_2, \ldots, x_n)$ be a quadratic form in n indeterminates over any finite field $\mathbb{F}_q$. Prove that $Q(x_1, x_2, \ldots, x_n)$ is not non-singular if and only if it is cogredient to a quadratic form in less than n indeterminates.

11.2 Reduce the following quadratic forms to normal forms:

(i) $x_1x_2 + x_3x_4 + x_5x_6$ over $\mathbb{F}_q$, where q is odd,

(ii) $x_1x_2 + x_2x_3 + x_3x_4$ over $\mathbb{F}_3$,

(iii) $x_1^2 + 2x_1x_2 + 3x_1x_4 + 4x_3x_4$ over $\mathbb{F}_5$.

11.3 Let q be a power of an odd prime number. Prove that the number of solutions of the equation $x^2 + \alpha y^2 = \beta$, where $\alpha, \beta \in \mathbb{F}_q^*$, is equal to $q + 1$ or $q - 1$ if the 2×2 diagonal matrix

$$\begin{pmatrix} 1 & 0 \\ 0 & \alpha \end{pmatrix}$$

is definite or not, respectively.

11.4 Let q be a power of an odd prime number. Prove that the number of solutions of the equation

$$x_1^2 + x_2^2 + \cdots + x_{2\nu}^2$$

in 2ν indeterminates $x_1, x_2, \ldots, x_{2\nu}$ is

$$q^{2\nu-1} + q^\nu - q^{\nu-1}, \text{ if } (-1)^\nu \in \mathbb{F}_q^{*2}$$
$$q^{2\nu-1} - q^\nu + q^{\nu-1}, \text{ if } (-1)^\nu \notin \mathbb{F}_q^{*2},$$

respectively.

11.5 Let S be an $n \times n$ nonsingular symmetric matrix over $\mathbb{F}_q$, where q is a power of an odd prime number. Let P be an m-dimensional subspace of $\mathbb{F}_q^n$ and $v_1, v_2, \ldots, v_m$ be a basis of P. The $n \times m$ matrix

$$(v_1, v_2, \ldots, v_m)$$

is of rank m and is called a matrix representation of the subspace P and denoted also by P. The subspace P is said to be *totally isotropic* with respect to S if $^t P S P = 0$. A totally isotropic subspace with respect to S is said to be maximal if it is not contained in any other totally isotropic subspace with respect to S. Prove that maximal totally isotropic subspaces with respect to S are of the same dimension and that the index of S is equal to the dimension of maximal totally isotropic subspaces with respect to S.

11.6 Reduce the following alternate forms to normal forms:

  (i) $x_1(y_2 + y_3 + y_4) - (x_2 + x_3 + x_4)y_1$ over $\mathbb{F}_3$,

  (ii) $\sum_{i=1}^{n-1} (x_i y_{i+1} + x_{i+1} y_i)$ over $\mathbb{F}_2$.

11.7 Prove that the determinant of a nonsingular alternate matrix is a square element.

11.8 Reduce the following quadratic forms to normal forms:

  (i) $x_1^2 + x_2^2 + x_3^2 + x_4^2$ over $\mathbb{F}_2$,
  (ii) $x_1^2 + x_2^2 + x_3^2 + x_4^2 + x_1 x_2 + x_1 x_3 + x_1 x_4 + x_2 x_3 + x_2 x_4 + x_3 x_4$ over $\mathbb{F}_{2^2}$.

11.9 Let G be an $n \times n$ nonsingular matrix over $\mathbb{F}_q$, where q is a power of 2. A subspace P of $\mathbb{F}_q^n$ is said to be *totally singular* with respect to G if $^t P G P \equiv 0$. A totally singular subspace with respect to S is said to be maximal if it is not contained in any other totally singular subspace with respect to G. Prove that maximal totally singular subspace with respect to G are of the same dimension and that the index of G is equal to the dimension of maximal totally singular subspaces with respect to G.

11.10 Prove that any quadratic form in three or more indeterminates over any finite field is necessarily indefinite.

Chapter 12

More Group Theory and Ring Theory

12.1 Homomorphisms of Groups, Normal Subgroups and Factor Groups

The concept of the isomorphism of groups can be generalized as follows.

Definition 12.1 *Let G and G' be groups. If σ is a map from G to G'*

$$\sigma : G \rightarrow G'$$
$$a \mapsto \sigma(a)$$

which preserves the group operation, i.e.,

$$\sigma(ab) = \sigma(a)\sigma(b) \quad for\ all\ a, b \in G,$$

then we say that σ is a homomorphic map or a homomorphism from G to G'. Moreover, if σ is surjective, we say that G' is a homomorphic image of G.

Let σ be a homomorphism from a group G to a group G'. If σ is a bijective map, then σ is an isomorphism from G to G'.

Example 12.1 Let m be a positive integer. Consider the map

$$- : \mathbb{Z} \rightarrow \mathbb{Z}_m$$
$$a \mapsto \bar{a} \quad for\ all\ a \in \mathbb{Z}.$$

Since $\overline{a + b} = \bar{a} + \bar{b}$ for all $a, b \in \mathbb{Z}$, $-$ is a homomorphism from $\mathbb{Z}$ to $\mathbb{Z}_m$. Moreover, $-$ is surjective. If $m = 0$, then $-$ is the identity map of $\mathbb{Z}$ and, hence, is an automorphism of $\mathbb{Z}$.

Example 12.2 Let S_4 be the symmetric group of degree 4 (see Example 2.6) and A_4 be the alternating group of degree 4, (see Example 2.11). Define a map

$$\phi : S_4 \;\to\; \mathbb{Z}_2$$

$$\pi \;\mapsto\; \phi(\pi) = \begin{cases} 0, & \text{if } \pi \in A_4, \\ 1, & \text{if } \pi \in S_4 \backslash A_4. \end{cases}$$

It can be verified that ϕ is a surjective homomorphism from S_4 to $\mathbb{Z}_2$.

Example 12.3 Let q be a prime power and n be a positive integer. By Theorem 7.12(i) and (ii) the map

$$\text{Tr} : \mathbb{F}_{q^n} \;\to\; \mathbb{F}_q$$

$$\alpha \;\mapsto\; \sum_{i=0}^{n-1} \alpha^{q^i}$$

is a homomorphism from the additive group of $\mathbb{F}_{q^n}$ to that of $\mathbb{F}_q$, and by Theorem 7.12(i′) and (ii′) the map

$$\text{N} : \mathbb{F}_{q^n}^* \;\to\; \mathbb{F}_q^*$$

$$\alpha \;\mapsto\; \prod_{i=0}^{n-1} \alpha^{q^i}$$

is a homomorphism from the multiplicative group $\mathbb{F}_{q^n}^*$ of the nonzero elements of $\mathbb{F}_{q^n}$ to that of $\mathbb{F}_q$.

Similar to Theorem 2.4 for the isomorphism of groups, for the homomorphism we have

Theorem 12.1 *Let σ be a homomorphism from a group G to a group G', then σ maps the identity e of G to that of G' and it also maps the inverse a^{-1} of an element a of G to the inverse $\sigma(a)^{-1}$ of the image $\sigma(a)$ of a.*

The proof of Theorem 2.4 can be carried over word by word and will not be repeated here.

Definition 12.2 *Let G and G' be groups and σ be a homomorphism from G to G'. The set of images $\sigma(a)$ for all $a \in G$ is called the image of σ and is denoted by $\text{Im}\,\sigma$. The complete inverse image of the identity e' of G' is called the kernel of σ and is denoted by $\text{Ker}\,\sigma$. More precisely,*

$$\text{Im}\,\sigma = \{\sigma(a) : a \in G\}$$

and

$$\text{Ker}\,\sigma = \{a \in G : \sigma(a) = e'\}.$$

Example 12.4 Consider the map $- : \mathbb{Z} \to \mathbb{Z}_m$ of Example 12.1. Clearly, $\operatorname{Im} - = \mathbb{Z}_m$ and $\operatorname{Ker} - = \{km : k \in \mathbb{Z}\} = (m)$. For any $\bar{a} \in \mathbb{Z}_m$, we also have $\phi^{-1}(\bar{a}) = \{a + km : k \in \mathbb{Z}\} = a + (m)$.

Example 12.5 For the map $\phi : S_4 \to \mathbb{Z}_2$ of Example 12.2, we have $\operatorname{Im} \phi = \mathbb{Z}_2$ and $\operatorname{Ker} \phi = A_4$.

Example 12.6 Consider the map Tr and N of Example 12.3 again. By Theorem 7.16(i), $\operatorname{Im} \operatorname{Tr} = \mathbb{F}_q$ and

$$\operatorname{Ker} \operatorname{Tr} = \{\beta - \beta^q : \beta \in \mathbb{F}_{q^n}\}.$$

By Theorem 7.16(ii), $\operatorname{Im} N = \mathbb{F}_q^*$ and

$$\operatorname{Ker} N = \{\beta^{1-q} : \beta \in \mathbb{F}_{q^n}^*\}.$$

Let G be a group and H be a subgroup of G. For an element $a \in G$ we have defined the left coset aH of a relative to H and the right coset Ha of a relative to H in Definitions 2.7 and 2.10, respectively. We mentioned in Section 2.2 that if G is not abelian, it may occur that $aH \neq Ha$ and we have Example 2.19 to illustrate this situation. Now we give the following definition.

Definition 12.3 *Let G be a group and N be a subgroup of G. If $aN=Na$ for all $a \in G$, N is called a normal subgroup of G. We use the notation $N \lhd G$ to denote that N is a normal subgroup of G. If $N \lhd G$, for any $a \in G$ aN is simply called the coset of a relative to N.*

Clearly, for any group G

$$G \lhd G \quad \text{and} \quad \{e\} \lhd G,$$

where $\{e\}$ is the subgroup of G consisting only of the identity of G. Moreover, if G is an abelian group, every subgroup of G is normal in G. However, if G is not abelian, there may be some subgroup of G which is not normal in G. For instance, by Example 2.19, the subgroup

$$H = \left\{ \begin{pmatrix} 1 & 2 & 3 \\ 1 & 2 & 3 \end{pmatrix}, \begin{pmatrix} 1 & 2 & 3 \\ 2 & 1 & 3 \end{pmatrix} \right\}$$

is not normal in S_3. But it can be verified directly that $A_4 \lhd S_4$ and $V_4 \lhd A_4$.

Theorem 12.2 *Let G be a group and N be a subgroup of G. Then the following conditions are equivalent:*

(i) $N \lhd G$, i.e., $aN{=}Na$ for all $a \in G$;

(ii) $a^{-1}Na = N$ for all $a \in G$;

(iii) $a^{-1}Na \subset N$ for all $a \in G$.

Proof. Clearly, (i) implies (ii) and (ii) implies (iii). Now suppose that (iii) holds. For any $a \in G$, we have $a^{-1} \in G$, then by (iii) $(a^{-1})^{-1}Na^{-1} \subset N$, i.e., $aNa^{-1} \subset N$. It follows that

$$N = a^{-1}(aNa^{-1})a \subset a^{-1}Na.$$

Hence $a^{-1}Na = N$ and $aN = Na$ for all $a \in G$.   □

Theorem 12.3 *Let G be a group and N be a normal subgroup of G. Denote the set of cosets of G relative to N by G/N and define the product of two cosets by*

$$(aN)(bN) = (ab)N \ \ for \ all \ a, \ b \in G. \tag{12.1}$$

Then this definition is well-defined and the set of cosets G/N forms a group with respect to the product of cosets so defined and with the coset N as its identity.

Proof. Assume that $aN = a'N$ and $bN = b'N$. Then $a = a'n_a$ and $b = b'n_b$ for some $n_a, n_b \in N$. Thus

$$(ab)N = ((a'n_a)(b'n_b))N = (a'(n_ab')n_b)N.$$

Since $N \lhd G$, $Nb' = b'N$. Hence there is an $n' \in N$ such that $n_ab' = b'n'$. Thus

$$\begin{aligned}
(a'(n_ab')n_b)N &= (a'(b'n')n_b)N \\
&= ((a'b')(n'n_b))N \\
&= ((a'b')N)((n'n_b)N) \\
&= ((a'b')N)(N) \\
&= (a'b')N.
\end{aligned}$$

This proves that the definition (12.1) is independent of the particular choice of the coset representatives; in other words, it is well-defined.

We shall prove that G/N is a group with respect to the product of cosets defined by (12.1).

Clearly, G/N is closed with respect to the multiplication.

Let aN, bN, and cN be any three cosets. By definition

$$((aN)(bN))(cN) = ((ab)N)(cN) = ((ab)c)N.$$

Similarly,

$$(aN)((bN)(cN)) = (a(bc))N.$$

Since $(ab)c = a(bc)$, we have

$$((aN)(bN))(cN) = (aN)((bN)(cN)).$$

This proves the associative law of multiplication.

For any coset aN, by definition

$$(aN)N = (aN)(eN) = (ae)N = aN.$$

Moreover, $a^{-1}N$ is also a coset and

$$(aN)(a^{-1}N) = (aa^{-1})N = eN = N.$$

Therefore G/N is a group. □

Corollary 12.4 *If G is an abelian group and N is a subgroup of G, then G/N is also an abelian group.*

Proof. For any two cosets aN and bN, we have

$$(aN)(bN) = (ab)N = (ba)N = (bN)(aN).$$

□

Definition 12.4 *Let G be a group and N be a normal subgroup of G. The group G/N defined in Theorem 12.3 is called the factor group of G relative to N.*

Example 12.7 We know that $A_4 \lhd S_4$ and $S_4 : A_4 = 2$. Therefore the factor group S_4/A_4 is a group of order 2.

Example 12.8 We know that $V_4 \lhd A_4$ and $A_4 : V_4 = 3$. Therefore the factor group A_4/V_4 is a group of order 3.

The following theorem, called the fundamental theorem of homomorphisms of groups, discloses the intimate relationship of homomorphisms, normal subgroups and factor groups.

Theorem 12.5 (i) *Let G be a group and N a normal subgroup of G. Then the map*

$$\phi: \begin{array}{ccc} G & \to & G/N \\ a & \mapsto & aN, \end{array}$$

is a homomorphism of groups, called the natural homomorphism of groups, and $\operatorname{Ker} \phi = N$, $\operatorname{Im} \phi = G/N$.

(ii) *Conversely, let $\phi:\; G \to G'$ be a homomorphism of groups, then* $\operatorname{Im}\phi$
is a subgroup of G', $\operatorname{Ker}\phi$ is a normal subgroup of G, and the map

$$\bar{\phi}:\; G/\operatorname{Ker}\phi \;\to\; \operatorname{Im}\phi$$
$$a\operatorname{Ker}\phi \;\mapsto\; \phi(a) \qquad\qquad (12.2)$$

is an isomorphism of groups.

Proof. (i) Clearly,

$$(aN)(bN) = (ab)N.$$

This proves that map ϕ is a homomorphism. Both $\operatorname{Ker}\phi = N$ and $\operatorname{Im}\phi = G/N$ are obvious.

(ii) For any $\phi(a), \phi(b) \in \operatorname{Im}\phi$, where $a, b \in G$, we have

$$\phi(a)\phi(b)^{-1} = \phi(a)\phi(b^{-1}) = \phi(ab^{-1}) \in \operatorname{Im}\phi.$$

By Theorem 2.7, $\operatorname{Im}\phi$ is a subgroup of G'.

Let $a, b \in \operatorname{Ker}\phi$, i.e., $\phi(a) = \phi(b) = e'$. Then

$$\phi(ab^{-1}) = \phi(a)\phi(b^{-1}) = \phi(a)\phi(b)^{-1} = e'e'^{-1} = e'.$$

This proves that $\operatorname{Ker}\phi$ is a subgroup of G. Let $a \in \operatorname{Ker}\phi$ and $x \in G$, then

$$\phi(x^{-1}ax) = \phi(x^{-1})\phi(a)\phi(x) = \phi(x)^{-1}e'\phi(x) = \phi(x)^{-1}\phi(x) = e'.$$

Thus $x^{-1}ax \in \operatorname{Ker}\phi$. Hence $x^{-1}\operatorname{Ker}\phi\, x \subset \operatorname{Ker}\phi$. By Theorem 12.2, $\operatorname{Ker}\phi \lhd G$.

Define a map $\bar{\phi}$ from $G/\operatorname{Ker}\phi$ to $\operatorname{Im}\phi$ by (12.2). Let $a, b \in G$ and assume $a\operatorname{Ker}\phi = b\operatorname{Ker}\phi$. Then there is an element $k \in \operatorname{Ker}\phi$ such that $a = bk$. Thus $\phi(a) = \phi(bk) = \phi(b)\phi(k) = \phi(b)e' = \phi(b)$. This proves that the definition (12.2) is independent of the particular choice of the coset representatives and, hence, is well-defined.

Let $a\operatorname{Ker}\phi$ and $b\operatorname{Ker}\phi$ be two elements of $G/\operatorname{Ker}\phi$. Then

$$\bar{\phi}(a\operatorname{Ker}\phi\; b\operatorname{Ker}\phi) = \bar{\phi}(ab\operatorname{Ker}\phi)$$
$$= \phi(ab)$$
$$= \phi(a)\phi(b)$$
$$= \bar{\phi}(a\operatorname{Ker}\phi)\bar{\phi}(b\operatorname{Ker}\phi).$$

This proves that $\bar{\phi}$ is a homomorphism. For any $\phi(a) \in \operatorname{Im}\phi$, we have a coset $a\operatorname{Ker}\phi$ and $\bar{\phi}(a\operatorname{Ker}\phi) = \phi(a)$. Therefore $\bar{\phi}$ is surjective. Finally, for any $a\mathrm{Ker}\,\phi$, $b\mathrm{Ker}\,\phi \in G/\mathrm{Ker}\,\phi$, assume that $\bar{\phi}(a\mathrm{Ker}\,\phi) = \bar{\phi}(b\mathrm{Ker}\,\phi)$, then $\phi(a) = \phi(b)$. Thus $\phi(ab^{-1}) = \phi(a)\phi(b^{-1}) = \phi(a)\phi(b)^{-1} = e'$, so $ab^{-1} \in \mathrm{Ker}\,\phi$ and, hence, $a\mathrm{Ker}\,\phi = b\mathrm{Ker}\,\phi$. This proves that the map $\bar{\phi}$ is injective. $\qquad\square$

Corollary 12.6 *Let G and G' be groups and ϕ is a homomorphism from G to G'. Then*

(i) *ϕ is surjective if and only if $\mathrm{Im}\,\phi = G$;*

(ii) *ϕ is injective if and only if $\mathrm{Ker}\,\phi = \{e\}$.*

Moreover, we have the following two isomorphism theorems of groups.

Theorem 12.7 *Let $\phi : G \to G'$ be a homomorphism of groups and assume that ϕ is surjective. Then the following holds.*

(i) *If N is a subgroup of G, $\phi(N)$ is subgroup of G'. If $N \lhd G$, $\phi(N) \lhd G'$.*

(ii) *If N' is a subgroup of G', $\phi^{-1}(N')$ is a subgroup of G, $\phi^{-1}(N') \supseteq \mathrm{Ker}\,\phi$ and $\phi^{-1}(N')/\mathrm{Ker}\,\phi \simeq N'$. If $N' \lhd G'$, $\phi^{-1}(N') \lhd G$.*

(iii) *If N is a normal subgroup of G and let $N' = \phi(N)$, then the map*

$$\tilde{\phi}: \begin{array}{ccc} G/N & \to & G'/N' \\ aN & \mapsto & \phi(a)N' \end{array}$$

is a homomorphism of groups. Assume further $N \supseteq \mathrm{Ker}\,\phi$, then $\phi^{-1}(\phi(N)) = N$ and the above map is an isomorphism of groups.

Proof. The verification of (i) and (ii) is trivial and is omitted.

Let us prove (iii). First we prove that the map $\tilde{\phi}$ is well-defined. Let $aN = bN$, where $a, b \in G$. Then $ab^{-1} \in N$, thus $\phi(a)\phi(b)^{-1} = \phi(ab^{-1}) \in \phi(N) = N'$, which implies $\phi(a)N' = \phi(b)N'$. Therefore $\tilde{\phi}(aN) = \phi(a)N' = \phi(b)N' = \tilde{\phi}(bN)$.

For any $a, b \in G$,

$$\tilde{\phi}(aN)\tilde{\phi}(bN) = \phi(a)N'\phi(b)N' = \phi(a)\phi(b)N'$$
$$= \phi(ab)N' = \tilde{\phi}(abN) = \tilde{\phi}(aNbN).$$

This proves that $\tilde{\phi}$ is a homomorphism of groups.

Assume that $N \supseteq \mathrm{Ker}\,\phi$. Clearly $\phi^{-1}(\phi(N)) \supseteq N$. Let $a \in \phi^{-1}(\phi(N))$, then $\phi(a) \in \phi(N)$. Thus there is an element $n \in N$ such that $\phi(a) =$

$\phi(n)$, which implies $\phi(an^{-1}) = \phi(a)\phi(n)^{-1} = e'$ and $an^{-1} \in \operatorname{Ker}\phi$. By assumption $\operatorname{Ker}\phi \subseteq N$, therefore $an^{-1} \in N$ and $a \in Nn = N$. This proves $\phi^{-1}(\phi(N)) \subseteq N$. Hence $\phi^{-1}(\phi(N)) = N$. Since ϕ is surjective, $\widetilde{\phi}$ is also surjective. To prove that $\widetilde{\phi}$ is injective, by Corollary 12.6(ii) it is enough to show that $\operatorname{Ker}\widetilde{\phi} = \{N\}$. Let $\widetilde{\phi}(aN) = N'$, where $a \in G$, then $\phi(a)N' = N'$, which implies $\phi(a) \in N' = \phi(N)$. Then $a \in \phi^{-1}(\phi(a)) \subseteq \phi^{-1}(\phi(N)) = N$. Thus $aN = N$ and $\operatorname{Ker}\widetilde{\phi} = \{N\}$. $\qquad\square$

Theorem 12.8 *Let G be a group, H and N be subgroups of G, and $N \lhd G$. Define $HN = \{hn : h \in H, n \in N\}$. Then*

(i) *HN is a subgroup of G, $N \lhd HN$, and $(H \cap N) \lhd H$.*

(ii) *The map*

$$\phi : \quad HN \quad \to \quad H/(H \cap N)$$
$$hn \quad \mapsto \quad h(H \cap N)$$

is well-defined and is a homomorphism of groups, the kernel of this homomorphism is N, and, consequently, the map

$$\bar{\phi} : \quad HN/N \quad \to \quad H/(H \cap N) \qquad (12.3)$$
$$hnN \quad \mapsto \quad h(H \cap N)$$

is an isomorphism of groups.

Proof. (i) can be proved by routine verification and is omitted.

Let us prove (ii). First we prove that the map ϕ is well-defined. Let $h_1, h_2 \in H$ and $n_1, n_2 \in N$. Assume that $h_1 n_1 = h_2 n_2$. Then $h_2^{-1} h_1 = n_2 n_1^{-1} \in H \cap N$. Thus $h_1(H \cap N) = h_2(H \cap N)$ and ϕ is well-defined.

Next, we show that ϕ is a homomorphism. We have

$$\phi(h_1 n_1)\phi(h_2 n_2) = h_1(H \cap N)h_2(H \cap N)$$
$$= h_1 h_2(H \cap N) \quad (\because (H \cap N) \lhd H).$$

Since $N \lhd G$, there is an element $n_1' \in N$ such that $h_2^{-1} n_1 h_2 = n_1'$. Then $n_1 h_2 = h_2 n_1'$ and

$$\phi(h_1 n_1 h_2 n_2) = \phi(h_1 h_2 n_1' n_2) = h_1 h_2(H \cap N).$$

Therefore

$$\phi(h_1 n_1 h_2 n_2) = \phi(h_1 n_1)\phi(h_2 n_2),$$

i.e., ϕ is a homomorphism.

Clearly, $N \subseteq \operatorname{Ker}\phi$. Conversely, let $hn \in \operatorname{Ker}\phi$. Then $\phi(hn) = h(H \cap N) = H \cap N$. Thus $h \in N$ and $hn \in N$. This proves $\operatorname{Ker}\phi \subseteq N$. Therefore $\operatorname{Ker}\phi = N$. By the fundamental theorem of group homomorphisms, (12.3) is an isomorphism of groups. $\qquad\square$

12.2 Direct Product Decomposition of Groups

Definition 12.5 *Let G be a group, H and K be subgroups of G. G is said to be the direct product of H and K, if the following conditions are fulfilled:*

(i) *$H \lhd G$, $K \lhd G$,*

(ii) *$G = HK$, where $HK = \{hk : h \in H, k \in K\}$,*

(iii) *$H \cap K = \{e\}$.*

Then we write $G = H \times K$ and say that $H \times K$ is a direct product decomposition of G.

Theorem 12.9 *Let G be a group, H and K be subgroups of G. Then G is the direct product of H and K if and only if*

(i) *$H \lhd G$, $K \lhd G$,*

(ii) *$G = HK$,*

(iii′) *Every element g of G can be written uniquely as a product $g = hk$, where $h \in H$, $k \in K$.*

Proof. Assume that (i), (ii), (iii) hold. Let us prove (iii′). Let $g = hk = h'k'$, where $h, h' \in H$, $k, k' \in K$. Then $h'^{-1}h = k'k^{-1} \in H \cap K = \{e\}$. Therefore $h'^{-1}h = e$ and $k'k^{-1} = e$. Hence $h = h'$ and $k = k'$.

Conversely, assume that (i), (ii), (iii′) hold. Let us prove (iii). Let $h \in H \cap K$, then $h = he$, where $h \in H$, $e \in K$, and $h = eh$, where $e \in H$, $h \in K$. By the uniqueness of the representation, $h = e$. Therefore $H \cap K = \{e\}$. □

We also have

Theorem 12.10 *Let G be a group, H and K be subgroups of G. Then G is the direct product of H and K if and only if*

(i′) *$hk = kh$ for all $h \in H$, $k \in K$,*

(ii) *$G = HK$,*

(iii) *$H \cap K = \{e\}$.*

Proof. Assume that (i), (ii), (iii) hold. Let us prove (i′). For any $h \in H$ and $k \in K$, $h^{-1}k^{-1}hk = (h^{-1}k^{-1}h)k \in K$, for by (i) $h^{-1}k^{-1}h \in K$. Similarly, $h^{-1}k^{-1}hk = h^{-1}(k^{-1}hk) \in H$. But by (iii) $H \cap K = \{e\}$. Therefore $h^{-1}k^{-1}hk = e$ and, hence, $hk = kh$.

That (i′), (ii), (iii) imply (i), (ii), (iii) is trivial. □

Example 12.9 Let $G = \langle a \rangle$ be a cyclic group of order 6 generated by a. Then $\langle a^3 \rangle$ and $\langle a^2 \rangle$ are subgroups of order 2 and 3, respectively. Let $H = \langle a^3 \rangle$ and $K = \langle a^2 \rangle$. Then (i') is fulfilled automatically. Any $a^i \in G$ can be expressed as $a^i = (a^3)^i \cdot (a^2)^{2i} \in HK$, thus (ii) holds. Suppose $a^j \in H \cap K$, then $a^j = (a^3)^h = (a^2)^k$ for some integer h, k. It follows that $a^{2j} = ((a^3)^h)^2 = a^{6h} = e$ and $a^{3j} = ((a^2)^k)^3 = e$, which implies $3 \mid j$ and $2 \mid j$, respectively. Therefore $6 \mid j$ and $a^j = e$, hence (iii) also holds. By Theorem 12.10, $G = \langle a^3 \rangle \times \langle a^2 \rangle$.

More generally, let $G = \langle a \rangle$ be a cyclic group of order mn, where $\gcd(m, n) = 1$, then $G = \langle a^n \rangle \times \langle a^m \rangle$.

Definition 12.5 can be generalized as follows.

Definition 12.6 *Let G be a group, $H_1, H_2, \ldots, H_r$ all be subgroups of G. G is said to be the direct product of $H_1, H_2, \ldots, H_r$ if the following conditions are fulfilled:*

(i) $H_i \lhd G$ *for* $i = 1, 2, \ldots, r$,

(ii) $G = H_1 H_2 \cdots H_r$, *where*

$$H_1 H_2 \cdots H_r = \{h_1 h_2 \cdots h_r : h_i \in H_i, i = 1, 2, \ldots, r\},$$

(iii) $H_i \cap (H_1 H_2 \cdots H_{i-1} H_{i+1} \cdots H_r) = \{e\}$ *for all* $i = 1, 2, \ldots, r$.

Then we write $G = H_1 \times H_2 \times \cdots \times H_r$ and say that $H_1 \times H_2 \times \cdots \times H_r$ is a direct product decomposition of G.

Theorems 12.9 and 12.10 can be generalized as follows.

Theorem 12.11 *Let G be a group, $H_1, H_2, \ldots, H_r$ all be subgroups of G. Then $G = H_1 \times H_2 \times \cdots \times H_r$ if and only if*

(i) $H_i \lhd G$ *for* $i = 1, 2, \ldots, r$,

(ii) $G = H_1 H_2 \cdots H_r$,

(iii') *Every element $g \in G$ can be written uniquely as a product*

$$g = h_1 h_2 \cdots h_r,$$

where $h_i \in H_i$, $i = 1, 2, \ldots, r$.

Theorem 12.12 *Let G be a group, $H_1, H_2, \ldots, H_r$ all be subgroups of G. Then $G = H_1 \times H_2 \times \cdots \times H_r$ if and only if* (i') *For every pair i and j, $i \neq j$, $h_i h_j = h_j h_i$ for all $h_i \in H_i$, $h_j \in H_j$,*

(ii) $G = H_1 H_2 \cdots H_r$,

(iii) $H_i \cap (H_1 H_2 \cdots H_{i-1} H_{i+1} \cdots H_r) = \{e\}$ *for all $i = 1, 2, \ldots, r$.*

The proof of Theorems 12.11 and 12.12 are similar to the proofs of Theorem 12.9 and 12.10, respectively, and are left as exercises.

Let G be an additive abelian group and $H_1, H_2, \ldots, H_k$ be subgroups of G. If G is the direct product of $H_1, H_2, \ldots, H_k$, we also write

$$G = H_1 \dotplus H_2 \dotplus \cdots \dotplus H_k$$

and call it the *direct sum decomposition* of G, or say that G is a *direct sum* of $H_1, H_2, \ldots, H_k$.

Example 12.10 Let $F[x]$ be a polynomial ring in an indeterminate x over a field F and let $f(x)$ be a polynomial of degree $n > 0$ over F. Consider the residue class ring $F[x]/(f(x))$. Denote the residue class of $x \bmod f(x)$ by $\bar{x}$. Then

$$F[x]/(f(x)) = \left\{ a_0 + a_1 \bar{x} + \cdots + a_{n-1} \bar{x}^{n-1} : a_0, a_1, \ldots, a_{n-1} \in F \right\}.$$

Let

$$G_i = \left\{ a_i \bar{x}^i : a_i \in F \right\},$$

then G_i's, $i = 0, 1, \ldots, n-1$, are subgroups of the additive group of the residue class ring $F[x]/(f(x))$ and

$$F[x]/(f(x)) = G_0 \dotplus G_1 \dotplus \cdots \dotplus G_{n-1}.$$

To illustrate the concept of the direct product decomposition of groups let us study finite abelian groups.

Let G be a finite abelian group. For any prime p, define

$$G_p = \{g \in G : \operatorname{ord}(g) = p^m \text{ for some integer } m \geq 0\}.$$

Then we have

Theorem 12.13 *Let G be a finite abelian group. Then for any prime p, G_p is a subgroup of G and there are only a finite number of primes p such that $G_p \neq \{e\}$. Let $p_1, p_2, \ldots, p_r$ be all the primes such that $G_{p_i} \neq \{e\}$, then $G = G_{p_1} \times G_{p_2} \times \cdots \times G_{p_r}$ and the direct product decomposition is unique up to a rearrangement of $G_{p_1}, G_{p_2}, \ldots, G_{p_r}$.*

Proof. Let g and g' be any two elements of G_p. Assume that $\mathrm{ord}(g) = p^m$ and $\mathrm{ord}(g') = p^{m'}$, where m and $m' > 0$. Then $(gg')^{p^{m+m'}} = e$ and $(g^{-1})^{p^m} = e$. Therefore both gg' and g^{-1} are elements of G_p. Hence G_p is a subgroup of G. Since G is a finite group, there are only a finite number of primes p such that $G_p \neq \{e\}$.

Let $p_1, p_2, \ldots, p_r$ be all the primes such that $G_{p_i} \neq \{e\}$. Let us verify that the condition (i)-(iii) of Definition 12.6 are fulfilled. Since G is abelian, (i) is automatically satisfied. For any i, $1 \leqslant i \leqslant r$, the order of any element of G_{p_i} is a power of p_i, while the order of any element of $G_{p_1} G_{p_2} \cdots G_{p_{i-1}} G_{p_{i+1}} \cdots G_{p_r}$ is of the form $p_1^{m_1} p_2^{m_2} \cdots p_{i-1}^{m_{i-1}} p_{i+1}^{m_{i+1}} \cdots p_r^{m_r}$. Since $\gcd(p_i, p_1 p_2 \cdots p_{i-1} p_{i+1} \cdots p_r) = 1$,

$$G_{p_i} \cap G_{p_1} G_{p_2} \cdots G_{p_{i-1}} G_{p_{i+1}} \cdots G_{p_r} = \{e\}.$$

Thus (iii) holds. To verify that (ii) holds it is enough to show that any element $g \in G$ can be written in the form $g = g_1 g_2 \cdots g_r$, where $g_i \in G_{p_i}$, $i = 1, 2, \ldots, r$. Let $g \in G$ and $\mathrm{ord}(g) = n$. If $n = 1$, then $g = e$ and

$$g = \underbrace{e\, e\, \cdots\, e}_{r \text{ in number}} \in G_{p_1} G_{p_2} \cdots G_{p_r}.$$

If $n > 1$, let p be a prime such that $p|n$, then $\mathrm{ord}(g^{n/p}) = p$, $g^{n/p} \in G_p$, and, hence, p is one of $p_1, p_2, \ldots, p_r$. So we can assume that $n = p_1^{e_1} p_2^{e_2} \cdots p_r^{e_r}$, where $e_i \geq 0$ for $i = 1, 2, \ldots, r$. Then $\gcd(n/p_1^{e_1}, n/p_2^{e_2}, \ldots, n/p_r^{e_r}) = 1$. By Theorem 1.4 there are integers $m_1, m_2, \ldots, m_r$ such that $m_1 \cdot n/p_1^{e_1} + m_2 \cdot n/p_2^{e_2} + \cdots + m_r \cdot n/p_r^{e_r} = 1$. Let $g_i = g^{m_i \cdot n/p_i^{e_i}}$, $i = 1, 2, \ldots, r$, then $g_i \in G_{p_i}$ and $g = g_1 g_2 \cdots g_r$. Thus (ii) is also verified. Therefore $G = G_{p_1} \times G_{p_2} \times \cdots \times G_{p_r}$. The uniqueness is clear, since each G_{p_i} is uniquely determined by G and p_i. □

By Theorem 12.13 the study of the structure of finite abelian groups is reduced to the study of the structure of finite abelian groups of prime power orders.

Lemma 12.14 *Let G be an abelian group of order p^n, where p is a prime. Then G is cyclic if and only if G has only one subgroup of order p.*

Proof. If G is cyclic, $G = \langle a \rangle$, then $\langle a^{p^{n-1}} \rangle$ is a cyclic subgroup of order p. Let $H = \langle b \rangle$ be any cyclic subgroup of order p. We can assume that $b = a^h$, $0 < h < p^n$. Then $e = b^p = a^{hp}$, thus $p^n | hp$, which implies $h = h_1 p^{n-1}$ where $p \nmid h_1$. Then $b = (a^{p^{n-1}})^{h_1} \in \langle a^{p^{n-1}} \rangle$ and $H = \langle b \rangle \subseteq \langle a^{p^{n-1}} \rangle$. But $|H| = p$ and $|\langle a^{p^{n-1}} \rangle| = p$, so $H = \langle a^{p^{n-1}} \rangle$. Therefore G has only one subgroup of order p.

Conversely, assume that G has only one subgroup P of order p. Since $|G| = p^n$, the order of any subgroup of G is a power of p. Then any subgroup of G, which is not $\{e\}$, has a subgroup of order p, which must be P. This proves that any subgroup of G, which is not $\{e\}$, has P as its only subgroup of order p. We apply induction on the order of G to prove that G is cyclic. Consider the map

$$\phi : G \;\to\; G \qquad\qquad (12.4)$$
$$g \;\mapsto\; g^p \quad \forall g \in G.$$

Clearly, ϕ is a homomorphism and $\operatorname{Ker} \phi$ consists of all elements of order p and the identity e. Since $\operatorname{Ker} \phi$ has only one subgroup of order p, $\operatorname{Ker} \phi = P$ and we have the isomorphism

$$\overline{\phi} : \; G/P \simeq \operatorname{Im} \phi.$$

Thus $G : \operatorname{Im} \phi = p$. If $\operatorname{Im} \phi = \{e\}$, $G = P$ is cyclic. If $\operatorname{Im} \phi \neq \{e\}$, by induction hypothesis $\operatorname{Im} \phi$ is cyclic. Let $\operatorname{Im} \phi = \langle a \rangle$ and assume that $g \in \phi^{-1}(a)$, then $\phi(g) = g^p = a$. Thus $\langle g \rangle : \operatorname{Im} \phi = p$. We have already proved $G : \operatorname{Im} \phi = p$. Therefore $G = \langle g \rangle$ is cyclic. $\qquad\square$

Lemma 12.15 *Let G be an abelian group of order p^n, where p is a prime, and assume that G is not cyclic. Let a be an element of the largest order in G. Then there is a subgroup H of G such that $G = \langle a \rangle \times H$.*

Proof. We apply induction on the order of G. Since G is not cyclic, G has at least two subgroups of order p. Then G has a subgroup of order p, not contained in $\langle a \rangle$. Denote this subgroup by P. Let $\overline{G} = G/P$. We assert that $\operatorname{ord}(aP) = \operatorname{ord}(a)$. Clearly, $\operatorname{ord}(aP) | \operatorname{ord}(a)$. Suppose $\operatorname{ord}(aP) = p^m$, then $(aP)^{p^m} = P$ which implies $a^{p^m} \in P$. Let $P = \langle b \rangle$, then $a^{p^m} = b^i$, $0 \leqslant i \leqslant p - 1$. If $i \neq 0$, there is an integer j such that $ij \equiv 1 \pmod{p}$, then $b = b^{ij} = a^{p^m j} \in \langle a \rangle$, which implies $P \subset \langle a \rangle$, a contradiction. Therefore $i = 0$, $a^{p^m} = e$, and $\operatorname{ord}(a) | \operatorname{ord}(aP)$. Hence $\operatorname{ord}(aP) = \operatorname{ord}(a)$, as we asserted. Then aP is an element of the largest order in $\overline{G}$. Since $|\overline{G}| < |G|$, by induction hypothesis there is a subgroup $\overline{H}$ of $\overline{G}$ such that $\overline{G} = \langle aP \rangle \times \overline{H}$. Let H be the complete inverse image of $\overline{H}$ under the homomorphism

$$- : \; G \to \overline{G} = G/P$$
$$g \mapsto gP.$$

By Theorem 12.7, $H \supseteq P$ and $\overline{H} = H/P$. From $\overline{G} = \langle aP \rangle \times \overline{H}$ and $H \supseteq P$ we deduce $G = \langle a \rangle H$ and $\langle a \rangle \cap H \subset P$. Thus $\langle a \rangle \cap H = P$ or $\{e\}$. By assumption, $P \subsetneq \langle a \rangle$. Therefore $\langle a \rangle \cap H = \{e\}$. Hence $G = \langle a \rangle \times H$. $\qquad\square$

Theorem 12.16 *Let G be an abelian group of order p^n, where p is a prime. Then G is a direct product of cyclic subgroups*

$$G = \langle a_1 \rangle \times \cdots \times \langle a_r \rangle \tag{12.5}$$

where r and the orders $\mathrm{ord}(a_i) = p^{e_i}, i = 1, 2, \ldots, r$, are uniquely determined by G.

Proof. The direct product decomposition (12.5) follows immediately from Lemma 12.15. Let us prove the uniqueness. Consider the group homomorphism (12.4). Clearly

$$\mathrm{Ker}\, \phi = \{g \in G : \ g^p = 1\}$$

and $|\mathrm{Ker}\, \phi|$ is independent of the direct product decomposition of G. For the direct product decomposition (12.5) we have

$$\mathrm{Ker}\, \phi = \langle a_1^{p^{e_1-1}} \rangle \times \cdots \times \langle a_r^{p^{e_r-1}} \rangle$$

and $|\mathrm{Ker}\, \phi| = p^r$, therefore r is uniquely determined by G. We also have

$$\mathrm{Im}\, \phi = \langle a_1^p \rangle \times \cdots \times \langle a_r^p \rangle.$$

We apply induction on $|G|$. Now $|\mathrm{Im}\, \phi| < |G|$, so $\mathrm{ord}(a_i^p) = p^{e_i-1}, i = 1, 2, \ldots, r$, are uniquely determined by $\mathrm{Im}\, \phi$, and hence by G. Therefore $\mathrm{ord}(a_i) = p^{e_i}, i = 1, 2, \ldots, r$, are uniquely determined by G.

In the decomposition (12.5) we usually assume that $e_1 \geqslant e_2 \geqslant \cdots \geqslant e_r$ and call $\{p^{e_1}, p^{e_2}, \ldots, p^{e_r}\}$ the *type* of the abelian group G.

12.3   Some Ring Theory

In this section we recapitulate some concepts and facts from ring theory, which will be needed in the study of Galois rings.

As before, by a ring we mean a commutative ring with identity. First let us introduce the concept of an ideal of a ring.

Definition 12.7 *Let R be a ring. A nonempty set I of R is called an ideal of R, if $a, b \in I$ and $r \in R$ imply $a + b \in I$ and $ra \in I$.*

Clearly, R is an ideal of R and the subset $\{0\}$ consisting of the zero of R alone is also an ideal of R. An ideal of R distinct from R is called a *proper ideal* of R. For instance, when $|R| > 1$, $\{0\}$ is a proper ideal of R.

An ideal is not necessarily a subring of R; for example, for any integer $m > 1$, $(m) = \{km : k \in \mathbb{Z}\}$ is an ideal of the ring $\mathbb{Z}$, but it is not a subring, since it does not have an identity. We recall that (m) was regarded as a subgroup of the additive group of $\mathbb{Z}$, but it is also an ideal of $\mathbb{Z}$. Conversely, a subring of R is not always an ideal of R; for example, let F be any field, then F is a subring of the polynomial ring $F[x]$, but it is not an ideal of $F[x]$.

Example 12.11 Let R be any ring and $a_1, a_2, \ldots, a_n$ be elements of R. The set

$$Ra_1 + Ra_2 + \cdots + Ra_n = \{r_1a_1 + r_2a_2 + \cdots + r_na_n : r_1, r_2, \ldots, r_n \in R\}$$

is an ideal of R, called the *ideal generated by* $a_1, a_2, \ldots, a_n$ and denoted by $(a_1, a_2, \ldots, a_n)$.

In particular, we have

Definition 12.8 *Let R be any ring and $a \in R$. The ideal $Ra = \{ra : r \in R\}$ of R is called a principal ideal or, more precisely, the principal ideal generated by a and denoted by (a).*

For example, $R = (1)$, $\{0\} = (0)$, and for $R = \mathbb{Z}$, $\mathbb{Z}m = (m)$.

Theorem 12.17 *Every ideal of the ring $\mathbb{Z}$ is principal.*

Proof. Let I be any ideal of $\mathbb{Z}$. If $I = \{0\}$, then $I = (0)$ is principal. Let $I \neq \{0\}$ and m be the least positive integer contained in I. For any $a \in I$, by the division algorithm, there are $q, r \in \mathbb{Z}$ such that

$$a = qm + r \quad \text{and} \quad 0 \le r < m.$$

Then $r = a - qm \in I$. Since m is the least positive integer contained in I and $0 \le r < m$, we must have $r = 0$. Thus $a = qm$. Hence $I = \mathbb{Z}m = (m)$. $\square$

Similarly, we have

Theorem 12.18 *For any field F, every ideal of $F[x]$ is principal.*

Definition 12.9 *Let D be an integral domain. If every ideal of D is principal, D is called a principal ideal domain.*

For example, both $\mathbb{Z}$ and $F[x]$ are principal ideal domains. They are also unique factorization domains. More generally, we have

Theorem 12.19 *Every principal ideal domain is a unique factorization domain.*

This theorem will not be used in the sequel, so we will not give the proof. Interested readers may consult Jacobson (1985).

Let I be an ideal of R. For any $a, b \in R$, define

$$a \equiv b \,(\text{mod}\, I) \text{ if and only if } a - b \in I. \tag{12.6}$$

We read $a \equiv b$ (mod I) as "a is congruent to b modulo I". For $R = \mathbb{Z}$ and $I = (m)$ where $m \in R$, $a \equiv b$ (mod I) is the same as $a \equiv b$ (mod m) introduced in Section 5.1. Thus $a \equiv b$ (mod I) is a generalization of $a \equiv b$ (mod m). It can be proved in the same way that the relation $a \equiv b$ (mod I) is an equivalence relation in R. An equivalence class is called a *residue class* mod I. For any $a \in R$, the residue class mod I containing a will be denoted by $\bar{a}$. It is easy to show that

$$\bar{a} = a + I = \{a + x : x \in I\}$$

and that $\bar{a} \cap \bar{b} = \phi$ if and only if $a \not\equiv b$ (mod I). Thus we have a partition of R into distinct residue classes mod I. Denote the set of distinct residue classes mod I by R/I. Clearly, when $I = (m)$, then $R/I = R/(m)$, where $R/(m)$ is introduced in Section 5.1. For any $\bar{a}, \bar{b} \in R/I$, where $a, b \in R$, define the addition and multiplication of $\bar{a}$ and $\bar{b}$ by

$$\bar{a} + \bar{b} = \overline{a + b} \tag{12.7}$$

and

$$\bar{a} \cdot \bar{b} = \overline{ab}, \tag{12.8}$$

respectively. As in Section 5.1 we can show that these two definitions are well-defined and we have

Theorem 12.20 *Let R be any ring and I be an ideal of R. Then R/I is a ring with respect to the addition and multiplication defined by (12.7) and (12.8), respectively.*

Definition 12.10 *Let R be any ring and I be an ideal of R. The ring R/I is called the residue class ring of R modulo the ideal I.*

Clearly, the zero of R/I is I.

We already studied many concrete residue class rings in Chapter 5.

Parallel to groups, the isomorphism of rings can also be generalized to homomorphism of rings.

Definition 12.11 *Let R and R' be rings. A map ϕ from R to R'*

$$\phi : \quad R \quad \to \quad R'$$
$$a \quad \mapsto \quad \phi(a)$$

is called a homomorphism of rings, if it preserves the addition and multiplication, i.e.,

$$\phi(a + b) = \phi(a) + \phi(b)$$

and

$$\phi(ab) = \phi(a)\phi(b)$$

for all $a, b \in R$.

Let $\phi : R \to R'$ be a homomorphism of rings. Clearly, $\phi(0) = 0'$ is the zero of R' and $\phi(-a) = -\phi(a)$ for all $a \in R$. Moreover, if ϕ is surjective, $\phi(1) = 1'$ is the identity of R'.

Definition 12.12 *Let $\phi : \ R \to R'$ be a homomorphism of rings. Define*

$$\operatorname{Im} \phi = \{\phi(a) : a \in R\}$$

and

$$\operatorname{Ker} \phi = \{a \in R : \phi(a) = 0\}.$$

$\operatorname{Im} \phi$ is called the image of ϕ and $\operatorname{Ker} \phi$ the kernel of ϕ.

Parallel to Theorem 12.5, we have the fundamental theorem of homomorphisms of rings.

Theorem 12.21   (i) *Let R be any ring and I be an ideal of R. Then the map*

$$\phi : \quad R \quad \to \quad R/I$$
$$a \quad \mapsto \quad \bar{a} = a + I$$

is a homomorphism of rings, called the natural homomorphism of rings, and $\operatorname{Ker} \phi = I$, $\operatorname{Im} \phi = R/I$.

(ii) *Let $\phi : \ R \to R'$ be a homomorphism of rings. Then $\operatorname{Im} \phi$ is a subring of R' and $\operatorname{Ker} \phi$ is an ideal of R. Moreover, the map*

$$\overline{\phi} : \quad R/\operatorname{Ker}\phi \quad \to \quad \operatorname{Im} \phi$$
$$\bar{a} = a + \operatorname{Ker}\phi \quad \mapsto \quad \phi(a)$$

is well-defined and is an isomorphism of rings.

The proof is similar to that of Theorem 12.5 and is left as an exercise.

From Theorem 12.21, we deduce the following two isomorphism theorems of rings.

Theorem 12.22 *Let $\phi : R \to R'$ be a homomorphism of rings and assume that ϕ is surjective. Then*

(i) *If I is an ideal or a subring of R, then $\phi(I)$ is an ideal or a subring, respectively, of R'.*

(ii) *If I' is an ideal or a subring of R', then $\phi^{-1}(I')$ is an ideal or a subring, respectively, of R, $\phi^{-1}(I') \supseteq \operatorname{Ker}\phi$ and $\phi^{-1}(I')/\operatorname{Ker}\phi \simeq I'$.*

(iii) *If I is an ideal of R and let $\phi(I) = I'$, then the map*

$$
\begin{array}{ccc}
R/I & \to & R'/I' \\
a + I & \mapsto & \phi(a) + I'
\end{array}
$$

is a homomorphism of rings. Assume further that $I \supseteq \operatorname{Ker}\phi$, then $\phi^{-1}(\phi(I)) = I$ and the above map is an isomorphism of rings.

Theorem 12.23 *Let R be a ring, R_0 a subring of R, I an ideal of R, and $R_0 + I = \{r_0 + a : r_0 \in R_0, a \in I\}$. Then*

(i) *$R_0 + I$ is a subring of R, I is an ideal of $R_0 + I$, and $R_0 \cap I$ is an ideal of R_0.*

(ii) *The map*

$$
\begin{array}{ccc}
R_0 + I & \to & R_0/(R_0 \cap I) \\
r_0 + a & \mapsto & r_0 + (R_0 \cap I)
\end{array}
$$

is a homomorphism of rings, the kernel of this homomorphism is $\{I\}$, and consequently

$$
(R_0 + I)/I \simeq R_0/(R_0 \cap I).
$$

The proofs of Theorems 12.22 and 12.23 are left as exercises.

Now let us introduce maximal ideals, prime ideals, and primary ideals of a ring.

Definition 12.13 *Let R be any ring.*

(i) *An ideal M of R is called a maximal ideal if $M \neq R$ and there is no other ideal not equal to R and containing M properly.*

(ii) *An ideal P of R is called a prime ideal if $P \neq R$ and $ab \in P$ implies $a \in P$ or $b \in P$.*

(iii) *An ideal Q of R is called a primary ideal if $Q \neq R$ and $ab \in Q$ implies $a \in Q$ or $b^n \in Q$ for some positive integer n.*

Theorem 12.24 *Let R be any ring and I be an ideal of R.*

(i) *I is a maximal ideal if and only if R/I is a field.*

(ii) *I is a prime ideal if and only if R/I is an integral domain.*

Proof. (i) Suppose that I is a maximal ideal. Let $\bar{a} \in R/I$ and $\bar{a} \neq \bar{0}$, then $a \notin I$. The set of elements

$$M = \{ra + m : r \in R, m \in I\}$$

is an ideal of R, $I \subset M$, and $a \in M$ but $a \notin I$. Therefore $M = R$. For $1 \in R$, we have $1 = r_0 a + m_0$, where $r_0 \in R, m_0 \in I$. Then $\bar{1} = \overline{r_0}\bar{a}$, i.e., $\bar{a}$ has an inverse in R/I. Hence R/I is a field.

Conversely, suppose that R/I is a field. Let M be an ideal of R such that $M \supseteq I$ and $M \neq I$. Then there is an element $a \in M$ and $a \notin I$. Thus $\bar{a} \neq \bar{0}$ in R/I. There is an element $\bar{b} \in R/I$ such that $\bar{a}\bar{b} = \bar{1}$. Then $1 = ab + m$ where $m \in I$. Hence $1 \in M$ and $M = R$. That is, I is maximal.

(ii) Suppose that I is a prime ideal. Let $\bar{a}, \bar{b} \in R/I$ be such that $\bar{a}\bar{b} = \bar{0}$. Then $ab \in I$, which implies $a \in I$ or $b \in I$. If $a \in I$ then $\bar{a} = 0$, and if $b \in I$ then $\bar{b} = 0$. Therefore R/I is an integral domain.

Conversely, suppose that R/I is an integral domain. Let $a, b \in R$ be such that $ab \in I$. Then $\bar{a}\bar{b} = \bar{0}$ in R/I. If follows that $\bar{a} = \bar{0}$ or $\bar{b} = \bar{0}$. If $\bar{a} = \bar{0}$ then $a \in I$, and if $\bar{b} = \bar{0}$ then $b \in I$. Therefore I is a prime ideal. $\square$

Corollary 12.25 *All maximal ideals of a ring are prime ideals.*

But conversely, there are prime ideals which are not maximal. For example, let $R = F[x, y]$, then $I = (x)$ is a prime ideal, but not maximal. However we have

Theorem 12.26 *The ideal (m) of the ring $\mathbb{Z}$ is a prime ideal if and only if m is a prime element or $m = 0$, and if and only if $|m|$ is a prime number or $m = 0$. Moreover, all prime ideals $\neq (0)$ of $\mathbb{Z}$ are maximal.*

Proof. Assume that $m = 0$. Then $\mathbb{Z}/(m) = \mathbb{Z}/(0) = \mathbb{Z}$ is an integral domain. By Theorem 12.24 (ii) $(m) = (0)$ is a prime ideal of $\mathbb{Z}$.

Assume that $|m|$ is a prime number. Let $a, b \in \mathbb{Z}$ and $ab \in (m)$. Then $ab = r|m|$, where $r \in \mathbb{Z}$. By the fundamental theorem of arithmetic (Theorem 1.5) $|m| \,|\, a$ or $|m| \,|\, b$. If $|m| \,|\, a$, $a \in (m)$, and if $|m| \,|\, b$, $b \in (m)$. Therefore (m) is a prime ideal.

Assume that $|m|$ is not a prime number. Then $|m| = m_1 m_2$, where $1 < m_1, m_2 < |m|$. Thus $m_1 m_2 \in (m)$, but $m_1 \notin (m)$ and $m_2 \notin (m)$. Hence (m) is not a prime ideal.

The last statement is an immediate consequence of Example 3.7.   □

Similarly, we have

Theorem 12.27 *Let F be any field, x be an indeterminate, and $f(x) \in F[x]$. The ideal $(f(x))$ of the ring $F[x]$ is a prime ideal if and only if $f(x)$ is a prime element or $f(x) = 0$, and if and only if $f(x)$ is irreducible or $f(x) = 0$. Moreover, all prime ideals $\neq \{0\}$ of $F[x]$ are maximal.*

From Definition 12.13 we deduce immediately

Theorem 12.28 *All prime ideals of a ring are primary.*

But primary ideals are not necessarily prime. For example, let p be a prime number, then (p^2) is a primary ideal of $\mathbb{Z}$, but not prime (Exercise 12.12). More generally, for any integer $s > 1$, (p^s) is also a primary ideal, but not prime. Similarly, for any irreducible polynomial $f(x)$ in $F[x]$, where F is a field, $(f(x)^s)$ is primary for any integer $s > 1$, but not prime.

Let F be any field and $f(x) \in F[x]$. It is clear that $f(x)$ is an irreducible polynomial in $F[x]$ if and only if the principal ideal $(f(x))$ is maximal. Now for $f(x) \neq 0$, we define $f(x)$ to be a *primary polynomial* if the principal ideal $(f(x))$ is primary. Clearly, in $F[x]$ primary polynomials are powers of irreducible polynomials and conversely (Exercise 12.13).

More generally, let R be any ring and $f(x) \in R[x]$, $f(x) \neq 0$. We define $f(x)$ to be a *primary polynomial* if the principal ideal $(f(x))$ is a primary ideal.

Now we define the radical of an ideal as follows.

Definition 12.14 *let R be any ring and I an ideal of R. Define*

$$\sqrt{I} = \{a \in R : a^n \in I \ for \ some \ integer \ n \geq 1\}$$

and call $\sqrt{I}$ the radical of I.

Theorem 12.29 $\sqrt{I}$ *is an ideal of R.*

The proof of Theorem 12.29 is left as an exercise.

Finally let us introduce the direct sum decomposition of rings.

Definition 12.15 *Let R be a ring, $R_1, R_2, \ldots, R_r$ all be subrings of R. R is said to be the direct sum of $R_1, R_2, \ldots, R_r$ if the following conditions are fulfilled.*

(i) *Each R_i is an ideal of R, $i = 1, 2, \ldots, r$.*

(ii) *$R = R_1 + R_2 + \cdots + R_r$, where $R_1 + R_2 + \cdots + R_r = \{a_1 + a_2 + \cdots + a_r : a_i \in R_i, \ i = 1, 2, \ldots, r\}$.*

(iii) *$R_i \cap (R_1 + \cdots + R_{i-1} + R_{i+1} + \cdots + R_r) = \{0\}$.*

Then we write $R = R_1 \dotplus R_2 \dotplus \cdots \dotplus R_r$ and say that $R_1 \dotplus R_2 \dotplus \cdots \dotplus R_r$ is a direct sum decomposition of R.

We also have

Theorem 12.30 *Let R be a ring, $R_1, R_2, \ldots, R_r$ all be subrings of R. Then R is the direct sum of $R_1, R_2, \ldots, R_r$ if and only if*

(i) *Each R_i is an ideal of R, $i = 1, 2, \ldots, r$.*

(ii) *$R = R_1 + R_2 + \cdots + R_r$.*

(iii) *Every element a of R can be written uniquely as a sum $a = a_1 + a_2 + \cdots + a_r$, where $a_i \in R_i$, $i = 1, 2, \ldots, r$.*

Using the concept of the direct sum decomposition of rings, we can formulate the Chinese Remainder Theorem (Theorem 2.26) as follows.

Theorem 12.31 *Let $m_1, m_2, \ldots, m_r$ be r integers greater than 1, which are pairwise coprime. Let $m = m_1 m_2 \cdots m_r$ and $\widehat{m}_i = m/m_i$ for each $i = 1, 2, \ldots, r$. Then there are integers $c_1, c_2, \ldots, c_r$ such that $c_1 \widehat{m}_1 + c_2 \widehat{m}_2 + \cdots + c_r \widehat{m}_r = 1$, and for each $i = 1, 2, \ldots, r$, the map*

$$\begin{array}{rcl}
\phi_i \quad \mathbb{Z}/(m_i) & \to & \mathbb{Z}/(m) \\
a + (m_i) & \mapsto & ac_i \widehat{m}_i + (m) \quad \text{for all } a \text{ with } 0 \le a \le m_i - 1
\end{array}$$

is an isomorphism of the ring $\mathbb{Z}/(m_i)$ into $\mathbb{Z}/(m)$. Moreover,

$$\mathbb{Z}/(m) = \phi_1(\mathbb{Z}/(m_1)) \dotplus \phi_2(\mathbb{Z}/(m_2)) \dotplus \cdots \dotplus \phi_r(\mathbb{Z}/(m_r)).$$

Similarly, Theorem 4.13 can be formulated as

Theorem 12.32 *Let F be a field and $f_1(x), f_2(x), \ldots, f_r(x)$ be r polynomials of degree ≥ 1 in $F[x]$, which are pairwise coprime. Let $f(x) =$*

$f_1(x)f_2(x)\cdots f_r(x)$ and $\widehat{f_i}(x) = f(x)/f_i(x)$ for each $i = 1, 2, \ldots, r$. Then there are polynomials $c_1(x), c_2(x), \ldots, c_r(x) \in F[x]$ such that

$$c_1(x)\widehat{f_1}(x) + c_2(x)\widehat{f_2}(x) + \cdots + c_r(x)\widehat{f_r}(x) = 1,$$

and for each $i = 1, 2, \ldots, r$, the map

$$\phi_i \quad F[x]/(f_i(x)) \ \to \quad F[x]/(f(x))$$
$$g(x) + (f_i(x)) \mapsto g(x)c_i(x)\widehat{f_i}(x) + (f(x)),$$

where $g(x) \in F[x]$ with $\deg g(x) < \deg f_i(x)$, is an isomorphism of the ring $F[x]/(f_i(x))$ into the ring $F[x]/(f(x))$. Moreover,

$$F[x]/(f(x)) = \phi_1(F[x]/(f_1(x)))\dot{+}\phi_2(F[x]/(f_2(x)))\dot{+}\cdots\dot{+}\phi_r(F[x]/(f_r(x))).$$

The proofs of Theorems 12.30 - 12.32 are left as exercises.

12.4 Modules

As a generalization of vector spaces over fields we have the concept of modules over rings.

Definition 12.16 *Let R be a commutative ring with identity 1. An R-module is an abelian group M together with a map*

$$R \times M \to M$$
$$(a, x) \mapsto ax$$

satisfying the following properties:

$M1$ $a(x + y) = ax + ay$,

$M2$ $(a + b)x = ax + bx$,

$M3$ $(ab)x = a(bx)$,

$M4$ $1x = x$

for $x, y \in M$, $a, b \in R$.

$M1, M2, M3$, and $M4$ are called the *axioms of R-modules*.

Example 12.12 Any abelian group M can be regarded as a $\mathbb{Z}$-module, if for any $x \in M$ and any positive integer n we define nx to be the sum of n x's, for any negative integer $-n$ we define $(-n)x = -(nx)$, and we define $0x = 0$, where the 0 on the left-hand side is the integer 0 and the 0 on the right-hand side is the zero element of M.

Example 12.13 A vector space V over a field F is an F-module. Moreover, let T be a linear transformation of V, i.e., $T\colon v \to T(v)$ is a map from V to itself satisfying

$$T(u + v) = T(u) + T(v) \text{ and } T(av) = aT(v)$$

for $u, v \in V$ and $a \in F$. Let $F[x]$ be the polynomial ring in an indeterminate x over the field F. For any $f(x) = a_0 + a_1 x + a_2 x^2 + \cdots + a_n x^n \in F[x]$, define a map

$$\begin{aligned} \mathbb{F}_q[x] \times V &\to V \\ (f(x), v) &\mapsto f(x)v, \end{aligned}$$

where

$$f(x)v = a_0 v + a_1 T(v) + a_2 T^2(v) + \cdots + a_n T^n(v)$$

for all $v \in V$. It is easy to verify that V is an $F[x]$-module.

Example 12.14 Let $\mathbb{F}_{q^n}$ and $\mathbb{F}_q$ be finite fields with q^n and q elements, respectively, and assume $\mathbb{F}_{q^n} \supset \mathbb{F}_q$. Let σ be the Frobenius automorphism of $\mathbb{F}_{q^n}$ over $\mathbb{F}_q$. Define a map

$$\begin{aligned} \mathbb{F}_q[x] \times \mathbb{F}_{q^n} &\to \mathbb{F}_{q^n} \\ (f(x), \beta) &\mapsto f(x)\beta, \end{aligned}$$

where $f(x) = a_0 + a_1 x + a_2 x^2 + \cdots + a_r x^r$, $a_i \in \mathbb{F}_q$ $(i = 0, 1, 2, \ldots, r)$ and

$$f(x)\beta = a_0 \beta + a_1 \sigma(\beta) + a_2 \sigma^2(\beta) + \cdots + a_r \sigma^r(\beta).$$

It is easy to verify that $\mathbb{F}_{q^n}$ is an $\mathbb{F}_q[x]$-module.

Clearly, the Frobenius automorphism σ of $\mathbb{F}_{q^n}$ over $\mathbb{F}_q$ is a linear transformation of the vector space $\mathbb{F}_{q^n}$ over $\mathbb{F}_q$. Hence Example 12.14 is a special case of Example 12.13.

Submodules, quotient modules, homomorphism, isomorphism, and automorphism of modules can be defined as for groups; the fundamental theorem of homomorphisms of modules and the isomorphism theorems of modules can be proved in a similar way. But we will not repeat the details. For latter purpose we need the following definition.

Definition 12.17 *Let R be a commutative ring with identity, M be an R-module, and $x_1, x_2, \ldots, x_n \in M$. If every element $x \in M$ can be expressed uniquely in the form*

$$x = a_1 x_1 + a_2 x_2 + \cdots + a_n x_n,$$

where $a_1, a_2, \ldots a_n \in R$, then M is called a free R-module of rank n and $\{x_1, x_2, \ldots, x_n\}$ is called a free basis of M.

Example 12.15 A vector space of finite dimension n over a field F is a free F-module of rank n.

Example 12.16 Let $R^n = \{{}^t(a_1, a_2, \ldots, a_n) : a_i \in R\}$ and define

$${}^t(a_1, a_2, \ldots, a_n) + {}^t(b_1, b_2, \ldots, b_n) = {}^t(a_1 + b_1, a_2 + b_2, \ldots, a_n + b_n),$$

$$a\,{}^t(a_1, a_2, \ldots, a_n) = {}^t(aa_1, aa_2, \ldots, aa_n)$$

for $a \in R$ and ${}^t(a_1, a_2, \ldots, a_n)$, ${}^t(b_1, b_2, \ldots, b_n) \in R^n$. Then R^n is a free R-module of rank n.

Theorem 12.33 *The rank of a free R-module M is uniquely determined by M.*

Proof. Let $\{e_i : 1 \leq i \leq n\}$ and $\{f_j : 1 \leq j \leq m\}$ be two free bases of M. Then we have

$$f_j = \sum_{i=1}^{n} a_{ij} e_i, \quad e_i = \sum_{j=1}^{m} b_{ji} f_j,$$

where $a_{ij}, b_{ij} \in R$. Substitution gives

$$f_j = \sum_{i=1}^{n} \sum_{j'=1}^{m} a_{ij} b_{j'i} f_{j'},$$

$$e_i = \sum_{j=1}^{m} \sum_{i'=1}^{n} b_{ji} a_{i'j} e_{i'}.$$

Since both $\{f_1, f_2, \ldots, f_m\}$ and $\{e_1, e_2, \ldots, e_n\}$ are bases, we have

$$\sum_{i=1}^{n} a_{ij} b_{j'i} = \delta_{jj'}, \tag{12.9}$$

$$\sum_{j=1}^{m} b_{ji} a_{i'j} = \delta_{ii'}. \tag{12.10}$$

Now suppose $m > n$. Let

$$A = \begin{pmatrix} a_{11} & a_{12} & \cdots & a_{1m} \\ \cdots & \cdots & \cdots & \cdots \\ a_{n1} & a_{n2} & \cdots & a_{nm} \\ 0 & 0 & \cdots & 0 \\ \cdots & \cdots & \cdots & \cdots \\ 0 & 0 & \cdots & 0 \end{pmatrix}$$

and

$$B = \begin{pmatrix} b_{11} & \cdots & b_{1n} & 0 \cdots 0 \\ b_{21} & \cdots & b_{2n} & 0 \cdots 0 \\ \cdots & \cdots & \cdots & \cdots \\ b_{m1} & \cdots & b_{mn} & 0 \ldots 0 \end{pmatrix}$$

both be $m \times m$ matrices. Then (12.9) is equivalent to $BA = I^{(m)}$. Since R is commutative, this implies $AB = I^{(m)}$ (Exercise 12.20). However, by (12.10)

$$AB = \begin{pmatrix} I^{(n)} & 0 \\ 0 & 0 \end{pmatrix}.$$

$\square$

Definition 12.18 *Let R be a commutative ring with identity and M be an R-module. For any $x \in M$, let*

$$\mathrm{ann}(x) = \{r \in R : rx = 0\}.$$

Clearly, $\mathrm{ann}(x)$ is an ideal of R. $\mathrm{ann}(x)$ is called the annihilator of x or the order ideal of x. If R is a principal ideal domain, there is an element $b \in R$ such that $\mathrm{ann}(x) = (b)$ and b is uniquely determined up to the units of R. b is called the order of x and is denoted by $\mathrm{ord}\,(x)$.

In Example 12.12 an abelian group M is regarded as a $\mathbb{Z}$-module. For an element $x \in M$, if we choose b to be the least positive integer such that $\mathrm{ann}\,(x) = (b)$, then b is the order of the group element $x \in M$.

For the $\mathbb{F}_q[x]$-module $\mathbb{F}_{q^n}$ introduced in Example 12.14, let σ be the Frobenius automorphism of $\mathbb{F}_{q^n}$ over $\mathbb{F}_q$ and α be an element of $\mathbb{F}_{q^n}$, then $(x^n - 1)\alpha = \sigma^n(\alpha) - \alpha = 0$. Thus $x^n - 1 \in \mathrm{ann}(\alpha)$ for every $\alpha \in \mathbb{F}_{q^n}$. Moreover, α is a normal basis generator of $\mathbb{F}_{q^n}$ over $\mathbb{F}_q$ if and only if $\mathrm{ann}(\alpha) = (x^n - 1)$, i.e., $\mathrm{ord}\,(\alpha) = x^n - 1$ (Exercise 12.20). More generally, we have

Theorem 12.34 *If we regard $\mathbb{F}_{q^n}$ as an $\mathbb{F}_q[x]$-module, then for any $\beta \in \mathbb{F}_{q^n}$, there is a uniquely determined monic polynomial $b(x)$ in $\mathbb{F}_q[x]$ such that $\mathrm{ann}(\beta) = (b(x))$, $\mathrm{ord}\,(\beta) = b(x)$ and $b(x)|(x^n - 1)$.*

Proof. We know that $\mathrm{ann}\,(\beta)$ is an ideal of $\mathbb{F}_q[x]$. Since $\mathbb{F}_q[x]$ is a unique factorization domain, there is a uniquely determined monic polynomial $b(x) \in \mathbb{F}_q[x]$ such that $\mathrm{ann}\,(\beta) = (b(x))$, then $\mathrm{ord}(\beta) = b(x)$. Dividing $x^n - 1$ by $b(x)$, we obtain $x^n - 1 = q(x)b(x) + r(x)$, where $q(x), r(x) \in \mathbb{F}_q[x]$ and $\deg r(x) < \deg b(x)$. Then $0 = (x^n - 1)\beta = q(x)b(x)\beta + r(x)\beta = r(x)\beta$, which implies $r(x) \in \text{ann}\,\beta = (b(x))$. But $\deg r(x) < \deg b(x)$, we must have $r(x) = 0$ and, hence, $b(x)|(x^n - 1)$. □

Theorem 2.21 can be generalized as

Lemma 12.35 *Lt* $\beta \in \mathbb{F}_{q^n}^*$ *and* $h(x) \in \mathbb{F}_q[x], h(x) \neq 0$. *Then*

$$\text{ord}\,(h(x)\beta) = \frac{\text{ord}\,(\beta)}{\gcd\,(\text{ord}\,(\beta),\,h(x))}.$$

The proof is similar to that of Theorem 2.21 and, hence, is omitted.

Definition 12.19 *Let R be a commutative ring with identity and M be an R-module. Let N be a submodule of M. If there exists an element $x \in N$ such that*
$$N = \{rx : r \in R\},$$
N is called a cyclic submodule and x is called a generator of N. In this case we write $N = Rx$. In particular, if M is a cyclic submodule, it is called a cyclic module.

From now on we concentrate our study on the $\mathbb{F}_q[x]$-module $\mathbb{F}_{q^n}$ introduced in Example 12.14, which will be used in the simpler proof of Theorem 8.34 below.

Theorem 12.36 *$\mathbb{F}_{q^n}$ is a cyclic $\mathbb{F}_q[x]$-module and any normal basis generator α of $\mathbb{F}_{q^n}$ over $\mathbb{F}_q$ is a generator of the cyclic module $\mathbb{F}_{q^n}$. More generally, every submodule of $\mathbb{F}_{q^n}$ is also cyclic and has a generator of the form $g(x)\alpha$, where $g(x) \in \mathbb{F}_q(x)$ and $g(x)|(x^n - 1)$.*

Proof. By Theorem 8.15, let α be a normal basis generator of $\mathbb{F}_{q^n}$ over $\mathbb{F}_q$, then
$$\mathbb{F}_{q^n} = \{a_0\alpha + a_1\sigma(\alpha) + \cdots + a_{n-1}\sigma^{n-1}(\alpha) : a_i \in \mathbb{F}_q\}.$$
Since $\sigma^n = 1$, we have
$$\mathbb{F}_{q^n} = \{f(x)\alpha : f(x) \in \mathbb{F}_q[x]\} = \mathbb{F}_q[x]\alpha.$$
Therefore $\mathbb{F}_{q^n}$ is a cyclic module with α as a generator.

Let N be a submodule. Then all elements of N are of the form $f(x)\alpha$. Let $g(x)\alpha$ be an element of N such that $\deg g(x)$ is minimum. Clearly, $\mathbb{F}_q[x]g(x)\alpha \subset N$. Let $f(x)\alpha$ be any element of N. Dividing $f(x)$ by $g(x)$, we have $f(x) = q(x)g(x) + r(x)$, where $q(x), r(x) \in \mathbb{F}_q[x]$ and $\deg r(x) < \deg g(x)$. Then $r(x)\alpha = f(x)\alpha - q(x)g(x)\alpha \in N$. Therefore $r(x) = 0$ and $f(x)\alpha = q(x)g(x)\alpha \in \mathbb{F}_q[x]g(x)\alpha$. Hence $N = \mathbb{F}_q[x]g(x)\alpha$ is cyclic and $g(x)\alpha$

is a generator of N. Moreover, dividing $x^n - 1$ by $g(x)$, we have $x^n - 1 = q_1(x)g(x) + r_1(x)$, where $q_1(x), r_1(x) \in \mathbb{F}_q[x]$ and $\deg r_1(x) < \deg g(x)$. Then $r_1(x)\alpha = (x^n - 1)\alpha - q_1(x)g(x)\alpha = -q_1(x)g(x)\alpha \in N$. Therefore $r_1(x) = 0$ and $g(x)|(x^n - 1)$. $\square$

Let $f(x)$ be a monic divisor of $x^n - 1$. Define

$$M_f = \{\beta \in \mathbb{F}_{q^n} : f(x)\beta = 0\}$$

and

$$M'_f = \{\beta \in \mathbb{F}_{q^n} : \mathrm{ord}\,(\beta) = f(x)\}.$$

For example, $M_{x^n-1} = \mathbb{F}_{q^n} = \mathbb{F}_q[x]\alpha$ for any normal basis generator α of $\mathbb{F}_{q^n}$ over $\mathbb{F}_q$. and M'_{x^n-1} consists of all normal basis generators of $\mathbb{F}_{q^n}$ over $\mathbb{F}_q$. In general, we have

Theorem 12.37 *Let $x^n - 1 = f(x)g(x)$ in $\mathbb{F}_q[x]$, where both $f(x)$ and $g(x)$ are monic, and let α be a normal basis generator of $\mathbb{F}_{q^n}$ over $\mathbb{F}_q$. Then M_f is a submodule of $\mathbb{F}_{q^n}$ and $M_f = \mathbb{F}_q[x]g(x)\alpha$, Moreover, $M'_f \neq \emptyset$ and M'_f consists of all generators of M_f.*

Proof. Clearly, M_f is a submodule of $\mathbb{F}_{q^n}$ and $\mathbb{F}_q[x]g(x)\alpha \subseteq M_f$. We are going to prove that $M_f \subseteq \mathbb{F}_q[x]g(x)\alpha$. Let $h(x)\alpha \in M_f$, where $h(x) \in \mathbb{F}_q[x]$, then $f(x)h(x)\alpha = 0$. Since $\mathrm{ord}\,(\alpha) = x^n - 1$, we have $(x^n - 1)|f(x)h(x)$. From the hypothesis $x^n - 1 = f(x)g(x)$ we deduce $g(x)|h(x)$. Let $h(x) = k(x)g(x)$ then $h(x)\alpha = k(x)g(x)\alpha \in \mathbb{F}_q[x]g(x)\alpha$. Thus $M_f \subseteq \mathbb{F}_q[x]g(x)\alpha$. Therefore $M_f = \mathbb{F}_q[x]g(x)\alpha$.

We know that $g(x)\alpha \in M_f$ and $\mathrm{ord}(\alpha) = x^n - 1 = f(x)g(x)$. It follows that $\mathrm{ord}(g(x)\alpha) = f(x)$. Thus $g(x)\alpha \in M'_f$. Hence $M'_f \neq \emptyset$.

It is clear that generators of M_f are elements of M'_f. Conversely, let $\beta \in M'_f$, then $\beta \in M_f$. But $M_f = \mathbb{F}_q[x]g(x)\alpha$, we can assume $\beta = k(x)g(x)\alpha$, where $k(x) \in \mathbb{F}_q[x]$. Since β, $g(x)\alpha \in M'_f$, we have $\mathrm{ord}(\beta) = \mathrm{ord}(g(x)\alpha) = f(x)$. By Lemma 12.35

$$f(x) = \mathrm{ord}(\beta) = \mathrm{ord}(k(x)g(x)\alpha) = \frac{\mathrm{ord}(g(x)\alpha)}{\gcd(\mathrm{ord}(g(x)\alpha, k(x))} = \frac{f(x)}{\gcd(f(x), k(x))},$$

thus $\gcd(f(x), k(x)) = 1$ and there exist $a(x), b(x) \in \mathbb{F}_q[x]$ such that $a(x)f(x) + b(x)k(x) = 1$. Then $g(x) = a(x)f(x)g(x) + b(x)k(x)g(x)$ and $g(x)\alpha = b(x)k(x)g(x)\alpha = b(x)\beta$. It follows that $M_f = \mathbb{F}_q[x]g(x)\alpha = \mathbb{F}_q[x]b(x)\beta \subset \mathbb{F}_q[x]\beta = \mathbb{F}_q[x]k(x)g(x)\alpha \subset \mathbb{F}_q[x]g(x)\alpha$. Thus $M_f = \mathbb{F}_q[x]\beta$ and β is a generatot of M_f. This proves that all elements of M'_f are genmerators of M_f. $\square$

Corollary 12.38 *Let $f(x)$ and $g(x)$ both be monic divisors of $x^n - 1$. Then $M_f \subset M_g$ if and only if $f(x) \mid g(x)$.*

Proof. Suppose $f(x) \mid g(x)$. We can assume $x^n - 1 = k_1(x)g(x)$ and $g(x) = k_2(x)f(x)$, where $k_1(x), k_2(x) \in \mathbb{F}_q[x]$. Then $x^n - 1 = k_1(x)k_2(x)f(x)$. Let α be a normal basis generator of $\mathbb{F}_{q^n}$ over $\mathbb{F}_q$. By Theorem 12.37,

$$M_f = \mathbb{F}_q[x]k_1(x)k_2(x)\alpha \subset \mathbb{F}_q[x]k_1(x)\alpha = M_g.$$

Conversely, suppose $M_f \subset M_g$. By Theorem 12.37 let $\beta \in M_f'$, then $\beta \in M_f \subset M_g$, which implies $g(x)\beta = 0$. Since $\mathrm{ord}\,(\beta) = f(x)$, $f(x) \mid g(x)$. $\square$

The following simple lemma will be used later.

Lemma 12.39 *Let $f(x)$ and $g(x)$ both be monic divisors of $x^n - 1$. If $M_f' \cap M_g' \neq \emptyset$, (in particular, $M_f' \subset M_g'$), then $f(x) = g(x)$.*

Proof. Let $\beta \in M_f' \cap M_g'$. Then $\mathrm{ord}\,(\beta) = f(x)$ and $\mathrm{ord}\,(\beta) = g(x)$. Therefore $f(x) = g(x)$. $\square$

Regarding $\mathbb{F}_{q^n}$ as an $\mathbb{F}_q[x]$-module, for any polynomial $h(x) \in \mathbb{F}_q[x]$ we have a map $\tau_h : \mathbb{F}_{q^n} \to \mathbb{F}_{q^n}$ defined by

$$\tau_h(\beta) = h(x)\beta \quad \text{for all } \beta \in \mathbb{F}_{q^n}.$$

τ_h is called the *polynomial map* associated with the polynomial $h(x)$, or, simply, a polynomial map. Let $f(x) = (x^n - 1)/h(x)$. It is easy to check that $\tau_h(\mathbb{F}_{q^n}) \subset M_f$.

Let σ be the Frobenius automorphism of $\mathbb{F}_{q^n}$ over $\mathbb{F}_q$. We have the trace map $\mathrm{Tr} : \mathbb{F}_{q^n} \to \mathbb{F}_q$, which is defined by

$$\mathrm{Tr}\,(\beta) = \beta + \sigma(\beta) + \sigma^2(\beta) + \cdots + \sigma^{n-1}(\beta) \quad \text{for all } \beta \in \mathbb{F}_{q^n},$$

(cf. Definition 7.5). Thus

$$\mathrm{Tr}\,(\beta) = (1 + \sigma + \sigma^2 + \cdots + \sigma^{n-1})(\beta) = \tau_{1+x+x^2+\cdots+x^{n-1}}(\beta).$$

Therefore the trace map is a polynomial map.

For monic divisors $f(x)$ and $g(x)$ of $x^n - 1$, satisfying $f(x) \mid g(x)$, let $h(x) = g(x)/f(x)$, then we have the polynomial map $\tau_h : \mathbb{F}_{q^n} \to \mathbb{F}_{q^n}$. For any $\beta \in M_g$, we have $g(x)\beta = 0$. Then $f(x)\tau_h(\beta) = f(x)h(x)\beta = g(x)\beta = 0$, i.e., $\tau_h(\beta) \in M_f$. Therefore $\tau_h(M_g) \subset M_f$. Denote the restriction of τ_h to M_g by τ_f^g. Then $\tau_f^g(M_g) \subset M_f$. Moreover, we have

Lemma 12.40 *Let $f(x)$ and $g(x)$ be monic divisors of $x^n - 1$, $f(x) \mid g(x)$, and $h(x) = g(x)/f(x)$. Then*

$$\tau_f^g : M_g \to M_f$$

is a homomorphism of $\mathbb{F}_q[x]$-modules, $\operatorname{Ker} \tau_f^g = M_h$, τ_f^g is surjective and $\tau_f^g(M_g') \subset M_f'$.

Proof. It is clear that τ_f^g is a homomorphism of $\mathbb{F}_q[x]$-modules. Let us prove that $\operatorname{Ker} \tau_f^g = M_h$. Assume $x^n - 1 = k(x)g(x)$, where $k(x) \in \mathbb{F}_q[x]$. Then $x^n - 1 = k(x)h(x)f(x)$. Let α be a normal basis generator of $\mathbb{F}_{q^n}$ over $\mathbb{F}_q$. Then $M_h = \mathbb{F}_q[x]k(x)f(x)\alpha$, $M_g = \mathbb{F}[x]k(x)\alpha$ and $M_f = \mathbb{F}[x]k(x)h(x)\alpha$. Clearly, $M_h \subset \operatorname{Ker} \tau_f^g$. Conversely, let $\beta \in M_g$ and $\beta \in \operatorname{Ker} \tau_f^g$. Then $\beta = a(x)k(x)\alpha$ for some $a(x) \in \mathbb{F}[x]$ and $h(x)\beta = 0$. Thus $a(x)k(x)h(x)\alpha = 0$. Since $\operatorname{ord}(\alpha) = x^n - 1$, $x^n - 1 = k(x)h(x)f(x) \mid a(x)k(x)h(x)$, which implies $a(x) = b(x)f(x)$ for some $b(x) \in \mathbb{F}_q[x]$. Then $\beta = b(x)f(x)k(x)\alpha \in M_h$. Therefore $\operatorname{Ker} \tau_f^g \subset M_h$. Hence $\operatorname{Ker} \tau_f^g = M_h$.

Then we prove that τ_f^g is surjective. Let $\beta \in M_f$. Then β can be expressed as $\beta = a(x)k(x)h(x)\alpha$ for some $a(x) \in \mathbb{F}_q[x]$. Clearly, $a(x)k(x)\alpha \in M_g$ and $\tau_f^g(a(x)k(x)\alpha) = \beta$.

It remains to prove that $\tau_f^g(M_g') \subset M_f'$. Let $\beta \in M_g'$. Then $\operatorname{ord}(\beta) = g(x)$ and $\tau_f^g(\beta) = h(x)\beta$. By Lemma 12.35,

$$\operatorname{ord}(h(x)\beta) = \operatorname{ord}(\beta)/\gcd(\operatorname{ord}(\beta), h(x)) = g(x)/\gcd(g(x), h(x)) = f(x).$$

Thus $\tau_f^g(\beta) \in M_f'$. Therefore $\tau_f^g(M_g') \subset M_f'$. $\qquad\square$

Note that

$$\operatorname{Tr} = \tau_{x-1}^{x^n-1}.$$

We call τ_f^g the *generalized trace map*.

We are interested in studying the originals of the generalized trace maps. At first we discuss two extremal cases in the following two lemmas.

Let $f(x) \in \mathbb{F}_q[x]$. Denote by $\nu(f)$ the set of pairwise coprime monic irreducible (or, prime) factors of $f(x)$.

Lemma 12.41 *Let $f(x)$ and $g(x)$ both be monic divisors of $x^n - 1$ in $\mathbb{F}_q[x]$ and $f(x) \mid g(x)$. Assume $\nu(f) = \nu(g)$. Then $(\tau_f^g)^{-1}(M_f') = M_g'$.*

Proof. Let $h(x) = g(x)/f(x)$, then $h(x) \in \mathbb{F}_q[x]$ and

$$\begin{aligned} \tau_f^g : \quad & M_g \to M_f \\ & \beta \mapsto h\beta \quad \text{for all } \beta \in M_g. \end{aligned}$$

By Lemma 12.40 $\tau_f^g(M_g') \subset M_f'$. Therefore $M_g' \subset (\tau_f^g)^{-1}(M_f')$.

Then we prove that $M_g' \supset (\tau_f^g)^{-1}(M_f')$ under the assumption of the present lemma. We apply induction on the number of irreducible factors of $h = g/f$. Consider first the case when h is irreducible. Let $\gamma \in M_f'$. By Lemma 12.40 τ_f^g is surjective, thus there is an element $\beta \in M_g$ such that $\gamma = \tau_f^g(\beta)$, then $\gamma = h\beta$. From $h \mid g$, $f \mid g$, and $\nu(f) = \nu(g)$ we deduce $h \mid f$. By Lemma 12.35,

$$\operatorname{ord}(\beta) = \operatorname{ord}(h\beta)/\gcd(\operatorname{ord}(\beta), h) = \operatorname{ord}(\gamma)\gcd(\operatorname{ord}(\beta), h).$$

From $\gamma \in M_f'$, i.e., $\operatorname{ord}(\gamma) = f$, we deduce $f \mid \operatorname{ord}(\beta)$. Then from $h \mid f$ and $f \mid \operatorname{ord}(\beta)$ we deduce $h \mid \operatorname{ord}(\beta)$, thus $\gcd(\operatorname{ord}(\beta), h) = h$. It follows that $\operatorname{ord}(\beta) = fh = g$, i.e., $\beta \in M_g'$. Therefore $(\tau_f^g)^{-1}(\gamma) \subset M_g'$. Hence $(\tau_f^g)^{-1}(M_f') \subset M_g'$.

Now let h be a product of $m > 1$ irreducible factors. Assume that our assertion is true when h is a product of $m-1$ irreducible factors. Let $h = lh_1$, where l is irreducible and h_1 is a product of $m - 1$ irreducible factors. By induction hypothesis, $(\tau_{lf}^g)^{-1}(M_{lf}') \subset M_g'$, and by the case $m = 1$ proved above, $(\tau_f^{lf})^{-1}(M_f') \subset M_{lf}'$. But $\tau_f^{lf} \circ \tau_{lf}^g = \tau_f^g$. Therefore

$$(\tau_f^g)^{-1}(M_f') = (\tau_{lf}^g)^{-1} \circ ((\tau_f^{lf})^{-1}(M_f')) = (\tau_{lf}^g)^{-1}(M_{lf}') = M_g'.$$

From $M_g' \subset (\tau_f^g)^{-1}(M_f')$ and $M_g' \supset (\tau_f^g)^{-1}(M_f')$ we conclude $(\tau_f^g)^{-1}(M_f') = M_g'$. $\qquad\square$

Lemma 12.42 *Let $f(x)$ and $g(x)$ both be monic divisors of $x^n - 1$ in $\mathbb{F}_q[x]$ and $f(x) \mid g(x)$. Let $h(x) = g(x)/f(x)$ and assume $\gcd(h(x), f(x)) = 1$. Then*

$$(\tau_f^g)^{-1}(M_f') = \bigcup_{a \mid h} M_{fa}' = M_f' + M_h.$$

Proof. First we assert that τ_f^g induces a module automorphism of M_f. By Lemma 12.40 τ_f^g is a module homomorphism from M_g to M_f, and by Lemma 12.38 $M_f \subset M_g$, thus τ_f^g induces a module homomorphism on M_f. Let $\beta \in M_f$, i.e. $f\beta = 0$. Suppose $\tau_f^g(\beta) = 0$, i.e., $h\beta = 0$. Since $\gcd(h, f) = 1$, there are polynomials $a, b \in \mathbb{F}_q[x]$ such that $ah + bf = 1$. Then $\beta = ah\beta + bf\beta = 0$. This proves that τ_f^g induces an injective map from M_f to itself. Since M_f is finite, τ_f^g induces an module automorphism of M_f.

Let $\beta \in M_f'$, i.e., $\operatorname{ord}(\beta) = f$. By Lemma 12.35

$$\operatorname{ord}(h\beta) = \operatorname{ord}(\beta)/\gcd(\operatorname{ord}(\beta), h) = f/\gcd(f, h) = f.$$

Therefore $h\beta \in M'_f$. Consequently, τ_f^g induces a permutation on M'_f. Then for each $\beta \in M'_f$, there is a $\beta' \in M'_f$ such that $\tau_f^g(\beta') = \beta$, thus $\beta' \in (\tau_f^g)^{-1}(\beta)$. By Lemma 12.40 Ker $\tau_f^g = M_h$. Therefore $(\tau_f^g)^{-1}(\beta) = \beta' + M_h$. Hence

$$(\tau_f^g)^{-1}(M'_f) = \bigcup_{\beta \in M'_f} (\beta' + M_h) = M'_f + M_h.$$

Clearly, $M_h = \bigcup_{a|h} M'_a$. Since $\gcd(h, f) = 1$, $M'_f + M'_a \subset M'_{fa}$ for every $a \mid h$. Conversely, suppose $\beta \in M'_{fa}$, where $a \mid h$. Then $\gcd(a, f) = 1$. There are polynomials b, $c \in \mathbb{F}_q[x]$ such that $ba + cf = 1$. Thus $\beta = ba\beta + cf\beta$. It is easy to check that $ba\beta \in M'_f$ and $cf\beta \in M'_a$. Therefore $M'_{fa} = M'_f + M'_a$. Consequently, $\bigcup_{a|h} M'_{fa} = \bigcup_{a|h} (M'_f + M'_a) = M'_f + M_h$. □

From the above two lemmas we deduce

Theorem 12.43 *Let $f(x)$ and $g(x)$ both be monic divisors of $x^n - 1$ in $\mathbb{F}_q[x]$ and $f(x) \mid g(x)$. Then there are polynomials $\hat{f}(x)$, $\bar{g}(x) \in \mathbb{F}_q[x]$ such that $g(x) = \hat{f}(x)\bar{g}(x)$, $\nu(f(x)) = \nu(\hat{f}(x))$, and $\gcd(\bar{g}(x), f(x)) = 1$. Furthermore,*

$$(\tau_f^g)^{-1}(M'_f) = \bigcup_{a|\bar{g}} M'_{fa} = M'_f + M_{\bar{g}}. \qquad (12.11)$$

Moreover, if τ_f^g is surjective, then for every $\beta \in M'_f$ there exists a $\gamma \in M'_g$ such that $\tau_f^g(\gamma) = \beta$.

Proof. Since $f \mid g$, we suppose $f = p_1^{e_1} \cdots p_r^{e_r}$ and $g = p_1^{e'_1} \cdots p_r^{e'_r} p_{r+1}^{e_{r+1}} \cdots p_{r+s}^{e_{r+s}}$, where $p_1, \ldots, p_{r+s}$ are pairwise distinct irreducible polynomials in $\mathbb{F}_q[x]$, $e_1, \ldots, e_{r+s}, e'_1, \ldots, e'_r$ are positive integers, and $e'_i \geq e_i$ for $i = 1, \ldots, r$. Let $\hat{f} = p_1^{e'_1} \cdots p_r^{e'_r}$ and $\bar{g} = p_{r+1}^{e_{r+1}} \cdots p_{r+s}^{e_{r+s}}$. Then $g = \hat{f}\bar{g}$, $\nu(f) = \nu(\hat{f})$, and $\gcd(\bar{g}, f) = 1$. We have $\tau_f^g = \tau_{\hat{f}}^g \circ \tau_f^{\hat{f}}$. Thus $(\tau_f^g)^{-1} = (\tau_{\hat{f}}^g)^{-1} \circ (\tau_f^{\hat{f}})^{-1}$. By Lemma 12.41,

$$(\tau_f^{\hat{f}})^{-1}(M'_f)) = M'_{\hat{f}}. \qquad (12.12)$$

By Lemma 12.42,

$$(\tau_{\hat{f}}^g)^{-1}(M'_{\hat{f}}) = \bigcup_{a|\bar{g}} M'_{\hat{f}a} = M'_{\hat{f}} + M_{\bar{g}}. \qquad (12.13)$$

Then (12.11) follows immediately from (12.12) and (12.13). □

After the foregoing preparation we are now ready to give the

Proof of Theorem 8.34 Let q be a power of a prime p and n be a positive integer. Assume every element $\alpha \in \mathbb{F}_{q^n}$ of degree n over $\mathbb{F}_q$ with

$\mathrm{Tr}_{\mathbb{F}_{q^n}/\mathbb{F}_q}(\alpha) \neq 0$ is a normal basis generator of $\mathbb{F}_{q^n}$ over $\mathbb{F}_q$. We are going to prove that either $n = p^e$ or n is a prime different from p having q as a primitive element modulo n.

If n is a power of p, it is all right. Now assume $n = p^e \bar{n}$, $\bar{n} > 1$, and $p \nmid \bar{n}$. Then

$$x^n - 1 = (x - 1)^{p^e} \Gamma(x),$$

where

$$\Gamma(x) = \prod_{d \mid \bar{n}, d \neq 1} \Phi_d(x)^{p^e}$$

and $\Phi_d(x)$ is the d-th cyclotomic polynomial (cf. Section 9.3). In the following $\mathrm{Tr}_{\mathbb{F}_{q^n}/\mathbb{F}_q}$ will simply be written as Tr. By Theorem 12.43,

$$\mathrm{Tr}^{-1}(M'_{x-1}) = (\tau^{x^n-1}_{x-1})^{-1}(M'_{x-1}) = \bigcup_{a \mid \Gamma} M'_{(x-1)^{p^e}a} = M'_{(x-1)^{p^e}} + M_\Gamma.$$

Thus $M'_{(x-1)^{p^e}\Phi_{\bar{n}}(x)}$ is a subset of $\mathrm{Tr}^{-1}(M'_{x-1})$. $\mathrm{Tr}^{-1}(M'_{x-1})$ is the set of elements $\beta \in \mathbb{F}_{q^n}$ such that $\mathrm{Tr}(\beta) \neq 0$. Therefore for all $\beta \in M'_{(x-1)^{p^e}\Phi_{\bar{n}}(x)}$, $\mathrm{Tr}(\beta) \neq 0$.

Recall that the period of a polynomial $f(x)$ of $\mathbb{F}_q[x]$ is denoted by $p(f(x))$ (cf. Section 9.4). By Theorem 9.24,

$$p((x-1)^{p^e}\Phi_{\bar{n}}(x)) = \mathrm{lcm}[p(x-1), p(\Phi_{\bar{n}}(x))]p^e = p(\Phi_{\bar{n}})(x)p^e = \bar{n}p^e = n.$$

Let $\beta \in \mathbb{F}_{q^n}$ and suppose $\mathrm{ord}(\beta) = (x-1)^{p^e}\Phi_{\bar{n}}(x)$, then n is the least positive integer such that $(x^n - 1)\beta = 0$, i.e., $\sigma^n(\beta) = \beta$, which implies that the degree of β over $\mathbb{F}_q$ is n. Thus all elements of $M'_{(x-1)^{p^e}\Phi_{\bar{n}}(x)}$ are of degree n over $\mathbb{F}_q$.

We proved that all elements of $M'_{(x-1)^{p^e}\Phi_{\bar{n}}(x)}$ have nonzero traces and are of degree n over $\mathbb{F}_q$. By the assumption of Theorem 8.34 all elements of $M'_{(x-1)^{p^e}\Phi_{\bar{n}}(x)}$ are normal basis generators. We know that M'_{x^n-1} consists of all normal basis generators (Exercise 12.20). Therefore $M'_{(x-1)^{p^e}\Phi_{\bar{n}}(x)} \subseteq M'_{x^n-1}$. By Lemma 12.39, $(x-1)^{p^e}\Phi_{\bar{n}} = x^n - 1$. It follows that $\Gamma(x) = \Phi_{\bar{n}}(x)$, which implies $p^e = 1$ and $\bar{n}$ is a prime. Thus $n = \bar{n}$ is a prime. Since $p \nmid \bar{n}$, $n = \bar{n}$ is a prime different from p. Moreover, we have $\Phi_n(x) = (x^n - 1)/(x - 1)$.

Suppose $\Phi_n(x)$ has a proper divisor $f(x) \in \mathbb{F}_q[x]$. Then $(x-1)f(x) \mid (x^n - 1)$. Let $\gamma \in M'_{(x-1)f(x)}$, i.e., $\gamma \in \mathbb{F}_{q^n}$ and is an element of order $(x-1)f(x)$. Since $(x-1)f(x) \mid (x^n-1)$, $(x^n-1)\gamma = 0$, i.e., $\sigma^n(\gamma) = \gamma$. Since n is a prime, n is the least positive integer such that $\sigma^n(\gamma) = \gamma$. Thus γ is of degree n over

$\mathbb{F}_q$. Therefore all elements of $M'_{(x-1)f(x)}$ are of degree n over $\mathbb{F}_q$. Moreover, by Lemma 12.39, from $x^n - 1 = (x - 1)\Phi_n(x)$ we deduce

$$\text{Tr}^{-1}(M'_{x-1}) = (\tau_{x-1}^{x^n-1})^{-1}(M'_{x-1}) = \bigcup_{a|\Phi_n} M'_{(x-1)a}.$$

It follows that $M'_{(x-1)f(x)}$ is a subset of $\text{Tr}^{-1}(M'_{x-1})$. Therefore for all $\beta \in M'_{(x-1)f(x)}$, $\text{Tr}(\beta) \neq 0$. By the assumption of Theorem 8.34 all elements of $M'_{(x-1)f(x)}$ are normal basis generators of $\mathbb{F}_{q^n}$ over $\mathbb{F}_q$. Thus $M'_{(x-1)f(x)} \subset M'_{x^n-1}$. By Lemma 12.39 $(x-1)f(x) = x^n - 1$, which implies $f(x) = \Phi_{\bar{n}}(x)$. Therefore $\Phi_n(x)$ is irreducible.

Let ξ be a root of $\Phi_n(x)$ in an extension of $\mathbb{F}_q$. Then $\xi^n = 1$ and $\xi, \xi^q, \xi^{q^2}, \ldots$, are roots of $\Phi_n(x)$ in the extension, too. Since $\Phi_n(x)$ is irreducible of degree $n - 1$, $\xi, \xi^q, \xi^{q^2}, \ldots, \xi^{q^{n-2}}$ are all the roots of $\Phi_n(x)$ and $\xi^{q^{n-1}} = \xi$. Thus $q^{n-1} \equiv 1 \pmod{n}$. This means that q is a primitive element modulo n.

The proof of Theorem 8.34 is now complete. $\square$

12.5 Exercises

12.1 Let G be a group and H be a subgroup of G. Prove that $H \lhd N_G(H)$. For the definition of $N_G(H)$, see Exercise 2.16.

12.2 Let G be a group and H be a subgroup of G. Prove that $H \lhd G$ if and only if H is the only subgroup of G conjugate to H.

12.3 Let ϕ be a surjective homomorphism of a group G to a group G'. Prove that the image of the center of G under ϕ is contained in the center of G'.

12.4 Prove that $V_4 \lhd A_4$ and establish an isomorphic map from A_4/V_4 to A_3.

12.5 Let $\mathbb{F}_q$ be a finite field of characteristic 2 and let $N = \{x^2+x : x \in \mathbb{F}_q\}$. Prove that the map $\mathbb{F} \to N : x \mapsto x^2 + x$ is a group homomorphism from the addition group of $\mathbb{F}_q$ to N. Determine the kernel of the homomorphism, then deduce that $\mathbb{F}_q : N = 2$ and $N^2 = N$.

12.6 Prove Theorems 12.11 and 12.12.

12.7 Let G be a finite abelian group, G_1 and G_2 be subgroups of G and $G = G_1 \times G_2$. Assume $\gcd(|G_1|, |G_2|) = 1$. Prove that G_i is the unique subgroup of order $|G_i| (i = 1, 2)$ of G.

12.8 Prove that a field F has only two ideals: the ideals (0) and (1), and $(0) = \{0\}$, $(1) = F$.

12.9 Prove Theorem 12.18.

12.10 Prove Theorem 12.20.

12.11 Fill in the proofs of Theorems 12.21–12.23.

12.12 Prove Theorem 12.29.

12.13   (i) Let p be a prime number and n be an integer ≥ 1. Prove that (p^n) is a primary ideal of $\mathbb{Z}$, and that when $n > 1$, (p^n) is not a prime ideal.

(ii) Let Q be a primary ideal of $\mathbb{Z}$ and $Q \neq \mathbb{Z}$. Prove that there exists a prime number p and an integer $n \geq 1$ such that $Q = (p^n)$.

12.14 Let F be a field. Prove that in $F[x]$ primary polynomials are powers of irreducible polynomials and conversely.

12.15 let $\phi : R \to R'$ be a homomorphism of rings and P' be a prime ideal of R'. Prove that $\phi^{-1}(P')$ is a prime ideal of R.

12.16 Prove Theorem 12.30.

12.17 Prove Theorems 12.31 and 12.32.

12.18 Let R be any ring and I be an ideal of R. Prove that I is the only maximal ideal of R if and only if all elements in $R \setminus I$ are units of R.

12.19 Let R be a commutative ring with identity. For any prime p, define

$$R_p = \{a \in R : p^r a = 0 \text{ for some integer } r \geq 0\}.$$

Prove that

(i) R_p is an ideal in R.

(ii) There is only a finite number of primes p such that $R_p \neq \{0\}$.

(iii) Let $p_1, p_2, \ldots, p_n$ be all the primes such that $R_i \neq \{0\}$, then $R = R_{p_1} \dot{+} R_{p_2} \dot{+} \cdots \dot{+} R_{p_n}$.

(iv) For any subring S of R, $S = \dot{+}_{i=1}^{n} S \cap R_i$.

(v) For any ideal I of R, $R/I \simeq \dot{+}_{i=1}^{n} R_i/R_i \cap I$.

12.20 Let R be a commutative ring with identity, and let A and B be two $n \times n$ matrices over R. Then $AB = I^{(n)}$ if and only if $BA = I^{(n)}$.

12.21 Regard $\mathbb{F}_{q^n}$ as an $\mathbb{F}_q[x]$-module as in Example 12.14. Prove that $\alpha \in \mathbb{F}_{q^n}$ is a normal basis generator of $\mathbb{F}_{q^n}$ over $\mathbb{F}_q$ if only if $\operatorname{ord}(\alpha) = x^n - 1$.

Chapter 13

Hensel's Lemma and Hensel Lift

13.1 The Polynomial Ring $\mathbb{Z}_{p^s}[x]$

Let p be any prime, s be a positive integer, and $\mathbb{Z}_{p^s}$ be the ring of integers modulo p^s. By the identification given in Example 2.13 we can assume that

$$\mathbb{Z}_{p^s} = \{0, 1, 2, \ldots, p^s - 1\}.$$

Clearly, all elements $a \in \mathbb{Z}_{p^s}$ which are coprime with p are units in $\mathbb{Z}_{p^s}$.

Theorem 13.1 *The principal ideals* $(1), (p), (p^2), \ldots, (p^{s-1}), (0)$ *are all the ideals of* $\mathbb{Z}_{p^s}$. (p) *is the unique maximal ideal of* $\mathbb{Z}_{p^s}$ *and* $\mathbb{Z}_{p^s}/(p) \simeq \mathbb{F}_p$.

Proof. Let I be any ideal of $\mathbb{Z}_{p^s}$. Suppose that $I \neq (0)$. Let m be the least positive integer of I. As in the proof of Theorem 12.17 we can show that $I = (m)$. Let $m = ap^i$, where $a \in \mathbb{Z}$, $a > 0$, $\gcd(a, p) = 1$ and $0 \leq i \leq s - 1$. There exist integers c, d such that $ca + dp^s = 1$. Then $ca = 1$ in $\mathbb{Z}_{p^s}$. Thus $p^i = cap^i = cm \in I$. By the minimality of m, $a = 1$. Then $I = (m) = (p^i)$. The first assertion is proved. Clearly, $(1) \supset (p) \supset (p^2) \supset \cdots \supset (p^{s-1}) \supset (0)$, thus (p) is the unique maximal ideal of $\mathbb{Z}_{p^s}$. That $\mathbb{Z}_{p^s}/(p) \simeq \mathbb{F}_p$ has already been proved in Theorem 5.3. $\quad\square$

Elements of $\mathbb{Z}_{p^s}$ can be expressed in the form

$$c_0 + c_1 p + \cdots + c_{s-1}p^{s-1}, \text{ where } 0 \leq c_i \leq p - 1. \quad (13.1)$$

We have the ring homomorphism

$$\mathbb{Z}_{p^s} \to \mathbb{Z}_{p^s}/(p) \simeq \mathbb{F}_p$$
$$c_0 + c_1 p + \cdots + c_{s-1}p^{s-1} \mapsto c_0 + (p), \quad (13.2)$$

which will be denoted by $-$, i.e.,

$$\overline{c_0 + c_1 p + \cdots + c_{s-1} p^{s-1}} = c_0 + (p).$$

The kernel of this homomorphism is the ideal (p).

Now we are going to study the polynomial ring $\mathbb{Z}_{p^s}[x]$.

The above homomorphism $-: \mathbb{Z}_{p^s} \to \mathbb{F}_p$ can be extended to a map from the polynomial ring $\mathbb{Z}_{p^s}[x]$ over $\mathbb{Z}_{p^s}$ to the polynomial ring $\mathbb{F}_p[x]$ over $\mathbb{F}_p$:

$$\mathbb{Z}_{p^s}[x] \to \mathbb{F}_p[x]$$
$$a_0 + a_1 x + \cdots + a_n x^n \mapsto \bar{a}_0 + \bar{a}_1 x + \cdots + \bar{a}_n x^n, \qquad (13.3)$$

where x is an indeterminate over $\mathbb{Z}_{p^s}$ and also over $\mathbb{F}_p$, and $a_0, a_1, \ldots, a_n \in \mathbb{Z}_{p^s}$. The extended map (13.3) will also be denoted by $-$ and the image of $f(x) \in \mathbb{Z}_{p^s}[x]$ under the map $-$ will be denoted by $\bar{f}(x)$. It can be easily verified that the extended map is a ring homomorphism from $\mathbb{Z}_{p^s}[x]$ onto $\mathbb{F}_p[x]$ with kernel (p), but now

$$(p) = \mathbb{Z}_{p^s}[x]p = \{f(x)p : f(x) \in \mathbb{Z}_{p^s}[x]\}$$

is the principal ideal generated by p in $\mathbb{Z}_{p^s}[x]$. More generally, for any $f(x) \in \mathbb{Z}_{p^s}[x]$ the principal ideal generated by $f(x)$ in $\mathbb{Z}_{p^s}[x]$ is

$$(f(x)) = \mathbb{Z}_{p^s}[x]f(x) = \{g(x)f(x) : g(x) \in \mathbb{Z}_{p^s}[x]\}.$$

In the following a polynomial $f(x) \in \mathbb{Z}_{p^s}[x]$ or $\mathbb{F}_p[x]$ will sometimes be denoted simply by f.

First we study the prime ideals of $\mathbb{Z}_{p^s}[x]$.

Theorem 13.2 *The ideal* $(p) = \mathbb{Z}_{p^s}[x]p$ *is a prime ideal of* $\mathbb{Z}_{p^s}[x]$, *every prime ideal of* $\mathbb{Z}_{p^s}[x]$ *contains* (p), *and every prime ideal of* $\mathbb{Z}_{p^s}[x]$ *containing* (p) *properly is maximal.*

Proof. Clearly, $\mathbb{Z}_{p^s}[x]/(p) \simeq \mathbb{F}_p[x]$ and $\mathbb{F}_p[x]$ is an integral domain. So (p) is a prime ideal of $\mathbb{Z}_{p^s}[x]$.

Let P be a prime ideal of $\mathbb{Z}_{p^s}[x]$. From $p \cdot p^{s-1} = p^s = 0 \in P$, we deduce $p \in P$ or $p^{s-1} \in P$. If $p^{s-1} \in P$ and $s > 2$, from $p \cdot p^{s-2} = p^{s-1} \in P$, we deduce $p \in P$ or $p^{s-2} \in P$. Proceeding in this way we obtain always $p \in P$. Therefore $(p) \subseteq P$.

Now let P be a prime ideal of $\mathbb{Z}_{p^s}[x]$ containing (p) properly. Denote the image of P under the homomorphism $- : \mathbb{Z}_{p^s}[x] \to \mathbb{F}_p[x]$ by $\overline{P}$. We assert that $\overline{P}$ is a prime ideal of $\mathbb{F}_p[x]$. Since $(p) = \mathrm{Ker}\,-$ and $P \supset (p)$, by Theorem 12.22(iii) the complete inverse image of $\overline{P}$ is P. Therefore $\overline{P} \neq \mathbb{F}_p[x]$. Let $f_0(x), g_0(x) \in \mathbb{F}_p[x]$ be such that $f_0(x)g_0(x) \in \overline{P}$, then there exist $f(x), g(x) \in \mathbb{Z}_{p^s}[x]$ such that $\overline{f(x)} = f_0(x), \overline{g(x)} = g_0(x)$. Hence $\overline{f(x)g(x)} \in \overline{P}$. Since the complete inverse image of $\overline{P}$ is P, $f(x)g(x) \in P$. But P is prime, so $f(x) \in P$ or $g(x) \in P$. If $f(x) \in P$, $f_0(x) \in \overline{P}$, and if $g(x) \in P$, $g_0(x) \in \overline{P}$. This proves that $\overline{P}$ is a prime ideal of $\mathbb{F}_p[x]$. Since P contains (p) properly, $\overline{P} \neq (0)$. By Theorem 12.27 $\overline{P}$ is a maximal ideal of $\mathbb{F}_p[x]$ and $\mathbb{F}_p[x]/\overline{P}$ is a field. By Theorem 12.22(iii)

$$\mathbb{Z}_{p^s}[x]/P \simeq (\mathbb{Z}_{p^s}[x]/(p))/(P/(p)) \simeq \mathbb{F}_p[x]/\overline{P}.$$

Hence P is a maximal ideal of $\mathbb{Z}_{p^s}[x]$. $\qquad\square$

Then we study the primary ideals of $\mathbb{Z}_{p^s}[x]$.

Theorem 13.3 *Let Q be an ideal of $\mathbb{Z}_{p^s}[x]$.*

(i) *If Q is primary then $\sqrt{Q}$ is prime.*

(ii) $\sqrt{Q} \supseteq (p)$.

(iii) *If $\sqrt{Q}$ is prime and contains (p) properly, then Q is primary.*

Proof. (i) Assume Q is primary. From $Q \neq \mathbb{Z}_{p^s}[x]$, we deduce immediately $\sqrt{Q} \neq \mathbb{Z}_{p^s}[x]$. Let $a, b \in \mathbb{Z}_{p^s}[x]$ be such that $ab \in \sqrt{Q}$. Then $(ab)^m \in Q$ for some positive integer m. Since Q is primary, we have $a^m \in Q$ or $(b^m)^n \in Q$ for some positive integer n. If $a^m \in Q$ then $a \in \sqrt{Q}$, and if $b^{mn} \in Q$ then $b \in \sqrt{Q}$. Therefore $\sqrt{Q}$ is prime.

(ii) Clearly, $p^s = 0 \in Q$. Therefore $p \in \sqrt{Q}$. Hence $(p) \subseteq \sqrt{Q}$.

(iii) Assume that $\sqrt{Q}$ is prime and contains (p) properly. By Theorem 13.2 $\sqrt{Q}$ is maximal. From $\sqrt{Q} \neq \mathbb{Z}_{p^s}[x]$, we deduce immediately $Q \neq \mathbb{Z}_{p^s}[x]$. Let $a, b \in \mathbb{Z}_{p^s}[x]$ be such that $ab \in Q$. Assume $b^m \notin Q$ for any positive integer m. Then $b \notin \sqrt{Q}$. Since $\sqrt{Q}$ is maximal, the ideal $(b, \sqrt{Q})$ generated by b and $\sqrt{Q}$ is $\mathbb{Z}_{p^s}[x]$. Then the identity 1 of $\mathbb{Z}_{p^s}[x]$ can be expressed as $1 = tb + r$, where $t \in \mathbb{Z}_{p^s}[x]$ and $r \in \sqrt{Q}$. There is a positive integer n such that $r^n \in Q$. Then

$$1 = 1^n = (tb + r)^n = yb + r^n,$$

where $y \in \mathbb{Z}_{p^s}[x]$. Multiplying the above equality by a, we obtain

$$a = yab + ar^n \in Q.$$

Hence Q is primary. $\qquad\square$

Lemma 13.4 *Let f be a polynomial in $\mathbb{Z}_{p^s}[x]$ and assume $\bar{f} = g^m$, where g is an irreducible polynomial in $\mathbb{F}_p[x]$ and m is a positive integer. Then f is a primary polynomial of $\mathbb{Z}_{p^s}[x]$.*

Proof. By Theorem 13.3(ii) $\sqrt{(f)} \supseteq (p)$. Since $f \in (f) \subseteq \sqrt{(f)}$ and $\bar{f} \neq 0$, $\sqrt{(f)}$ contains (p) properly. By Theorem 13.3 (iii) it is enough to prove that $\sqrt{(f)}$ is prime. Since $1 \notin (f)$, we have also $1 \notin \sqrt{(f)}$. Thus $\sqrt{(f)} \neq \mathbb{Z}_{p^s}[x]$. Let $a, b \in \mathbb{Z}_{p^s}[x]$ and assume that $ab \in \sqrt{(f)}$. Then there is a positive integer n such that $(ab)^n \in (f)$. It follows that $(\bar{a}\bar{b})^n \in (\bar{f}) = (g^m)$. By the unique factorization theorem of $\mathbb{F}_p[x]$, $g \mid \bar{a}$ or $g \mid \bar{b}$. If $g \mid \bar{a}$, then $\bar{f} \mid \bar{a}^m$. There are polynomials $c, d \in \mathbb{Z}_{p^s}[x]$ such that $a^m = cf + pd$. Then $a^{mp^s} = c^{p^s} f^{p^s} \in (f)$, which implies $a \in \sqrt{(f)}$. If $g \mid \bar{b}$ we can prove in a similar way that $b \in \sqrt{(f)}$. Therefore $\sqrt{(f)}$ is prime. $\qquad\square$

13.2   Hensel's Lemma

Let f_1 and f_2 be two polynomials in $\mathbb{Z}_{p^s}[x]$. We recall from Section 3.5 that, f_1 and f_2 are said to be coprime in $\mathbb{Z}_{p^s}[x]$, if there are polynomials λ_1 and λ_2 in $\mathbb{Z}_{p^s}[x]$ such that

$$\lambda_1 f_1 + \lambda_2 f_2 = 1,$$

or, equivalently,

$$\mathbb{Z}_{p^s}[x] f_1 + \mathbb{Z}_{p^s}[x] f_2 = \mathbb{Z}_{p^s}[x].$$

The coprimeness of two polynomials in $\mathbb{F}_p[x]$ is defined in a similar way. It is well-known that two polynomials in $\mathbb{F}_p[x]$ are coprime in $\mathbb{F}_p[x]$ if and only if they have no common divisors of degree ≥ 1 (cf. Corollary 4.9).

Lemma 13.5 *Let f_1 and $f_2 \in \mathbb{Z}_{p^s}[x]$. Then f_1 and f_2 are coprime in $\mathbb{Z}_{p^s}[x]$ if and only if $\bar{f}_1$ and $\bar{f}_2$ are coprime in $\mathbb{F}_p[x]$.*

Proof. When $s = 1$, the lemma trivially holds. Now let $s > 1$. Assume that $\bar{f}_1$ and $\bar{f}_2$ are coprime in $\mathbb{F}_p[x]$. Then there are λ_1 and $\lambda_2 \in \mathbb{Z}_{p^s}[x]$ such that

$$\bar{\lambda}_1 \bar{f}_1 + \bar{\lambda}_2 \bar{f}_2 = 1.$$

Thus

$$\lambda_1 f_1 + \lambda_2 f_2 = 1 + pk, \text{ where } k \in \mathbb{Z}_{p^s}[x]. \tag{13.4}$$

Let

$$l = \sum_{i=0}^{s-1} (-pk)^i.$$

Multiplying (13.4) by l, we obtain

$$l\lambda_1 f_1 + l\lambda_2 f_2 = 1. \tag{13.5}$$

Therefore f_1 and f_2 are coprime in $\mathbb{Z}_{p^s}[x]$.

The converse part is trivial. $\qquad\qquad\qquad\qquad\qquad\qquad\qquad\qquad\square$

Lemma 13.6 (Hensel's Lemma): *Let f be a monic polynomial in $\mathbb{Z}_{p^s}[x]$ and assume that*

$$\bar{f} = g_1 g_2 \text{ in } \mathbb{F}_p[x],$$

where g_1 and g_2 are coprime monic polynomials in $\mathbb{F}_p[x]$. Then there exist coprime monic polynomials f_1 and f_2 in $\mathbb{Z}_{p^s}[x]$ such that

$$f = f_1 f_2 \text{ in } \mathbb{Z}_{p^s}[x]$$

and

$$\bar{f}_1 = g_1, \ \bar{f}_2 = g_2.$$

Proof. We construct by mathematical induction, for each $n = 0, 1, 2, \ldots$ a pair of monic polynomials $f_1^{(n)}$ and $f_2^{(n)} \in \mathbb{Z}_{p^s}[x]$ and a polynomial $k^{(n)} \in \mathbb{Z}_{p^s}[x]$ of degree $< \deg f$ such that

$$f = f_1^{(n)} f_2^{(n)} + p^{2^n} k^{(n)} \text{ in } \mathbb{Z}_{p^s}[x] \tag{13.6}$$

and

$$\overline{f_i^{(n)}} = g_i, \ i = 1, 2. \tag{13.7}$$

When $n = 0$, let $f_1^{(0)}$ and $f_2^{(0)}$ be monic polynomials over $\mathbb{Z}_{p^s}$ such that $\overline{f_i^{(0)}} = g_i$, $i = 1, 2$. Clearly, $f - f_1^{(0)} f_2^{(0)} \in (p)$ and $\deg(f - f_1^{(0)} f_2^{(0)}) < \deg f$. Thus there is a polynomial $k^{(0)}$ over $\mathbb{Z}_{p^s}$ such that $f = f_1^{(0)} f_2^{(0)} + p k^{(0)}$ and $\deg k^{(0)} < \deg f$.

Let $n \geq 0$ and assume that monic polynomials $f_1^{(n)}$ and $f_2^{(n)}$ over $\mathbb{Z}_{p^s}$ and a polynomial $k^{(n)}$ over $\mathbb{Z}_{p^s}$ of degree $< \deg f$ have been constructed such that (13.6) and (13.7) hold. By hypothesis g_1 and g_2 are coprime in $\mathbb{F}_p[x]$, so by Lemma 13.5, $f_1^{(n)}$ and $f_2^{(n)}$ are coprime in $\mathbb{Z}_{p^s}[x]$. Then there exist $\lambda_1^{(n)}$ and $\lambda_2^{(n)} \in \mathbb{Z}_{p^s}[x]$ such that

$$\lambda_1^{(n)} f_1^{(n)} + \lambda_2^{(n)} f_2^{(n)} = k^{(n)}. \tag{13.8}$$

Dividing $\lambda_1^{(n)}$ by $f_2^{(n)}$, we obtain

$$\lambda_1^{(n)} = q_1^{(n)} f_2^{(n)} + r_1^{(n)}, \tag{13.9}$$

where $q_1^{(n)}, r_1^{(n)} \in \mathbb{Z}_{p^s}[x]$ and $\deg r_1^{(n)} < \deg f_2^{(n)}$. Similarly,

$$\lambda_2^{(n)} = q_2^{(n)} f_1^{(n)} + r_2^{(n)}, \tag{13.10}$$

where $q_2^{(n)}, r_2^{(n)} \in \mathbb{Z}_{p^s}[x]$ and $\deg r_2^{(n)} < \deg f_1^{(n)}$. Substituting (13.9) and (13.10) into (13.8) and rearranging, we obtain

$$(q_1^{(n)} + q_2^{(n)}) f_1^{(n)} f_2^{(n)} = k^{(n)} - r_1^{(n)} f_1^{(n)} - r_2^{(n)} f_2^{(n)}.$$

The right-hand side of the above equation is a polynomial of degree $< \deg f$ and the left-hand side is one of degree $\geq \deg f$, if $q_1^{(n)} + q_2^{(n)} \neq 0$. Therefore we must have $q_1^{(n)} + q_2^{(n)} = 0$ and, consequently, $r_1^{(n)} f_1^{(n)} + r_2^{(n)} f_2^{(n)} = k^{(n)}$. Let $f_1^{(n+1)} = f_1^{(n)} + p^{2^n} r_2^{(n)}$, $f_2^{(n+1)} = f_2^{(n)} + p^{2^n} r_1^{(n)}$, and $k^{(n+1)} = -r_1^{(n)} r_2^{(n)}$. Then $f_1^{(n+1)}$ and $f_2^{(n+1)}$ are monic polynomials over $\mathbb{Z}_{p^s}$, $k^{(n+1)}$ is a polynomial over $\mathbb{Z}_{p^s}$ of degree $< \deg f$, $f = f_1^{(n+1)} f_2^{(n+1)} + p^{2^{n+1}} k^{(n+1)}$, and $\overline{f_i^{(n+1)}} = \overline{f_i^{(n)}} = g_i$, $i = 1, 2$. The induction construction is now complete.

Let l be an integer such that $2^l \geq s > 2^{l-1}$ and let $f_i = f_i^{(l)}$, $i = 1, 2$, then f_1 and f_2 are monic polynomials over $\mathbb{Z}_{p^s}$, $f = f_1 f_2$ in $\mathbb{Z}_{p^s}[x]$, and $\overline{f_i} = \overline{f_i^{(l)}} = g_i$, $i = 1, 2$. By Lemma 13.5 f_1 and f_2 are coprime in $\mathbb{Z}_{p^s}[x]$. $\square$

By mathematical induction, Lemma 13.6 can be generalized to r factors, where $r \geq 2$.

Lemma 13.7 (Hensel's Lemma): *Let f be a monic polynomial over $\mathbb{Z}_{p^s}$ and assume that $\bar{f} = g_1 g_2 \cdots g_r$, where $g_1, g_2, \ldots, g_r$ are pairwise coprime monic polynomials over $\mathbb{F}_p$. Then there exist pairwise coprime monic polynomials $f_1, f_2, \ldots, f_r$ over $\mathbb{Z}_{p^s}$ such that $f = f_1 f_2 \cdots f_r$ in $\mathbb{Z}_{p^s}[x]$ and $\overline{f_i} = g_i$, $i = 1, 2, \ldots, r$.*

13.3   Factorization of Monic Polynomials in $\mathbb{Z}_{p^s}[x]$

We have the following unique factorization theorem.

Theorem 13.8 *Let f be a monic polynomial of degree ≥ 1 in $\mathbb{Z}_{p^s}[x]$. Then*

(i) *f can be factorized into a product of some number, say r, of pairwise coprime monic primary polynomials $f_1, f_2, \ldots, f_r$ of $\mathbb{Z}_{p^s}[x]$:*

$$f = f_1 f_2 \cdots f_r \tag{13.11}$$

and for each $i = 1, 2, \ldots, r$, $\overline{f_i}$ is a power of a monic irreducible polynomial over $\mathbb{F}_p$.

(ii) *Let*
$$f = f_1 \cdots f_r = h_1 \cdots h_t \qquad (13.12)$$
be two factorizations of f into products of pairwise coprime monic primary polynomials of $\mathbb{Z}_{p^s}[x]$, then $r = t$ and after renumbering, $f_i = h_i$, $i = 1, 2, \ldots, r$.

Proof. (i) By the unique factorization theorem of $\mathbb{F}_p[x]$ we can assume that
$$\bar{f} = g_1^{e_1} \cdots g_r^{e_r},$$
where $g_1, \ldots, g_r$ are distinct monic irreducible polynomials over $\mathbb{F}_p$ and $e_1, \ldots, e_r$ are positive integers. By Hensel's Lemma, there exist pairwise coprime monic polynomials $f_1, \ldots, f_r$ over $\mathbb{Z}_{p^s}$ such that (13.11) holds and $\bar{f}_i = g_i^{e_i}$, $i = 1, \ldots, r$. By Lemma 13.4 f_i ($i = 1, \ldots, r$) are primary polynomials of $\mathbb{Z}_{p^s}[x]$.

(ii) From (13.12) we deduce $f_1 \cdots f_r \in (h_i)$ for all $i = 1, \ldots, t$. Since h_i is primary, there is an integer $k_i, 1 \le k_i \le r$, and a positive integer n_i such that $f_{k_i}^{n_i} \in (h_i)$. We assert that k_i is uniquely determined. Assume that there is another $k_i' \ne k_i$ and n_i' such that $f_{k_i'}^{n_i'} \in (h_i)$, since f_{k_i} and $f_{k_i'}$ are coprime in $\mathbb{Z}_{p^s}[x]$, there are $a, b \in \mathbb{Z}_{p^s}[x]$ such that $1 = af_{k_i} + bf_{k_i'}$. Then
$$1 = 1^{n_i + n_i' - 1} = (af_{k_i} + bf_{k_i'})^{n_i + n_i' - 1} \in (h_i),$$
a contradiction.

Similarly, for each $j = 1, \ldots, r$, there is a uniquely determined integer l_j, $1 \le l_j \le t$ and a positive integer m_j such that $h_{l_j}^{m_j} \in (f_j)$. Then for every i, $1 \le i \le t$, $h_{l_{k_i}}^{m_{k_i} n_i} \in (h_i)$, thus $\bar{h}_{l_{k_i}}^{m_{k_i} n_i} \in (\bar{h}_i)$. Since h_i and h_j are coprime for $i \ne j$, $\bar{h}_i$ and $\bar{h}_j$ are coprime for $i \ne j$ and we must have $l_{k_i} = i$ for every $i = 1, \ldots, t$. It follows that the map
$$\{1, \ldots, t\} \to \{1, \ldots, r\}$$
$$i \to k_i$$
is a well-defined injective map. Thus $t \le r$. Similarly, $r \le t$. Hence $r = t$. After renumbering we can assume that $k_i = i$ for $i = 1, \ldots, r$. Then $l_i = i$ for $i = 1, \ldots, r$. Thus $f_i^{n_i} \in (h_i)$ and $h_i^{m_i} \in (f_i)$ for $i = 1, \ldots, r$.

For $j \ne 1$, f_j and f_1 are coprime. By Lemma 13.5, $\bar{f}_j$ and $\bar{f}_1$ are coprime, which implies $\bar{f}_j$ and $\bar{f}_1^{n_1}$ are coprime. Hence $\bar{f}_2 \cdots \bar{f}_r$ and $\bar{f}_1^{n_1}$ are coprime. By Lemma 13.5 again, $f_2 \cdots f_r$ and $f_1^{n_1}$ are coprime. Since $f_1^{n_1} \in (h_1)$, $f_2 \cdots f_r$ and h_1 are coprime. Then there exist $c, d \in \mathbb{Z}_{p^s}[x]$ such that
$$cf_2 \cdots f_r + dh_1 = 1.$$

Multiplying both sides of the above equality by f_1, we obtain

$$f_1 = cf_1f_2\cdots f_r + df_1h_1 = ch_1h_2\cdots h_r + df_1h_1,$$

which implies $h_1\,|\,f_1$. Similarly, $f_1\,|\,h_1$. Since both f_1 and h_1 are monic, $f_1 = h_1$. Similarly, $f_i = h_i$, $i = 2,\dots,r$.   □

13.4 Basic Irreducible Polynomials and Hensel Lift

Let $f(x)$ be a monic polynomial of degree $m \geq 1$ in $\mathbb{Z}_{p^s}[x]$. If $\bar{f}(x)$ is irreducible (or primitive) in $\mathbb{F}_p[x]$, $f(x)$ is called a *monic basic irreducible* (or *monic basic primitive*, respectively,) polynomial in $\mathbb{Z}_{p^s}[x]$. Clearly, monic basic primitive polynomials in $\mathbb{Z}_{p^s}[x]$ are monic basic irreducible.

Theorem 13.9 *For any integer $m \geq 1$ there exist monic basic irreducible (and monic basic primitive) polynomials of degree m over $\mathbb{Z}_{p^s}$ and dividing $x^{p^m-1} - 1$ in $\mathbb{Z}_{p^s}[x]$.*

Proof. By Corollary 6.13 (and Theorem 7.7) there exist monic irreducible polynomials (and monic primitive polynomials, respectively,) of degree m over $\mathbb{F}_p$. By Lemma 6.9 such a polynomial divides $x^{p^m-1} - 1$ in $\mathbb{F}_p[x]$. Let $f_p(x)$ be one of them and let $g_p(x) = (x^{p^m-1} - 1)/f_p(x)$. Then

$$x^{p^m-1} - 1 = f_p(x)g_p(x) \text{ in } \mathbb{F}_p[x].$$

Since $x^{p^m-1} - 1$ has no multiple roots, $f_p(x)$ and $g_p(x)$ are coprime in $\mathbb{F}_p[x]$. By Hensel's Lemma, there exist monic polynomials $f(x)$ and $g(x)$ in $\mathbb{Z}_{p^s}[x]$ such that

$$x^{p^m-1} - 1 = f(x)g(x) \text{ in } \mathbb{Z}_{p^s}[x]$$

and $\bar{f}(x) = f_p(x), \bar{g}(x) = g_p(x)$. Then $\deg f(x) = \deg f_p(x) = m$. Hence $f(x)$ is a monic basic irreducible (or monic basic primitive) polynomial of degree m over $\mathbb{Z}_{p^s}$ and dividing $x^{p^m-1} - 1$ in $\mathbb{Z}_{p^s}[x]$.   □

The monic basic irreducible (or monic basic primitive) polynomial $f(x)$ over $\mathbb{Z}_{p^s}$ in the proof of Theorem 13.9 satisfies not only the condition that $\bar{f}(x)$ is monic irreducible (or monic primitive, respectively,) over $\mathbb{F}_p$, but also the condition that $f(x)\,|\,(x^{p^m-1} - 1)$ in $\mathbb{Z}_{p^s}[x]$. This suggests the following

Definition 13.1 *Let $g(x)$ be a monic polynomial over $\mathbb{F}_p$. A monic polynomial $f(x)$ over $\mathbb{Z}_{p^s}$ is called a Hensel lift of $g(x)$ if $\bar{f}(x) = g(x)$ and there is a positive integer n not divisible by p such that*

$$f(x)\,|\,(x^n - 1) \text{ in } \mathbb{Z}_{p^s}[x].$$

For Hensel lift, cf. Wan(2002a).

Thus the monic basic irreducible (or monic basic primitive) polynomial $f(x)$ over $\mathbb{Z}_{p^s}$ in the proof of Theorem 13.9 is the Hensel lift of the monic irreducible (or monic primitive, respectively,) polynomial $\bar{f}(x) = f_p(x)$ over $\mathbb{F}_p$. But not every monic basic irreducible polynomial over $\mathbb{Z}_{p^s}$ is a Hensel lift; for example, $x - 3$ is a monic basic irreducible polynomial over $\mathbb{Z}_{2^2}$, but it is not a Hensel lift of $\overline{x-3} = x + 1$ over $\mathbb{F}_2$ (Exercise 13.3).

In the following we will prove that a monic polynomial $g(x)$ over $\mathbb{F}_p$ has a Hensel lift if and only if $g(x)$ has no multiple roots and $x \nmid g(x)$ in $\mathbb{F}_p[x]$ and that the Hensel lift of $g(x)$ is unique if it has one.

Theorem 13.10 *Let s be an integer ≥ 1. Then a monic polynomial $g(x)$ over $\mathbb{F}_p$ has a Hensel lift $f(x)$ over $\mathbb{Z}_{p^s}$ if and only if $g(x)$ has no multiple roots and $x \nmid g(x)$ in $\mathbb{F}_p[x]$.*

Proof. Let $f(x)$ be a Hensel lift of $g(x)$ over $\mathbb{Z}_{p^s}$. Then $\bar{f}(x) = g(x)$ and there is a positive integer n not divisible by p such that $f(x) \mid (x^n - 1)$ in $\mathbb{Z}_{p^s}[x]$. From $f(x) \mid (x^n - 1)$ in $\mathbb{Z}_{p^s}[x]$, we deduce $\bar{f}(x) \mid (x^n - 1)$ in $\mathbb{F}_p[x]$, i.e., $g(x) \mid (x^n - 1)$ in $\mathbb{F}_p[x]$. Since $p \nmid n$, $x^n - 1$ has no multiple roots and $g(x)$ has also no multiple roots. Since $x \nmid (x^n - 1)$ in $\mathbb{F}_p[x]$, we also have $x \nmid g(x)$ in $\mathbb{F}_p[x]$.

Conversely, assume that $g(x) \in \mathbb{F}_q[x]$ has no multiple roots and that $x \nmid g(x)$ in $\mathbb{F}_p[x]$. Since $x \nmid g(x)$ in $\mathbb{F}_p[x]$, by Definition 9.2 and Theorem 9.18, there is a positive integer n such that $g(x) \mid (x^n - 1)$ in $\mathbb{F}_p[x]$. Suppose $p \mid n$, let $n = p^e n_1$, where $\gcd(p, n_1) = 1$, then $(x^n - 1) = (x^{n_1} - 1)^{p^e}$. Since $g(x)$ has no multiple roots, from $g(x) \mid (x^n - 1)$ we deduce $g(x) \mid (x^{n_1} - 1)$. Therefore we can assume $p \nmid n$. Let $g_0(x) \in \mathbb{F}_p[x]$ be such that

$$x^n - 1 = g(x)g_0(x).$$

Since $p \nmid n$, $x^n - 1$ has no multiple roots. It follows that $g(x)$ and $g_0(x)$ are coprime in $\mathbb{F}_p[x]$. By Hensel's Lemma, there are monic polynomials $f(x)$, $f_0(x) \in \mathbb{Z}_{p^s}[x]$ such that

$$x^n - 1 = f(x)f_0(x) \in \mathbb{Z}_{p^s}[x],$$

$$\overline{f}(x) = g(x), \ \overline{f}_0(x) = g_0(x),$$

and $f(x)$, $f_0(x)$ are coprime in $\mathbb{Z}_{p^s}[x]$.

Since $f(x)$ is a monic polynomial over $\mathbb{Z}_{p^s}$, $f(x)$ divides $x^n - 1$ in $\mathbb{Z}_{p^s}[x]$, where $p \nmid n$, and $\overline{f}(x) = g(x)$, $f(x)$ is a Hensel lift of $g(x)$. $\square$

Theorem 13.11 *Let s be an integer ≥ 1. Let $g(x)$ be a monic polynomial over $\mathbb{F}_p$ without multiple roots and $x \nmid g(x)$ in $\mathbb{F}_p[x]$. Then $g(x)$ has a unique Hensel lift in $\mathbb{Z}_{p^s}[x]$.*

Proof. By Theorem 13.10, $g(x)$ has a Hensel lift in $\mathbb{Z}_{p^s}[x]$. Let $f^{(1)}(x)$ and $f^{(2)}(x)$ be two Hensel lifts of $g(x)$ in $\mathbb{Z}_{p^s}[x]$. Then both of them are monic polynomials over $\mathbb{Z}_{p^s}$, $\overline{f^{(1)}}(x) = \overline{f^{(2)}}(x) = g(x)$, and there are positive integers n_1 and n_2, both not divisible by p, such that $f^{(1)}(x) \mid (x^{n_1} - 1)$ and $f^{(2)}(x) \mid (x^{n_2} - 1)$, respectively, in $\mathbb{Z}_{p^s}[x]$. Thus $g(x) \mid (x^{n_1} - 1)$ and $g(x) \mid (x^{n_2} - 1)$ in $\mathbb{F}_p[x]$. We distinguish the following two cases.

(a) $n_1 = n_2$. For simplicity, let $n = n_1 = n_2$. Since $p \nmid n$, $x^n - 1$ has no multiple roots. Let

$$x^n - 1 = g_1(x)g_2(x)\cdots g_r(x) \tag{13.13}$$

be the unique factorization of $x^n - 1$ into a product of distinct monic irreducible polynomials $g_i(x)$ $(i = 1, 2, \ldots, r)$ over $\mathbb{F}_p$. By Hensel's Lemma, there are monic polynomials $f_1(x), f_2(x), \ldots, f_r(x) \in \mathbb{Z}_{p^s}[x]$ such that

$$x^n - 1 = f_1(x)f_2(x)\cdots f_r(x) \text{ in } \mathbb{Z}_{p^s}[x], \tag{13.14}$$

$$\overline{f_i}(x) = g_i(x), \ i = 1, 2, \ldots, r, \tag{13.15}$$

and $f_1(x), f_2(x), \ldots, f_r(x)$ are pairwise coprime in $\mathbb{Z}_{p^s}[x]$. Since $g(x) \mid (x^n - 1)$ in $\mathbb{F}_p[x]$, by the unique factorization theorem in $\mathbb{F}_p[x]$, we can assume that there is an integer t, where $1 \leq t \leq r$, such that

$$g(x) = g_1(x)g_2(x)\cdots g_t(x) \tag{13.16}$$

up to a rearrangement of $g_1(x), g_2(x), \ldots, g_r(x)$. Let

$$f(x) = f_1(x)f_2(x)\cdots f_t(x).$$

Clearly, $f(x)$ is a Hensel lift of $g(x)$.

We have $\overline{f^{(1)}}(x) = g(x)$. By Hensel's Lemma, from the factorization (13.16) we have pairwise coprime monic polynomials $h_1(x), h_2(x), \ldots, h_t(x)$ such that

$$f^{(1)}(x) = h_1(x)h_2(x)\cdots h_t(x)$$

and

$$\overline{h_i}(x) = g_i(x), \ i = 1, 2, \ldots, t.$$

Since $f^{(1)}(x) \mid (x^n - 1)$ in $\mathbb{Z}_{p^s}[x]$, by Theorem 13.8 all $h_1(x), h_2(x), \ldots, h_t(x)$ appear in $\{f_1(x), f_2(x), \ldots, f_r(x)\}$. Since $\overline{h_i}(x) = g_i(x), i = 1, 2 \ldots, t$ and

$g_1(x)$, $g_2(x)$, $\ldots$, $g_t(x)$ are distinct in pairs, we must have $h_i(x) = f_i(x)$, $i = 1, 2, \ldots, t$. Consequently, $f^{(1)}(x) = f(x)$. Similarly, $f^{(2)}(x) = f(x)$.

(b) $n_1 \neq n_2$. Let $n = \operatorname{lcm}[n_1, n_2]$, then $p \nmid n$ and $(x^{n_1} - 1) \mid (x^n - 1)$ and $(x^{n_2} - 1) \mid (x^n - 1)$ in $\mathbb{Z}_{p^s}[x]$. Then $f^{(1)}(x) \mid (x^n - 1)$ and $f^{(2)}(x) \mid (x^n - 1)$ in $\mathbb{Z}_{p^s}[x]$. By Case (a), we must have $f^{(1)}(x) = f^{(2)}(x)$.

This proves the uniqueness of the Hensel lift. $\qquad\square$

Let $f_2(x)$ be a polynomial over $\mathbb{Z}_2$ without multiple roots and not divisible by x. Its Hensel lift can be calculated using Graeffe's method for finding a polynomial whose roots are squares of the roots of $f_2(x)$, as the following theorem shows.

Theorem 13.12 *Let $f_2(x)$ be a polynomial over $\mathbb{Z}_2$ without multiple roots and not divisible by x. Write $f_2(x) = e(x) - d(x)$, where $e(x)$ contains only even power terms and $d(x)$ only odd power terms. Then $e(x)^2 - d(x)^2$, computed in $\mathbb{Z}_4[x]$, is a polynomial having only even power terms and of degree $2 \deg f_2(x)$. Let $f(x^2) = \pm(e(x)^2 - d(x)^2)$, where we take the $+$ or $-$ sign if $\deg e(x) > \deg d(x)$ or $\deg d(x) > \deg e(x)$, then $f(x)$ is the Hensel lift of $f_2(x)$.*

Proof. The first statement is clear. By the choice of $\pm$ sign, $f(x^2)$ is monic, hence $f(x)$ is monic. We have

$$f(x^2) \equiv e(x^2) - d(x^2) = f_2(x^2) \pmod{2},$$

which implies $\bar{f}(x) = f_2(x)$. We also have

$$f(x^2) = \pm f_2(x) f_2(-x),$$

computed in $\mathbb{Z}_4[x]$. There is an odd positive integer n such that $f_2(x) \mid (x^n - 1)$ in $\mathbb{Z}_2[x]$. Computed in $\mathbb{Z}_4[x]$,

$$x^n - 1 = f_2(x)a(x) + 2b(x),$$

where $a(x), b(x) \in \mathbb{Z}_4[x]$. Then

$$(-x)^n - 1 = f_2(-x)a(-x) + 2b(-x)$$

and

$$
\begin{aligned}
x^{2n} - 1 &= (x^n - 1)(x^n + 1) \\
&= -f_2(x)f_2(-x)a(x)a(-x) \\
&\quad + 2[f_2(x)a(x)b(-x) + f_2(-x)a(-x)b(x)] \\
&= \mp f(x^2)a(x)a(-x) + 2[f_2(x)a(x)b(-x) + f_2(-x)a(-x)b(x)].
\end{aligned}
$$

Let $\deg(f_2(x)a(x)b(-x)) = l$ and write

$$f_2(x)a(x)b(-x) = a_0 + a_1x + a_2x^2 + \cdots + a_lx^l, \ a_i \in \mathbb{Z}_4.$$

We see that

$$2[f_2(x)a(x)b(-x) + f_2(-x)a(-x)b(x)]$$
$$= 2(2a_0 + 2a_2x^2 + \cdots) = 0 \ \text{ in } \mathbb{Z}_4[x].$$

Therefore $f(x^2) \mid (x^{2n} - 1)$ in $\mathbb{Z}_4[x]$. Hence $f(x) \mid (x^n - 1)$ in $\mathbb{Z}_4[x]$. We conclude that $f(x)$ is the Hensel lift of $f_2(x)$. □

Example 13.1 Let $f_2(x) = x^2 + x + 1 \in \mathbb{Z}_2[x]$. Write $f_2(x) = e(x) - d(x)$, where $e(x) = x^2 + 1$ and $d(x) = -x$. Then

$$e(x)^2 - d(x)^2 = x^4 + x^2 + 1.$$

Hence $f(x) = x^2 + x + 1$ is the Hensel lift of $x^2 + x + 1$. □

Example 13.2 Let $f_2(x) = x^3 + x + 1 \in \mathbb{Z}_2[x]$. Write $f_2(x) = e(x) - d(x)$, where $e(x) = 1$ and $d(x) = -x^3 - x$. We have

$$-e(x)^2 + d(x)^2 = x^6 + 2x^4 + x^2 - 1.$$

Then $f(x) = x^3 + 2x^2 + x - 1$ is the Hensel lift of $x^3 + x + 1$. □

13.5   Exercises

13.1 Show that $x^2 - 2$ has 0 as a multiple root in $\mathbb{Z}_2$, but no roots in $\mathbb{Z}_8$.

13.2 Factorize the following polynomials over $\mathbb{Z}_4$:

  (i) $x^5 + 3x^4 + 2x^3 + 1$.

  (ii) $x^6 + 3x^5 + x^4 + x^3 + x^2 + 3x + 1$.

13.3 Prove that the monic basic irreducible polynomial $x - 3$ over $\mathbb{Z}_{2^2}$ is not a Hensel lift of $\overline{x - 3} = x + 1$ over $\mathbb{F}_2$.

13.4 Find the Hensel lifts in $\mathbb{Z}_4[x]$ of the following polynomials in $\mathbb{F}_2[x]$:

  (i) $x^3 + x + 1$.

  (ii) $x^4 + x^3 + x^2 + x + 1$.

  (iii) $x^5 + x^2 + 1$.

13.5 Let $f(x) \in \mathbb{Z}_{p^s}[x]$ and assume $\overline{f}(x)$ has a nonzero constant term. Prove that there is a smallest positive integer T such that $f(x) \mid (x^T - 1)$ in $\mathbb{Z}_{p^s}[x]$. (T is called the *period* of $f(x)$ over $\mathbb{Z}_{p^s}$.)

Chapter 14

Galois Rings

14.1 Examples of Galois Rings

The theory of Galois Rings was first developed by W. Krull (1924).

Definition 14.1 *A Galois ring is defined to be a finite ring with identity 1 such that the set of its zero divisors with 0 added forms a principal ideal $(p1)$ for some prime number p.*

We recall that in the above definition

$$p1 = \underbrace{1 + 1 + \cdots + 1}_{p}.$$

Example 14.1 Consider the ring $\mathbb{Z}_{p^s}$, where p is a prime number and s is a positive integer. As before we identify $n1$ with n for every $n \in \mathbb{Z}_{p^n}$. Clearly, 1 is the identity of $\mathbb{Z}_{p^s}$ and the set of its zero divisors with 0 added forms the principal ideal (p). Therefore $\mathbb{Z}_{p^s}$ is a Galois ring with p^s elements. By Theorem 13.1, (p) is the unique maximal ideal of $\mathbb{Z}_{p^s}$.

When $s = 1$, $\mathbb{Z}_{p^s} = \mathbb{F}_p$ is the field with p elements and $(p) = (0)$ is the zero ideal.

Example 14.2 Let $h(x)$ be a monic basic irreducible polynomial of degree m in $\mathbb{Z}_{p^s}[x]$. Consider the residue class ring

$$\mathbb{Z}_{p^s}[x]/(h(x)).$$

The residue classes

$$a_0 + a_1 x + \cdots + a_{m-1} x^{m-1} + (h(x)),$$

where $a_0, a_1, \ldots, a_{m-1} \in \mathbb{Z}_{p^s}$, are all the distinct elements of $\mathbb{Z}_{p^s}[x]/(h(x))$. Therefore $|\mathbb{Z}_{p^s}[x]/(h(x))| = p^{sm}$. It is evident that $1 + (h(x))$ is the identity of $\mathbb{Z}_{p^s}[x]/(h(x))$ and $(h(x))$ is its zero. Hence $\mathbb{Z}_{p^s}[x]/(h(x))$ is a finite commutative ring with identity. Clearly, $p[1 + (h(x))] = p + (h(x))$ and $(p[1 + (h(x))]) = (p + (h(x)))$. For any element

$$[a_0 + a_1 x + \cdots + a_{m-1} x^{m-1} + (h(x))][p + (h(x))]$$

of the principal ideal $(p[1 + (h(x))]) = (p + (h(x)))$, we have

$$[p^{s-1} + (h(x))][a_0 + a_1 x + \cdots + a_{m-1} x^{m-1} + (h(x))][p + (h(x))] = (h(x)).$$

Thus, all elements of the principal ideal $(p + (h(x)))$ are either zero divisors or zero.

We have defined a ring homomorphism (13.2)

$$- : \quad \mathbb{Z}_{p^s} \to \mathbb{Z}_{p^s}/(p) \simeq \mathbb{F}_p$$
$$c_0 + c_1 p + \cdots + c_{s-1} p^{s-1} \mapsto c_0 + (p),$$

where $0 \le c_i \le p - 1$, whose kernel is the principal ideal (p) of the ring $\mathbb{Z}_{p^s}$. The ring homomorphism (13.2) has been extended to the ring homomorphism (13.3)

$$- : \quad \mathbb{Z}_{p^s}[x] \to \mathbb{F}_p[x]$$
$$a_0 + a_1 x + \cdots + a_n x^n \mapsto \overline{a}_0 + \overline{a}_1 x + \cdots + \overline{a}_n x^n,$$

where $a_0, a_1, \ldots, a_n \in \mathbb{Z}_{p^s}$ and we know that the kernel of (13.3) is the principal ideal (p) generated by p in $\mathbb{Z}_{p^s}[x]$. Clearly, the image of the ideal $(h(x))$ is the ideal $(\overline{h}(x))$. By Theorem 12.22 the ring homomorphism (13.3) induces a ring homomorphism

$$\mathbb{Z}_{p^s}[x]/(h(x)) \to \mathbb{F}_p[x]/(\overline{h}(x)) \tag{14.1}$$
$$a_0 + \cdots + a_{m-1} x^{m-1} + (h(x)) \mapsto \overline{a}_0 + \cdots + \overline{a}_{m-1} x^{m-1} + (\overline{h}(x)),$$

where $a_0, \ldots, a_{m-1} \in \mathbb{Z}_{p^s}$. This ring homomorphism will also be denoted by $-$.

It is clear that the kernel of the ring homomorphism (14.1) is the principal ideal $(p + (h(x)))$ generated by $p + (h(x))$ in $\mathbb{Z}_{p^s}[x]/(h(x))$. By the fundamental theorem of homomorphisms of rings

$$(\mathbb{Z}_{p^s}[x]/(h(x)))/(p + (h(x))) \simeq \mathbb{F}_p[x]/(\overline{h}(x)).$$

Since $h(x)$ is a monic basic irreducible polynomial of degree m over $\mathbb{Z}_{p^s}$, $\overline{h}(x)$ is a monic irreducible polynomial of degree m over $\mathbb{F}_p$ and $\mathbb{F}_p[x]/(\overline{h}(x))$ is the Galois field $\mathbb{F}_{p^m}$. It follows that the principal ideal $(p + (h(x)))$ generated by $p + (h(x))$ in $\mathbb{Z}_{p^s}[x]/(h(x))$ is a maximal ideal.

Let $a(x) + (h(x)) \in \mathbb{Z}_{p^s}[x]/(h(x))$ and $a(x) + (h(x)) \notin (p + (h(x)))$. Then the ideal $(a(x) + (h(x)), p + (h(x)))$ generated by $a(x) + (h(x))$ and the principal ideal $(p + (h(x)))$ is the whole ring $\mathbb{Z}_{p^s}[x]/(h(x))$. Thus, there are elements $b(x) + (h(x))$ and $c(x) + (h(x)) \in \mathbb{Z}_{p^s}[x]/(h(x))$ such that

$$1 + (h(x)) = [b(x) + (h(x))][a(x) + (h(x))] + [c(x) + (h(x))][p + (h(x))].$$

Then
$$1 + (h(x)) = b(x)a(x) + pc(x) + (h(x)).$$

We recall that $1 + (h(x))$ is the identity element of $\mathbb{Z}_{p^s}[x]/(h(x))$, therefore

$$1 + (h(x)) = [1 + (h(x))]^p = [b(x)a(x)]^p + p^2 c_1(x) + (h(x)),$$

where $c_1(x) \in \mathbb{Z}_{p^s}[x]$,

$$1 + (h(x)) = [1 + (h(x))]^{p^2} = [b(x)a(x)]^{p^2} + p^3 c_2(x) + (h(x)),$$

where $c_2(x) \in \mathbb{Z}_{p^s}[x]$, etc., and finally

$$
\begin{aligned}
1 + (h(x)) &= [b(x)a(x)]^{p^{s-1}} + (h(x)) \\
&= [a(x) + (h(x))][b(x)^{p^{s-1}} a(x)^{p^{s-1}-1} + (h(x))].
\end{aligned}
$$

Thus $a(x) + (h(x))$ is a unit of $\mathbb{Z}_{p^s}[x]/(h(x))$. Hence the principal ideal $(p + (h(x)))$ consists of all zero divisors and 0 of the ring $\mathbb{Z}_{p^s}[x]/(h(x))$. This proves that $\mathbb{Z}_{p^s}[x]/(h(x))$ is a Galois ring. Moreover, since all elements of $[\mathbb{Z}_{p^s}[x]/(h(x))] \setminus (p + (h(x)))$ are units, $(p + (h(x)))$ is the only maximal ideal of $\mathbb{Z}_{p^s}[x]/(h(x))$. We write also the principal ideal $(p + (h(x)))$ simply as (p).

$\mathbb{Z}_{p^s}[x]/(h(x))$ contains a subring R_0 consisting of elements of the form $a + (h(x))$, where $a \in \mathbb{Z}_{p^s}$. Clearly, the map

$$a \mapsto a + (h(x))$$

is an isomorphism of $\mathbb{Z}_{p^s}$ to R_0. Then we identify $a \in \mathbb{Z}_{p^s}$ with $a + (h(x)) \in \mathbb{Z}_{p^s}[x]/(h(x))$ and regard $\mathbb{Z}_{p^s}$ as a subring of $\mathbb{Z}_{p^s}[x]/(h(x))$.

For simplicity, we write $\xi = x + (h(x))$. Then $h(\xi) = 0$, i.e., ξ is a root of $h(x)$ and

$$a_0 + a_1 x + \cdots + a_{m-1} x^{m-1} + (h(x)) = a_0 + a_1 \xi + \cdots + a_{m-1} \xi^{m-1}.$$

Therefore
$$\mathbb{Z}_{p^s}[x]/(h(x)) = \mathbb{Z}_{p^s}[\xi],$$

and all elements of $\mathbb{Z}_{p^s}[x]/(h(x))$ can be expressed uniquely in the form

$$a_0 + a_1 \xi + \cdots + a_{m-1} \xi^{m-1}, \ a_i \in \mathbb{Z}_{p^s} \ (i = 0, 1, \ldots, m-1), \qquad (14.2)$$

which is called the *additive representation* of the element of the Galois ring $\mathbb{Z}_{p^s}[\xi]$. $\mathbb{Z}_{p^s}[\xi]$ is a free $\mathbb{Z}_{p^s}$-module of rank m and $\{1, \xi, \xi^2, \ldots, \xi^{m-1}\}$ is a free basis. The multiplication of $\mathbb{Z}_{p^s}[\xi]$ is performed as follows. Let $a(\xi), b(\xi) \in \mathbb{Z}_{p^s}[\xi]$, then

$$a(\xi)b(\xi) = r(\xi),$$

where $r(\xi)$ is obtained by substituting $x = \xi$ in the remainder $r(x)$ of dividing $a(x)b(x)$ by the monic polynomial $h(x)$ in $\mathbb{Z}_{p^s}[x]$.

Denote the image of $\xi = x + (h(x))$ under the map (14.1) by $\bar{\xi}$, i.e., $\bar{\xi} = x + (\bar{h}(x))$, then $\bar{\xi}$ is a root of the monic irreducible polynomial $\bar{h}(x)$ over $\mathbb{F}_p$ and

$$\mathbb{F}_p[x]/(\bar{h}(x)) = \mathbb{F}_p[\bar{\xi}] \simeq \mathbb{F}_{p^m}.$$

Moreover, (14.1) can be written as

$$- : \quad \mathbb{Z}_{p^s}[\xi] \to \mathbb{F}_p[\bar{\xi}]$$
$$a_0 + a_1\xi + \cdots + a_{m-1}\xi^{m-1} \mapsto \bar{a}_0 + \bar{a}_1\bar{\xi} + \cdots + \bar{a}_{m-1}\bar{\xi}^{m-1}. \quad (14.3)$$

Obviously, the following diagram is commutative.

$$\begin{array}{ccc} \mathbb{Z}_{p^s}[x] & \overset{-}{\longrightarrow} & \mathbb{F}_p[x] \\ \downarrow & & \downarrow \\ \mathbb{Z}_{p^s}[\xi] & \overset{-}{\longrightarrow} & \mathbb{F}_p[\xi] \end{array}$$

If the monic basic irreducible polynomial $h(x)$ is of degree 1, then $\mathbb{Z}_{p^s}[x]/(h(x)) \simeq \mathbb{Z}_{p^s}$.

If $s = 1$, $\mathbb{Z}_{p^s}[x]/(h(x)) = \mathbb{F}_p[x]/(h(x)) \simeq \mathbb{F}_{p^m}$, where $m = \deg h(x)$.

Definition 14.2 *For a commutative ring with identity 1, the order of 1 in the additive group of the ring is called the characteristic of the ring.*

Clearly, both the ring $\mathbb{Z}_{p^s}$ and the ring $\mathbb{Z}_{p^s}[x]/(h(x))$, where $h(x)$ is a monic basic irreducible polynomial over $\mathbb{Z}_{p^s}$, are of characteristic p^s.

Denote the Galois ring $\mathbb{Z}_{p^s}[x]/(h(x))$, where $h(x)$ is a monic basic irreducible polynomial of degree m over $\mathbb{Z}_{p^s}$, by $GR(p^s, p^{sm})$, where p^s is its characteristic and p^{sm} its cardinality. In particular, $\mathbb{Z}_{p^s} = GR(p^s, p^s)$. The notation $GR(p^s, p^{sm})$ being independent of $h(x)$ will be justified later by Theorem 14.6.

We summarize the results obtained in Example 14.2 in the following theorem.

Theorem 14.1 *Let $h(x)$ be a monic basic irreducible polynomial of degree m over $\mathbb{Z}_{p^s}$. Then the residue class ring $\mathbb{Z}_{p^s}[x]/(h(x))$ is a Galois ring of characteristic p^s and cardinality p^{sm}, it contains $\mathbb{Z}_{p^s}$ as a subring, and is denoted by $GR(p^s, p^{sm})$. Write $\xi = x + (h(x))$, then $h(\xi) = 0$, all elements of $\mathbb{Z}_{p^s}[x]/(h(x))$ can be expressed uniquely in the form (14.2), $\mathbb{Z}_{p^s}[x]/(h(x)) = \mathbb{Z}_{p^s}[\xi]$, and $\mathbb{Z}_{p^s}[\xi]$ is a free $\mathbb{Z}_{p^s}$-module of rank m with $\{1, \xi, \xi^2, \ldots, \xi^{m-1}\}$ as a free basis. The ideal (p) of $\mathbb{Z}_{p^s}[\xi]$ consists of all zero divisors together with the zero of $\mathbb{Z}_{p^s}[\xi]$ and is the only maximal ideal of $\mathbb{Z}_{p^s}[\xi]$. Moreover, denote the image of ξ under the ring homomorphism (14.1) by $\overline{\xi}$, i.e., $\overline{\xi} = x + (\overline{h}(x))$, then $\overline{h}(\overline{\xi}) = 0$, $\mathbb{F}_p[x]/(\overline{h}(x)) = \mathbb{F}_p[\overline{\xi}] = \mathbb{F}_{p^m}$, and (14.1) can be written as (14.3). Finally, the following diagram*

$$
\begin{array}{ccc}
\mathbb{Z}_{p^s}[x] & \overset{-}{\longrightarrow} & \mathbb{F}_p[x] \\
\downarrow & & \downarrow \\
\mathbb{Z}_{p^s}[\xi] & \overset{-}{\longrightarrow} & \mathbb{F}_p[\xi]
\end{array}
$$

is commutative.

14.2 Structure of Galois Rings

Lemma 14.2 *Let R be a Galois ring whose zero divisors together with 0 form a principal ideal $(p1)$ for a prime number p. Then $(p1)$ is the only maximal ideal of R, $R/(p1)$ is a finite field $\mathbb{F}_{p^m}$ for some positive integer m, and the characteristic of R is a power of p.*

Proof. In a finite ring every nonzero element which is not a zero divisor is a unit (Exercise 3.16). Therefore $(p1)$ is the only maximal ideal of R and $R/(p1)$ is a finite field. Denote the natural homomorphism $R \to R/(p1)$ by $-$ and the image of $r \in R$ in $R/(p1)$ by $\bar{r}$. Let n be any positive integer and $a \in R$ or $R/(p1)$, denote

$$
\underbrace{a + a + \cdots + a}_{n}
$$

by na. Then $p\bar{1} = \overline{p1} = \bar{0}$. Therefore $R/(p1)$ is a finite field of characteristic p. Let $R/(p1) \simeq \mathbb{F}_{p^m}$ for some positive integer m.

Let k be the characteristic of R. From $k1 = 0$ we deduce $k\bar{1} = \overline{k1} = \bar{0}$. Therefore $p \mid k$. Assume that $k = p^s l$, where s, l are positive integers and $\gcd(p, l) = 1$. If $l > 1$, then $a = p^s 1$ and $b = l1$ are nonzero elements of R and $ab = 0$. It follows that $l1 \in (p1)$ and $l\bar{1} = \overline{l1} = \bar{0}$ in $R/(p1)$. But $R/(p1)$ is of characteristic p, so $p \mid l$, which contradicts $\gcd(p, l) = 1$. Therefore $l = 1$ and $k = p^s$. $\square$

Lemma 14.3 *Let R be a Galois ring of characteristic p^s, where p is a prime, and let 1 be the identity of R. Then the zero divisors of R together with 0 form the principal ideal $(p1)$, $\{r1 : r \in \mathbb{Z}_{p^s}\}$ is a subring of R and is isomorphic to $\mathbb{Z}_{p^s}$.*

Proof. By Definition 14.1 let the zero divisors of R together with 0 form a principal ideal $(q1)$ for some prime number q. By Lemma 14.2, the characteristic of R is a power of q. Therefore $q = p$.

It is easy to verify that the map

$$\mathbb{Z}_{p^s} \to R$$
$$r \mapsto r1$$

is an isomorphism of $\mathbb{Z}_{p^s}$ into R and its image is $\{r1 : r \in \mathbb{Z}_{p^s}\}$. $\quad\square$

Let R be a Galois ring of characteristic p^s. We usually identify $r \in \mathbb{Z}_{p^s}$ with the element $r1 \in R$ for all $r \in \mathbb{Z}_{p^s}$ and regard $\mathbb{Z}_{p^s}$ as a subring of R. In particular, $p \in \mathbb{Z}_{p^s}$ is identified with $p1 \in R$. The principal ideal $(p1)$ is usually written as (p) or Rp.

Lemma 14.4 *Let R be a Galois ring of characteristic p^s, where p is a prime number, and let (p) be the principal ideal consisting of the zero divisors of R and 0. Then $R/(p) \simeq \mathbb{F}_{p^m}$ for some positive integer m, the principal ideal (p^i), $0 \le i \le s$, has cardinality $p^{(s-i)m}$, and in particular, $|R| = p^{sm}$.*

Proof. By Lemma 14.2 (p) is the only maximal ideal of R, then $R/(p) \simeq \mathbb{F}_{p^m}$ for some positive integer m. Regard R as an additive group, the principal ideals (p^i), $0 \le i \le s - 1$, as its subgroups, and consider the map

$$R \to (p^i)/(p^{i+1})$$
$$r \mapsto p^i r + (p^{i+1}).$$

Clearly, this is a homomorphism of groups, it is surjective, and the kernel is an ideal of R and includes (p). Suppose 1 belongs to the kernel, then $p^i \in (p^{i+1})$ and $p^i = cp^{i+1}$, where $c \in R$. Thus $p^{s-1} = p^{s-1-i}p^i = p^{s-1-i}cp^{i+1} = cp^s = 0$, which is a contradiction. Thus 1 does not belong to the kernel. Since (p) is a maximal ideal of R, the kernel must be (p). By the fundamental theorem of homomorphisms of groups we have the isomorphism of additive groups

$$R/(p) \simeq (p^i)/(p^{i+1}).$$

It follows that

$$|R/(p)| = |(p)/(p^2)| = \cdots = |(p^{s-2})/(p^{s-1})| = |(p^{s-1})|.$$

But $R/(p) \simeq \mathbb{F}_{p^m}$, thus $|R/(p)| = p^m$. Therefore

$$|(p^i)| = |(p^i)/(p^{i+1})||(p^{i+1})/(p^{i+2})| \cdots |(p^{s-1})| = |(p^{(s-i)m})|.$$

In particular, $|R| = |(1)| = |(p^0)| = p^{sm}$. $\qquad\qquad\square$

Let R be a Galois ring of characteristic p^s. The natural homomorphism $R \to R/Rp$ will be denoted by $-$ and the image of $r \in R$ in R/Rp by $\bar{r}$. We have regarded $\mathbb{Z}_{p^s}$ as a subring of R. Then $-$ induces a homomorphism $\mathbb{Z}_{p^s} \to (\mathbb{Z}_{p^s} + Rp)/Rp$. By Theorem 12.23,

$$(\mathbb{Z}_{p^s} + Rp)/Rp \simeq \mathbb{Z}_{p^s}/(\mathbb{Z}_{p^s} \cap Rp)$$

and it is easy to verify that $\mathbb{Z}_{p^s} \cap Rp = \mathbb{Z}_{p^s}p$. We know that $\mathbb{Z}_{p^s}/\mathbb{Z}_{p^s}p \simeq \mathbb{F}_p$, which can be regarded as a subfield of $R/Rp \simeq \mathbb{F}_{p^m}$. Therefore $-$ induces a surjective homomorphism $\mathbb{Z}_{p^s} \to \mathbb{F}_p$, which coincides with (13.2). We have also the extended homomorphism (13.3)

$$\mathbb{Z}_{p^s}[x] \to \mathbb{F}_p[x]$$

and the homomorphim

$$R[x] \to R/(p)[x]$$
$$a_0 + a_1 x + \cdots + a_n x^n \mapsto \bar{a}_0 + \bar{a}_1 x + \cdots + \bar{a}_n x^n,$$

where $a_0, a_1, \ldots, a_n \in R$. Both these homomorphisms are also denoted by $-$. The image of $f(x) \in \mathbb{Z}_{p^s}[x]$ or $R[x]$ under $-$ is denoted by $\bar{f}(x)$.

Lemma 14.5 *Let R be a Galois ring of characteristic p^s and cardinality p^{sm}, where p is a prime number and s and m are positive integers. Let $f(x)$ be a polynomial over $\mathbb{Z}_{p^s}$ and assume that $\bar{f}(x)$ has a root $\bar{\beta}$ in $R/(p) \simeq \mathbb{F}_{p^m}$ and that $\bar{f}'(\bar{\beta}) \neq \bar{0}$. Then there exists a unique root $\alpha \in R$ of the polynomial $f(x)$ such that $\bar{\alpha} = \bar{\beta}$.*

Proof. Let β be an original of $\bar{\beta}$ under the map $- : R \longrightarrow R/(p)$. Let us construct a sequence of elements $\alpha_0, \alpha_1, \ldots, \alpha_{s-1} \in R$ such that $\bar{\alpha}_i = \bar{\beta}$ and $f(\alpha_i) \in (p^{i+1})$ for $i = 0, 1, \ldots, s - 1$. First, let $\alpha_0 = \beta$, then $\bar{\alpha}_0 = \bar{\beta}$ and $f(\alpha_0) = f(\beta) \in (p)$, since $\bar{f}(\bar{\beta}) = \bar{0}$. Assume that the element α_i has already been constructed and satisfies $\bar{\alpha}_i = \bar{\beta}$ and $f(\alpha_i) \in (p^{i+1})$. Then $\bar{f}'(\bar{\alpha}_i) = \bar{f}'(\bar{\beta}) \neq \bar{0}$, thus $f'(\alpha_i)$ is a unit of R. Let $\alpha_{i+1} = \alpha_i - f'(\alpha_i)^{-1}f(\alpha_i) \in R$, then $\bar{\alpha}_{i+1} = \bar{\alpha}_i$ and by Taylor's formula (Exercise 14.1)

$$f(\alpha_{i+1}) = f(\alpha_i) + \tfrac{f'(\alpha_i)}{1!}(-f'(\alpha_i)^{-1}f(\alpha_i))$$
$$+ \tfrac{f''(\alpha_i)}{2!}(-f'(\alpha_i)^{-1}f(\alpha_i))^2 + \cdots \in (p^{i+2}).$$

By mathematical induction, finally $\alpha_{s-1} \in R$ is constructed and satisfies $\overline{\alpha_{s-1}} = \overline{\beta}$ and $f(\alpha_{s-1}) \in (p^s) = (0)$, which implies $f(\alpha_{s-1}) = 0$. Let $\alpha = \alpha_{s-1}$, then $f(\alpha) = 0$ and $\overline{\alpha} = \overline{\beta}$.

Let α' be any root of $f(x)$ such that $\overline{\alpha}' = \overline{\beta}$. Then $\overline{\alpha}' = \overline{\alpha}$. If $\alpha' \neq \alpha$, there is an integer i, $1 \leq i \leq s - 1$, and an element $r \in R$ such that $\alpha' = \alpha + p^i r$ and $p^i r \notin (p^{i+1})$. Again by Taylor's formula

$$f(\alpha') = f(\alpha) + \frac{f'(\alpha)}{1!}(p^i r) + \frac{f''(\alpha)}{2!}(p^i r)^2 + \cdots .$$

Since $f(\alpha') = f(\alpha) = 0$, we have $f'(\alpha)p^i r \in (p^{2i}) \subseteq (p^{i+1})$. But $\overline{f'}(\overline{\alpha}) = \overline{f}'(\overline{\beta}) \neq \overline{0}$, thus $f'(\alpha)$ is a unit of R. It follows that $p^i r \in (p^{i+1})$, a contradiction. □

Theorem 14.6 *Let R be a Galois ring of characteristic p^s and cardinality p^{sm}, where p is a prime and s and m are positive integers. Then R is isomorphic to the ring $\mathbb{Z}_{p^s}[x]/(h(x))$ for any monic basic irreducible polynomial $h(x)$ of degree m over $\mathbb{Z}_{p^s}$.*

Proof. Let $h(x)$ be any monic basic irreducible polynomial of degree m over $\mathbb{Z}_{p^s}$. Then $\overline{h}(x)$ is monic irreducible over $\mathbb{F}_p$, $\deg \overline{h}(x) = m$, and $\overline{h}(x)$ has a root in $\mathbb{F}_{p^m}$. By Lemma 14.4 $R/(p) \simeq \mathbb{F}_{p^m}$. Then $\overline{h}(x)$ has a root in $R/(p)$. Let it be $\overline{\beta}$, then $\overline{h}(\overline{\beta}) = 0$. Since $\overline{h}(x)$ is irreducible, $\overline{h}(x)$ has no multiple root. Therefore $\overline{h}'(\beta) \neq 0$. By Lemma 14.5 there exists a unique root $\alpha \in R$ of the polynomial $h(x)$ such that $\overline{\alpha} = \overline{\beta}$. Consider the map

$$\mathbb{Z}_{p^s}[x]/(h(x)) \to R$$
$$a_0 + \cdots + a_{m-1}x^{m-1} + (h(x)) \mapsto a_0 + \cdots + a_{m-1}\alpha^{m-1}, \quad (14.4)$$

where $a_0, \ldots, a_{m-1} \in \mathbb{Z}_{p^s}$. Clearly, it is a well-defined ring homomorphism. Let us prove that it is injective. Assume that $a_0 + \cdots + a_{m-1}\alpha^{m-1} = 0$, then $\overline{a}_0 + \cdots + \overline{a}_{m-1}\overline{\alpha}^{m-1} = 0$. But $\overline{\alpha} = \overline{\beta}$ and $\overline{\beta}$ is a root of the irreducible polynomial $\overline{h}(x)$ of degree m, so $\overline{a}_0 = \overline{a}_1 = \cdots = \overline{a}_{m-1} = 0$, then we may write $a_i = pb_i^{(1)}$, where $b_i^{(1)} \in \mathbb{Z}_{p^s}$, $i = 0, 1, \ldots, m - 1$. It follows that $p(b_0^{(1)} + b_1^{(1)}\alpha + \cdots + b_{m-1}^{(1)}\alpha^{m-1}) = 0$. Then $b_0^{(1)} + b_1^{(1)}\alpha + \cdots + b_{m-1}^{(1)}\alpha^{m-1}$ is either a zero divisor or zero, $b_0^{(1)} + b_1^{(1)}\alpha + \cdots + b_{m-1}^{(1)}\alpha^{m-1} \in (p)$, and $\overline{b_0^{(1)}} + \overline{b_1^{(1)}}\overline{\alpha} + \cdots + \overline{b_{m-1}^{(1)}}\overline{\alpha}^{m-1} = 0$. As before, we have $\overline{b_0^{(1)}} = \overline{b_1^{(1)}} = \cdots = \overline{b_{m-1}^{(1)}} = 0$, then we may write $b_i^{(1)} = pb_i^{(2)}$, where $b_i^{(2)} \in \mathbb{Z}_{p^s}$, $i = 0, 1, \ldots, m - 1$. Proceeding in this way, we obtain successively, for $j = 1, 2, \ldots, s-1$, $b_i^{(j)} = pb_i^{(j+1)}$, where $b_i^{(j)}, b_i^{(j+1)} \in \mathbb{Z}_{p^s}$. Then

$$a_i = pb_i^{(1)} = p^2 b_i^{(2)} = \cdots = p^s b_i^{(s)} = 0, \quad i = 0, 1, \ldots, m - 1.$$

Therefore the map (14.4) is injective. By Theorem 14.1 $|\mathbb{Z}_{p^s}[x]/(h(x))| = p^{sm}$ and by hypothesis $|R| = p^{sm}$. Therefore the map (14.4) is also surjective. Hence $\mathbb{Z}_{p^s}[x]/(h(x)) \simeq R$. $\square$

Corollary 14.7 *Any two Galois rings of the same characteristic and the same cardinality are isomorphic.* $\square$

Therefore we can use the notation $GR(p^s, p^{sm})$ to denote any Galois ring of characteristic p^s and cardinality p^{sm}.

14.3 The p-adic Representation

We know that every element of $\mathbb{Z}_{p^s} = GR(p^s, p^s)$ can be expressed in the form (13.1)

$$c_0 + c_1 p + \cdots + c_{s-1} p^{s-1}, \quad \text{where } 0 \le c_i \le p - 1.$$

This can be generalized to general Galois rings.

Theorem 14.8 (i) *In the Galois ring $GR(p^s, p^{sm})$ there exists a nonzero element ξ of order $p^m - 1$, which is a root of a monic basic primitive polynomial $h(x)$ of degree m over $\mathbb{Z}_{p^s}$ and dividing $x^{p^m - 1} - 1$ in $\mathbb{Z}_{p^s}[x]$, and every element of $GR(p^s, p^{sm})$ can be expressed uniquely in the form* (14.2)

$$a_0 + a_1 \xi + \cdots + a_{m-1} \xi^{m-1}, \ a_i \in \mathbb{Z}_{p^s} \ (i = 0, 1, \ldots, m-1).$$

Thus

$$GR(p^s, p^{sm}) = \mathbb{Z}_{p^s}[\xi] = \{a_0 + a_1 \xi + \cdots + a_{m-1} \xi^{m-1} : a_0, a_1, \ldots, a_{m-1} \in \mathbb{Z}_{p^s}\}.$$
(14.5)
Moreover, $h(x)$ is the unique monic polynomial of degree $\le m$ over $\mathbb{Z}_{p^s}$ and having ξ as a root.

(ii) *Let*

$$\mathcal{T} = \{0, 1, \xi, \xi^2, \ldots, \xi^{p^m - 2}\},$$

then every element $c \in GR(p^s, p^{sm})$ can be written uniquely as

$$c = c_0 + c_1 p + \cdots + c_{s-1} p^{s-1}, \tag{14.6}$$

where $c_0, c_1, \ldots, c_{s-1} \in \mathcal{T}$. Moreover, c is a unit if and only of $c_0 \ne 0$, and c is a zero divisor or 0 if and only if $c_0 = 0$.

Proof. (i) By Lemma 13.9 there is a monic basic primitive polynomial $h(x)$ of degree m over $\mathbb{Z}_{p^s}$ and dividing $x^{p^m - 1} - 1$ in $\mathbb{Z}_{p^s}[x]$. Then $\bar{h}(x)$ is a monic primitive polynomial of degree m over $\mathbb{F}_p$. Let ξ_p be a root of $\overline{h}(x)$ in $\mathbb{F}_{p^m}$, then ξ_p is a primitive element of $\mathbb{F}_{p^m}$. By Lemma 14.5 the polynomial $h(x)$ has a unique root $\xi \in GR(p^s, p^{sm})$ such that $\overline{\xi} = \xi_p$. Since $h(x)$ divides $x^{p^m-1} - 1$, $\xi^{p^m-1} - 1 = 0$. Since $\overline{\xi} = \xi_p$ is of order $p^m - 1$, ξ is also of order $p^m - 1$.

By the proof of Theorem 14.6, the map

$$\mathbb{Z}_{p^s}[x]/(h(x)) \to GR(p^s, p^{sm})$$
$$a_0 + \cdots + a_{m-1}x^{m-1} + (h(x)) \mapsto a_0 + \cdots + a_{m-1}\xi^{m-1}, \quad (14.7)$$

where $a_0, \ldots, a_{m-1} \in \mathbb{Z}_{p^s}$, is a ring isomorphism. Therefore every element of $GR(p^s, p^{sm})$ can be expressed uniquely in the form (14.2) and we have (14.5).

Now let $f(x)$ be a monic polynomial of degree $\leq m$ over $\mathbb{Z}_{p^s}$ and having ξ as a root. If $\deg f(x) < m$, by the ring isomorphism (14.7) $f(x) \in (h(x))$, which is impossible. If $\deg f(x) = m$, then $\deg(f(x) - h(x)) < m$, but $f(\xi) - h(\xi) = 0$, by the isomorphism (14.7) we must have $f(x) - h(x) = 0$ and, hence, $f(x) = h(x)$.

(ii) We observe that for $i = 0, 1, \ldots, p^m - 2$,

$$\xi^i \xi^{p^m - 1 - i} = 1.$$

Thus all ξ^i, $0 \leq i \leq p^m - 2$, are units of $GR(p^s, p^{sm})$. Moreover, all $1 - \xi^j$, $0 < j \leq p^m - 2$, are also units of $GR(p^s, p^{sm})$; for, suppose $1 - \xi^j \in (p)$, then $\overline{\xi}^j = 1$, which contracts the order of $\overline{\xi}$ being $p^m - 1$. It follows that for $0 \leq i < j \leq p^m - 2$, all $\xi^i - \xi^j$ are units of $GR(p^s, p^{sm})$.

We know that $|GR(p^s, p^{sm})| = p^{sm}$. If we can show that all the p^{sm} elements of the form (14.6) are distinct, then the first statement of (ii) will be proved. Assume that

$$c_0 + c_1 p + \cdots + c_{s-1}p^{s-1} = c_0' + c_1'p + \cdots + c_{s-1}'p^{s-1}, \quad (14.8)$$

where $c_0, c_1, \ldots, c_{s-1}, c_0', c_1', \ldots, c_{s-1}' \in \mathcal{T}$. Multiplying (14.8) by p^{s-1}, we obtain $c_0 p^{s-1} = c_0' p^{s-1}$. Clearly $c_0 = 0$ if and only if $c_0' = 0$. Assume that both c_0 and c_0' are nonzero, we have $(c_0 - c_0')p^{s-1} = 0$. If $c_0 - c_0' \neq 0$, by what we proved in the proceeding paragraph $c_0 - c_0'$ is a unit of $GR(p^s, p^{sm})$. Multiplying $(c_0 - c_0')p^{s-1} = 0$ by the inverse of $c_0 - c_0'$, we obtain $p^{s-1} = 0$, which is impossible. Therefore $c_0 = c_0'$. Then we have

$$c_1 p + \cdots + c_{s-1}p^{s-1} = c_1'p + \cdots + c_{s-1}'p^{s-1}. \quad (14.9)$$

Multiplying (14.9) by p^{s-2} we obtain $c_1 p^{s-1} = c_1' p^{s-1}$. Reasoning as above, we obtain $c_1 = c_1'$. Then we have

$$c_2 p^2 + \cdots + c_{s-1}p^{s-1} = c_2'p^2 + \cdots + c_{s-1}'p^{s-1}.$$

Proceeding as above, we obtain successively $c_2 = c_2', \ldots, c_{s-1} = c_{s-1}'$.

Finally, it is clear that the element (14.6) belongs to (p) if and only if $c_0 = 0$. By Lemma 14.3, (p) consists of all zero divisors and the zero of $GR(p^s, p^{sm})$. Therefore the last assertion of (ii) is proved. $\square$

The representation (14.6) of an element of $GR(p^s, p^{sm})$ is called the *p-adic representation* of the element.

Example 14.3 Let $h(x) = x^3 + 2x^2 + x - 1 \in \mathbb{Z}_{2^2}$. By Example 13.2, $h(x)$ is the Hensel lift of $\bar{h}(x) = x^3 + x + 1 \in \mathbb{F}_2[x]$. $\bar{h}(x)$ is a primitive polynomial in $\mathbb{F}_2[x]$, so $h(x)$ is a monic basic primitive polynomial in $\mathbb{Z}_{2^2}[x]$. By Lemma 14.5, let ξ be a root of $h(x)$ in $GR(2^2, 2^6) = \mathbb{Z}_{2^2}[x]/(h(x))$ such that $\bar{\xi}$ is a root of $\bar{h}(x)$ in $\mathbb{F}_{2^3}$. Then ξ is of order 7. We have

$$\xi^0 = 1, \ \xi^1 = \xi, \ \xi^2 = \xi^2,$$
$$\xi^3 = 2\xi^2 + 3\xi + 1,$$
$$\xi^4 = 3\xi^2 + 3\xi + 2,$$
$$\xi^5 = \xi^2 + 3\xi + 3,$$
$$\xi^6 = \xi^2 + 2\xi + 1.$$

Therefore

$$\mathcal{T} = \{0, 1, \xi, \xi^2, 2\xi^2 + 3\xi + 1, 3\xi^2 + 3\xi + 2, \xi^2 + 3\xi + 3, \xi^2 + 2\xi + 1\}.$$

Corollary 14.9 *Express every element $c \in GR(p^s, p^{sm})$ in the form (14.6)*

$$c = c_0 + c_1 p + \cdots + c_{s-1} p^{s-1},$$

where $c_0, c_1, \ldots, c_{s-1} \in \mathcal{T}$. Then

(i) *All the elements c with $c_0 \neq 0$ are units and they form a multiplicative group of order $(p^m - 1)p^{(s-1)m}$, which is a direct product $\langle \xi \rangle \times \mathcal{E}$, where $\langle \xi \rangle$ is the cyclic group of order $p^m - 1$ generated by ξ and $\mathcal{E} = \{1 + \pi : \pi \in (p)\}$ is a group of order $p^{(s-1)m}$.*

(ii) *The order of ξ^i, $0 \leq i \leq p^m - 2$, is a divisor of $p^m - 1$ and the order of $1 + \pi$, where $\pi \in (p)$, is a divisor of p^{s-1}. An element of order a divisor of $p^m - 1$ in $GR(p^s, p^{sm})$ is of the form ξ^i, where $0 \leq i \leq p^m - 2$. In particular, an element of order $p^m - 1$ is of the form ξ^i, where $0 \leq i \leq p^m - 2$ and $\gcd(i, p^m - 1) = 1$, and is a root of a basic primitive polynomial of degree m over $\mathbb{Z}_{p^s}$ and dividing $x^{p^m - 1} - 1$ in $\mathbb{Z}_{p^s}[x]$.*

(iii) *All the elements c with $c_0 = 0$ are either zero divisors or zero, and they form the ideal (p) of $GR(p^s, p^{sm})$.*

(iv) *Every nonzero element $c \in GR(p^s, p^{sm})$ can be expressed uniquely in the form $c = up^r$, where u is a unit and $0 \leq r < s$.*

14.4   The Group of Units of a Galois Ring

Let $R = GR(p^s, p^{sm})$ and denote the group of units of R by R^*. By Corollary 14.9(i), $R^* = \langle \xi \rangle \times \mathcal{E}$, where $\langle \xi \rangle$ is the cyclic group of order $p^m - 1$ generated by ξ and $\mathcal{E} = \{1 + \pi : \pi \in (p)\}$ is a subgroup order $p^{(s-1)m}$. We shall explore further the structure of $\mathcal{E}$. We need the following lemma.

Lemma 14.10 *Let X be an indeterminate, p be an odd prime number, and a_t, b_t, c_t be the coefficients of X^t in the expansions of $(1 + pX)^N, (1 + 2X)^N, (1 + 4X)^N$, respectively.*

(i) *If $p^e \mid N$, then $p^{e+1} \mid a_1$ and $p^{e+2} \mid a_t$ for $t \geq 2$.*

(ii) *If $2^e \mid N$, then $2^{e+1} \mid b_t$ for $t = 1, 2$ and $2^{e+2} \mid b_t$ for $t \geq 3$.*

(iii) *If $2^e \mid N$, then $2^{e+2} \mid c_1$ and $2^{e+3} \mid c_t$ for $t \geq 2$.*

(iv) *$4 \mid c_t$ for all $t \geq 1$.*

Proof. (i) We have $a_1 = Np$. By assumption $p^e \mid N$, so $p^{e+1} \mid a_1$.

Now let $t \geq 2$. Then

$$a_t = \binom{N}{t} p^t = \frac{N}{t} \binom{N-1}{t-1} p^t.$$

Both

$$\binom{N}{t} \text{ and } \binom{N-1}{t-1}$$

are integers. Suppose $p^f \| t$. By assumption $p^e \mid N$, so

$$p^{e-f} \Big| \binom{N}{t},$$

which implies $p^{e-f+t} \mid a_t$. If $f = 0$, clearly $t \geq f + 2$. If $f > 0$, then $p^f \geq f + 2$ and thus $t \geq p^f \geq f + 2$. In both cases we have $-f + t \geq 2$. Consequently, $p^{e+2} \mid p^{e-f+t}$. Hence $p^{e+2} \mid a_t$.

(ii) and (iii) can be proved in the same way as (i).

(iv) is trivial. $\square$

Theorem 14.11 *Let $R = GR(p^s,\ p^{sm})$ and denote by R^* the group of units of R. Then*

$$R^* = G_1 \times G_2,$$

where G_1 is a cyclic group of order $p^m - 1$ and G_2 is a group of order $p^{(s-1)m}$ such that

(i) *If p is odd or if $p = 2$ and $s \leq 2$, then G_2 is a direct product of m cyclic groups each of order p^{s-1}.*

(ii) *If $p = 2$ and $s \geq 3$, then G_2 is a direct product of a cyclic group of order 2, a cyclic group of order 2^{s-2} and $m - 1$ cyclic groups each of order 2^{s-1}.*

Proof. When $s = 1$, $R = GR(p, p^m) = \mathbb{F}_{p^m}$. By Theorem 6.3 $R^* = \langle \xi \rangle$ is a cyclic group of order $p^m - 1$ generated by a primitive element ξ of $\mathbb{F}_{p^m}$. In this case $G_1 = R^*$ and $G_2 = 1$. Hence our theorem holds.

Now consider the case $s \geq 2$. By Theorem 14.8, $R = \mathbb{Z}_{p^s}[\xi]$, where ξ is an element of order $p^m - 1$ and is a root of a basic primitive polynomial $h(x)$ of degree m over $\mathbb{Z}_{p^s}$ and dividing $x^{p^m - 1} - 1$ in $\mathbb{Z}_{p^s}[x]$. Then by Corollary 14.9 $R^* = \langle \xi \rangle \times \mathcal{E}$, where $\langle \xi \rangle$ is a cyclic group of order $p^m - 1$ generated by ξ, and $\mathcal{E} = \{1 + \pi : \pi \in (p)\}$ is a subgroup of order $p^{(s-1)m}$. Let $G_1 = \langle \xi \rangle$ and $G_2 = \mathcal{E}$, then $R^* = G_1 \times G_2$. We have to determine the structure of G_2. We distinguish the following cases.

Case 1 p is odd. Clearly, $1 + p\xi^i \in G_2$ and by Lemma 14.10(i) $(1 + p\xi^i)^{p^{s-1}} = 1$ $(i = 0, 1, 2, \ldots m - 1)$. We claim that if

$$\prod_{i=0}^{m-1} (1 + p\xi^i)^{n_i} = 1, \tag{14.10}$$

where $n_0, n_1, \ldots, n_{m-1}$ are positive integers $\leq p^{s-1}$, then $n_i = p^{s-1}$ for all $i = 0, 1, \ldots, m - 1$.

Expanding (14.10), we obtain

$$p(\sum_{i=0}^{m-1} n_i \xi^i + pa) = 0 \text{ for some } a \in R,$$

which implies $\sum_{i=0}^{m-1} n_i \xi^i + pa \in (p)$. Then

$$\sum_{i=0}^{m-1} \overline{n_i} \overline{\xi}^i = \overline{0}.$$

$\overline{\xi}$ is a primitive element of $\mathbb{F}_{p^m}$, thus the minimal polynomial of $\overline{\xi}$ over $\mathbb{F}_p$ is of degree m and $1, \overline{\xi}, \ldots, \overline{\xi}^{m-1}$ are linearly independent over $\mathbb{F}_p$. Therefore $\overline{n_i} = 0$ for all $i = 0, 1, \ldots, m - 1$. That is, $p \,|\, n_i$, $i = 0, 1, \ldots, m - 1$. Let e be the integer $0 \leq e \leq s - 2$ such that p^{e+1} is the highest power of p dividing each of the integers $n_0, n_1, \ldots, n_{m-1}$. We proceed to show that

$e = s - 2$. Let $n_i = p^{e+1} r_i$ for $i = 0, 1, \ldots, m - 1$. Then at least one of $r_0, r_1, \ldots, r_{m-1}$ is not divisible by p. Expanding (14.10) and using Lemma 14.10(i), we obtain

$$p^{e+2}\left(\sum_{i=0}^{m-1} r_i \xi^i + pc\right) = 0 \text{ for some } c \in R.$$

If $e + 2 < s$, then $\sum_{i=1}^{m-1} r_i \xi^i + pc \in (p)$. Thus

$$\sum_{i=0}^{m-1} \overline{r}_i \overline{\xi}^i = \overline{0},$$

which leads to a contradiction. Therefore $e = s - 2$ and our claim is proved.

Let
$$H_i = \langle 1 + p\xi^i \rangle, \qquad i = 0, 1, \ldots, m - 1,$$

then each H_i is of order p^{s-1}, and $H_0 \times H_1 \times \cdots \times H_{m-1}$ is a direct product and is of order $p^{(s-1)m}$. Therefore $G_2 = H_0 \times H_1 \times \cdots \times H_{m-1}$.

Case 2 $p = 2$ and $s = 2$. In this case the square of every element of G_2 equals 1 so G_2 is a direct product of m cyclic group of order 2 (Exercise 14.5).

Case 3 $p = 2$ and $s \geq 3$. By Lemma 11.15, $N = \{\overline{a}^2 + \overline{a} : \overline{a} \in \mathbb{F}_{2^m}\}$ is an additive subgroup of index 2 in $\mathbb{F}_{2^m}$. Hence there is an element $\overline{b} \in \mathbb{F}_{2^m} \setminus N$. Let $b \in R$ be an original of $\overline{b}$ under the map $-$. By Theorem 14.1 every element of R can be expressed uniquely in the form (14.2)

$$a_0 + a_1 \xi + \cdots + a_{m-1} \xi^{m-1}, \ a_i \in \mathbb{Z}_{2^s} \ (i = 0, 1 \ldots, m - 1).$$

Clearly, $1 + 2 + 2^2 + \cdots + 2^{s-2}$, $1 + 2\xi^i \ (i = 1, 2, \ldots, m - 1)$, $1 + 4b \in G_2$. It can be verified by direct multiplication that $(1 + 2 + 2^2 + \cdots + 2^{s-2})^2 = 1$ and by Lemma 14.10(ii) and (iii) that $(1 + 2\xi^i)^{2^{s-1}} = 1$, $(1 + 4b)^{2^{s-2}} = 1$.

We claim that if

$$(1 + 2 + 2^2 + \cdots + 2^{s-2})^{n_0} \prod_{i=1}^{m-1} (1 + 2\xi^i)^{n_i} (1 + 4b)^n = 1, \qquad (14.11)$$

where $n_0, n_1, n_2, \ldots, n_{m-1}, n$ are positive integers with

$$n_0 \leq 2, \quad n_1, n_2, \ldots, n_{m-1} \leq 2^{s-1}, \text{ and } n \leq 2^{s-2},$$

then
$$n_0 = 2, \quad n_1 = n_2 = \cdots = n_{m-1} = 2^{s-1}, \text{ and } n = 2^{s-2}.$$

Suppose $n_0 = 1$. Expanding (14.11) and using Lemma 14.10(ii) we obtain

$$2(1 + \sum_{i=1}^{m-1} n_i \xi^i + 2a) = 0 \text{ for some } a \in R.$$

Thus $1 + \sum_{i=1}^{m-1} n_i \xi^i + 2a \in (2)$. Then $1 + \sum_{i=1}^{m-1} \overline{n_i} \overline{\xi}^i = 0$, which contradicts that $1, \overline{\xi}, \overline{\xi}^2, \ldots, \overline{\xi}^{m-1}$ is a basis of $\mathbb{F}_{2^m}$ over $\mathbb{F}_2$. Hence $n_0 = 2$ and (14.11) reduces to

$$\prod_{i=1}^{m-1} (1 + 2\xi^i)^{n_i} (1 + 4b)^n = 1. \tag{14.12}$$

By the above argument we can prove that $n_1, n_2, \ldots, n_{m-1}$ are all even.

Let e be the integer $0 \le e \le s - 2$ such that 2^{e+1} is the highest power of 2 dividing each of the integers $2n, n_1, n_2, \ldots, n_{m-1}$. We proceed to show that $e = s - 2$. Let $n = 2^e r$, $n_i = 2^{e+1} r_i$ for $i = 1, 2, \ldots, m - 1$. Then at least one of $r, r_1, r_2, \ldots, r_{m-1}$ is odd. Expanding (14.12) and using Lemma 14.10(ii) (with $e + 1$ in place of e) and (iii), we obtain

$$2^{e+2} (\sum_{i=1}^{m-1} (r_i \xi^i + r_i (r_i 2^{e+1} - 1)\xi^{2i}) + rb + 2c) = 0 \text{ for some } c \in R.$$

If $e + 2 < s$, then

$$\sum_{i=1}^{m-1} (\overline{r}_i \overline{\xi}^i + \overline{r}_i \overline{\xi}^{2i}) + \overline{r}\overline{b} = \overline{0}.$$

Since $\overline{r}_i^2 \equiv \overline{r}_i \pmod{2}$, $\sum_{i=1}^{m-1} \overline{r}_i \overline{\xi}^i + \overline{r}_i \overline{\xi}^{2i} \in N$. By the choice of $\overline{b}$, $\overline{b} \notin N$, hence r must be even and $\sum_{i=1}^{m-1} (\overline{r}_i \overline{\xi}^i + \overline{r}_i \overline{\xi}^{2i}) = \overline{0}$. Then one of r_i $(i = 1, 2, \ldots, m-1)$ must be odd, and

$$\sum_{i=1}^{m-1} \overline{r}_i \overline{\xi}^i = \overline{0} \text{ or } \sum_{i=1}^{m-1} \overline{r}_i \overline{\xi}^i = \overline{1},$$

both of which lead to contradictions. Therefore $e = s - 2$. Our claim is proved.

Let

$$H_0 = \langle 1 + 2 + 2^2 + \cdots + 2^{s-2} \rangle,$$
$$H_i = \langle 1 + 2\xi^i \rangle, \quad i = 1, 2, \ldots, m - 1,$$
$$H_r = \langle 1 + 4b \rangle.$$

Then H_0, H_i $(i = 1, 2, \ldots, r-1)$, and H_r are cyclic groups of order 2, 2^{s-1} and 2^{s-2} respectively. Our claim implies that we have a direct product $H_0 \times H_1 \times \cdots \times H_r$. But $|H_0 \times H_1 \times \cdots \times H_r| = |G_2|$. Therefore $G_2 = H_0 \times H_1 \times \cdots \times H_r$. □

Theorem 14.11 is due to Raghavendran (1970).

Corollary 14.12 *Let p be a prime and s be a positive integer. Then*

$$\mathbb{Z}_{p^s}^* = G_1 \times G_2,$$

where G_1 is a cyclic group of order $p - 1$ and G_2 is a group of order p^{s-1} such that

(i) *If p is odd or if $p = 2$ and $s \leq 2$, then G_2 is a cyclic group of order p^{s-1}.*

(ii) *If $p = 2$ and $s \geq 3$, then G_2 is a direct product of a cyclic group of order 2 and a cyclic group of order 2^{s-2}.*

14.5   Extension of Galois Rings

Theorem 14.13 *Let $R = GR(p^s, p^{sm})$ and $R' = GR(p^s, p^{sn})$. If R' is an extension ring of R, then $m \mid n$.*

Proof. The principal ideal (p) of R can be written as Rp and, similarly, the principal ideal (p) of R' can be written as $R'p$. Consider the map

$$\begin{aligned} - : \quad & R' \to R'/R'p \\ & r' \mapsto r' + R'p. \end{aligned}$$

By Lemma 14.4, $R'/R'p \simeq \mathbb{F}_{p^n}$. Clearly, $R + R'p$ is a subring of R' and $R'p$ is an ideal of $R + R'p$. Therefore $(R + R'p)/R'p$ is a subring of $R'/R'p$. Thus

$$(R + R'p)/R'p \subseteq R'/R'p \simeq \mathbb{F}_{p^n}.$$

We assert that $R \cap R'p = Rp$. Clearly, $R \cap R'p \supseteq Rp$. For any $r \in R \cap R'p$, we have $r \in R$ and $r \in R'p$. There is an $r' \in R'$ such that $r = r'p$. Then $rp^{s-1} = r'p^s = 0$. Thus r is either 0 or a zero divisor of R, which implies $r \in Rp$. Therefore $R \cap R'p \subseteq Rp$. Hence $R \cap R'p = Rp$. By Theorem 12.23, we have an isomorphism of rings

$$(R + R'p)/R'p \simeq R/Rp \simeq \mathbb{F}_{p^m}.$$

Therefore $\mathbb{F}_{p^m} \subseteq \mathbb{F}_{p^n}$. By Theorem 6.17 $m \mid n$. □

Remark: We have the commutative diagram

$$- : \quad GR(p^s, p^{sn}) \to \mathbb{F}_{p^n}$$
$$\quad\quad \Big\uparrow \quad\quad\quad\quad \Big\uparrow$$
$$- : \quad GR(p^s, p^{sm}) \to \mathbb{F}_{p^m}$$

where $\Big\uparrow$ denotes the inclusion.

Conversely, we have

Theorem 14.14 *Let $R = GR(p^s, p^{sm})$ and $m \mid n$. Then there is a Galois ring $R' = GR(p^s, p^{sn})$ which contains R as a subring.*

To prove this theorem we have to generalize the construction of the Galois ring extension $\mathbb{Z}_{p^s}[\xi] = \mathbb{Z}_{p^s}[x]/(h(x))$ of $\mathbb{Z}_{p^s}$ to that of $GR(p^s, p^{sm})$. The construction is almost the same, so we omit the proofs.

In the following let $R = GR(p^s, p^{sm})$ and denote the natural homomorphism $R \to R/Rp$ by $-$. Then $\overline{R} = R/Rp \simeq \mathbb{F}_{p^m}$.

First, the properties of $\mathbb{Z}_{p^s}$ obtained in Chapter 13 are extended to general Galois rings $GR(p^s, p^{sm})$. Parallel to Theorems 13.1, 13.2, 13.3 and Lemma 13.4 we have

Theorem 14.15 *The principal ideals $(1), (p), (p^2), \ldots, (p^{s-1}), (0)$ are all the ideals of $R = GR(p^s, p^{sm})$, (p) is the unique maximal ideal of R and $R/(p) \simeq \mathbb{F}_{p^m}$.*

Theorem 14.16 *The ideal $(p) = R[x]p$ is a prime ideal of $R[x]$, every prime ideal of $R[x]$ contains (p), and every prime ideal of $R[x]$ containing (p) properly is maximal.*

Theorem 14.17 *Let Q be an ideal of $R[x]$.*

(i) *If Q is primary then $\sqrt{Q}$ is prime.*

(ii) *$\sqrt{Q} \supseteq (p)$.*

(iii) *If $\sqrt{Q}$ is prime and contains (p) properly, then Q is primary.*

Lemma 14.18 *Let f be a polynomial in $R[x]$ and assume that $\bar{f} = g^l$, where g is an irreducible polynomial in $\mathbb{F}_{p^m}[x]$ and l is a positive integer. Then f is a primary polynomial of $R[x]$.*

We have the coprimeness of polynomials over R. Let $f, g \in R[x]$. They are said to be *coprime* in $R[x]$ if and only if $R[x]f + R[x]g = R[x]$. Then we have

Lemma 14.19 *Let f_1, $f_2 \in R[x]$. Then f_1 and f_2 are coprime in $R[x]$ if and only if $\overline{f_1}$ and $\overline{f_2}$ are coprime in $\mathbb{F}_{p^m}[x]$.*

Hensel's Lemma (Lemma 13.7) can also be generalized as follows.

Lemma 14.20 (Hensel's Lemma): *Let f be a monic polynomial in $R[x]$ and $g_1, g_2, \ldots, g_r$ be pairwise coprime monic polynomials in $\overline{R}[x]$. Assume that $\overline{f} = g_1 g_2 \ldots g_r$ in $\overline{R}[x]$. Then there exist pairwise coprime monic polynomials $f_1, f_2, \ldots, f_r$ in $R[x]$ such that $f = f_1 f_2 \ldots f_r$ and $\overline{f_i} = g_i$ for $i = 1, 2, \ldots, r$.*

Parallel to Theorem 13.8, we have also the following unique factorization theorem.

Theorem 14.21 *Let f be a monic polynomial of degree ≥ 1 in $R[x]$. Then*

(i) *f can be factorized into a product of some number, say r, of pairwise coprime monic primary polynomials $f_1, f_2, \ldots, f_r$ over R:*

$$f = f_1 f_2 \cdots f_r$$

and for each $i = 1, 2, \ldots, r$ $\overline{f_i}$ is a power of a monic irreducible polynomial over $\mathbb{F}_{p^m}$.

(ii) *Let*

$$f = f_1 \cdots f_r = h_1 \cdots h_t$$

be two factorizations of f into products of pairwise coprime monic primary polynomials over R, then $r = t$ and after renumbering, $f_i = h_i$, $i = 1, 2, \ldots, r$.

Let f be a monic polynomial of degree ≥ 1 in $R[x]$. If $\overline{f}$ is irreducible (or primitive) in $\overline{R}[x]$, f is called a *monic basic irreducible* (or *monic basic primitive*, respectively,) *polynomial* over R. Clearly, monic basic primitive polynomials over R are monic basic irreducible. Parallel to Theorem 13.9 we have

Theorem 14.22 *For any integer $l \geq 1$ there exist monic basic irreducible (and monic basic primitive) polynomials of degree l over R and dividing $x^{p^{ml}-1} - 1$ in $R[x]$.*

Finally, Theorem 14.1 can be generalized as follows:

Theorem 14.23 *Let $h(x)$ be a monic basic irreducible polynomial of degree l over R. Then the residue class ring $R[x]/(h(x))$ is a Galois ring of characteristic p^s and cardinality p^{sml} and it contains R as a subring. Thus*

$$R[x]/(h(x)) = GR(p^s, p^{sml}).$$

Write $\xi = x + (h(x))$, then $h(\xi) = 0$, all elements of $R[x]/(h(x))$ can be expressed uniquely in the form

$$a_0 + a_1\xi + \cdots + a_{l-1}\xi^{l-1}, \quad \text{where } a_0, a_1, \ldots, a_{l-1} \in R, \quad (14.13)$$

$R[x]/(h(x)) = R[\xi]$ and $R[\xi]$ is a free R-module of rank l with $\{1, \xi, \xi^2, \ldots, \xi^{l-1}\}$ as a free basis. The ideal (p) of $R[\xi]$ consists of all zero divisors together with the zero of $R[\xi]$ and is the only maximal ideal of $R[\xi]$. Moreover, denote the image of $\xi = x + (h(x))$ under the ring isomorphism

$$R[x]/(h(x)) \to \mathbb{F}_{p^m}[x]/(\overline{h}(x)) \quad (14.14)$$
$$a_0 + \cdots + a_{l-1}x^{l-1} + (h(x)) \mapsto \overline{a}_0 + \cdots + \overline{a}_{l-1}x^{l-1} + (\overline{h}(x)),$$

where $a_0, \ldots, a_{l-1} \in R$, by $\overline{\xi}$. That is, $\overline{\xi} = x + \overline{h}(x)$, then $\overline{h}(\overline{\xi}) = 0$, $R[x]/(h(x)) \simeq \mathbb{F}_{p^m}[x]/(\overline{h}(x)) \simeq \mathbb{F}_{p^m}[\overline{\xi}] = \mathbb{F}_{p^{ml}}$, and (14.14) can be written as

$$- : \quad R[\xi] \to \mathbb{F}_{p^m}[\overline{\xi}]$$
$$a_0 + a_1\xi + \cdots + a_{l-1}\xi^{l-1} \mapsto \overline{a}_0 + \overline{a}_1\overline{\xi} + \cdots + \overline{a}_{l-1}\overline{\xi}^{l-1}.$$

Proof of Theorem 14.14. It follows from Theorems 14.22 and 14.23. □

Moreover, we have

Theorem 14.24 *Let $R' = GR(p^s, p^{sn})$ and m be a positive divisor of n. Then R' contains a unique Galois ring of characteristic p^s and cardinality p^{sm} as its subring.*

Proof. Let $R = GR(p^s, p^{sm})$ be a Galois ring of characteristic p^s and cardinality p^{sm}. Let $n = ml$. By Theorem 14.14 there is a Galois ring $R'' = GR(p^s, p^{sml})$ containing R as a subring. By Corollary 14.7 $R'' = GR(p^s, p^{sml}) \simeq GR(p^s, p^{sn}) = R'$. Under this isomorphism, the subring $R = GR(p^s, p^{sm})$ of $R'' = GR(p^s, p^{sml})$ corresponds to a subring R_0 of R' and R_0 is a Galois ring of characteristic p^s and cardinality p^{sm}.

Now let us prove the uniqueness. By Theorem 14.8 there is a nonzero element η of order $p^m - 1$ in R_0, which is a root of a monic basic primitive polynomial of degree m over $\mathbb{Z}_{p^s}$ and dividing $x^{p^m-1} - 1$ in $\mathbb{Z}_{p^s}[x]$, and

$$R_0 = \{c_0 + c_1p + \cdots + c_{s-1}p^{s-1} : c_0, c_1, \cdots, c_{s-1} \in \mathcal{T}_0\},$$

where
$$\mathcal{T}_0 = \{0, 1, \eta, \eta^2, \ldots, \eta^{p^m-2}\}.$$

Let R_1 be any Galois ring of characteristic p^s and cardinality p^{sm} contained in R' as a subring. Then there is a nonzero element ζ of order $p^m - 1$ in R_1, which is a root of a basic primitive polynomial of degree m over $\mathbb{Z}_{p^s}$ and dividing $x^{p^m-1} - 1$ in $\mathbb{Z}_{p^s}[x]$. Let

$$\mathcal{T}_1 = \{0, 1, \zeta, \zeta^2, \ldots, \zeta^{p^m-2}\}.$$

Then

$$R_1 = \{c_0 + c_1 p + \cdots + c_{s-1} p^{s-1} : c_0, c_1, \cdots, c_{s-1} \in \mathcal{T}_1\}.$$

Denote the group of units of R' by R'^*. By Corollary 14.9(i), Theorem 12.13 and also Exercise 12.7, R'^* has a unique cyclic subgroup G_1 of order $p^n - 1$. By Theorem 2.28(ii) G_1 has a unique cyclic subgroup of order $p^m - 1$. It follows that R'^* has a unique cyclic subgroup of order $p^m - 1$. Hence $\mathcal{T}_1 = \mathcal{T}_0$ and $R_1 = R_0$. $\square$

We have also the following analogues of Lemma 14.5, Theorems 14.6 and 14.8.

Lemma 14.25 *Let $R = GR(p^s, p^{sm})$, $R' = GR(p^s, p^{sml})$ and R be a subring of R'. Let $f(x)$ be a polynomial over R, and assume that $\overline{f}(x)$ has a root $\overline{\beta}$ in $\overline{R'} = \mathbb{F}_{p^{ml}}$ and that $\overline{f}'(\overline{\beta}) \neq \overline{0}$. Then there is a unique root $\alpha \in R'$ of the polynomial $f(x)$ such that $\overline{\alpha} = \overline{\beta}$.*

Theorem 14.26 *Let R' be a Galois ring of characteristic p^s and cardinality p^{sml}, where p is a prime and s, m, l are positive integers, and let $R = GR(p^s, p^{sm})$ be a subring of R'. Then R' is isomorphic to the residue class ring $R[x]/(h(x))$ for any monic basic irreducible polynomial $h(x)$ of degree l over R.*

Theorem 14.27 (i) *In the Galois ring $GR(p^s, p^{sml})$ there exists a nonzero element ξ of order $p^{ml} - 1$, which is a root of a monic basic primitive polynomial $h(x)$ of degree l over $GR(p^s, p^{sm})$ and dividing $x^{p^{ml}-1} - 1$ over $GR(p^s, p^{sm})$, and every element of $GR(p^s, p^{sml})$ can be expressed uniquely in the form (14.13). Thus*

$$\begin{aligned}
GR(p^s, p^{sml}) &= GR(p^s, p^{sm})[\xi] \\
&= \{c_0 + c_1 \xi + \cdots + c_{l-1} \xi^{l-1} : c_0, c_1, \ldots, c_{l-1} \in GR(p^s, p^{sm})\}.
\end{aligned}$$

Moreover, $h(x)$ is the unique monic polynomial over $GR(p^s, p^{sm})$ of degree $\leq l$ and having ξ as a root.

(ii) *Let*

$$T = \{0, 1, \xi, \xi^2, \ldots, \xi^{p^{ml}-2}\},$$

then any element $c \in T$ *can be written uniquely as*

$$c = c_0 + c_1 p + \cdots + c_{s-1} p^{s-1}, \quad where \quad c_0, c_1, \ldots, c_{s-1} \in T. \tag{14.15}$$

Moreover, c is invertible (or a zero divisor or 0) if and only if $a_0 \neq 0$ *(or* $a_0 = 0$, *respectively).*

Corollary 14.28 *Under the hypothesis of Theorem 14.27, let*

$$\eta = \xi^{(p^{ml}-1)/(p^m-1)},$$

then η *is of order* $p^m - 1$. *Let*

$$T' = \{0, 1, \eta, \eta^2, \ldots, \eta^{p^m-2}\},$$

then those elements of the form (14.15), where $c_0, c_1, \cdots, c_{l-1} \in T'$ *form the Galois ring* $GR(p^s, p^{sm})$.

Proof. Similar to the proof of the uniqueness part of Theorem 14.24. $\square$

14.6 Automorphisms of Galois Rings

Let $R = GR(p^s, p^{sm})$, $R' = GR(p^s, p^{sml})$, and R be a subring of R'. By Theorem 14.27, there is a nonzero element ξ of order $p^{ml} - 1$ in R', which is a root of a monic basic primitive polynomial $h(x)$ of degree l over R and dividing $x^{p^{ml}-1} - 1$ over R, and $R' = R[\xi]$.

First we prove the following lemma.

Lemma 14.29 *Let* R' *contains* R *as a subring and* $R' = R[\xi]$, *where* ξ *is a nonzero element of order* $p^{ml} - 1$ *in* R' *and is a root of a monic basic primitive polynomial* $h(x)$ *of degree* l *over* R *and dividing* $x^{p^{ml}-1} - 1$ *over* R. *Then* $\xi, \xi^{p^m}, \cdots, \xi^{p^{m(l-1)}}$ *are all the roots of* $h(x)$.

Proof. Since $h(x)$ is a monic basic primitive polynomial of degree l over R and has ξ as one of its roots, $\overline{h}(x)$ is a primitive polynomial of degree l over $\overline{R} = \mathbb{F}_{p^m}$ and has $\overline{\xi} \in \mathbb{F}_{p^{ml}}$ as one of its roots. Therefore $\overline{\xi}$ is of order $p^{ml} - 1$ and $\overline{\xi}, \overline{\xi}^{p^m}, \ldots, \overline{\xi}^{p^{m(l-1)}}$ are all the roots of $\overline{h}(x)$ in $\overline{R'} = \mathbb{F}_{p^{ml}}$. By Lemma 14.25 for any $\overline{\xi}^{p^{mi}}$, $0 \leq i \leq l-1$, the polynomial $h(x)$ has a unique root $\alpha_i \in R'$ such that $\overline{\alpha}_i = \overline{\xi}^{p^{mi}}$. Since $h(x) \mid (x^{p^{ml}-1} - 1)$ over R, we have $\alpha_i^{p^{ml}-1} - 1 = 0$ and $\alpha_i^{p^{ml}} = \alpha_i$. By Corollary 14.9(ii) $\alpha_i = \xi^{l_i}$, for some integer l_i, $0 \le l_i \le p^{ml} - 2$. Then $\overline{\alpha_i} = \overline{\xi}^{l_i}$, which together with $\overline{\alpha}_i = \overline{\xi}^{p^{mi}}$ implies $l_i = p^{mi}$. Therefore $\alpha_i = \xi^{p^{mi}}$. Hence $\xi, \xi^{p^m}, \ldots, \xi^{p^{m(l-1)}}$ are l roots of $h(x)$ in $R' = R[\xi]$. By the unique factorization theorem (Theorem 14.21) they are all the roots of $h(x)$. $\square$

Theorem 14.30 *Let* $R' = GR(p^s, p^{sml})$ *contain* $R = GR(p^s, p^{sm})$ *as a subring,* $R' = R[\xi]$, *where* ξ *is an element of order* $p^{ml} - 1$ *in* R' *and is a root of a monic basic primitive polynomial* $h(x)$ *of degree* l *over* R *and dividing* $x^{p^{ml}-1} - 1$ *in* $R[x]$. *Define a map* $\phi : R' \to R'$ *by*

$$\phi(a_0 + a_1\xi + \cdots + a_{l-1}\xi^{l-1}) = a_0 + a_1\xi^{p^m} + \cdots + a_{l-1}\xi^{p^{m(l-1)}} \quad (14.16)$$

for all $a_0, a_1, \ldots, a_{l-1} \in R$. *Then* ϕ *is an automorphism of* R' *leaving* R *fixed elementwise. Moreover, for* $\alpha \in R'$ $\phi(\alpha) = \alpha$ *if and only if* $\alpha \in R$.

Proof. Clearly, ϕ is additive. Thus, to prove that ϕ is injective, it is enough to show that $\mathrm{Ker}\,\phi = \{0\}$. Assume $\phi(a_0 + a_1\xi + \cdots + a_{l-1}\xi^{l-1}) = 0$, where $a_0, a_1, \ldots, a_{l-1} \in R$, i.e., $a_0 + a_1\xi^{p^m} + \cdots + a_{l-1}\xi^{p^{m(l-1)}} = 0$. Let $g(x) = a_0 + a_1 x + \cdots + a_{l-1}x^{l-1}$, then $g(\xi^{p^m}) = 0$ and, hence, $\overline{g}(\overline{\xi}^{p^m}) = 0$. By Lemma 14.29 ξ^{p^m} is a root of $h(x)$, which implies $\overline{\xi}^{p^m}$ is a root of $\overline{h}(x)$ over $\overline{R}$. We have $\deg \overline{g}(x) < l$ and $\deg \overline{h}(x) = l$. Since $\overline{h}(x)$ is the minimal polynomial of $\overline{\xi}^{p^m}$ over R, we must have $\overline{g}(x) = 0$, i.e., $\overline{a}_0 = \overline{a}_1 = \cdots = \overline{a}_{l-1} = 0$. Then we may write $a_i = pb_i^{(1)}$, where $b_i^{(1)} \in R$, $i = 0, 1, \ldots, l-1$. It follows that $p(b_0^{(1)} + b_1^{(1)}\xi^{p^m} + \cdots + b_{l-1}^{(1)}\xi^{p^{m(l-1)}}) = 0$. Then $b_0^{(1)} + b_1^{(1)}\xi^{p^m} + \cdots + b_{l-1}^{(1)}\xi^{p^{m(l-1)}} \in (p)$ and $\overline{b}_0^{(1)} + \overline{b}_1^{(1)}\overline{\xi}^{p^m} + \cdots + \overline{b}_{l-1}^{(1)}\overline{\xi}^{p^{m(l-1)}} = 0$. As before, we have $\overline{b}_0^{(1)} = \overline{b}_1^{(1)} = \cdots = \overline{b}_{l-1}^{(1)} = 0$, then we may write $b_i^{(1)} = pb_i^{(2)}$, where $b_i^{(2)} \in R$, $i = 0, 1, \ldots, l-1$. Proceeding in this way, we obtain successfully, for $j = 1, 2, \ldots, s-1$, $b_i^{(j)} = pb_i^{(j-1)}$, where $b_i^{(j)}, b_i^{(j-1)} \in R$. Then

$$a_i = pb_i^{(1)} = p^2 b_i^{(2)} = \cdots = p^s b_i^{(s)} = 0, \; i = 0, 1, \ldots, m-1.$$

Now let us show that ϕ preserves multiplication. Let $a_0 + a_1\xi + \cdots + a_{l-1}\xi^{l-1}$, $b_0 + b_1\xi + \cdots + b_{l-1}\xi^{l-1} \in R[x]$, and

$$a(x) = a_0 + a_1 x + \cdots + a_{l-1}x^{l-1}, \; b(x) = b_0 + b_1 x + \cdots + b_{l-1}x^{l-1}.$$

Dividing $a(x)b(x)$ by the monic polynomial $h(x)$, we have

$$a(x)b(x) = q(x)h(x) + r(x),$$

where $q(x)$, $r(x) \in R[x]$ and $\deg r(x) < \deg h(x)$. Substituting ξ and ξ^{p^m} into the above equation, we obtain

$$a(\xi)b(\xi) = r(\xi) \text{ and } a(\xi^{p^m})b(\xi^{p^m}) = r(\xi^{p^m}),$$

respectively. Then

$$\phi(a(\xi)b(\xi)) = \phi(r(\xi)) = r(\xi^{p^m}) = a(\xi^{p^m})b(\xi^{p^m}) = \phi(a(\xi))\phi(b(\xi)).$$

This proves that ϕ is an automorphism of R'.

Moreover, let α be an element of R' and assume that $\phi(\alpha) = \alpha$. We express α in the p-adic representation as follows.

$$\alpha = c_0 + c_1 p + \cdots + c_{l-1}p^{l-1},$$

where $c_0, c_1, \cdots, c_{l-1} \in \mathcal{T}$ and

$$\mathcal{T} = \{0, 1, \xi, \xi^2, \ldots, \xi^{p^{ml}-2}\}.$$

Since $\phi(\xi) = \xi^{p^m}$,

$$\phi(\alpha) = c_0^{p^m} + c_1^{p^m}p + \cdots + c_{l-1}^{p^m}p^{l-1}.$$

Thus $\phi(\alpha) = \alpha$ if and only if $c_i^{p^m} = c_i$ for $i = 0, 1, \ldots, l-1$. Since $c_i \in \mathcal{T}$, $c_i = 0$ if and only if $c_i^{p^m} = 0$. If $c_i \neq 0$, $c_i = \xi^j$ for some j, $0 \le j \le p^{ml} - 2$. Then $\xi^{jp^m} = \xi^j$ and $\xi^{j(p^m-1)} = 1$. But $\langle \xi \rangle$ has a unique subgroup $\langle \eta \rangle$ of order $p^m - 1$, where $\eta = \xi^{(p^{ml}-1)/(p^m-1)}$. Therefore we always have $c_i \in \mathcal{T}'$, where

$$\mathcal{T}' = \{0, 1, \eta, \eta^2, \ldots, \eta^{p^m-2}\}.$$

By Corollary 14.28 $\alpha \in R$. $\qquad\square$

An automorphism of R' which leaves every element of R fixed is called an *automorphism* of R' over R. For example, the map ϕ defined by (14.16) is an automorphism of R' over R. We call ϕ the *generalized Frobenius automorphism* of R' over R. Define

$$\phi^0 = 1,$$

where 1 denotes the identity map of R', and

$$\phi^{i+1} = \phi^i \circ \phi, \ i = 1, 2, \ldots.$$

Then all ϕ^i, $i = 0, 1, 2, \ldots$, are automorphisms of R' over R. Clearly,

$$\phi^i(a_0 + a_1\xi + \cdots + a_{l-1}\xi^{l-1}) = a_0 + a_1\xi^{p^{mi}} + \cdots + a_{l-1}\xi^{p^{mi(l-1)}}$$

for all $a_0, a_1, \ldots, a_{l-1} \in R$. Since ξ is of order $p^{ml} - 1$, l is the least positive integer such that $\phi^l = 1$. Then $\phi^0 = 1$, ϕ, ϕ^2, $\ldots$, ϕ^{l-1} form a cyclic group $\langle \phi \rangle$ of order l generated by ϕ.

The set of automorphisms of R' over R forms a group with respect to the composition of maps, which is called the *Galois group* of R' over R and is denoted by $\mathrm{Gal}(R'/R)$. We have

Theorem 14.31 *Under the hypothesis of Theorem* 14.30, $\mathrm{Gal}(R'/R) = \langle \phi \rangle$.

Proof. Let $\tau \in \mathrm{Gal}(R'/R)$. Since $h(\xi) = 0$, we have $h(\tau(\xi)) = 0$. By Lemma 14.30 $\tau(\xi) = \xi^{p^{mi}}$ for some i with $0 \leq i \leq l - 1$. Then $\tau(\xi) = \phi^i(\xi)$, which implies

$$\tau(a_0 + a_1\xi + \cdots + a_{l-1}\xi^{l-1}) = \phi^i(a_0 + a_1\xi + \cdots + a_{l-1}\xi^{l-1}),$$

where $a_0, a_1, \ldots, a_{l-1} \in R$. Thus $\tau(\alpha) = \phi^i(\alpha)$ for all $\alpha \in R'$. Therefore $\tau = \phi^i$. Hence $\mathrm{Gal}(R'/R) = \langle \phi \rangle$. □

Theorem 14.32 *Let* $\tau \in \mathrm{Gal}(R'/R)$. *For any* $a \in \mathbb{F}_{p^{ml}}$, *there is an element* $\alpha \in R'$ *such that* $\overline{\alpha} = a$, *we define*

$$\overline{\tau} : \mathbb{F}_{p^{ml}} \to \mathbb{F}_{p^{ml}}$$
$$a \mapsto \overline{\tau}(a) = \overline{\tau(\alpha)}.$$

Then

(i) $\overline{\tau}$ *is well-defined and* $\overline{\tau} \in \mathrm{Gal}(\mathbb{F}_{p^{ml}}/\mathbb{F}_{p^m})$.

(ii) *The map*

$$\mathrm{Gal}(R'/R) \to \mathrm{Gal}(\mathbb{F}_{p^{ml}}/\mathbb{F}_{p^m})$$
$$\tau \mapsto \overline{\tau}$$

is an isomorphism of groups.

(iii) *The generalized Frobenius automorphism* ϕ *of* R' *over* R *corresponds to the Frobenius automorphism* σ *of* $\mathbb{F}_{p^{ml}}$ *over* $\mathbb{F}_{p^m}$, *i.e.,* $\overline{\phi} = \sigma$.

Proof. Let $\alpha, \alpha' \in R'$ be such that $\overline{\alpha} = \overline{\alpha'}$. Then $\alpha - \alpha' = p\pi$, where $\pi \in R'$. Thus $\overline{\tau(\alpha)} - \overline{\tau(\alpha')} = \overline{\tau(\alpha - \alpha')} = \overline{p\tau(\pi)} = 0$, i.e. $\overline{\tau(\alpha)} = \overline{\tau(\alpha')}$. Therefore the map $\overline{\tau}$ is well-defined. For any $a, b \in \mathbb{F}_{p^{ml}}$, there are elements $\alpha, \beta \in R'$ such that $\overline{\alpha} = a, \overline{\beta} = b$. Then $\overline{\alpha + \beta} = a + b$ and $\overline{\alpha\beta} = ab$. Thus

$$\overline{\tau}(a + b) = \overline{\tau(\alpha + \beta)} = \overline{\tau(\alpha)} + \overline{\tau(\beta)} = \overline{\tau}(a) + \overline{\tau}(b),$$

$$\overline{\tau}(ab) = \overline{\tau(\alpha\beta)} = \overline{\tau(\alpha)\tau(\beta)} = \overline{\tau}(a)\overline{\tau}(b).$$

This proves that $\overline{\tau}$ is a ring homomorphism. Suppose $\overline{\tau}(a) = 0$. Let $\alpha \in R'$ be such that $\overline{\alpha} = a$, then $\overline{\tau(\alpha)} = 0$, i.e. $\tau(\alpha) = p\pi$, where $\pi \in R'$. Thus $\tau(\alpha^s) = \tau(\alpha)^s = p^s\pi^s = 0$. It follows that $\alpha^s = 0$, which implies $\alpha \in (p)$ and $a = \overline{\alpha} = \overline{0}$. Therefore Ker $\overline{\tau} = \{\overline{0}\}$ and $\overline{\tau}$ is an automorphism of $\mathbb{F}_{p^{ml}}$. For $a \in \mathbb{F}_{p^m}$ there is an $\alpha \in R$ such that $\overline{\alpha} = a$. Then $\overline{\tau}(a) = \overline{\tau(\alpha)} = \overline{\alpha} = a$. Therefore $\overline{\tau} \in \text{Gal}(\mathbb{F}_{p^{ml}}/\mathbb{F}_{p^m})$.

The last two statements are clear. $\square$

Corollary 14.33 *Let $n = mld$ and ϕ be the generalized Frobenius automorphism of $R' = GR(p^s, p^{sn})$ over $R = GR(p^s, p^{sm})$. Then*

(i) ϕ^d is the generalized Frobenius automorphism of $R' = GR(p^s, p^{sn})$ over $R'' = GR(p^s, p^{smd})$ and $\text{Gal}(R'/R'') = \langle \phi^d \rangle$ is a cyclic group of order l.

(ii) For all $\alpha \in R''$, $\phi(\alpha) \in R''$. Denote the restriction of ϕ to R'' by $\phi|_{R''}$, then the map

$$\phi|_{R''} : \quad R'' \to R''$$
$$\alpha \mapsto \phi(\alpha)$$

is well-defined, $\phi|_{R''}$ is the generalized Frobenius automorphism of R'' over R and $\text{Gal}(R''/R) = \langle \phi|_{R''} \rangle$ is a cyclic group of order d.

14.7 Generalized Trace and Norm

Let $R = GR(p^s, p^{sm})$ be a subring of $R' = GR(p^s, p^{sml})$ and ϕ be the generalized Frobenius automorphism of R' over R. For any $\alpha \in R'$, define

$$\mathrm{T}_{R'/R}(\alpha) = \alpha + \phi(\alpha) + \cdots + \phi^{l-1}(\alpha)$$

and

$$\mathrm{N}_{R'/R}(\alpha) = \alpha\phi(\alpha)\cdots\phi^{l-1}(\alpha)$$

to be the *generalized trace* and *norm* of $\alpha \in R'$ relative to R, respectively. If R' and R are clear from the context, we write T and N simply for $\mathrm{T}_{R'/R}$ and $\mathrm{N}_{R'/R}$, respectively.

Theorem 14.34 *For α, $\beta \in R'$ and $a \in R$ we have*

(i) $\mathrm{T}(\alpha) \in R$, (i') $\mathrm{N}(\alpha) \in R$;

(ii) $\mathrm{T}(\alpha + \beta) = \mathrm{T}(\alpha) + \mathrm{T}(\beta)$, (ii') $\mathrm{N}(\alpha\beta) = \mathrm{N}(\alpha)\mathrm{N}(\beta)$;

(iii) $\mathrm{T}(a\alpha) = a\mathrm{T}(\alpha)$ and, (iii') $\mathrm{N}(a\alpha) = a^l\,\mathrm{N}(\alpha)$ and,
in particular, $\mathrm{T}(a) = la$, *in particular*, $\mathrm{N}(a) = a^l$;

(iv) $\mathrm{T}(\phi(\alpha)) = \mathrm{T}(\alpha)$, (iv') $\mathrm{N}(\phi(\alpha)) = \mathrm{N}(\alpha)$.

The proof is similar to that of Theorem 7.12 and is omitted.

Theorem 14.35 *Let $R = GR(p^s, p^{sm})$ be a subring of $R' = GR(p^s, p^{sml})$ and $R' = GR(p^s, p^{sml})$ be a subring of $R'' = GR(p^s, p^{smld})$. Then for all $\alpha \in R''$*

$$\mathrm{T}_{R''/R}(\alpha) = \mathrm{T}_{R'/R}(\mathrm{T}_{R''/R'}(\alpha)), \quad \mathrm{N}_{R''/R}(\alpha) = \mathrm{N}_{R'/R}(\mathrm{N}_{R''/R'}(\alpha)).$$

The proof is similar to that of Theorem 7.13 and is omitted.

Theorem 14.36 *Let $R = GR(p^s, p^{sm})$ be a subring of $R' = GR(p^s, p^{sml})$ and let $v_1, v_2, \ldots, v_l$ be a free basis of R' over R. For any $\alpha \in R'$, we have*

$$\alpha v_j = \sum_{i=1}^{l} a_{ij} v_i, \ a_{ij} \in R, \ j = 1, 2, \ldots, l,$$

then

$$M(\alpha) = (a_{ij})_{1 \leq i,j \leq n}$$

is an $l \times l$ matrix over R and

$$\mathrm{N}(\alpha) = \det M(\alpha), \quad \mathrm{T}(\alpha) = \mathrm{Tr} M(\alpha).$$

The proof is left as an exercise.

We have also a generalization of Hilbert's Theorem 90.

Theorem 14.37 *Let $R = GR(p^s, p^{sm})$ be a subring of $R' = GR(p^s, p^{sml})$.*

(i) *The map $\mathrm{T} : R' \mapsto R$ is surjective. For $\alpha \in R'$, $\mathrm{T}(\alpha) = 0$ if and only if there exists an element $\beta \in R'$ such that $\alpha = \beta - \phi(\beta)$.*

(ii) *The map $\mathrm{N} : R'^* \to R^*$ is surjective. For $\alpha \in R'^*$, $\mathrm{N}(\alpha) = 1$ if and only if there exists an element $\beta \in R'^*$ such that $\alpha = \beta\phi(\beta)^{-1}$.*

The proof is also left as an exercise.

14.8 Exercises

14.1 Let R be any ring and $f(x) = a_0 x^n + a_1 x^{n-1} + \cdots + a_n$ be a polynomial over R. Define

$$f'(x) = na_0 x^{n-1} + (n-1)a_1 x^{n-2} + \cdots + a_{n-1},$$
$$f^{(m)}(x) = (f^{(m-1)}(x))' \qquad \text{for } m \geq 2.$$

Prove Taylor's formula

$$f(x + a) = f(x) + \frac{f'(x)}{1!}a + \frac{f''(x)}{2!}a^2 + \cdots + \frac{f^{(n)}(x)}{n!}a^n.$$

14.2 Prove Theorem 14.15.

14.3 Let (p^r) be an ideal in the Galois ring $GR(p^s, p^{sm})$, where $1 \leq r < s$. Prove that $GR(p^s, p^{sm})/(p^r) \simeq GR(p^r, p^{rm})$.

14.4 Prove Corollary 14.9.

14.5 Complete the proof of Lemma 14.10.

14.6 Let G be a finite group and assume that all elements of G except the identity are of order 2. Prove that G is abelian and is direct product of cyclic groups of order 2.

14.7 Determine the number of basic irreducible polynomials and the number of basic primitive polynomials over $\mathbb{Z}_{p^s}$. Generalize to the Galois ring $GR(p^s, p^{sm})$.

14.8 Let A be an $n \times n'$ matrix over the Galois ring $GR(p^s, p^{sm})$. Prove that there is an $n \times n$ invertible matrix P and an $n' \times n'$ invertible matrix Q, both over $GR(p^s, p^{sm})$ such that PAQ is of the form

$$\begin{pmatrix} \mathrm{diag}[p^{d_1}, p^{d_2}, \ldots, p^{d_r}] & 0^{(r, n'-r)} \\ 0^{(n-r, r)} & 0^{(n-r, n'-r)} \end{pmatrix},$$

where $0 \leq r \leq \min\{m, n\}$, and $0 \leq d_1 \leq d_2 \leq \cdots \leq d_r \leq s$. Moreover, the numbers r and $d_1, d_2, \ldots, d_r$ are uniquely determined by A.

14.9 Let A be an $n \times n$ matrix over the Galois ring $GR(p^s, p^{sm})$ and assume $A^2 = A$. Prove that there is an $n \times n$ invertible matrix P over $GR(p^s, p^{sm})$ such that $P^{-1}AP$ is of the form

$$\begin{pmatrix} I^{(r)} & 0 \\ 0 & 0 \end{pmatrix},$$

where r is uniquely determined by A.

14.10 Prove that any subring of a Galois ring, with the identity of the Galois ring as its identity, is a Galois ring.

14.11 Prove that any Galois ring is a homomorphic image of a principal ideal domain.

14.12 Prove Theorem 14.36.

14.13 Prove Theorem 14.37.

Bibliography

[1] Agou, S. (1977). Irréductibilité des polynôme $f(X^{p^r} - aX)$ sur un corps fini $\mathbb{F}_{p^s}$, *J. Reine Angew. Math.*, **202**, 191-195.

[2] Agou, S. (1978). Irréductibilité des polynôme $f(X^{p^{2r}} - aX^{p^r} - bX)$ sur un corps fini $\mathbb{F}_{p^s}$, *J. Number Theory*, **10**, 64-69; **11**, 20.

[3] Agou, S. (1979). Irréductibilité des polynôme $f(\sum_{i=0}^{m} a_i X^{p^i})$ sur un corps fini $\mathbb{F}_{p^s}$, *Canad. Math. Bull.*, **23**, 207-212.

[4] Berlekamp, E. R. (1968). *Algebraic Coding Theory*, McGraw-Hill, New York.

[5] Berelekamp, E. R. (1970). Factorizing polynomials over large finite fields, *Math. Comp.*, **24**, 713-735.

[6] Berlekamp, E. (1982). Bit-serial Reed-Solomon encoder, *IEEE Trans. Info. Th.*, **28**, 868-874.

[7] Brawley, J. V. and Carlitz, L. (1987). Irreducibles and the composed product for polynomials over a finite field, *Discrete Math.*, **65**, 115-139.

[8] Brawley, J. V., Gao, S. and Mills, D. (1999). Computing composed products of polynomials, *Finite Fields: Theory, Applications and Alorithms, Comtempory Mathematics* **225**, 1-16.

[9] Blake, I., Gao, S. and Mullin, R. (1991). Factorization of polynomials of type $f(x^t)$, presented at the International Conference on Finite Fields, Coding Theory, and Advanes in Comm. and Computing, Las Vegas, Aug. 1991.

[10] Blake, I., Gao, S. and Mullin, R. (1993). Explicit factorization of $x^{2^k} + 1$ over $\mathbb{F}_p$ with $p \equiv 3 \,(mod\,4)$, *Applicable Algebra in Engineering, Communication and Computing* **4**, 89-94.

[11] Caldwell, C. K. (2009). Mersenne primes: history, theorems and lists, 2009, caldwell@utm.edu

[12] Chang, Y., Truong, T. K. and Reed, I. S. (1999). Normal bases for $GF(q)$, *Journal of Algebra*, **241**, 89-101.

[13] Cohen, S. (1969). On irreducible polynomials of certain types in finite fields, *Proc. Camb. Phil., Soc.,* **66**, 335-344.

[14] Cohen, S. (1989). The irreducibility theorem for linearized polynomials over finite fields, *Bull. Austral. Math. Soc.,* **49**, 407-412.

[15] Cohen, S. (1992). The explicit construction of irreducible polynomials over finite fields, *Designs, Codes and Cryptography,* **2**, 169-174.

[16] Dickson, L. E. (1900). *Linear Groups,* Teubner, Leipzig.

[17] Hachenberger, D. (2004). Characterizing normal basis via the trace map, *Communications in Algebra,* **32** 269-277.

[18] Imamura, I. (1983). On Self-complementary bases of GF(q^n) over GF(q), *Trans. IECE Japan (Section E),* **66**, 717-721.

[19] Jacobson, N. (1985). *Basic Algebra,* vol. I, 2nd ed., Freeman.

[20] Jungnickel, D. (1992). *Finite Fields: Structure and Arithmetics,* Wissenschaftsverlag, Mannheim.

[21] Krull, W. (1924). Algebraische Theorie der Ringe, *Math. Annalen,* **92**, 183-213.

[22] Lidl, R. and Niederreiter, H. (1983). *Finite Fidlds,* Addison-Wesley, Reading, Massachussetts.

[23] MacWilliams, F. J. (1969). Orthogonal matrices over finite fields, *Amer. Math. Monthly,* **76**, 152-164.

[24] McDonald, B. R. (1974). *Finite Rings with Identity,* Marcel Dekker, New York.

[25] Menezes, A. J. (Editor) (1993), Blake, I. F., Gao, X., Mullen, K. C., Vanstone, S.A. and Yaghoobian, T. *Applications of Finite Fields,* Kluwer, Boston.

[26] Omura, J. and Massey, J. (1986). *Computational method and apparatus for finite field arithmetic,* U. S. patent ♯ 4,587,627.

[27] Pei, D., Wang, C. and Omura, J. (1986). Normal bases of finite field GF(2^m), *IEEE Transaction on Info. Th.* **32**, 285-287.

[28] Pellet, A. E. (1870). Sur les fonctions irréductibles suivant un module premier et une fonctionaire, *C. R. Acad. Sci. Paris*, **70**, 328-330.

[29] Perlis, S. (1942). Normal basis of cyclic fields of prime-power degree, *Duke math. J.*, **9**, 507-517.

[30] Pincin, A. (1989). Bases for finite fields and a canonical decomposition for a normal basis generator, *Communications in Algebra*, **17**, 1337-1352.

[31] Raghavendran, R. (1970). A class of finite rings, *Compositio Math.*, **22**, 49-57.

[32] Schwarz, S. (1988). Construction of normal bases in cyclic extensions of a field, *Czechslovak Math. J.*, **38**, 291-312.

[33] Serret, J. A. (1866). *Cours d'algèbre supériears*, 3rd ed., Gauthicr-Villars, Paris.

[34] Serret, J. A. (1866a). Mémoire sur la théorie des congruences suivant un module premier et suivant une fonction modulaire irréductible, *Mém. Acad. Sci. Inst. de France*, **35**, 617-688.

[35] Shparlinski, I. E. (1999). *Finite Fields: Theory and Computation*, Kluwer, Dordrecht.

[36] Shoup, V. (1990). New algorithms for finding irreducible polynomials over finite fields, *Math. Comp.*, **54**, 435-447.

[37] Varshamov, R. (1984). A general method of synthesizing irreducible polynomials over Galois fields, *Soviet Math. Dokl.*, **29**, 334-336.

[38] Wan, Z.-X. (2002). *Geometry of Classical Groups over Finite Fields*, 2nd ed., Science Press, Beijing.

[39] Wan, Z.-X. (2002a). On the Hensel lift of a polynomial, *Differential Geometry and Related Topics*, World Scientific, Singapore, 250-256.

[40] Wan, Z.-X. and Zhou, K. (2007). On the complexity of the dual basis of a Type I optimal normal basis, *Finite Fields and Their Applications*, **13**.

Index